ARCHITECTURE OF SYSTEMS PROBLEM SOLVING

ARCHITECTURE OF SYSTEMS PROBLEM SOLVING

GEORGE J. KLIR

State University of New York at Binghamton
Binghamton, New York

SPRINGER SCIENCE+BUSINESS MEDIA, LLC

Library of Congress Cataloging in Publication Data

Klir, George J., 1932–.
Architecture of systems problem solving.

Bibliography: p.
Inlcudes indexes.
1. System theory. I. Title.
Q295.K55 1985 003 85-9283
ISBN 978-1-4757-1170-7 ISBN 978-1-4757-1168-4 (eBook)
DOI 10.1007/978-1-4757-1168-4

Originally published by Plenum Press, New York in 1985
Softcover reprint of the hardcover 1st edition 1985
A Division of Plenum Publishing Corporation
233 Spring Street, New York, N.Y. 10013

To the Memory of

W. ROSS ASHBY

What is the hardest thing
of all to see: that which
lies before your eyes.

—GOETHE

PREFACE

One criterion for classifying books is whether they are written for a single purpose or for multiple purposes. This book belongs to the category of multipurpose books, but one of its roles is predominant—it is primarily a textbook. As such, it can be used for a variety of courses at the first-year graduate or upper-division undergraduate level. A common characteristic of these courses is that they cover fundamental systems concepts, major categories of systems problems, and some selected methods for dealing with these problems at a rather general level.

A unique feature of the book is that the concepts, problems, and methods are introduced in the context of an architectural formulation of an expert system—referred to as the general systems problem solver or GSPS—whose aim is to provide users of all kinds with computer-based systems knowledge and methodology. The GSPS architecture, which is developed throughout the book, facilitates a framework that is conducive to a coherent, comprehensive, and pragmatic coverage of systems fundamentals—concepts, problems, and methods.

A course that covers systems fundamentals is now offered not only in systems science, information science, or systems engineering programs, but in many programs in other disciplines as well. Although the level of coverage for systems science or engineering students is surely different from that used for students in other disciplines, this book is designed to serve both of these needs.

A course in systems science or engineering programs would normally cover the whole text, including the various appendices. This material will provide the students with a broad base for further studies. When equipped with such a base, the student will maintain an overall perspective during his studies of more advanced and specialized topics. He will be able to recognize the role of each topic within the overall GSPS architecture, to see from this larger perspective how the various topics are interrelated, and to use this knowledge in developing a meaningful program of study for himself.

When used in a course offered to students in the various traditional disciplines, some parts of the text need not be covered. Specific passages which can be skipped without jeopardizing the intelligibility of subsequent parts of the book are marked by symbols ► and ◄ at the beginning and end, respectively. They include some instances of a mathematical presentation of material that for general understanding is adequately covered at the conceptual level and sufficiently illustrated by examples. Also marked are passages in which a specific methodological alternative is described that is not essential for a general comprehension of systems fundamentals. Depending on the course objectives (or study objectives of an individual reader), the marked passages can be either totally excluded from study, or only omitted during a preliminary reading of the book.

In addition to its primary function as a text, the book is also intended for practicing scientists and professionals in various subject areas. An increasing number of them are becoming interested in learning about modern developments in systems science, which may be utilized in their own work. The book is obviously of particular significance to those specialists who are involved in multidisciplinary team projects.

It is also expected that the book will serve as a useful reference for researchers as well as practitioners in systems science and related fields on one side, and the area of expert systems on the other side. Systems science researchers will find in the book a rich source of underdeveloped research areas. Practitioners, on the other hand, will find in it some general methodological tools of considerably broad applicability.

While most expert systems described in the literature are designed to provide the user with expertise in a traditional discipline (such as a specific subject area of medicine, geology, chemistry, or law), the role of the GSPS is to assist the user in dealing with systems problems. Its expertise is thus systems knowledge and methodology and, consequently, its utility transcends boundaries between the traditional disciplines. In this sense, the book should be a useful reference for designers of expert systems and, in fact, also for computer systems architects—it is the computer architecture that ought to reflect the underlying systems problem—solving architecture and not the other way around.

Prerequisite dependencies between individual chapters and sections of this book are well defined and are expressed by the diagram in Figure P.1. Since they do not form a linear ordering, there are several alternative ways of studying the material. Chapter 1, which represents an overall introduction, must always be read first. Chapters 2–5 form a core of the book and are dependent on each other as shown in the diagram. All fundamental types of systems and key categories of systems problems are introduced in these chapters. One way of studying the material is to read all these core chapters before proceeding to the remaining chapters. Another alternative is to follow the prerequisite dependencies and proceed to relevant sections in Chapter 7 (Goal-Oriented Systems) and Chapter 8 (Systems Similarity) after completing the study of each of the core chapters. Chapter 6, which is devoted to systems complexity, can be read in virtually any order. The last chapter, Chapter 9, which overviews the whole GSPS, should also be the last one read.

Mathematical prerequisites are restricted to the material covered normally in a one-semester course in finite mathematics. Some knowledge of calculus is useful, but it is not necessary. Special mathematical concepts, such as the concepts of the Shannon entropy, fuzzy measure, or metric distance, are introduced in the book before they are used. For a quick reference, lists of relevant mathematical symbols and a glossary of all mathematical terms employed in the book are given in Appendices A and B, respectively.

In order to minimize interruptions in the main text, almost all bibliographical, historical, terminological, and other remarks are included in the Notes that

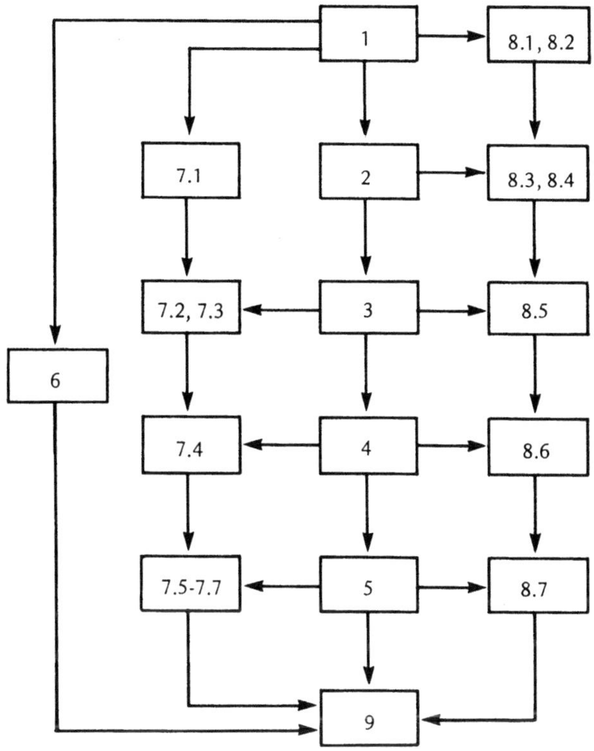

Figure P. 1. Prerequisite dependencies of this book.

accompany individual chapters. They are placed at the end of each chapter, numbered, and occasionally referred to in the main text.

One additional feature of this book should be mentioned. Each of its chapters and sections is introduced by a quote that depicts the essence of the material covered in it. The aim of these carefully selected quotes is to appeal to the right hemisphere of the readers's brain to communicate the key ideas advanced in this book. I expect that in some instances the quotes will help the reader to understand the material presented under it, while in other instances the studied material will help him to properly understand and appreciate the quote. In any case, I am confident that the quotes will reinforce the learning process and will make it a little more enjoyable.

GEORGE J. KLIR
Binghamton, New York

NOTE TO THE READER

Passages that are set off by symbols ▶ and ◀ at the beginning and at the end, respectively, can be omitted without jeopardizing the intelligibility of the remaining text.

References to literature are denoted by brackets and contain the first two letters of the author's last name and the reference number. For instance, [AS1] means the first reference to Ashby found in the References at the end of the book.

When confused with mathematical notation, the reader should consult the *List of Symbols* in Appendix A. When uncertain about the meaning of a mathematical term, he should consult the *Glossary of Relevant Mathematical Concepts* in Appendix B.

CONTENTS

ACKNOWLEDGMENTS

This book was made possible primarily by the generous support of the Netherlands Institute for Advanced Studies (NIAS) in Wassenaar, The Netherlands, by which I was invited to be a Fellow in Residence during the academic years 1975–1976 and 1982–1983. The key ideas presented in the book were created during my first NIAS residence; the actual writing took place during my second residence.

When I came to NIAS in August 1982 for my second residence, almost all material for the book was already well developed, scrutinized, and tested in a classroom environment. Most research work associated with the GSPS was done at the Department of Systems Science, School of Advanced Technology,* SUNY-Binghamton. Some of it, related particularly to Chapter 4, was supported by the National Science Foundation under Research grants ENG-78-18954 and ECS-80-06590.

A number of graduate students in systems science at SUNY-Binghamton were directly involved in the research, most notably Roger Cavallo, Iris Chang, Douglas Elias, Robert Gerardy, Abdul Hai, Masahiko Higashi, Matthew Mariano, Behzad Parviz, Michael Pittarelli, Arthur Ramer, Scott Sancetta, and Hugo Uyttenhove. Some of them were supported by the National Science Foundation, some by SUNY. Other students contributed through their term papers or workshop projects.

The development of the key ideas presented in the book was greatly influenced by Antonin Svoboda, my intellectual mentor and close friend, as well as by personal contacts with many colleagues all over the world, particularly W. Ross Ashby, Gerrit Broekstra, Brian R. Gaines, Yoichi Kaya, Lars Lofgren, Walter Lowen, Robert A. Orchard, Franz Pichler, Robert Rosen, Lotfi Zadeh, and Bernard P. Zeigler.

I am very grateful to two persons who were most instrumental when I was working on the manuscript, Michael Pittarelli and Marina H. M. Voerman. Michael read carefully the whole manuscript, checked all mathematical formulas and examples, and solved all exercises. His scrutiny of the original version of the manuscript resulted in many changes by which the manuscript was greatly improved. Marina typed for me virtually the whole manuscript during my stay at NIAS in 1982–1983. Not only was the typing superb, but she provided me with daily encouragement in my writing.

The book contains many excellent quotes and I am grateful to the copyright owners for permitting me to use the materials. They are Academic Press, Addison-Wesley, American Scientist, Cornell University Press, Entropy Limited, Estate of

* In summer 1983, the School of Advanced Technology became integrated into the Thomas J. Watson School of Engineering, Applied Science, and Technology.

Buckminister Fuller, Gordon and Breach, Institute of Electrical and Electronics Engineers, International Institute for Applied Systems Analysis, John Wiley, Alfred A. Knopf, Longmans, Green and Co., Nature, North-Holland, Ohio University Press, Pattern Recognition, Pepperdine University Press, Philosophy of Science Association, Physica-Verlag, Reidel, Society for General Systems Research, Southwest Journal of Philosophy, and The University of Massachusetts Press.

Last, but not least, I would like to thank my wife Milena and my children John and Jane for their encouragement.

1

INTRODUCTION

> *If one does not begin with a right attitude, there is little hope for a right ending.*
>
> —Kung Fu Meditation

1.1. SYSTEMS SCIENCE

> *We must stop acting as though nature were organized into disciplines in the same way that universities are.*
>
> —RUSSELL L. ACKOFF

The evolution of a highly complex hierarchy of disciplinary specializations has been one of the major characteristics of the history of science. The ancient scientist/philosopher such as Aristotle, who was able to comprehend almost all knowledge available in his time, has gradually been replaced by generations of scientists with ever increasing depth of knowledge and narrowness of interest and competence.

Limitations of the human mind seem to be the primary reason for this trend of fragmenting science into narrow specializations. Once the amount of knowledge becomes greater than what the human mind is able to comprehend, any increase in the knowledge necessarily means that the human comprehends a smaller fraction of it. The more in-depth this knowledge is, the narrower it must be.

The evolution of disciplinary specialization is not unique to science. Other areas of human endeavor, such as engineering, medicine, humanitites, or the arts, have been going through a similar evolution. Engineering, for instance, has evolved from one discipline (the classical civil engineering) into a spectrum of engineering branches, such as mechanical, electrical, chemical, or nuclear engineering, each of them being further divided into many specializations.

One of the major characteristics of science in the second half of this century is the emergence of a number of related intellectual areas such as cybernetics, general systems research, information theory, control theory, mathematical systems theory, decision theory, operations research, and artificial intelligence. All those areas, whose appearance and development are strongly correlated with the origins and advances of computer technology, have one thing in common: they deal with such systems problems in which informational, relational, or structural aspects predominate, whereas the kind of entities which form the system is considerably less significant. It has increasingly been recognized that it is useful to view these interrelated intellectual developments as parts of a larger field of inquiry, usually referred to as *systems science*.

If systems science is a science in the usual sense, then three basic components should be distinguished in it:

i. a *domain* of inquiry,
ii. a body of *knowledge* regarding the domain,
iii. a *methodology* (a coherent collection of methods) for the acquisition of new knowledge within the domain as well as utilization of the knowledge for dealing with problems relevant to the domain.

It is the purpose of this introductory section to characterize these three components—the domain, knowledge, and methodology—of systems science. Moreover, it is argued that systems science is not directly comparable with the other sciences; that, instead, it is more appropriate to view it as a new dimension in science.

It is fair to say that the domain of each scientific discipline is a particular class of systems. Indeed, the term "*system*" is unquestionably one of the most widely used terms in describing activities in the various disciplines of science, particularly in recent times. It has become, unfortunately, a highly overworked term which enjoys different meanings under different circumstances and for different people.

Looking up the term "system" in a standard dictionary, one is likely to find that it is defined as "*a set or arrangement of things so related or connected as to form a unity or organic whole*" (*Webster's New World Dictionary*), although different dictionaries may contain stylistic variations of this formulation.

To follow the common definition, the term "system" stands, in general, for a set of some *things* and a *relation* among the things. The term "relation" is used here in a broad sense to encompass the whole set of kindred terms such as "constraint," "structure," "information," "organization," "cohesion," "interaction," "coupling," "linkage," "interconnection," "dependence," "correlation," "pattern," and the like. A system, say system S, is thus an ordered pair $S = (A, R)$, where A denotes a set of relevant things and R denotes a relation among the things in set A. Such a conception of a system is too general and, consequently, of little pragmatic value. To make it pragmatically useful, it has to be refined in the sense that specific classes of ordered pairs (A, R), relevant to recognized problems, must be introduced. Such classes can basically be introduced by one of two fundamentally different criteria:

a. by a restriction to systems which are based on certain kinds of things;
b. by a restriction to systems which are based on certain kinds of relations.

Classification criteria (a) and (b) can be viewed as orthogonal. Criterion (a) is exemplified by the traditional classification of science and technology into disciplines and specializations, each focusing on the study of certain kinds of things (physical, chemical, biological, political, economical, etc.) without committing to any particular kind of relations. Since different kinds of things require different experimental (instrumentation) procedures for data acquisition, this classification is essentially *experimentally based.*

Criterion (b) leads to fundamentally different classes of systems, each characterized by a specific kind of relations with no commitment to any particular kind of things on which the relations are defined. This classification is related primarily to data processing rather than data acquisition and, as such, it is predominantly *theoretically based.*

As discussed later in more detail, the largest classes of systems based on criterion (b) are those which characterize various *epistemological levels*, i.e., levels of knowledge regarding the phenomena under consideration. They are further refined by various *methodological distinctions.* Each class of systems defined by a particular epistemological level and specific methodological distinctions is then divided into still smaller classes. Each of these classes consists of systems that are equivalent with regard to some specific, pragmatically relevant aspects of their relations. Such equivalence is usually called isomorphism and classes based on it are called *isomorphic classes.*

Depending on the relational aspects in which systems are required to be isomorphic, some isomorphic classes are subsets of others. The smallest isomorphic classes are obviously those in which systems are isomorphic with respect to all aspects of their relations.

Although systems in each particular isomorphic class are equivalent in at least some aspects of their relations, they may be based on completely different kinds of things. To deal solely with relational aspects of systems, it is sufficient to replace each isomorphic class of systems by a single system chosen as its representative. Although the choice of these representatives is arbitrary, in principle, it is important that the same selection criteria be used for all isomorphic classes. Otherwise, the representatives would not be compatible and, consequently, it would be methodologically rather difficult to deal with them. For our purpose, let the representatives be defined as systems whose sets of things are some comparable abstract (interpretation-free) sets and whose relations are described in some convenient standard form.

Let representatives of isomorphic classes that satisfy these characteristics, under some specific meaning given to the term "standard," be called general systems. Hence, a *general system is a standard and interpretation-free system chosen to represent a class of systems equivalent (isomorphic) with respect to some relational aspects that are pragmatically relevant.* The term "standard" is used in this definition to refer to a description which satisfies certain conventions, influenced primarily by the use of the system; some convenient form by which the system is represented on a computer, for example, may be accepted as a standard description.

The orthogonality of classification criteria (a) and (b) is illustrated by Figure 1.1. Classes of systems based on the kind of things involved (set A) are characterized by the vertical strips; classes of systems based on relations involved (set R) are characterized by the horizontal strips.

While systems classification based on criterion (b) is foreign to traditional science, its significance has increasingly been recognized. All activities involved in the study of those properties of systems and relevant problems which emanate from this classification are now becoming identified with the general name "*systems science.*" In this sense, "systems science" is a name for scientific activities which are predominantly

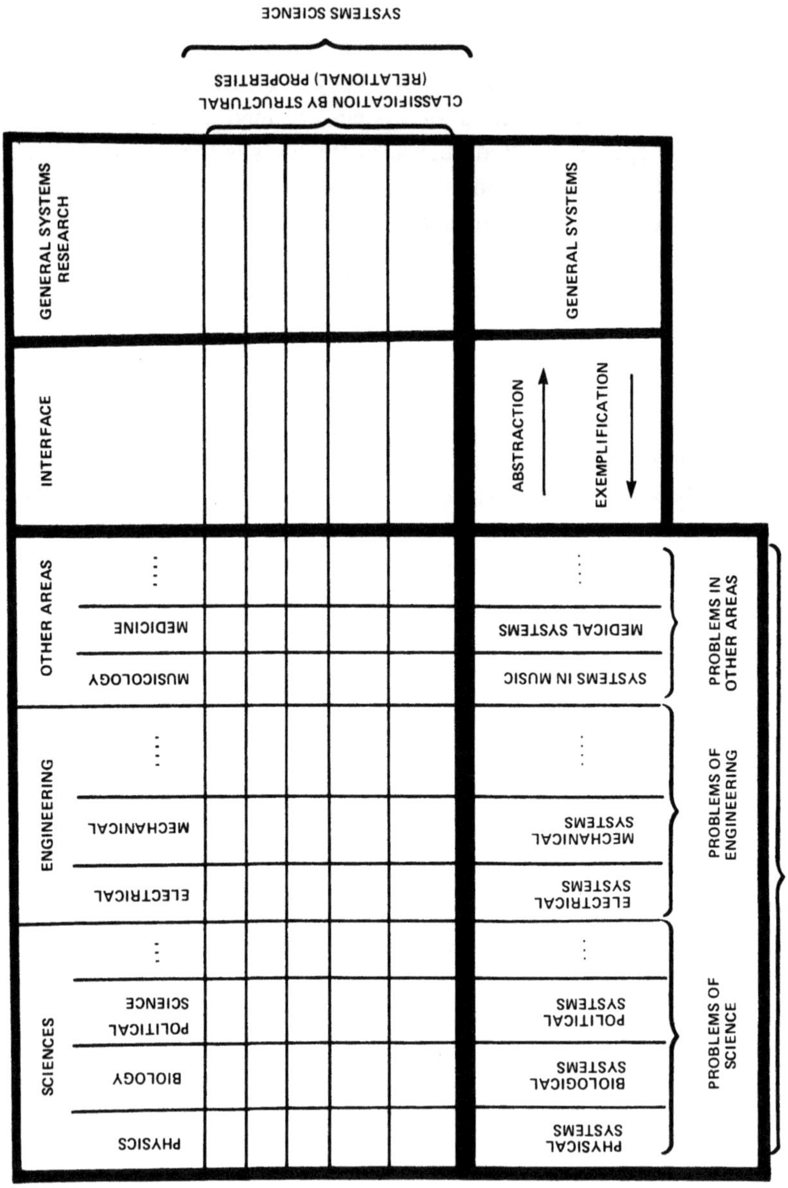

Figure 1.1. Two ways of classifying systems.

theoretically based and, hence, complementary to the experimentally based activities of the traditional science.

The *domain of systems science* consists thus of all kinds of relational properties which are valid for particular classes of systems, or, in some rare instances, are valid for all systems. The chosen relational classification of systems determines the way in which the domain of systems is divided into subdomains, in a similar fashion as the domain of the traditional science has been divided into subdomains of the various disciplines and specializations.

The *knowledge of systems science*, i.e., knowledge regarding the various classes of relational properties of systems, can be obtained either mathematically or through experiments with systems simulated on a computer. Examples of mathematically derived knowledge in systems science are the Ashby law of requisite variety [AS2, AS3], the principles of maximum entropy and minimum cross-entropy [CH5, JA2], or the various laws of information which govern systems [CO4]. As far as the experimentally derived knowledge is concerned, it is the computer which represents the *laboratory for systems science*. It allows the systems scientist to perform experiments in exactly the same way other scientists do in their laboratories, although the experimental entities he deals with are abstract structural properties (simulated on the computer) rather than specific properties of the real world. Some instances of systems knowledge obtained by computer experimentation are described later in this book.

The third component of systems science—*systems methodology*—is a coherent collection of methods for studying relational properties of various classes of systems and for solving systems problems, i.e., problems which deal with the relational aspects of systems. A useful classification of systems from the relational point of view is the kernel of systems methodology. When properly developed, the classification is a basis for a comprehensive description and taxonomy of systems problems. The ultimate goal of systems methodology is to provide potential users in various disciplines and problem areas with methodological tools for all the recognized types of systems problems.

From the standpoint of the disciplinary classification of traditional science, systems science is clearly cross-disciplinary. There are at least two implications of this fact. Firstly, systems science knowledge and methodology are directly applicable, at least in principle, in virtually all disciplines of traditional science. Secondly, systems science has the flexibility to study relational properties of such systems and the associated problems which include aspects derived from any number of different disciplines and specializations of traditional science. Such cross-disciplinary systems and problems can thus be studied as wholes rather than collections of the disciplinary subsystems and subproblems.

It follows from the previous discussion that systems science, like any other science, has a specific domain of inquiry, body of knowledge, and methodology. And yet, it is not a science in the traditional sense. While the traditional science is oriented to the study of various categories of phenomena, systems science is oriented to the study of various classes of relations. As such, it should be viewed as a new dimension in science rather than a new science comparable with the other sciences.

The two dimensions of science, which reflect the two-dimensional classification of systems symbolized by Figure 1.1, are complementary. When combined in scientific inquiries, they are more powerful than either of them alone. The traditional dimension of science provides a meaning and context to each inquiry. The systems dimension, on the other hand, provides a means for dealing with any desirable system, regardless of whether or not it is restricted to a traditional discipline of science.

It seems that three major periods in human history can naturally be recognized with regard to the character of science:

i. *prescientific period* (until about the sixteenth century)—characterized by common sense, speculation, the method of trial and error, craft skills, deductive reasoning, and the emphasis on tradition;
ii. *one-dimensional science* (the period from the seventeenth century until about the middle of this century)—characterized by the integration of speculation, deductive reasoning, and experimentation, with a particular emphasis on the latter, which gives rise to the various experimentally based disciplines and specializations of science; they emerge primarily due to differences in experimental (instrumentation) procedures rather than differences in the relational properties of the investigated systems;
iii. *two-dimensional science* (developing since about the middle of this century)—characterized by the emergence of systems science, which focuses on the relational rather than experimental aspects of the investigated systems, and its integration with the experimentally based (traditional) disciplines of science.

In summary, it is reasonable to characterize the development of science during the second half of this century in terms of a major transition from a one-dimensional science—primarily experimentally based—into a two-dimensional science, in the course of which systems science—primarily relationally based—gradually enters as the second dimension. The significance of this radically new paradigm of science—the two-dimensional science—has not been fully realized as yet, but its implications for the future seem to be quite profound.

1.2. SYSTEMS PROBLEM SOLVING

> *Only by means of a full understanding of the tasks may we find means relevant to their solution. It is more important for the result to put correct questions than to give correct answers to wrong questions.*
>
> CHRISTIAN NORBERG-SCHULZ

The notion of systems problem solving, as a central theme of this book, raises three questions regarding its meaning and significance:

1. Can systems problems be recognized as instances of a special and well-defined class of general problems?
2. Can the class of systems problems be operationally described to make it possible to develop a comprehensive methodology for solving problems in this class?
3. Is the class of systems problems of sufficient practical significance to warrant the development of a systems problem-solving methodology?

In my opinion, the answers to these questions are positive. Although the rationale of this opinion can be fully comprehended only after reading this book, let me offer the following remarks as a brief preview.

As argued in Section 1.1, the concept of general systems, as standard representatives of pragmatically significant equivalence classes of systems, emerges naturally from the two-dimensional classification of systems illustrated by Figure 1.1. Although it is quite clear that general systems enjoy infinite variety, this variety can be adequately captured by a *finite number of types of general systems*, each characterized by a particular epistemological level and a finite set of relevant and desirable methodological distinctions.

Once desirable types of general systems are defined, they form a space within which types of systems problems can be defined. Such a space is usually called a *problem space*. Each problem type is defined in terms of an ordered connection in the problem space, from some initial systems type to a terminal systems type, and a set of requirement types that are compatible with the two systems types involved.

Requirements can be objectives or constraints of particular problems. Although the variety of actual requirements applicable to any nonempty problem space is infinite, it can be adequately represented by a finite number of types, as already indicated in the previous paragraph. Each of the problem types is thus characterized by the two types of systems involved and a finite set of specified requirement types.

A problem type becomes a particular problem when particular requirements of all specified types are given and, depending on the requirement types, either only a particular initial system of the specified type is given or particular systems are given for both of the specified types. In the former case, the initial system represents the initial problem situation; the problem solution (or goal problem situation) consists of one or more particular terminal systems of the required type. In the latter case, the initial problem situation is represented by two particular systems and the solution is some relationship between them.

It follows from this characterization of systems problems that, indeed, systems problems form a special and well-defined class of general problems. The fact that their infinite variety is reducible into a finite number of well-defined problem types makes it certainly possible to develop a methodology for this class of problems. Hence, the answers to our first and second questions are clearly positive. The third question needs some additional discussion.

Systems problem solving, as conceptualized in this book, is restricted to problems in which problem situations are represented by general systems of well-defined types. As such, it is concerned solely with those aspects of overall problems that are interpretation free and context independent. The use of systems problem-solving methodology is thus based on the assumption that interpretation-free and context-independent subproblems can be extracted from the individual overall problems.

Is it meaningful and useful to divide overall problems in this way? I would like to argue that it is. Indeed, we all employ this division in solving simple everyday problems when we use arithmetic, for example. Bernard Zeigler expresses this point quite well in the Preface to his book *Theory of Modelling and Simulation* [ZE2]:

> Nobody questions the role of arithmetic in the sciences, engineering, and management. Arithmetic is all pervasive, yet it is a mathematical discipline having its own axioms and logical structure. Its content is not specific to any other disciplines but is directly applicable to them all. Thus students of biology and engineering are not taught how to add differently—the different training comes in what to add, when to do it, and why.
>
> The practice of modelling and simulation too is all pervasive. However it has its own concepts of model description, simplification, validation, simulation, and exploration, which are not specific to any particular discipline. These statements would be agreed to by all. Not everyone, however, would say that the concepts named can be isolated and abstracted in a generally useful form.

Although Zeigler emphasizes modelling and simulation, his observations are applicable equally well to other classes of problems such as systems design, analysis, identification, reconstruction, control, performance evaluation, testing, etc. Sophisticated methodological tools can be developed for many subproblems of these various overall problems in terms of the relevant general systems, i.e., without regard to any interpretation or context. Such tools introduce great efficiency and unification into the methodological process of solving complex problems, in the same way as arithmetic does for very simple problems.

Let a conceptual framework through which types of systems problems are defined together with methodological tools for solving problems of these types be called a *general systems problem solver* (or GSPS, in abbreviation).

In different problem-solving contexts associated with the various traditional disciplines of science, engineering, medicine, and other areas, as well as cross-disciplinary studies, the GSPS should be primarily viewed as a methodological resource, presumably computer based. When available, its service can be utilized whenever systems problems arise in the process of dealing with some overall problem.

Figure 1.2 illustrates the role of the GSPS as an aid to scientific investigation in different disciplines of science. Two levels are distinguished in its operation:

1 (Represented by the inner rectangles). The investigator is familiar enough with the basic language of the GSPS to be able to formulate an *interpretation* of a systems problem within his own discipline. In this case the investigator, or *user*, maps the interpretation to GSPS formulation (as described later); GSPS solves the problem and

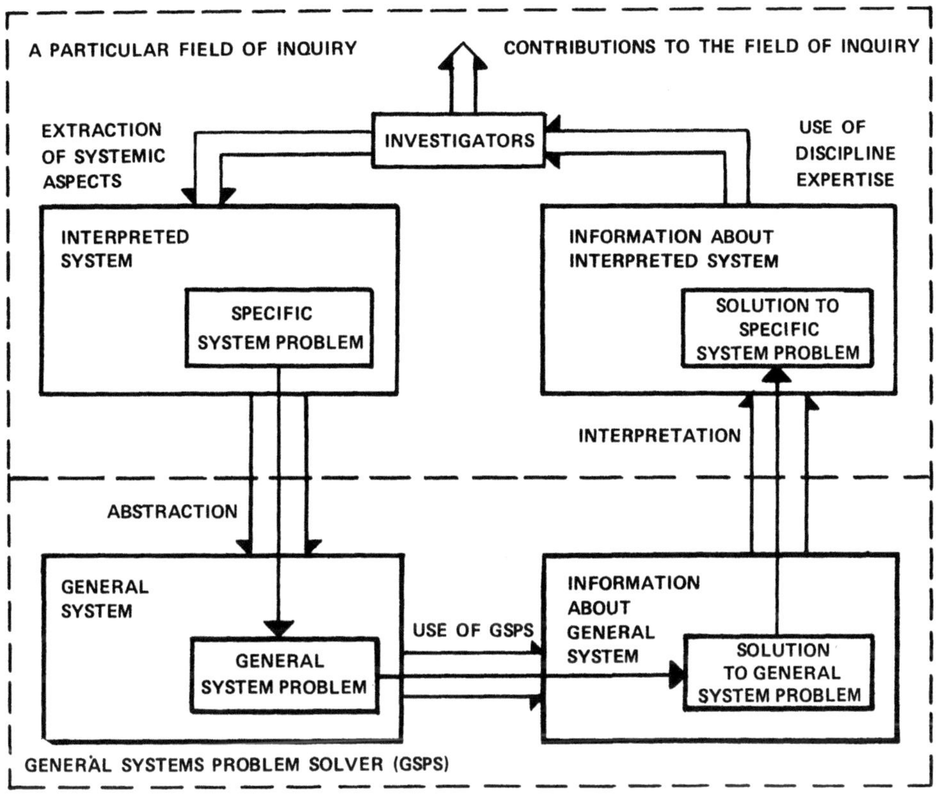

Figure 1.2. The role of GSPS as a methodological resource in science.

maps the solution to the interpreted system. This situation can also arise through the development of procedures in the form of simple questions put to the user, the answers to which identify the systems problem applicable to the situation.

2 (Outer rectangles). Many systems investigations are of sufficient complexity that the investigator could make meaningful use of more information than that provided by the solution to a particular systems problem. In this case, also, procedures can be developed through which a transformation from interpreted system to general system can be effected. Based on the information supplied through this transformation, GSPS can translate new information about the general system back to the interpreted system. The investigator is thus given new knowledge about the interpreted system.

The utilization of the GSPS, or similar developments in systems science, for the study of particular systems requires thus an interface with the disciplines involved. Such an interface consists of two dual and alternately used processes—*abstraction* and *interpretation*. In a scientific investigation, the application of these processes has, generally, an on-going and nonterminating character. This vital characteristic of science is well depicted by G. Spencer Brown [BR9]:

> Science is a continuous living process; it is made up of activities rather than records; and if the activities cease it dies. Science differs from mere records in much the same way as a teacher differs from a library. . . . Scientific knowledge, like negative entropy, tends constantly to diminish. It is prevented from dwindling completely into anecdote only by the attitude which seeks to repeat experiments and confirm results without end. . . . Science is a significant game: one player tries to reduce significance to insignificance by asking more questions, while another seeks to counter his activities by doing more experiments. The scientist, like the chess-enthusiast, often plays both sides himself. . . . Repetitions of scientific results serve two purposes. First, they inhibit alternative questions which would tend to reduce their significance; and secondly, each successful repetition tends to increase the significance which such question might reduce. We thus have a race between the questions and the results.

The scheme in Figure 1.2 is not restricted to science. It is applicable equally well to other areas such as engineering, medicine or management. Although problems in these areas (e.g., systems design, testing, diagnosis, decision making, etc.) are different from the problems involved in scientific inquiry, the role of the GSPS in assisting the various users to deal with their systems subproblems is essentially the same.

Systems problem solving, as represented by the GSPS, is thus applicable only in combination with the traditional disciplines of science and other areas, within which overall problems arise in specific contexts. To be practically useful, the GSPS must cover as large a class of systems problems as possible, particularly those systems problems that are common to many disciplines. Hence, the GSPS conceptual framework should be derived by a process of abstracting and organizing systems conceptions and problems from as many disciplines as possible and supplementing them, wherever desirable, with new conceptions and problems to form a coherent whole. The GSPS framework that is described in this book has actually been developed in this way over a period of almost 20 years.

The current version of the GSPS, as well as any of its future versions, should be viewed as provisional. Given a particular framework, it is always likely that sooner or later, some new systems concepts and problems will be discovered in the process of using the GSPS that have no meaning within the framework. While some of them might be too specialized and limited in applications, others may be of sufficiently broad applicability to warrant their integration into the framework and a development of associated methods. The GSPS is thus evolving through its interactions with users and its scope of applicability steadily increases during this evolutionary process. From this point of view, the GSPS may be also viewed as an ongoing research program.

From the remarks made in this section, we may now conclude that systems problems are meaningful subproblems of overall problems that arise in the various traditional disciplines of science and other areas of human endeavor, that these subproblems can be operationally described, and that their methodology is an important resource for the traditional disciplines as well as for dealing with cross-disciplinary problems.

1.3. HIERARCHY OF EPISTEMOLOGICAL LEVELS OF SYSTEMS

> *Epistemology, or the theory of knowledge, is that branch of philosophy which is concerned with the nature and scope of knowledge, its presuppositions and basis, and the general reliability of claims to knowledge.*
>
> —THE ENCYCLOPEDIA OF PHILOSOPHY

The skeleton of the GSPS taxonomy of systems is a *hierarchy of epistemological levels of systems.* It seems that such hierarchy is vital, in one form or another, to the development of any organized package of methodological tools for systems problem solving. Although the individual epistemological levels of systems are described in great detail in Chapters 2–5, a simple characterization of the whole hierarchy is presented in this section as a preview.

The hierarchy is derived from some primitive notions: an *investigator* (observer) and his environment, an investigated (observed) *object* and its environment; and an *interaction* between the investigator and object.

At the lowest level in the hierarchy, denoted as level 0, *a system is what is distinguished as a system by the investigator.* That is to say, the investigator makes a choice regarding the manner in which he wants to interact with the investigated object. His choice is not completely arbitrary in most instances; it is at least partially determined by the purpose of his investigation, investigative constraints (availability of measuring instruments, financial and time limitations, legal restrictions, etc.), and available knowledge relevant to the investigation. The following quote from a recent paper by Brian Gaines is an engaging discussion of this very general notion of the term "system" [GA4]:

> *Definition*: A system is what is distinguished as a system. At first sight this looks to be a nonstatement. Systems are whatever we like to distinguish as systems. Has anything been said? Is there any possible foundation here for a systems science? I want to answer both these questions affirmatively and show that this definition is full of content and rich in its interpretation.
>
> Let me first answer one obvious objection to the definition above and turn it to my advantage. You may ask, "What is peculiarly systemic about this definition"? "Could I not equally well apply it to all other objects I might wish to define?" i.e., A rabbit is what is distinguished as a rabbit. "Ah, but," I shall reply, "my definition is adequate to define a system but yours is not adequate to define a rabbit." In this lies the essence of systems theory: that to distinguish some entity as being a system is a necessary and sufficient criterion for its being a system, and this is uniquely true for systems. Whereas to distinguish some entity as being anything else is a necessary criterion to its being that something but not a sufficient one.
>
> More poetically we may say that the concept of a system stands at the supremum of the hierarchy of being. That sounds like a very important place to be.

> Perhaps it is. But when we realize that getting there is achieved through the rather negative virtue of not having any further distinguishing characteristics, then it is not so impressive a qualification. I believe this definition of a system as being that which uniquely is defined by making a distinction explains many of the virtues, and the vices, of systems theory. The power of the concept is its sheer generality; and we emphasize this naked lack of qualification in the term general systems theory, rather than attempt to obfuscate the matter by giving it some respectable covering term such as mathematical systems theory. The weakness, and paradoxically the prime strength of the concept is in its failure to require further distinctions. It is a weakness when we fail to recognize the significance of those further distinctions to the subject matter in hand. It is a strength when those further distinctions are themselves unnecessary to the argument and only serve to obscure a general truth through a covering of specialist extremes of vilification and praise.

The form of the interaction with the object can be described in a number of alternative ways. In the GSPS framework, a system at the epistemological level 0 is defined by a set of variables, a set of potential states (values) recognized for each variable, and some operational way of describing the meaning of their states in terms of the manifestations of the associated attributes of the object. The term "*source system*" has been used for systems defined at this level to indicate that such a system is, at least potentially, a source of empirical data. Other names used in the literature are "primitive system" and "dataless system," which suggest that a system at this level represents a primitive stage in the process of systems investigation with no data regarding the variables available.

The set of variables is usually partitioned into two subsets, referred to as *basic and supporting variables*. Aggregate states of all supporting variables form a *support set* (also called a *parameter set*), within which changes in states of the individual basic variables occur. The most frequent examples of supporting variables are time, space, and various populations of individuals of the same kind (social groups, sets of countries, manufactured products of the same kind, etc.).

Source systems can usefully be classified by various criteria through which methodologically significant special properties of the variables or state sets are distinguished. According to one such criterion, the basic variables may be partitioned into *input and output variables*. Under such a partition, states of input variables are viewed as conditions which affect the output variables. Input variables are not the subject of inquiry but are viewed as being determined by some agent which is not part of the system under consideration. Such an agent is referred to as an *environment* of the system; it includes, in many cases, the investigator. It is important that the notion of input variables not be confused with the notion of independent variables.

Systems whose variables are classified into input and output variables are called *directed systems*; those for which no such classification is given are called *neutral systems*. A number of additional distinctions are recognized for state sets associated with the involved variables (basic or supporting) and provide a basis for further methodological classification of source systems. They include, for instance, the distinctions between crisp and fuzzy variables, discrete and continuous variables, and variables of different scales.

Systems at different higher epistemological levels are distinguished from each other by the level of knowledge regarding the variables of the associated source system. A higher-level system entails all knowledge of the corresponding systems at any lower level and contains some additional knowledge which is not available at the lower levels. Hence, the source system is included in all of the higher-level systems.

When the source system is supplemented by data, i.e., by actual states of the basic variables within the defined support set, we view the new system (a source system with data) as a system defined at epistemological level 1. Systems at this level are called *data systems*. Depending on the problem, data may be obtained by observation or measurement (as in the problem of systems modelling) or are defined as desirable states (as in the problem of systems design).

Higher epistemological levels involve knowledge of some *support-invariant relational characteristics* of the variables involved through which the data can be generated for appropriate initial or boundary conditions. The data generation may be exact (deterministic) or approximate in some specific fashion (stochastic, fuzzy).

At level 2, the support invariance is represented by one overall characterization of the constraint among a set of basic variables within the support set. The set of basic variables includes those defined by the associated source system and, possibly, some additional basic variables. Each of the additional variables is defined in terms of a specific *translation rule* in the support set, applied either to a basic variable of the source system or to a hypothetical (unobserved) variable, introduced by the user (modeler, designer) and usually referred to as an *internal variable*. Each translation rule is basically a one-to-one function by which each element of the support set is assigned another (unique) element of the same support set.

Since the aim of the support-invariant constraint characterization is to describe a process by which states of the basic variables can be generated within the support set for each initial or boundary condition, systems at level 2 are called *generative systems*.

At epistemological level 3, the system is defined in terms of a set of generative systems (or, sometimes, lower-level systems), referred to as *subsystems* of the overall system. The subsystems may be coupled in the sense that they share some variables, or may interact in some other way. Systems at this level are called *structure systems*.

At epistemological level 4, the system consists of a set of systems, defined at some lower levels, and some support-invariant *metacharacterization* (a rule, relation, procedure) by which changes in the lower-level systems are described; the lower-level systems are required to share the same source system and are defined at levels 1, 2, or 3. Systems defined in this way are called *metasystems*. At level 5, the metacharacterization is allowed to change in the support set according to higher-level support-invariant characterization or meta-metacharacterization; such systems are called *meta-metasystems* or *metasystems of the second order*. Metasystems of higher orders are defined similarly.

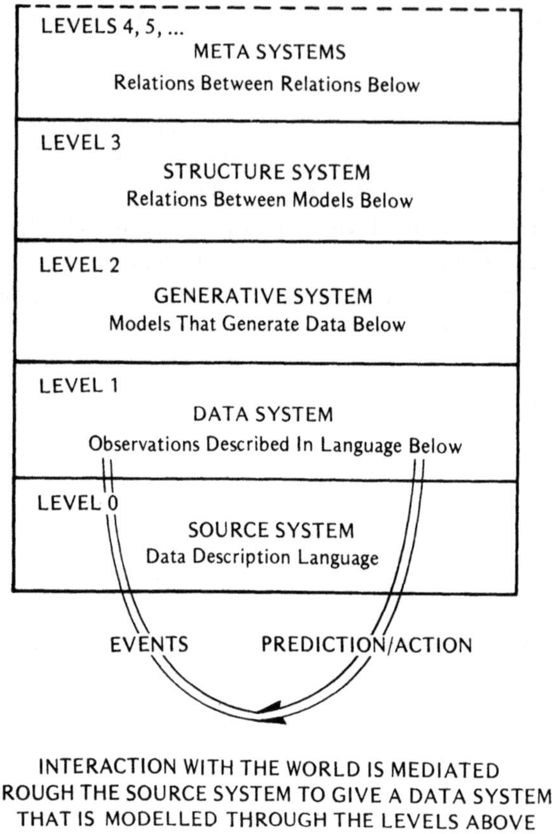

Figure 1.3. Hierarchy of epistemological levels of systems: a simplified overview.

For easy reference, a simplified overview of the epistemological systems hierarchy is given in Figure 1.3.

1.4. THE ROLE OF MATHEMATICS

> *Don't mistake a solution method for a problem definition—especially if it's your own solution method.*
>
> —DONALD GAUSE AND GERALD M. WEINBERG

As discussed in Section 1.2, the notion of systems problem solving emerges naturally from the two-dimensional classification of systems expressed by Figure 1.1. Its aim is to deal with relational aspects of systems in an interpretation-free and context-independent fashion. This, however, is also what mathematics is supposed to do. What, then, is the difference between the two?

Mathematics can roughly be divided into pure and applied. Pure mathematics is basically oriented to the development of various axiomatic theories, regardless of whether or not they have any real world meaning. The proper activity of the pure mathematician is thus to derive theorems from postulated assumptions (axioms), and it is not his concern to determine whether there is some interpretation of the theory in which the assumptions are true. This "l'art pour l'art" attitude, which has been increasingly influencial in mathematics since the nineteenth century, is even emphasized by some mathematicians as crucial for mathematics. In spite of this attitude, however, many mathematical theories have various degrees of relevance to the real world. It may be a lucky accident for a mathematical theory when any such relevance is discovered. More frequently, however, it seems to be a result of either some unconscious process in the mind of the mathematician (intuition, insight) or his conscious effort (often hidden or at least unreported) to abstract and formalize some aspects of reality.

It should be mentioned at this point that there are some inherent limitations in the axiomatic formalization. It was discovered by Kurt Gödel in 1931 that some axiomatic theories (e.g., any axiomatic theory of ordinary arithmetic) are such that their internal consistency (i.e., that no mutually contradictory theorems can be deduced from the axioms) cannot be proven. More precisely, the consistency of an axiomatic theory cannot be proven within its own rules of inference. A consistency proof based on more powerful rules of inference may exist but, then, the consistency of the assumptions in these new rules has to be proven. This may require the use of still more powerful rules of inference. This argument can be repeated to show that the consistency question can never be fully answered for some mathematical theories. Gödel also showed, and this is even more important, that if some of the axiomatic theories whose consistency is not provable were consistent, then they would not be complete (i.e., some true statements of the theories would not be derivable from their axioms). Hence, there are mathematical theories that are either inconsistent or incomplete, and it cannot be decided to which of the two categories each of them belongs.

The role of applied mathematics is to search for practical interpretations of the various mathematical theories and, when such interpretations are found, to further develop the theories into useful methodological tools for dealing with the interpreted systems and associated problems. As such, applied mathematics is oriented to the development of methods based on specific mathematical theories and their use in as many interpreted areas as possible. It is, of course, subject to the fundamental limitations of mathematical theories exposed by Gödel. Moreover, each mathematical theory is derived from some specific assumptions (axioms) and, consequently, the use of any methodology based on the theory is restricted to problems which conform to these assumptions. If a problem does not conform to them and an applied mathematician trained in the methodology still wants to use it, he has to adjust (reformulate) the problem to make it fit the assumptions. This means, however, that a different problem is now solved. The problem adjustment is often not stated explicitly and, as a consequence, an impression is created that the original problem was solved while, in fact, it was not.

Applied mathematics provides thus the various users (scientists, engineers, etc.) with a set of methodological tools, each derived from a mathematical theory which, in turn, is based on a specific set of assumptions. Mathematical theories are most frequently developed for assumptions that are interesting or convenient from the mathematical point of view. As a consequence, they produce methods which cover rather small and scattered parts of the whole spectrum of systems problems. In some sense, the idea of systems problem solving is a reaction to this unsatisfactory situation.

In contrast with applied mathematics, systems problem solving is committed to the investigation of the domain of systems problems as a coherent whole. In particular, it attempts to identify pragmatically rich subproblems, i.e., subproblems that occur in as many genuine systems problems as possible. This emphasis of systems problem solving on comprehensiveness and pragmatic significance in pursuing methodological research is quite different from the usual emphasis in mathematics to pursue research of methodological areas based on convenient (and often arbitrary) mathematical properties.

Hence, the *primacy of problems* in systems problem solving is in sharp contrast with the *primacy of methods* in applied mathematics. It is the most fundamental commitment of systems problem solving to develop methods for solving systems problems in their natural formulation with no simplifying assumptions imposed upon the solution at all or, if unavoidable, with assumptions that make the problem manageable but at the same time distort it as little as possible. The methodological tools for solving the problems are of secondary importance and are chosen in such a way as to best fit the problem rather than the other way around. Moreover, the tools need not be only mathematical in nature but may consist of a combination of mathematical, computational, heuristic, experimental, or any other desirable aspects.

In order to manage the complexity involved in the solution process, systems problems can rarely be handled without any simplifying assumptions. However, simplifying assumptions can be introduced in each problem in many different ways. Each set of assumptions reduces, in a particular manner, the range of possible solutions and, at the same time, reduces the complexity of the solution process.

Given a particular systems problem, a set of assumptions regarding its solutions is referred to as a *methodological paradigm*. When a problem is solved within a particular methodological paradigm, the solution does not contain any features inconsistent with the paradigm.

It is reasonable to view a paradigm which represents a proper subset of assumptions of another paradigm as a generalization of the latter. Given the set of all assumptions which are considered for a problem type, the relation "paradigm A is more general than paradigm **B**" (i.e., A contains a subset of the assumptions contained in **B**) forms a partial ordering among all meaningful paradigms associated with the problem type. The term "meaningful paradigm" may be used in a strong sense to characterize sets of assumptions that guarantee that all particular problems of the given problem type are solvable; alternatively, it may be used in a weak sense to require only that some particular problems of the given problem type are solvable.

The most general paradigm for each problem type is unique: it is the assumption-free paradigm. On the other hand, there are usually several least general paradigms which are meaningful for a given problem type.

Paradigm generalization is a current trend stimulated primarily by the advances in computer technology. Any generalization of a paradigm extends the set of possible solutions to the problem and makes it possible in many cases to reach a better solution. At the same time, however, it usually requires a solution procedure with greater complexity. The study of the relationship between possible methodological paradigms and classes of systems problems is a subject of *systems metamethodology*. This is an important new area of research in which little has been accomplished as yet. The central issue of systems metamethodology is to determine those paradigms, for various classes of problems and the current state of computer technology, which represent the best compromise between the two conflicting criteria—the quality of the solution and the complexity of the solution procedure. The main difficulty in this investigation is that there are usually many alternative solution procedures which can be developed for a given problem under the same methodological paradigm.

Another issue of systems metamethodology is the determination and characterization of clusters of systems paradigms that usefully complement each other and may thus be effectively used in parallel for dealing with the same problem. Together, they may give the investigator much better insight than any one of them could provide alone.

Every mathematical theory that has some meaning in terms of a systems problem-solving framework (such as that of the GSPS) is actually a methodological paradigm. It is associated with a problem type and represents a local frame within which methods can be developed for solving particular problems of this type. One of the roles of systems metamethodology is to compile relevant mathematical theories and identify their place in the overall problem space. Another of its roles is to propose new meaningful paradigms; the ultimate goal is to characterize and order all possible paradigms for each problem type. Since the recognition of a new paradigm is an impetus for developing a new mathematical theory, comprehensive investigations in systems metamethodology will undoubtedly be a tremendous stimulus for basic mathematical research of great pragmatic significance. Mathematics is thus a contributor to systems problem solving as well as a beneficiary of the latter.

1.5. THE ROLE OF COMPUTER TECHNOLOGY

> *Symbiosis has played an important role in evolution; . . . it may have played a decisive role. . . . within the last generation man has acquired an important symbiote. Man's new partner is the high-speed computer.*
>
> —John G. Kemeny

Systems science is strongly dependent on the computer, which is its laboratory as well as the most important methodological tool. It is thus not surprising that modern systems ideas began to emerge shortly after the first fully automatic digital computers were built in the late 1940s and early 1950s. Systems science and computer technology have been developing side by side and have been influencing each other since that time.

Advances in computer technology, together with developments in the area of artificial intelligence, have opened new methodological possibilities, have helped to clarify or sharpen the formulation of some fundamental philosophical problems, have made many speculative ideas increasingly operational, and have made it possible to implement some simple functions of the human mind on the computer. However, the aim of systems problem solving is not to replace the human mind by a machine, but rather to supplement it, in a symbiotic fashion, by a computer equipped with an organized package of appropriate methodological tools. This view is based on the recognition that the human mind, when encountering very complex systems, has certain faculties which make it superior even to the most sophisticated methods applied on the most advanced computers. The current understanding of these faculties is rather rudimentary and certainly unsatisfactory. In spite of the progress made by artificial intelligence, together with neurophysiology, psychology, and other relevant areas, it is reasonable to expect that there are abilities of the human mind which will never be fully understood operationally.

Intuition, insight, and the ability of global comprehension are possibly the most valuable assets of the human mind, particularly one that is appropriately trained. However, complex systems frequently possess properties which are counterintuitive and resistant to global comprehension. As such, they represent traps for the human mind in the sense that they may guide it into illusory insights. To discover such traps, it is usually unavoidable to perform the tedious work of detailed analysis of the system at hand. While the human mind is weak and severely limited in this respect, it is exactly this domain of detailed analysis where the computer, equipped with appropriate methodology, excels. This ability gives the computer an important role as an *intuition safeguard* and *intuition amplifier*.

The symbiosis of the human being (scientist, decision maker, designer, and the like) with the methodologically equipped computer, such as the GSPS, makes it possible to invent and utilize new approaches to various intellectual tasks, far superior to those applicable by either of them alone. The strength of the human being is his experience in the area of study, understanding and taking advantage of the context of investigation, intuition, global comprehension, feeling for the right solution, visual and auditory capabilities, creativity, and the like. The strength of the computer lies in its computational power, the ease with which it can handle a tremendous number of operations, far exceeding the human capability in this respect. It is the computational power which, when properly utilized, can significantly enhance the human intellectual qualities by providing the human being with desirable detailed analyses and, as already mentioned, help him to avoid the many counterintuitive traps associated with complex systems.

One such counterintuitive trap lies in the assumption, often taken for granted, that properties of overall systems can be reconstructed from knowledge of corresponding properties associated with their subsystems. For instance, it has often been assumed in interdisciplinary societal projects that the whole system is understood when we understand its economic, legal, political, ecological, and other relevant subsystems. Such an assumption is unfortunately warranted only rarely and, even if it is warranted, its validity depends on the chosen subsystems. There is no reason to believe that the "natural" subsystems (economic, political, etc.) are adequate in the sense that they contain enough information to allow a fairly accurate reconstruction (understanding) of the overall system. If the assumption of the ability to reconstruct an overall system from its specific subsystems is not warranted, the various conclusions about the overall system obtained from the subsystems may be incorrect and vastly misleading. Although information about the reconstruction possibilities is implicitly included in data regarding the overall system, their explicit determination requires a detailed analysis of the data. Methods for performing such analysis, referred to as reconstructability analysis, have been under development during the last few years and are described in Chapter 4. While it is virtually impossible for the human mind to perform reconstructability analysis, except for minuscule systems, the computer has a great potential for extending it to systems of practical significance.

Reconstructability analysis is just one example of an important methodological area which would have no practical significance without the aid of sophisticated computer technology. Such examples are not rare in systems problem solving; on the contrary, they are rather typical.

The use of the computer as an intuition safeguard and amplifier in systems problem solving is only one of its two major roles in systems science. The other one is its use as the systems science laboratory. In this latter role, the computer is used for experimenting with systems simulated on it. At least three distinct aims of this experimentation can be recognized:

1. *The traditional use of computer simulation.* A system that models relevant aspects of some object of investigation is simulated on the computer for the purpose of generating scenarios under various assumptions regarding the environment of the system as well as various parameters of the system itself. Some of the best-known examples in this category are in the areas of industrial and world dynamics initiated by Jay Forrester [FO2, FO3].

2. *Discovery or validation of systems science laws.* Experiments of some kind are performed on the computer with many different systems of the same class. The aim of this experimentation is to discover useful properties characterizing the class of systems under investigation or, alternatively, to validate some postulated hypotheses regarding the class.

One of the most exemplary experiments of this kind was performed by Gardner and Ashby [GA6]. Their objective was to determine the effect of the size of a system (the number of variables involved) and its connectance (the number of dependencies between the variables) on the probability of stability in a particular class of systems. Gardner and Ashby restricted their investigation to a very special class of systems (linear dynamical systems described by simultaneous first-order and linear differential equations with constant coefficients). Among other results, their study led to the discovery of a critical connectance and the following statistical law valid for the investigated class of systems: If a linear dynamical system (as described above) is sufficiently large (consists of 10 variables or more) and its connectance (percentage of nonzero off-diagonal entries in the matrix describing the system) is smaller than 13% (critical connectance), then it is almost certain that it is stable; if its connectance is greater than 13%, then it is almost certain that it is not stable; a 2% deviation either way from the critical connectance is sufficient to convert the answer to the stability question from "almost certainly stable" to "almost certainly unstable" (Figure 1.4). Makridakis and Faucheux continued the same kind of experimental investigations for a more general class of dynamic systems described by nonlinear and time-varying differential equations [MA3]. They expressed some of their results for various circumstances by mathematical formulas. For instance, the probability of stability $p(n)$ of a given system with n variables from the above-described class of systems (chosen randomly) is given by the function

$$p(n) = e^{1-1.1n},$$

which fits the experimental data extremely well. It is certainly quite reasonable to view this function as a law of systems science.

Somewhat similar simulation studies, biologically motivated, were conducted by Kauffman [KA4] for systems represented by interconnected logic elements.

Walker, Ashby, and Gelfand [GE1, WA1, WA2] experimentally investigated systems built up from functionally identical elements representing a simple finite state

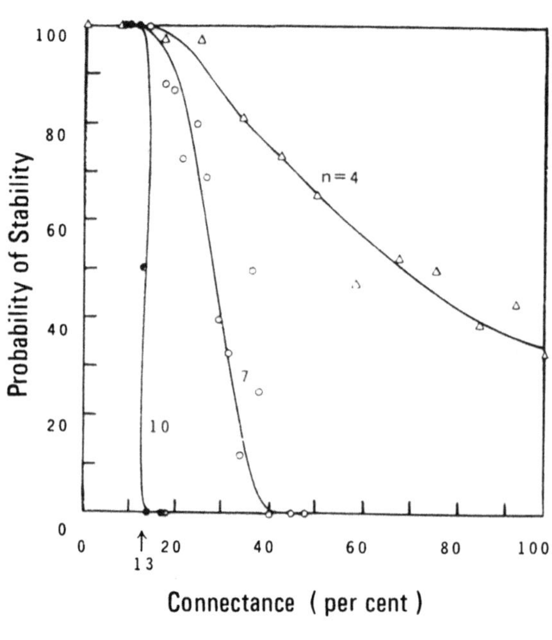

Figure 1.4. Variation of stability with connectance (experimental results obtained by Gardner and Ashby).

machine with two inputs and two internal states. The aim of these studies was to determine the dependence of cycle length and other behavioral characteristics on the size of the system for various types of finite-state machines.

As another example, of a rather different kind of experimental investigation of certain systems properties, let us mention an empirical formula

$$C = \frac{K_1(K_2)^n gh}{g+h}$$

for the average cost C of a two-level memoryless logic network (switching circuit) implementing a single Boolean function with n independent variables, g "one" vertices, and h "zero" vertices. The formula was determined by Kellerman [KE1] on the basis of a large number of computer experiments. K_1 and K_2 are constants that depend, generally, on the technology used, available types of components, and the definition of "cost." Kellerman also determined values of these constants for some cases of practical significance. This formula can help the logic designer in getting rough estimates of several different designs and throw some light upon the general question: "Using the same technology, objective criteria, and constraints, is a switching circuit with a inputs and b outputs cheaper or more expensive (on the average) than a switching circuit with c inputs and d outputs, when either $a < c$ and $b > d$ or $a > c$ and $b < d$?"

3. *Experimental characteristics of methodological tools.* A problem situation whose solution is known is simulated on the computer. A relevant methodological tool under investigation (usually one designed for a problem of a nondeductive nature) is used to solve the problem. The result obtained is then compared with the known solution. This is repeated a sufficient number of times for different problem situations of the investigated class with the aim of determining useful characteristics of the methodological tool involved. Such characteristics are very important for users of the various methodological tools, as they allow them to properly interpret the results obtained and make appropriate decisions. Experiments of this kind are described in Chapter 4 for one of the problems of reconstructability analysis.

The relationship between systems problem solving and computer technology is thus very rich. It is safe to say that systems problem solving, as understood and described in this book, would have virtually no practical value without the support of powerful computer technology.

1.6. ARCHITECTURE OF SYSTEMS PROBLEM SOLVING

> *Good architecture should be a projection of life itself and that implies an intimate knowledge of biological, social, technical and artistic problems.*
>
> —Walter Gropius

Although it is quite clear that systems problem solving enjoys infinite variety, it has increasingly been recognized that the infinite variety can quite satisfactorily be captured by a finite number of categories. Such categories of mutually interrelated systems problems result from some underlying principles by which all recognized systems are conveniently classified and organized.

Systems problem solving can be studied and developed at various levels of generality and detail. At the highest level of generality, the emphasis is on the development of pragmatically sound principles for organizing systems and on capturing a comprehensive view of systems problem-solving processes. Such general aspects of systems problem solving will be referred to as *systems problem solving architecture*. Professionals with competence in systems problem solving at this general level will be called *systems architects*.

Architecture is one of the oldest professions. Indeed, it was already well developed and recognized as a profession in Ancient Greece more than 2,000 years ago. One of the best characterizations of architecture remains the famous book *De Architectura*, written in Latin by Marcus Vitruvius Pollio, a Roman architect and engineer, in the first century B.C. [VI1]. The following quote from this book captures his views quite well:

> The architect should be equipped with knowledge of many branches of study and varied kinds of learning, for it is by his judgement that all work done by other arts is put to test. This knowledge is the child of practice and theory.

In its long and interesting history, architecture has been almost exclusively associated with the design of buildings. It has only recently been argued that certain general principles of architecture are not restricted to buildings, but are equally relevant and important to other areas of design.

The need for architecture has been recognized in the area of computer design since the early 1960s. It seems that the term "computer architecture" was coined by Fred Brooks in association with the development of the IBM computer STRETCH. In his article entitled "Architectural Philosophy," he introduced the following definition [BR6]:

> Computer architecture, like any other architecture, is the art of determining the needs of the user of a structure and then designing it to meet those needs as effectively as possible within the economic and technological constraints.

This view of computer architecture, which emerged through the experiences in designing STRETCH, was then consciously followed in the development of the IBM System/360, a family of mutually compatible computers designed in a common style that reflected users' needs as well as economic and technological realities. The result of this innovative approach to computer design is described in a paper entitled "Architecture of the IBM System/360" by its three architects, Amdahl, Blaauw, and Brooks [AM1].

After these beginnings, computer architecture quickly became recognized as an important part of computer design. It is now a standard subject in virtually every curriculum in computer science or engineering, and it is also well covered in computer literature.

The recognition of computer architecture was the first step in extending the concept of architecture beyond its traditional meaning as architecture of buildings. This book proceeds one step further and extends the concept of architecture to all systems. It is not the first instance when such a generalization is suggested. For example, Herbert Simon discussed this idea under the name "architecture of complexity" as early as 1962 [SI1], Heinz Zemanek has repeatedly argued the significance of generalizing the notion of architecture to all systems and introduced the names "generalized architecture" [ZE6] and "abstract architecture" [ZE7], and George Towner expressed similar sentiments in his book entitled '*The Architecture of Knowledge*' [TO2]. None of these suggestions, however, is oriented to systems problem solving. In that sense, this book is unique.

The current trend of extending the scope of architecture beyond its traditionally narrow domain of buildings (including, perhaps, other related constructions such as bridges or ships) is by no means in conflict with the usual common dictionary definition of architecture. For instance, architecture is defined in the *Oxford English Dictionary* as "the art or science of building or constructing edifices of any kind for human use" or "the action or process of building" or "a building or structure." We can thus see that (i) the term "architecture" has three distinct connotations: a particular discipline, a particular type of human activity, and a particular result of that activity, and (ii) architecture is not restricted only to buildings in any of these connotations.

Two key characteristics of architecture can clearly be extracted from the common dictionary definition. First, architecture is associated with design, construction, building, and the like, i.e., processess of creating artificial objects. Second, it is concerned with human use of the created objects. Let me elaborate on both of these characteristics.

Although architecture is oriented to designing, constructing, and building, it does not cover the full range of these activities. The architect is thus a designer whose work has to be completed by other people. His role is to oversee the design at a global level, focusing on those aspects that involve each interface with the user. The remaining aspects, which are not necessary from user's point of view, should be left open in any good architectural design. In his considerations, however, the architect must be aware of the technological possibilities and economic restrictions to make sure that his architectural design can be completed without any great difficulties.

The aim of architectural design is to prepare overall specifications, derived from the needs and desires of the user, for subsequent design and construction stages. The first task for the architect in each design project is thus to determine what the real needs and desires of the user are, as beautifully expressed by Julien Guadet, a French architect: "The architect must first of all determine the content, from which he can then derive the container."

The process of good architectural design disregards almost all details of the prospective final construction, thus leaving enough freedom for further design and construction work, but it does contain specifications of all features that are significant for the user. As such, it represents an overall description of the envisioned construction, observed from an appropriate distance. The ability to choose the right distance, from which all user-significant properties are still well distinguished while all other properties vanish, characterizes a good architect more than anything else. More poetically, this point can also be expressed by the following quote from the book *Tao Te Ching* by a famous Chinese philosopher, Lao Tsu (sixth century B.C.):

> Greatness means vanishing;
> Vanishing means distance;
> Distance means return to greatness.

Our previous remarks can be now well summarized by a statement made by Heinz Zemanek [ZE6]:

> Architectural design is design from top to bottom, making every detail a function of the whole. In this view, architectural design becomes complementary to formal definition: only if the method of description provides full freedom to omit details and to speak about the wanted properties of the full system before starting any work combining the building parts, can one derive the details from the general structure.

According to Gerrit Blaauw [BL1], one of the architects of the IBM System/360, three main levels are characteristic of the top-down design of any system: architecture, implementation, and realization. The *architecture* of a system consists of the functional appearance of the system to the user; the *implementation* is concerned with the inner structure, considered from a logical point of view, which makes the required functions possible; and *realization* is a physical embodiment of the implementation.

Several principles of good architecture are generally recognized. They are well described by Blaauw in his paper [BL1]. The following list of principles is extracted from this paper:

1. *Consistency*. Good architecture is consistent. That is, with a partial knowledge of the system the remainder of the system can be predicted.
2. *Orthogonality*. This principle requires that functions which are independent of each other are kept separate in their specification.
3. *Propriety*. According to this principle, only functions that are proper to the essential requirements of the system should be accepted; in other words, good architecture does not contain any unnecessary functions.
4. *Parsimony*. No function in the architectural description should be repeated in different forms.
5. *Transparency*. Functions introduced in the process of implementation should not be imposed upon the user.

6. *Generality.* If a function has to be introduced, it should be introduced in such a way that it can be used for as many purposes as possible.

7. *Open-endedness.* Freedom should be provided for the use of a function in other ways than those envisioned during the design.

8. *Completeness.* The introduced functions should satisfy the needs and desires of the user as completely as possible within the technological and economic constraints.

The architecture of systems problem solving, to which this book is primarily devoted, should follow the general aims and principles of any architecture, as outlined in this section. First of all, it should be user-oriented, i.e., it should cover all types of systems problems with which the envisioned users deal. Although the notion of the user should be given as broad interpretation as possible, the primary focus is on scientists, engineers, and professionals of all sorts. The various types of systems problems are thus predominantly extracted from systems problems recognized in various branches of science and engineering, as well as professions such as medicine, management, or law.

At the architectural level, systems problem solving should be seen and described from a proper distance to recognize its overall structure without being distracted by details. The actual architectural design for systems problem solving, which is the GSPS mentioned in Section 1.2, should be developed in an appropriate top-down manner and should reflect the various principles of good architecture. This is illustrated in Chapter 9, after all relevant functions of systems problem solving are determined and described in Chapter 2–8. As fas as the implementation of the GSPS is concerned, only some aspects of it are covered in the book, primarily associated with those problems which are epistemologically significant and which are based on highly general methodological distinctions. No realization of the GSPS is described in the book, but references are made to two partial realizations which are currently available.

NOTES

1.1. It seems that the terms "general system" and "general systems theory" are due to Ludwig von Bertalanffy. Although he introduced them orally in the 1930s, the first written presentations appeared only after World War II [BE6–8]. In his view, "general systems theory is a logicomathematical field whose task is the formulation and derivation of those general principles that are applicable to 'systems' in general." Von Bertalanffy was not only the originator of the idea of general systems theory but also one of the major organizers of the general systems movement, represented primarily by the Society for General Systems Research. The Society was founded in 1954 with the following objectives:

(1) to investigate the isomorphy of concepts, laws, and models from various fields, and to help in useful transfers from one field to another;

(2) to encourage development of adequate theoretical models in fields which lack them;
(3) to minimize the duplication of theoretical effort in different fields; and
(4) to promote the unity of science through improving communication among specialists.

Among early proponents of the general systems movement, Anatol Rapoport and Kenneth E. Boulding have probably been the most influential. Boulding envisioned general systems theory as "a level of theoretical model-building which lies somewhere between the highly generalized constructions of pure mathematics and the specific theories of the specialized disciplines" [BO2].

Ideas quite similar to those associated with general systems research, although focusing on information processes in systems, such as communication and control, were proposed, in the late 1940s and early 1950s, under the name "cybernetics." The most influential in this direction were the classic books by Norbert Wiener [WI1] and W. Ross Ashby [AS2]. Cybernetics, defined by Wiener as the study of "control and communication in the animal and machine," is based on the recognition that information related problems can be meaningfully and beneficially studied, at least to some extent, independently of any specific context. This view was considerably reinforced by a successful mathematical treatment given to the concept of information by Claude C. Shannon [SH3, SH4], a nucleus from which a mathematical theory of information emerged [FE1, GU1, KH1, WA6].

Later, in the 1960s, several efforts were made to formulate and develop various mathematical systems theories at high levels of generality. One of these theories, initiated mainly by Mihajlo D. Mesarovic, is based on the assumption that every system can be represented as a relation defined on a family of sets [ME1]. More mathematical structure is then introduced in various ways to study systems with certain specific properties. A good coverage of the theory can be found in a book by Windeknecht [WI3] and a more recent book by Mesarovic and Takahara [ME3]. Other mathematical systems theories were motivated by the desire to subsume theories of systems described by differential equations and finite state automata under one mathematical theory. The most successful of these theories turned out to be those developed by A. Wayne Wymore [WY1] and Michael A. Arbib [AR1]. Still other mathematical systems theories were developed from electric circuit theory through generalized circuit theory [ZA1] or from other background areas [BA2, HA7, RO4, ZA6].

The three intellectual areas—general systems research, cybernetics, and mathematical systems theories—together with computer technology, are the key elements of systems science. More details about the history and significance of these areas, as well as relevant aspects of some related areas (operations research, decision theories, artificial intelligence, etc.) can be found in a number of references [CA4, FL1, GA3, HA5, HA12, KL4, KL8, KL9, RO7].

1.2. It is interesting to note that the three periods of science suggested in Section 1.1 are characteristic of the three levels of societies that are usually recognized—the preindustrial, industrial, and postindustrial societies [BE1, GE5]. The preindustrial society is basically prescientific; the industrial society is characterized by the one-dimensional science; and the postindustrial society, which seems better characterized by the name "information society" [NA1], is associated with the two-dimensional science.

1.3. The idea of systems problem solving developed in this book occurred to me in the mid-1960s, when I became interested in methodological issues of general systems research. It was first hinted in one of my papers [KL2]. Later, I explained the idea, which was still half baked, a little more in one of my books [KL3]. The first sketch of the GSPS architecture was published in 1978 in a paper I coauthored with Roger Cavallo [CA4]. The same year, Hugo Uyttenhove completed his dissertation at the Department of Systems Science of SUNY-Binghamton that was devoted to the implementation and realization of some parts of the GSPS [UY1]. This initial realization has been considerably extended since 1978 and this process is still ongoing. A software package, which represents a subset of the GSPS, is now commercially available [UY2].

1.4. Gödel published his proof that the axiomatic method has certain inherent limitations in 1931 in German in *Monatshefte für Mathematik und Physik*, vol. 38, pp. 173–198. A popular exposition is included in *Gödel's Proof* by E. Nagel and J. R. Newman (Routledge & Kegan Paul, London, 1959).

1.5. Figure 1.3 was prepared by Brian Gaines and Mildred Shaw to characterize my hierarchy of epistemological levels of systems; it is used in this book with their permission.

2

SOURCE AND DATA SYSTEMS

A man said to the universe
"Sir, I exist!"
"However," replies the universe
"The fact has not created in me
A sense of obligation."

—STEPHEN CRANE

2.1. OBJECTS AND OBJECT SYSTEMS

> *Flower in the crannied wall,*
> *I pluck you out of the crannies,*
> *Hold you here, root and all, in my hand,*
> *Little flower—but if I could understand*
> *What you are, root and all, and all and all,*
> *I should know what God and man is.*
>
> —ALFRED TENNYSON

As human beings, we are able to distinguish ourselves from our environment. Our immediate awareness of the environment is a result of our perception. We have also the ability to store, process, and utilize information received from the environment, and this, in turn, reinforces our perception. These abilities are fundamental for our survival and well-being. They allow us to make decisions and act appropriately.

In daily life, when we interact with various objects in our environment, the interaction is usually restricted to just a few representative properties of each object. Although the interaction with an object may become increasingly richer when one becomes more and more acquainted with the object, it is always considerably restricted by the limited scope of human perception and ability to act, as well as all other limits of the human scale. In situations such as scientific inquiry, engineering design, medical diagnosis, criminal investigation, or artistic creation, the interactions with the objects of interest are considerably more pronounced and often extended beyond the limits of the human scale.

People trained and actively involved in each of the traditional disciplines of science, engineering, or other areas (medicine, law, etc.) are interested, in their professional work, in rather specific kinds of objects. For instance, objects of interest for ecologists include lakes, rivers, and forests; musicologists are interested in musical compositions or composers; psychologists study human individuals or small social groups; engineers are interested in all kinds of man-made objects such as power stations, cars, airplanes, computers, and the like; physicians deal with human patients and veterinarians with sick animals; criminologists are trained to investigate crimes; and biologists study all sorts of phenomena associated with living things.

For our further considerations, let the term "object" be defined as a part of the world that is distinguishable as a single entity from the rest of the world for an appreciable length of time. According to this definition, objects can be either material or abstract. *Material objects* can further be classified into *natural* (such as a piece of rock, a biological cell, the sun, or a group of animals) and *man-made* (such as an airport, a computer center, New York City, or a hospital). *Abstract objects* are usually man-made (such as a musical composition, a poem, or the U.S. Constitution), but some may be viewed as natural, at least to some degree (e.g. English or any other natural language).

In most cases, objects consist of a virtually unlimited number of properties, each of which can meaningfully be investigated. As a consequence, it is almost always practically impossible to study objects completely. This means that a selection of a limited (and, usually, rather small) number of attributes, which best characterize the studied phenomenon of the object, has to be made. When such a selection is made, a measurement (observation) procedure has to be defined for each attribute which, in turn, defines an abstract variable that represents our image of the corresponding attribute.

We say that a *system* is defined on the object of interest by selecting a set of relevant attributes on the object and by assigning a variable, in a particular way (through a measurement procedure), to each of them. The term "system" is thus always viewed as an abstraction—or an image—of some aspects of the object and not a real thing. This important distinction between the notions of object and system is well characterized by Ross Ashby in his book on cybernetics [AS2]:

> At this point we must be clear about how a "system" is to be defined. Our first impulse is to point at the pendulum and to say "the system is that thing there." This method, however, has a fundamental disadvantage: every material object contains no less than an infinity of variables and therefore of possible systems. The real pendulum, for instance, has not only length and position; it has also mass, temperature, electric conductivity, crystalline structure, chemical impurities, some radio-activity, velocity, reflecting power, tensile strength, a surface film of moisture, bacterial contamination, an optical absorption, elasticity, shape, specific gravity, and so on and on. Any suggestion that we should study "all" the facts is unrealistic and actually the attempt is never made. What is necessary is that we should pick out and study the facts that are relevant to some main interest that is already given . . . The system now means, not a thing, but a list of variables.

As already hinted, the term "variable" is used in this book for an abstract image of an attribute. Hence, to be able to define it properly, we have to elaborate on the underlying concept of an attribute first.

We observe that each attribute is associated with a set of possible appearances (manifestations). For instance, if the attribute is the relative humidity at a certain place on the Earth, the set of appearances consists of all possible values of relative humidity (defined in some specific way) in the range from 0% to 100%; if the attribute is the amount of estrogen hormone in a cm^3 of the blood of a woman, each particular amount of the hormone is an appearance of this attribute; if the attribute is defined by the color of the light at an intersection controlling traffic moving in a particular direction, the appearances are normally red, yellow, and green.

In a single observation, the observed attribute takes on a particular appearance. To be able to determine possible changes in its appearance, multiple observations of the attribute must be made. This requires, however, that the individual observations of the same attribute, performed according to exactly the same observation procedure, must be distinguished from each other in some way. Let any underlying property *that is actually used* to distinguish different observations of the same attribute be called a *backdrop*. The choice of this term, which may seem peculiar, is motivated by the recognition that the distinguishing property, whatever it is, is in fact some sort of background against which the attribute is observed.

A typical backdrop that is applicable to virtually every attribute is *time*. In this case, different observations of the same attribute are distinguished by being made at different times. For example, the relative humidity measured at one particular place allows multiple measurements provided they are made at different times, e.g., on every hour; similarly, different measurements of the amount of estrogen in a cm^3 of the blood of the same patient in a hospital can be made at different times, say at 8:00 a.m. and 8:00 p.m. every day, during her treatment period.

In some cases, different observations of the same attribute are not distinguished by time (i.e., are made at the same time or time is not relevant at all), but by different locations in *space* at which they are made. For example, various attributes characterizing acoustic quality can be observed at the same time at different locations of a concert hall. Space as a backdrop is particularly significant in some disciplines, e.g., crystallography, civil and optical engineering, the fine arts of painting and sculpture, and anatomy. The term "space" is not limited only to one, two, or three-dimensional Euclidean space. For example, space represented by points on a sphere, which is associated with the Riemannian geometry, may be an appropriate backdrop for some attributes (e.g., geologic, climatologic, or geographic attributes defined on the Earth). The order of words in a text of some sort can also be viewed as one-dimensional (abstract) space; attributes such as the positions and functions of words in each sentence, numbers of letters in the individual words, etc. can be observed at each point (word) of the text.

Time and space are not the only possible backdrops. Multiple observations regarding the same attribute can also be distinguished by individuals of some *population* on which the attribute is defined, such as a social group, a set of manufactured products of the same kind, the set of words in a particular poem or story, a set of countries, a group of laboratory mice, etc. For example, in any census, attributes such as the age, sex, income, occupation, academic degrees, etc., are observed at the same time for each individual of the total population of a country.

The three basic kinds of backdrops—time, space, population—can also be combined. Although all combinations are possible in principle, time–space and time–population are especially important and frequently used combinations. Let me illustrate each of them by some examples.

Time–space. This combination is best exemplified by any motion picture film, especially when it is used in research (growth of plants, study of microbioloical processes, or traffic situations at busy intersections and the like); most meteorologic attributes (relative humidity, temperature, wind velocity and direction, types of clouds, etc.) are observed at many places on the Earth as well as in time; a series of chess board situations of a particular chess game is another example of this combination.

Time–population. Attributes which characterize economic, political, and social situations are observed each year for different countries by various organizations such as the United Nations; a population of laboratory mice whose various physiological or behavioral attributes as well as attributes under the control of the investigator (stimuli, drugs, surgical treatments) are observed on a daily basis; attributes such as the number of published books or journals in certain categories, average book price and journal subscription rate, average size of books and journals, total income, and many others, can be observed annually for a population of publishers over a period of time.

Time, space, and population, which have special significance as backdrops, may also be used as attributes. For instance, when sunrise and sunset times are observed each day at various places on the Earth, the attribute is time and its backdrops are time and space; record times in some sport event, say swimming 400 m free style, are observations of an attribute distinguished by time (i.e., dates when they were accomplished); a location of a vehicle as an attribute can be observed in time as a backdrop; time needed to complete a problem can be observed for a population of computer programs used on the same computer.

As illustrated by this spectrum of examples, the selection of proper backdrops is quite flexible, but it is not completely arbitrary. Constraints on this selection are adequately expressed in terms of the following requirements for properly chosen backdrops; they can be used as guiding principles in the process of defining a system on an object of interest.

First, *backdrops must be applicable to all attributes in the system* for which they are defined. For instance, neither time nor space can be used to distinguish the same tests performed on manufactured products of some kind (it does not matter when and where the tests are made); no space or population are applicable to attributes that represent a musical composition; neither time nor a population is applicable for describing a mosaic.

Second, *backdrops of a system must be compatible with the purpose for which the system is defined.* For instance, appropriate attributes are usually observed on a patient for the purpose of monitoring his or her recovery after surgery and, if necessary, making desirable interventions; clearly, the only backdrop compatible with this purpose is time. On the other hand, if the purpose were to develop a medical data base, then the same attributes would be distinguished not only by time, but also by names or other identifiers of individuals in the same recovery stage whose data are to be included in the data base; hence, both time and a population are compatible with the purpose in this case.

Third, *observations of all attributes in a system must be uniquely distinguished by the backdrops of the system*, i.e., each element of the backdrop set (a particular time instant, point in space, individual of a population, or an appropriate combination of these) yields one and only one appearance for each of the attributes. For example, when attributes of words in a text of some sort are investigated (their position and function in each sentence, the number of letters in them, etc.), the population of all words in the text seems a reasonable choice of a backdrop. It is clearly applicable to the attributes as well as compatible with the purpose of investigation. However, it does not satisfy the requirement of unique distinguishability of observations. Indeed, the same word may have the same position and function in several sentences in the text and, of course, it has always the same number of letters. To distinguish each observation, we have to resort in this case to one-dimensional abstract space—the location in the text.

Once the meaning of attributes, backdrops, and their relationship is properly understood, it is easy to introduce a formal definition of a system defined on an object—or *object system*. It is a set of attributes, each associated with a set of appearances, and a set of backdrops, each associated with a set of its elements.

▶ Formally,

$$\mathbf{O} = (\{(a_i, A_i) | i \in N_n\}, \{(b_j, B_j) | j \in N_m\}) \tag{2.1}$$

where $N_n = \{1, 2, \ldots, n\}$ and $N_m = \{1, 2, \ldots, m\}$ (N subscripted by a positive integer is always used in this book to denote the set of positive integers from 1 through the value of the subscript); a_i, A_i denote an attribute and a set of its appearances, respectiely; b_j, B_j denote a backdrop and a set of its elements, respectively; and $\mathbf{O}$ denotes an object system.

The sets A_i and B_j in Eq. (2.1) are well defined for some attributes and backdrops. For example, when a population is used as a backdrop b_j (a social group or a group of animals, manufactured products of some kind, or a population of countries), the set B_j is usually well defined. Similarly, appearance sets A_i are well defined for attributes a_i such as the monthly income of a person, the colors of a traffic light, the number of letters in a word, or number of passengers on a flight. There are many cases in science, however, in which the sets are not known and cannot be determined without resorting to metaphysics. Nevertheless, independently of what is the case, they can be related to some well-defined sets by specific observation or measurement procedures. The latter sets are thus images of the set A_i and B_j in terms of which knowledge about the attributes is formulated. ◀

The very existence of attributes, backdrops, and the associated sets A_i, B_j, as properties of natural objects, is a subject of philosophical controversy. Various views range from naive realism, which fully accepts their existence, to the extreme form of operationalism (or instrumentalism), which rejects their existence and maintains that the meaning of any scientific concept is fully and exclusively determined by the specification of a measurement procedure.

This controversy is irrelevant to our aims. However, we have to be aware that, independent of the position one takes on the existence issue, the sets A_i and B_j are often unknown. In such cases, the object system is vacuous and can be given some meaning only in association with specific observation or measurement procedures through which images of the attributes are created. Hence, the object system has to be viewed as a component of a larger system; it is of little or no use when considered alone.

2.2. VARIABLES AND SUPPORTS

> *Nature does not declare herself freely, but only speaks when spoken to . . .*
>
> —Francis Bacon

The term "*variable*" is used in this book for an operational representation of an attribute, i.e., for an image of an attribute defined in terms of a specific measurement or observation procedure. Each variable has a name (label), which distinguishes it from all other variables under consideration, and is associated with a particular set of entities through which it manifests itself. These entities are usually referred to as *states* (or values) of the variable; the whole set is called a *state set.*

In a similar way, the term "*support*" is used for an operational representation of a backdrop. Each support has a unique name and is accompanied by a particular set; let this set be called a *support set* and its elements *support instances.*

In analogy with attributes and backdrops, it is assumed that different observations of the same variable are distinguished by instances of the supports involved. If two or more supports are used, their overall support set is the Cartesian product of the individual support sets. It is required that each particular support instance (of the overall support set) identify one and only one observation of the associated variables.

Some mathematical properties, such as an ordering or distance, may be defined on the individual state sets or support sets. As far as they are embedded in the relevant measurement procedures, they reflect some underlying properties of the corresponding attributes and backdrops. Differences in these properties among variables or supports, which have important methodological implications, are referred to as *methodological distinctions.* They are discussed in Section 2.3.

In addition to *specific variables and supports*, each of which represents a particular attribute or backdrop, respectively, we also recognize *general variables and supports.* The latter are abstract entities, i.e., they have no meaning in terms of some attributes or backdrops. Their state or support sets and the various properties defined on these sets are represented in some convenient standard manner.

A general variable is given an interpretation when elements of its state set are assigned by an isomorphic mapping (a one-to-one mapping that preserves all relevant mathematical properties defined on the set) to elements of the state set of a specific variable; the same holds for general and specific supports when applied to their support sets. Let any isomorphic mapping of this kind be called an *exemplification* of the general variable (or general support), and let the inverse of any of these mappings be called an *abstraction* of the specific variable (or specific support) involved.

► To formalize the concepts of general and specific variables and their supports, let the following symbols be added to those introduced in the last section: v_i, V_i, $\mathscr{V}_i$: a general variable, its state set, and a set of mathematical properties defined on V_i, respectively; $\dot{v}_i$, $\dot{V}_i$, $\dot{\mathscr{V}}_i$: the same aspects of a specific variable which is an exemplification of v_i; w_j, W_j, $\mathscr{W}_j$: a general support, its support set, and a set of mathematical properties defined on W_j, respectively; $\dot{w}_j$, $\dot{W}_j$, $\dot{\mathscr{W}}_j$: the same aspects of a specific support which is an exemplification of w_j.

Let any operation by which a specific variable is introduced as an image of an attribute be called an *observation channel*. The observation channel through which attribute a_i is represented by variable $\dot{v}_i$ realizes a function

$$o_i\colon A_i \to \dot{V}_i \tag{2.2}$$

that is homomorphic with regard to the presumed relevant properties of A_i and those in $\dot{\mathscr{V}}_i$. A similar function, say

$$\omega_j\colon B_j \to \dot{W}_j, \tag{2.3}$$

expresses a representation of backdrop b_j by support $\dot{w}_j$; it must be homomorphic with respect to the presumed relevant properties of the backdrop (say time) and those in set $\dot{\mathscr{W}}_j$. ◄

In some cases of attributes and backdrops, the observation channels may consist of explicit definitions of the functions o_i and ω_j. In other cases, however, when sets A_i and B_j are not known, no explicit definitions of the functions are possible without the use of some metaphysical assumptions. In such cases, the representations of attributes and backdrops are introduced physically (operationally) rather than by mathematical definitions.

Except for the trivial cases, when functions o_i and ω_j are explicitly defined, an observation channel consists of a physical device and a procedure describing its use. The device is usually called a measuring instrument or meter. The procedure is a set of instructions which specify how to use the instrument under various conditions.

The term "measuring instrument" should be given a broad interpretation. In some areas, such as psychology, social sciences, or ethology, the investigator himself (or his team) functions as the instrument or, alternatively, questionnaires or tests are used to measure attributes such as opinions, attitudes, or abilities of people. Any measuring instrument must be able to interact with the measured attribute and must convert this interaction into some form which directly represents states of the corresponding variable (e.g., a pointer on a scale, digital display, or graphical record).

Although measuring instruments and procedures that form observation channels must satisfy some general principles of measurement, they are considerably dependent on what is actually measured. As such, their study, design, and use are predominantly organized along the traditional disciplines of science. While both the theory and practice of measurement are crucial for the traditional disciplines of science, as well as such professions as engineering, medicine, management, etc., it is outside the scope of systems problem solving. Hence, observation channels are only acknowledged in the GSPS framework as components that are necessary for a full definition of any particular real world system, but they are not included in the GSPS proper. It is recognized that, owing to their intimate association with specific phenomena, observation channels must be studied and developed within the traditional disciplines.

►The GSPS deals only with general variables and supports. Given a general variable v_i, it is exemplifiable by a specific variable $\dot{v}_i$ if and only if a function

$$e_i\colon V_i \to \dot{V}_i \tag{2.4}$$

exists that is isomorphic with respect to the mathematical properties in $\mathscr{V}_i$. Similarly, a general support w_j is exemplifiable by a specific support $\dot{w}_j$ if and only if a function

$$\varepsilon_j\colon W_j \to \dot{W}_j \tag{2.5}$$

exists that is isomorphic with respect to $\mathscr{W}_j$. Each particular isomorphic function e_i (or ε_j) defines an exemplification of v_i by $\dot{v}_i$ (or w_j by $\dot{w}_j$, respectively). Inverse functions of e_i, ε_j, i.e.,

$$e_i^{-1}\colon \dot{V}_i \to V_i, \tag{2.6}$$

$$\varepsilon_j^{-1}\colon \dot{W}_j \to W_j, \tag{2.7}$$

define abstractions of $\dot{v}_i$ and $\dot{w}_j$, respectively. ◄

Example 2.1. To illustrate the concepts introduced, let a_i be the adjusted annual income of a U.S. taxpayer for the last year, as reported in his income tax form this year. Then, A_i consists of all possible amounts of money in U.S. dollars, from zero to the largest recognized amount, say \$100,000.00. The set is finite since the smallest unit of currency is 1c. We also recognize that it is totally (linearly) ordered. For calculating the income tax, it is sufficient to recognize only some ranges of the taxable income, each associated with some percentage of income that is to be paid as income tax. For the sake of simplicity, let these ranges be \$0–4,999.99, 5,000.00–9,999.99, . . . , 90,000.00–94,999.99, 95,000.00–100,000.00 and let the state set $\dot{V}_i$ of a specific variable $\dot{v}_i$, which is to be used to represent attribute a_i, be the set of minimal values in these ranges. A meaningful representation of a_i by $\dot{v}_i$ can be introduced by function o_i which assigns to all values in each of the defined ranges the minimum values in that range, e.g., $o_i(\$\,52{,}357) = \$\,50{,}000$ or $o_i\ (\$\,796) = \$\,0$. Function o_i is clearly homomorphic with respect to the total ordering in A_i since for each pair $\alpha, \beta \in A_i$, if $\alpha \leq \beta$, then $o_i(\alpha) \leq o_i(\beta)$. For methodological convenience, a general variable v_i can be defined for the specific variable $\dot{v}_i$ by an abstraction function $e_i^{-1}\colon \dot{V}_i \to V_i$. This function has to be isomorphic with respect to the ordering in $\dot{V}_i$. Assume that V_i is required to be a set of integers. Then e_i^{-1} can be defined, perhaps in the most natural way, by the equation

$$e_i^{-1}(\$\,5{,}000\ \mathrm{k}) = k \qquad (k = 0, 1, \ldots, 19).$$

The backdrop in this example is the set of all U.S. taxpayers in some category, say residents of the State of New York. No mathematical property is recognized in this set. Hence, ω_j: $B_j \rightarrow \dot{W}_j$ can be any one-to-one function by which a unique identifier, say social security number, is assigned to each taxpayer. For methodological convenience, an abstraction ε_j^{-1}: $\dot{W}_j \rightarrow W_j$ can be introduced by any one-to-one function that, e.g., assigns integers in the set N_n, where n is the number of taxpayers in the population, to the individual social security numbers.

► Let me elaborate now a little more on the concept of observation channel. Thus far, it has been defined in terms of functions o_i and ω_j given by (2.2) and (2.3), respectively. These functions induce partitions on sets A_i and B_j, say partitions A_i/o_i and B_j/ω_j. Elements in each block of these partitions are viewed as equivalent in the sense that they are not distinguished by the observation procedure involved. Each block of these partitions is thus represented as a whole by one state of variable $\dot{v}_i$ or one instance of support $\dot{w}_j$. When an observation of attribute a_i is made at some support instance, the observed attribute assumes a particular appearance from set A_i. This appearance is a member of exactly one block of the partition A_i/o_i. It is assigned by o_i to a particular state of variable $\dot{v}_i$. It is thus assumed that each observation permits us to recognize that block of A_i/o_i to which the actual appearance belongs, even though it does not allow us to identify the individual appearance itself.

The assumption that blocks of A_i/o_i can be distinguished by observations is warranted only when no observation errors are involved. Such situations do exist, as illustrated by Example 2.1, but they are relatively rare. Nevertheless, the assumption may be accepted as practically reasonable in other cases too, provided that the blocks of A_i/o_i are substantially larger than the estimated ranges of systematic observation errors. In such cases, a block of A_i/o_i can be correctly identified in each observation unless the actual appearance is close to a boundary between two blocks, i.e., within the range of observation errors. Since attributes (at least some) are not under the control of the investigator, he cannot prevent them from assuming appearances undesirably close to the boundaries between the blocks in A_i/o_i and, consequently, the possibility of identifying incorrect blocks by observations can only be reduced by an appropriate choice of the observation channel o_i, but it cannot be completely avoided.

As a result of the possibility of measurement errors, appearances near boundaries between the blocks of A_i/o_i have an associated observation uncertainty. This uncertainty can be viewed in either of the following two ways:

1. The blocks of a partition defined on A_i are viewed as sets without sharp boundaries. Using the terminology of fuzzy set theory, the blocks are fuzzy subsets of the set A_i. It is assumed that A_i is crisp (i.e., nonfuzzy). Each element of A_i belongs to each individual fuzzy subset with some membership grade. According to this view, the subsets are not defined by the function o_i, but solely by the membership grades.

2. The partition on A_i is defined by the function o_i. It is thus the same partition A_i/o_i that was considered previously. Its blocks are obviously crisp (nonfuzzy) subsets

of A_i. Given an element of A_i, it is generally uncertain to which block of A_i/o_i it belongs. This uncertainty can be expressed by a function that assigns a real number (usually between 0 and 1) to each pair consisting of an element of A_i and a block of A_i/o_i. This number is assigned in such a way that it expresses, according to the given context, the degree of certainty that the element belongs to the block.

In our further considerations, the second alternative is followed. That requires that function o_i, as given by (2.2), be defined first. Once defined, it imposes the partition A_i/o_i upon A_i. A function

$$\tilde{o}_i: A_i \times A_i/o_i \rightarrow [0,1] \tag{2.8}$$

is then defined, where $\tilde{o}_i(x, y)$ expresses the degree of certainty that x belongs to y. Since, however, each block of A_i/o_i is uniquely represented (labeled) by a state in set $\dot{V}_i$ (according to the function o_i), function $\tilde{o}_i$ can be redefined in a more convenient form

$$\tilde{o}_i: A_i \times \dot{V}_i \rightarrow [0,1], \tag{2.9}$$

where $\tilde{o}_i(x, y)$ expresses the degree of certainty that x belongs to that block of A_i/o_i which is represented by state y of variable $\dot{v}_i$.

Function $\tilde{o}_i$, defined by (2.9), characterizes observations of attribute a_i subject to uncertainty. It can also be viewed as a membership grade function that defines a fuzzy relation on the Cartesian product $A_i \times \dot{V}_i$. It is thus reasonable to call $\tilde{o}_i$ a *fuzzy observation channel*. Whenever confusion might arise, function o_i will be referred to as a *crisp observation channel*.

It is obvious that the crisp observation channel o_i is a prerequisite for defining the fuzzy observation channel. The crisp observation channel may also be viewed as a special case of the fuzzy observation channel. Indeed, when

$$\tilde{o}_i(x, y) = \begin{cases} 1, & \text{if } o_i(x) = y, \\ 0, & \text{otherwise,} \end{cases}$$

$\tilde{o}_i$ defines a crisp function from A_i into $\dot{V}_i$ that is identical with function o_i.

As far as backdrops are concerned, we can introduce a function

$$\tilde{\omega}_j: B_j \times \dot{W}_j \rightarrow [0,1], \tag{2.10}$$

analogous to (2.9) and based on (2.3), where $\tilde{\omega}_j(x, y)$ expresses the degree of certainty that x belongs to that block of the partition B_j/ω_j which is represented by instance y of support $\dot{w}_j$. However, there is virtually no use for this function. Indeed, when B_j is a population, function ω_j is one-to-one and no observation uncertainty is usually involved. When B_j is time or space, the actual observation is under the control of the investigator, i.e., he decides when to make observations or where to make them. This control over the actual observations, together with his considerable freedom to properly define function ω_j, enables the investigator to avoid any uncertainty in spite of unavoidable errors in time or space measurements. For example, if he decides that temperature, relative humidity, etc., be recorded at some meteorological station 24 times each day, 30 minutes after every hour, he may define function ω_j in such a way that each block of the resulting partition B_j/ω_j is a 1-hour time interval: [0–1 a.m.), [1–2 a.m.), . . . , [11 p.m.–0 a.m.). When a particular measurement of the observed attributes is taken, say at 1:30 a.m., there is clearly no uncertainty that this measurement represents the block [1–2 a.m.) under normal circumstances. Singular cases may occur (such as a gross violation of the rules or a malfunction of the used clock), but such cases are beyond the scope of what is usually considered a normal observation channel.

The various issues associated with observation channels can now be summarized as follows. For all practical purposes the crisp observation channel ω_j is sufficient for any backdrop, be it a population, time, or space. For attributes, however, both crisp and fuzzy observation channels (o_i and $\tilde{o}_i$) are useful and either type may be more appropriate under different circumstances.

Example 2.2. Let attribute a_i be the age of a person in a population B_j. Let elements of A_i be numbers of years in the range from 0 to 100. Let

$$\dot{V}_i = \{\text{very young, young, middle-aged, old, very old}\}$$

and let o_i be such that the one-to-one function $A_i/o_i \rightarrow \dot{V}_i$ is defined as follows:

$\{0, 1, \ldots, 14\}$—very young,
$\{15, 16, \ldots, 29\}$—young,
$\{30, 31, \ldots, 49\}$—middle-aged,
$\{50, 51, \ldots, 74\}$—old,
$\{75, 76, \ldots, 100\}$—very old.

When the crisp observation channel o_i is used, it leads to a rather poor characterization of persons whose age is close to the boundaries between the blocks of A_i/o_i. For instance, a 49-year-old person is labeled as middle-aged while a 50-year-old one is labeled old. When a fuzzy observation channel $\tilde{o}_i$ is used, for example the one characterized in Figure 2.1, the description is more satisfactory since it does not contain such abrupt changes. It is important to observe that a fuzzy observation channel does not produce one state of $\dot{V}_i$ for one observation, as the crisp channel does, but a tuple of values $\tilde{o}_i(x, y)$ for all $y \in \dot{V}_i$. For example, when observing the age of a person 25 years old, the following 5-tuple would be obtained via our fuzzy channel:

$\tilde{o}_i(25, \text{very young}) = 0.1$
$\tilde{o}_i(25, \text{young}) = 0.97$
$\tilde{o}_i(25, \text{middle age}) = 0.3$
$\tilde{o}_i(25, \text{old}) = 0$
$\tilde{o}_i(25, \text{very old}) = 0.$ ◀

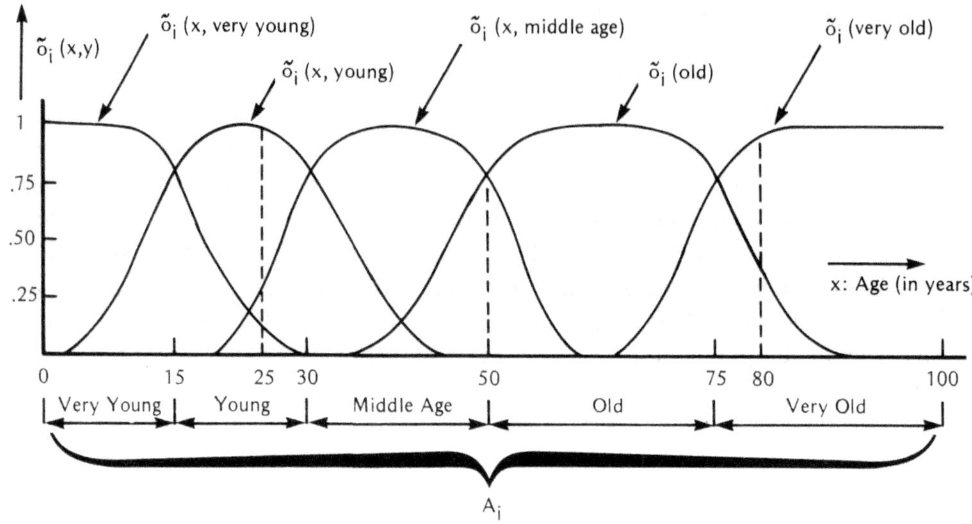

Figure 2.1. Crisp and fuzzy observation channels of the totally ordered attribute "age of a person."

2.3. METHODOLOGICAL DISTINCTIONS

> *If we say that a property has a distinct structure we mean a structure determined by empirical relations between empirical objects. . . . The more relations are taken into account in the definition of a scale, the more do the scale values tell us about reality. . . . It would be a waste of information to construct a scale which is a homomorphism with respect to an order relation and to neglect, say, an additive relation, if one can be empirically defined.*
>
> —J. PFANZAGL

The term "methodological distinction" is used in this book for characteristics of systems problems by which different problem types are distinguished within each epistemological problem category. Methodological distinctions involve either systems or requirements or both. Such changes as introducing a new methodological distinction into a system, excluding one, or replacing one by another do not change the epistemological type of the system. However, they may affect sets of methodological distinctions applicable to various requirements. Hence, methodological distinctions for systems must be chosen prior to those for requirements.

As the name suggests, problem types that differ from each other solely in some methodological distinctions require different methods, but they have exactly the same epistemological status in terms of the epistemological hierarchy of systems.

Methodological distinctions are thus criteria for a secondary classification of systems problems. They are applied in addition to the primary classification, the one based on epistemological criteria. Epistemological problem types are too broad for any specific methodological treatment. The aim of methodological distinctions is thus to introduce refined types of problems that are methodologically tractable.

The set of methodological distinctions recognized for systems, within a particular conceptual framework, is related to the set of recognized epistemological types of systems by the relation of "being applicable to"; a relation of the same kind exists also for requirements. While some methodological distinctions are applicable only to certain epistemological types, others are applicable to all types.

In this section, methodological distinctions applicable to variables and their supports are discussed. Since variables and supports are parts of every system, regardless of its epistemological type, these methodological distinctions are applicable to all epistemological types of systems.

Methodological distinctions regarding variables or supports involve properties that are recognized in their state sets or support sets, respectively. When a variable (or a support) represents an attribute (or a backdrop), the properties cannot be artitrary. Properties that are obviously not satisfied by the attribute or backdrop sets must not be recognized in the corresponding state or support sets. On the other hand, some presumed properties of the attribute (or backdrop), which are not relevant to the problem of concern, need not be recognized in the corresponding variable (or support).

To avoid possible confusion, the following remark is needed in order to clarify the meaning of methodological distinctions at the lowest epistemological level—the level of attributes, backdrops, and their abstract counterparts (variables and supports). For purposes of systems problem solving, methodological distinctions are defined for variables and supports (specific as well as general) and not for the corresponding attributes and backdrops. Methodological distinctions are thus defined at the lowest epistemological level solely in terms of the mathematical properties of the state sets and support sets involved. It must be ensured, of course, that the recognized properties do reflect some underlying properties of the corresponding attributes and backdrops. This, however, is an empirical issue, associated primarily with the methodology of measurement, which is beyond the scope of systems problem solving.

Each variable is associated with one or more supports and it is the overall support set within which changes in states of the variable occur. Hence, it is the combination of properties recognized in the state set and those in the overall support set which represents the most elementary kind of methodological distinctions.

When more than one support is involved, the overall support set is the Cartesian product of the individual support sets. Properties recognized in each individual support set have to be properly combined to express recognizable properties of this Cartesian product; these properties of the overall support set (the Cartesian product) are then used in characterizing, together with properties of the associated state set, an elementary methodological distinction. If the same properties are recognized in each of the individual support sets, it is easy to combine them, and the derived overall properties are homogeneous over the whole Cartesian product. The situation becomes more difficult when properties recognized in the individual support sets are not the same. In such cases, there are at least some overall properties that do not extend over the whole Cartesian product.

For the sake of simplicity, let us initially assume that we deal with one support set, regardless of whether it is an individual support set or a Cartesian product of several support sets, and that the properties recognized in it extend over the entire set.

The possibility that *no mathematical property* is recognized in a state set as well as the associated support set must be considered as one of the fundamental methodological distinctions. It is an extreme case, which would be a poor choice for any variable (or support) that is supposed to represent an attribute (or a backdrop) with some clearly recognizable and relevant properties. In many cases, however, this extreme methodological distinction is perfectly appropriate or even necessary. For example, variables such as marital status (single, married, divorced, widow), political affiliation (democrat, republican, independent), blood type (A, B, O, AB), or sex (female, male), each defined on individuals of some social population, illustrate the significance of this methodological distinction. In the literature devoted to measurement scales, variables of this kind are usually referred to as *nominal scale variables.*

The most fundamental property recognized in state or support sets is an *ordering.* From the methodological point of view, two kinds of orderings must be distinguished: partial orderings and linear orderings. A *partial ordering* is a binary relation on a set (a state or support set in our case) that is reflexive, antisymmetric, and transitive. A *linear ordering* is stronger; it is a partial ordering that is connected (i.e., each pair of distinct elements in the set is ordered either one way or the other).

►Formally, a partial ordering Q on a set, say a state set V_i, is a binary relation

$$Q \subset V_i \times V_i$$

that satisfies the following requirements:

1. $(x, x) \in Q$ (*reflexivity*);
2. if $(x, y) \in Q$ and $(y, x) \in Q$ then $x = y$ (*antisymmetry*);
3. if $(x, y) \in Q$ and $(y, z) \in Q$, then $(x, z) \in Q$ (*transitivity*).

If $(x, y) \in Q$, then x is called a *predecessor* of y and y is called a *successor* of x. If $(x, y) \in Q$ and there is no $z \in V_i$ such that $(x, z) \in Q$ and $(z, y) \in Q$, then x is called an *immediate predecessor* of y and y is called an *immediate successor* of x. In addition to the requirements of reflexivity, antisymmetry, and transitivity, a linear ordering satisfies the following requirement of *connectivity* for all $x, y \in A_i$: if $x \neq y$, then either $(x, y) \in Q$ or $(y, x) \in Q$. ◄

Examples of variables with partially ordered state sets are seniority or educational background of a person (defined, e.g., on a population of governmental employees). Examples of variables with linearly ordered state sets are the well-known Mohs' scale of hardness of solids, pitch as a characteristic of tones, or examination grades defined on a population of students. The ordering of support sets is best exemplified by the ordering of any time set. Although in most cases the ordering is linear, partially ordered time sets are also meaningful, e.g., in the study of spatially separated distinct processes (such as distributed computers) which communicate with one another and in which the transmission delays in the communication are not negligible compared with the time between changes in states of variables in the individual processes. Useful orderings can also be recognized in some populations. For example, human populations can be ordered by relations such as being older than, being a descendant of, or having a higher job position. They are usually partial orderings and their relevance depends on the kind of the population and the overall problem context. Variables with linearly ordered state sets are frequently referred to as *ordinal scale variables.*

In addition to the partial or linear orderings, there are other mathematical properties whose recognition in state or support sets is useful in many instances. One of the most important of these is a measure of distance between pairs of elements of the set involved. Such a measure is defined by a function that assigns to each pair of elements in the set a number which expresses how far apart the two elements are with respect to some underlying ordering.

►Given a set, say a state set V_i, a distance is thus defined by a function δ of the form

$$\delta: V_i \times V_i \rightarrow \mathbb{R}.$$

However, to qualify as an intuitively acceptable measure of distance, the function must satisfy the following requirements for all $x, y, z \in V_i$:

(δ1) $\delta(x, y) \geq 0$ (requirement of nonnegativity);
(δ2) $\delta(x, y) = 0$ iff $x = y$ (requirement of zero distance, also referred to as nondegeneracy requirement);
(δ3) $\delta(x, y) = \delta(y, x)$ (symmetry requirement);
(δ4) $\delta(x, z) \leq \delta(x, y) + \delta(y, z)$ (triangle inequality requirement).

Any function that satisfies requirements (δ1)–(δ4) is said to define a *metric distance* on set V_i. The pair (V_i, δ) is then called a *metric space.* Metric distances can, of course, be defined on both state sets and support sets. ◄

Examples of variables with recognizable and useful metric distances are almost all variables used in physics such as length, mass, pressure, electric current, voltage, or sound intensity, but there are many examples outside physics as well, e.g., variables whose state sets are money amounts, production amounts, numbers of defects, numbers of accidents, etc. It is quite obvious that space as well as time are supports for which the notion of metric distance is quite naturally applicable. On the other hand, rarely are useful metric distances recognized in populations. However, one such example is a population of students who are linearly ordered by their performance, and the distance is then defined for each pair of students by the absolute value of the difference between their positions in the ordered set. Variables whose state sets are associated with a metric distance are often called *metric variables.*

One additional property of state or support sets, which is of sufficient methodological significance to be recognized as a fundamental methodological distinction in our framework, is *continuity.* This is a concept well known from calculus and, consequently, it is not necessary to develop it here. Nevertheless, a few remarks regarding some aspects of continuity, which are relevant to later discussions, are appropriate.

►First, a necessary condition for continuity in a set is that the set is ordered. Since linear ordering is a special case of partial ordering, it is preferable to define continuity in terms of partial ordering. This can be done in a number of different manners. One definition of continuous partial ordering is based on the notion of a cut of a partially ordered set, which is defined as follows: A *cut of a partially ordered set,* say a state set V_i, is a partition of the set into two nonempty subsets, say X and $Y = V_i - X$, such that either no element of X is a predecessor (according to the partial ordering defined on V_i) of any element in Y and some element in Y is a predecessor of some element in X or no element in X is a successor of any element in Y and some element in Y is a successor of some element in X. A *continuous partial ordering* in V_i is then defined as a partial ordering in which any cut X, Y of V_i characterized by some element in X being a predecessor of some element in Y is such that either X has a largest upper bound in Y or Y has a greatest lower bound in X. ◄

The notion of continuous partial ordering is best exemplified by the ordering relation of "less than or equal to" defined on the set of real numbers or Cartesian products of this set. As a matter of fact, the very notion of a *continuous variable or support* is based on the requirement that the associated state or support set be isomorphic to the set of real numbers.

It follows from the previous remark that state or support sets of any continuous variables or supports, respectively, are uncountably infinite. As such, continuous variables and supports contrast with those which are associated with finite sets or, possibly, countably infinite sets. These latter are usually called *discrete variables or supports.*

While continuous variables and supports are represented by real numbers, their discrete counterparts can conveniently be represented by integers. This is particularly important when state or support sets of discrete variables or supports are linearly ordered and, hence, isomorphic to appropriate sets of integers. A natural metric distance defined by the absolute value of differences between integers, as well as by integer arithmetic, can also be used in dealing with some discrete variables or supports.

For our purposes the properties of ordering, metric distance, and continuity in state or support sets are considered a basis from which the most significant methodological distinctions are derived at the level of variables and supports. The following is a list of alternatives for each of these three properties in which the individual alternatives are assigned integer identifiers:

Ordering:	0—no ordering
	1—partial ordering
	2—linear ordering
Distance:	0—not recognized
	1—recognized
Continuity:	0—discrete
	1—continuous

The status of each variable (or support) with respect to these three properties can be characterized uniquely by a triple

(ordering, distance, continuity),

where a specific alternative (or its identifier) is entered for each of the properties. Thus, for example, (2, 1, 0) characterizes a discrete variable with a linearly ordered state set in which a metric distance is recognized.

Although 12 combinations of the three properties can be formed, the following three combinations are not meaningful: (0, 0, 1), (0, 1, 0), (0, 1, 1). Indeed, when no ordering is recognized in a set, no meaningful metric distance can be recognized in it and neither can it be viewed as continuous. Hence, there are nine meaningful combinations. Let these meaningful combinations be called *methodological types* of variables or supports. They can be partially ordered by a relation of "being methodologically more special than." The Hasse diagram of this partial ordering, which forms a lattice, is presented in Figure 2.2a. A simplified lattice is shown in Figure 2.2b for a framework in which the properties of ordering and distance are recognized, but not continuity.

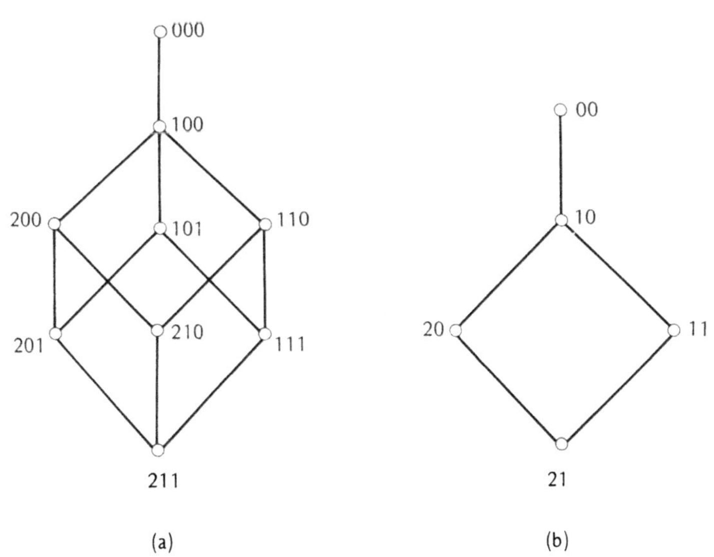

Figure 2.2. Lattices of methodological types of variables or supports.

At the level of variables and supports, each methodological distinction for a single variable is a combination of methodological types of that variable and the associated supports. There are 9 types of each. Hence, if only one support is involved or if it is required that all supports in a combination be of the same methodological types (the most common case), then the number of methodological distinctions is 81 (as methodological types of variables and supports do not restrict each other). If, in addition, the framework were restricted to discrete variables and supports, whose methodological types are summarized in Figure 2.2b, the number of methodological distinctions would be reduced to 25. The lattice of methodological distinctions for this case is described in Table 2.1.

Assume now that there are two or more supports, say m supports. They can all be of one type, two types (regardless of order), three types (regardless of order), etc. Assuming that $m \leq 9$ (which is a realistic assumption), the total number of methodological types of

TABLE 2.1.
Lattice of Methodological Distinctions for Discrete Variables and Supports

Methodological distinction	Immediate successors in the lattice
00/00	10/00 00/10
00/10	10/10 00/20 00/11
00/20	10/20 00/21
00/11	10/11 00/21
00/21	10/21
10/00	20/00 11/00 10/10
10/10	20/10 11/10 10/20 10/11
10/20	20/20 11/20 10/21
10/11	20/11 11/11 10/21
10/21	20/21 11/21
20/00	21/00 20/10
20/10	21/10 20/20 20/11
20/20	21/20 20/21
20/11	21/11 20/21
20/21	21/21
11/00	11/10 21/00
11/10	11/20 11/11 21/10
11/20	11/21 21/20
11/11	11/21 21/11
11/21	21/21
21/00	21/10
21/10	21/20 21/11
21/10	21/21
21/11	21/21
21/21	none

the overall support is given by the sum

$$\binom{9}{1}+\binom{9}{2}+\cdots+\binom{9}{m}.$$

When this sum is combined with the nine methodological types of a variable, we obtain the total number of possible methodological distinctions for one variable and its support; the number is given by the formula

$$9\times\sum_{i=1}^{m}\binom{9}{i}.$$

2.4. DISCRETE VERSUS CONTINUOUS

> *It is common for models of a theory to contain continuous functions or infinite sequences although the confirming data are highly discrete and finitistic in character.*
>
> —PATRICK SUPPES

As we saw in the previous section, the dichotomy of discrete sets versus continuous sets is included in the formulation of methodological distinctions at the level of variables and supports. Both discrete and continuous variables and supports are recognized in the GSPS framework. However, as far as the GSPS implementation is concerned, the book is almost exclusively restricted to discrete systems, i.e., systems with discrete variables and discrete supports. Continuous systems are covered only by occasional remarks and references to relevant literature.

There are several reasons why it was decided to restrict this book to discrete systems. The primary reason is that the scope of the GSPS is so large that it is not feasible to cover all aspects of its implementation in a book of reasonable size. In any event, the principal aim of this book is to describe the GSPS architecture and not its implementation. The GSPS implementation should thus be covered not for its own sake, but for the purpose of reinforcing the architectural description. In this sense, it is preferable, in my opinion, to include in the book a thorough description of a possible implementation of some meaningful and coherent subset of the GSPS rather than to cover the whole implementation spectrum superficially. The class of discrete systems and the associated problems represent such a desirable coherent subset.

Although continuous systems may seem equally suitable for illustrating the GSPS architecture, I would like to argue that discrete systems are preferable for a number of reasons, including the following:

(i) Regardless of whether one believes the world is basically discrete, continuous, or hybrid and regardless of the quality of the instruments used, the fact remains that most, if not all, observations are associated with some unavoidable finite error. The value of the error imposes some specific finite upper bound upon the resolution level involved in data gathering through a particular observation channel. This implies that data are always discrete regardless of philosophical beliefs or the state of technology.

(ii) When empirical considerations, as described in (i), are not dominating and the use of a continuous variable is desirable, an appropriate finite resolution level can always be chosen which defines a discrete variable approximating the desirable continuous variable as closely as one wishes; the same discrete approximations are, of course, possible for continuous supports as well. This point has clearly been demonstrated by Greenspan [GR1–5], who shows that classical as well as relativistic physics can be fully reformulated in terms of discrete variables and that this new formulation yields results that can be made as close as desirable to the results obtained through the traditional formulation based on continuous variables and differential equations.

(iii) While discrete variables (and supports) can always be defined in such a way as to approximate continuous variables to any accuracy desired, continuous variables are applicable only to certain kinds of attributes. In particular, sets of appearances of the applicable attributes must have structures isomorphic with the set of real numbers. This is an extremely severe restriction. The applicability of discrete variables and supports is thus considerably broader than that of their continuous counterparts.

(iv) If a real-world phenomenon can be described by continuous variables and supports, usually in terms of a set of differential equations, it is rare that methods of continuous mathematics can actually be used to handle such descriptions. Differential equations describing real-world phenomena are usually such that either they cannot be solved analytically (e.g., most nonlinear differential equations) or their analytical solution is difficult. Hence, it is either necessary or convenient to use numerical methods and digital computers for their solution; this obviously requires that the continuous variables and supports be converted to appropriate discrete counterparts. This is discussed by Greenspan [GR1]:

> It is usual, first, in the development of scientific knowledge, to have experimentation, which results in discrete sets of data. Theoreticians then analyze these data and, in the classical spirit, infer continuous models. Should the equations of these models be nonlinear, these would be solved today on computers by numerical methods, which results again in discrete data. Philosophically, the middle step of the activity sequence is inconsistent with the other two steps. Indeed, it would be simpler and more consistent to replace the continuous model inference by a discrete model inference, and this can be accomplished by denying the concept of infinity . . . . The concept of infinity and the consequential concepts of limit, derivative, and integral

are reasonable for the pure mathematical study of real numbers and real functions, but are *not* reasonable for the modeling of physical concepts and phenomena.

(v) Continuous variables and supports involve a number of purely mathematical difficulties and restrictions which not only result in rather high pedagogical demands but, more importantly, obscure the real issues.

(vi) It can clearly be observed that the dominance of the methods of continuous mathematics, characterizing the precomputer era, has been declining steadily since the appearance of the first commercial general-purpose digital computers in the 1950s. The analytic power of the calculus has become more and more outweighed by the steadily increasing computational power of digital computers. This will likely continue and, as a result, systems based on discrete variables will eventually dominate the field of systems research.

(vii) While the precision of continuous man-made systems (such as analog computers or regulators) is limited and cannot be increased beyond a certain level by any means, the precision of discrete man-made systems (digital computers, regulators, communication systems, etc.), is basically a matter of cost.

(viii) While man-made discrete systems may, by their very nature, be designed with various self-correcting features, no self-correction is possible for continuous man-made systems.

(ix) Discrete functions (expressing, e.g., dependencies of variables on their supports or on other variables) are more flexible than their continuous counterparts in their mode of representation. This issue is well argued by Andrew Barto [BA3]:

> . . . it is perfectly feasible to use symbolic expressions to define discrete functions. . . . It is also possible to define operators on discrete functions in terms of symbolic manipulations of these formulae. Thus turning to discrete functions one does not give up the possibility of concise symbolic expression. One gains, however, the advantage that using symbolic expression is not the *only* means of completely specifying functions as it is in the continuous case. Discrete functions can be completely defined by listing their values, e.g., storing the values in a computer so that "addresses" correspond to function arguments and "contents" correspond to function values, or by providing an algorithm whose input is a function argument and whose output is the corresponding value. . . . The primitive operations used in specifying algorithms (e.g., looping and conditional branching) permit the concise definition of functions which are impossible or very awkward to express by conventional algebraic means.

In addition to all these essential reasons for choosing discrete systems to illustrate some aspects of the GSPS implementation, there is also one practical reason for this choice. Continuous systems are much better developed and represented in the literature than discrete systems and it is thus more efficient to cover them indirectly by bibliographical notes and references.

2.5. IMAGE SYSTEMS AND SOURCE SYSTEMS

With only a single meter, there cannot be any science at all . . .
—Y. M. M. Bishop, S. E. Fienberg, and P. W. Holland

Attributes, specific variables and general variables, as well as backdrops, specific supports and general supports are components of three primitive systems—an object system, specific image system, and general image system, respectively—that form, together with the relationship among them, a source system. One of the three primitive systems—the object system—is conceptually introduced in Section 2.1 and formally defined by Eq. (2.1). The remaining two primitive systems have the same form as the object system, but their components are variables and supports rather than attributes and backdrops.

► Let $\dot{\mathbf{I}}$ and $\mathbf{I}$ denote a *specific image system* and a *general image system*, respectively. Then,

$$\dot{\mathbf{I}} = (\{(\dot{v}_i, \dot{V}_i) \mid i \in N_n\}, \{(\dot{w}_j, \dot{W}_j) \mid j \in N_m\}), \tag{2.11}$$

$$\mathbf{I} = (\{(v_i, V_i) \mid i \in N_n\}, \{(w_j, W_j) \mid j \in N_m\}), \tag{2.12}$$

where all symbols have the same meaning as defined in Section 2.2

A relationship among the three primitive systems—$\mathbf{O}$, $\dot{\mathbf{I}}$, $\mathbf{I}$—must now be defined. For the sake of notational simplicity, let us introduce a convention that, for each $i \in N_n$ and each $j \in N_m$, attribute a_i and variables $\dot{v}_i$, v_i correspond to each other and, similarly, backdrop b_j and supports $\dot{w}_j$, w_j correspond to each other.

The relationship between the object system and the specific image system is expressed by an overall observation channel that consists of individual observation channels, one for each attribute or backdrop in the object system. Let $\mathcal{O}$ denote a *crisp overall observation channel*. Then,

$$\mathcal{O} = (\{(A_i, \dot{V}_i, o_i) \mid i \in N_n, o_i \text{ is defined by Eq. (2.2) and must be } \textit{homomorphic} \text{ with respect to properties in } A_i, \dot{V}_i\}, \{(B_j, \dot{W}_j, \omega_j) \mid j \in N_m, \omega_j \text{ is defined by Eq. (2.3) and must be } \textit{homomorphic} \text{ with respect to properties in } B_j, W_j\}), \tag{2.13}$$

where all symbols have the same meaning as defined in Section 2.2.

A *fuzzy overall observation channel*, say $\tilde{\mathcal{O}}$, is obtained when o_i in Eq. (2.13) is replaced by $\tilde{o}_i$ defined by Eq. (2.9). Functions ω_j may also be replaced by $\tilde{\omega}_j$ defined by Eq. (2.10), but this is not considered in the GSPS framework for reasons discussed in Section 2.2.

The relationship between the specific and general image systems is expressed by a set of exemplifications/abstractions, one for each variable and support in the image systems. Let this set be called an *exemplification/abstraction channel* and let it be denoted by $\mathscr{E}$. Then,

$$\mathscr{E} = (\{(\dot{V}_i, V_i, e_i) \mid i \in N_n, e_i \text{ is defined by Eq. (2.4) and must be } \textit{isomorphic} \text{ with respect to properties in } \dot{V}_i, V_i\}, \{(\dot{W}_j, W_j, \varepsilon_j) \mid j \in N_m, \varepsilon_j \text{ is defined by Eq. (2.5) and must be } \textit{isomorphic} \text{ with respect to properties in } \dot{W}_j, W_j\}). \tag{2.14}$$

An observation channel from the object system directly to the general image system can also be considered. This channel, however, can be derived from the two channels defined by Eqs. (2.13) and (2.14). It consists of triples $(A_i, V_i, o_i \circ e_i^{-1})$ and $(B_j, W_j, \omega_j \circ \varepsilon_j^{-1})$, where the symbol $\circ$ denotes the operation of composition. ◀

The *source system* can now be defined as the quintuple

$$\mathbf{S} = (\mathbf{O}, \dot{\mathbf{I}}, \mathbf{I}, \mathscr{O}, \mathscr{E}). \tag{2.15}$$

These five components of source systems are shown in Figure 2.3, together with their relationships to the premethodological considerations (the investigator, object and purpose of investigation, etc.) and to epistemologically higher types of systems. The figure also summarizes the main methodological issues associated with the source system:

1. On one side, the source system represents interactions with the real world. They are mediated through the object system **O** and observation channel $\mathscr{O}$. On the other side, the source system is linked to the GSPS through the general image system **I** and exemplification/abstraction channel $\mathscr{E}$. These two components (**I** and $\mathscr{E}$) represent the interface between a particular discipline and the GSPS (as discussed in Section 1.2 and illustrated in Figure 1.2). This interface, which occurs at the lowest epistemological level of systems, is important since any interface at a higher epistemological level is based on it.

2. The GSPS conceptual framework is basically a language that is tailored to the description of significant systems problems. In its own domain, the GSPS is restricted to *syntactic aspects* of systems problem solving. They are represented by the notion of general image systems of various methodological distinctions and their epistemologically higher counterparts. The GSPS implementation can thus be developed and described solely in terms of general image systems and their various extensions at higher epistemological levels. When the GSPS is used in a particular inquiry or some other activity, the relevant *semantic aspects* are introduced through $\dot{\mathbf{I}}$, **O**, $\mathscr{O}$, and $\mathscr{E}$ of the source system. They consist of the dual processes of *abstraction* and *interpretation*. Abstraction is associated with functions o_i, ω_j, e_i^{-1}, and ε_j^{-1}; interpretation is characterized by functions e_i, ε_j, and partitions $o_i^{-1} = A_i/o_i$, $\omega_j^{-1} = B_j/\omega_j$. *Pragmatic aspects* are introduced at the premethodological level. They include the purpose and constraints of particular activities (scientific inquiries, systems design activities, etc.). Some of these pragmatic aspects are reflected in the extracted formulation in the GSPS language.

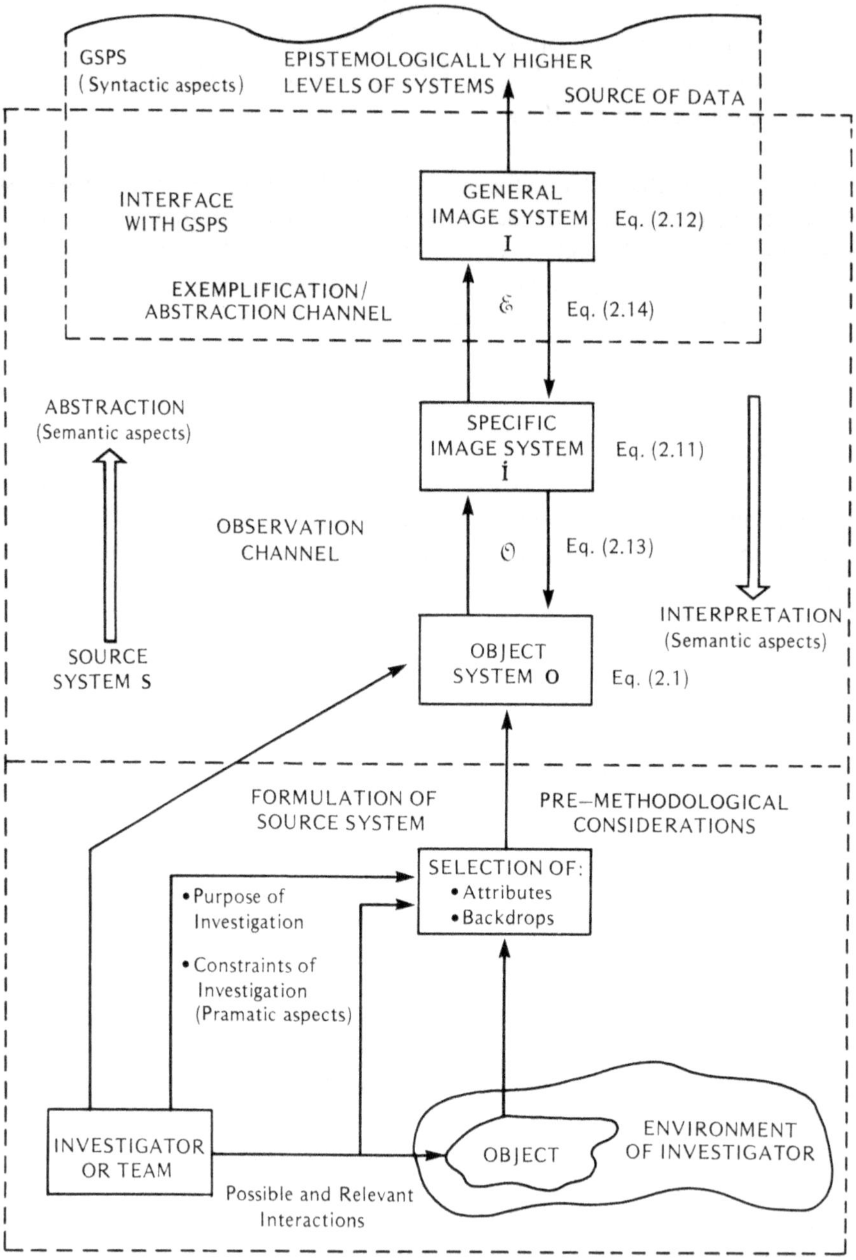

Figure 2.3. Conceptual elements involved in defining a source system on an object.

3. The source system, as the name suggests, is a source in two different respects. In one respect, exemplified by scientific investigations, it is a *source of empirical data*, i.e., a source of abstract images, expressed in the GSPS language, of some real-world phenomena. In another respect, exemplified by activities such as engineering design, it is a *source of interpretations of abstract data* that are either defined by the user or derived from within the GSPS.

As hinted in Section 1.3, two kinds of variables (or the corresponding attributes) may be usefully distinguished when a system is defined on an object. They are referred to as input and output variables (or attributes). In the following characterization of the difference between input and output variables, all aspects discussed are equally applicable to specific and general variables as well as to the corresponding attributes.

The dichotomy of input variables versus output variables emerges from pragmatic considerations. It expresses basically the user's point of view, which, in turn, is influenced or, in some cases, determined by the purpose for which the system is defined. *Output variables* of a source system are viewed by the user as variables whose states at the relevant support instances are determined from within the system, while the *input variables* are viewed as those determined from without. All factors that contribute to the determination of input variables are usually covered under the name "*environment of the system.*"

►Let systems with input and output variables be called *directed systems* and let those whose variables are not classified in this way be called *neutral systems*. According to this terminology, the source system, defined by Eq. (2.15), as well as its three primitive systems (**O**, **İ**, **I**), are neutral systems. To modify them into directed systems, each variable (or attribute) in their definitions must be declared either as an input variable or as an output variable (input or output attribute). Let this be done, say for system **I**, by defining a function

$$u : N_n \to \{0, 1\},$$

such that $u(i) = 0$ or $u(i) = 1$ are taken as declarations that variable v_i is an input variable or output variable, respectively. Let each n-tuple

$$\mathbf{u} = (u(1), u(2), \ldots, u(n)),$$

by which a particular input/output status is declared for each variable in the system, be called an *input/output identifier*. There are clearly 2^n possible input/output declarations for n variables, each represented by one input/output identifier $\mathbf{u}$.

The same input/output identifier is, of course applicable to variables $\dot{v}_i$ and attributes a_i as well. The definition of each of the three primitive systems—**O**, **İ**, **I**—can thus be easily modified to a definition of its directed counterpart by adding to it a particular input/output identifier. Let symbols of neutral systems be modified for their directed counterparts by adding a caret to each of them. Then,

$$\hat{\mathbf{O}} = (\{(a_i, A_i) | i \in N_n\}, \mathbf{u}, \{(b_j, B_j) | j \in N_m\}), \tag{2.16}$$

$$\hat{\dot{\mathbf{I}}}, = (\{(\dot{v}_i, \dot{V}_i) | i \in N_n\}, \mathbf{u}, \{(\dot{w}_j, \dot{W}_j) | j \in N_m\}), \tag{2.17}$$

$$\hat{\mathbf{I}} = (\{(v_i, V_i) | i \in N_n\}, \mathbf{u}, \{(w_j, W_j) | j \in N_m\}), \tag{2.18}$$

where $\hat{\mathbf{O}}, \hat{\dot{\mathbf{I}}}, \hat{\mathbf{I}}$ are the directed counterparts of the primitive neutral systems $\mathbf{O}, \dot{\mathbf{I}}, \mathbf{I}$, respectively. The *directed source system* is then defined by the quintuple

$$\hat{\mathbf{S}} = (\hat{\mathbf{O}}, \hat{\dot{\mathbf{I}}}, \hat{\mathbf{I}}, \mathcal{O}, \mathcal{E}). \blacktriangleleft \tag{2.19}$$

The difference between input and output variables is not well exhibited at the level of source systems. It becomes more apparent at higher epistemological levels, where various kinds of relationships among the variables are described. Since input variables are not considered as being determined from within the system, their states are viewed as conditions that are determined by the environment and have some influence upon the output variables. Relationships among variables of directed systems are thus expressed in terms of conditional propositions of the general form "if x, then y," where x is an overall state of input variables (determined by the environment) and y describes some property of the system. This contrasts with neutral systems, whose relationships among variables are described by simple propositions of the general form "it is true that y," where y again describes a property of the system. Input variables of directed systems may be influenced by their output variables as well, but such influence, if any, is not mediated through the system. It is mediated through the environment, as illustrated in Figure 2.4a. Properties of input variables of a directed system are thus not a subject of investigation within the context of that system.

There are two kinds of *degenerate directed systems*:

1. Directed systems with no output variables (Figure 2.4b), i.e., systems with $\mathbf{u} = (0, 0, \ldots, 0)$. These systems are methodologically vacuous. Indeed, each such system contains only input variables which, by definition, are fully determined by its environment and thus their properties cannot be expressed and investigated within the system itself. Hence, there is nothing in the system to be described and investigated. Any proposition that can be formulated within the system is meaningless since it contains only a condition but no consequent. Directed systems of this kind are excluded from the GSPS framework as methodologically meaningless. There are thus only $2^n - 1$ meaningful input/output declarations for n variables.

2. Directed systems with no input variables (Figure 2.4c), i.e., systems with $\mathbf{u} = (1, 1, \ldots, 1)$. These systems are methodologically sound since meaningful propositions can be formulated within them. However, the propositions cease to be conditional because there are no input variables in the systems that would provide the conditions. As such, the propositions become essentially the same as those formulated within the comparable neutral systems, i.e., neutral systems with the same sets of

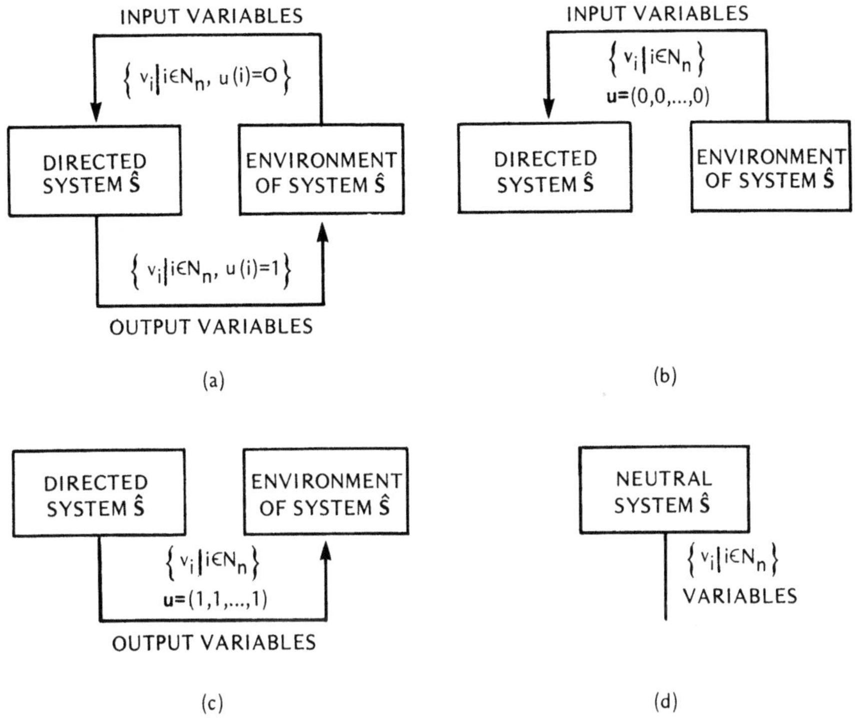

Figure 2.4. Methodological distinction of directed versus neutral source systems.

variables, supports, and the other components. However, there is a subtle difference between the two kinds of systems, as illustrated in Figures 2.4c, and 2.4d. While all variables in these degenerate directed systems are declared as output variables, those in the corresponding neutral systems are undeclared. As such, they are initially left uncommited and, if necessary or desirable, can be converted into appropriate directed systems at some later stage of their investigation. Note that, although these two kinds of systems are conceptually distinct, they are equivalent in the sense that either can be converted into the other simply by including or excluding **u**.

Properties of the environment of a directed system may not be known and, consequently, states of its input variables may be unpredictable within the support set. However, this lack of knowledge has no effect on the system itself since states of input variables participate in propositions about the system only as conditions. In some cases of directed systems, the determination of states of their input variables by the environment may be completely known. Such knowledge still has no effect on the description of the system, but it can be utilized in some problems that involve the system. There are also cases in which the input variables are completely controlled by the investigator, i.e., he represents the environment.

No environment is recognized for a neutral systems (Figure 2.4d). When a neutral system is replaced by a directed system, an environment is introduced and, provided that $\mathbf{u} \neq (1, 1, \ldots, 1)$, some information included in the neutral system is moved into the environment. Hence the resulting directed system contains less information than the original neutral system. A neutral system with n variables can be replaced by 2^n-1 directed systems, each represented by one n-dimensional input/output identifier u, but the one with no input variables (Figure 2.4c) is a rather trivial replacement. We may thus conclude that there are 2^n-2 nontrivial ways in which a neutral system can be replaced by a directed system.

The distinctions between neutral and directed systems and between crisp and fuzzy observation channels are two additional methodological distinctions recognized for source systems. They are independent of each other and of the distinctions based on properties in state and support sets. Each source system can be either neutral or directed, and observation channels of its variables can be all crisp, all fuzzy, or mixed. Hence, the new distinctions introduce $2 \times 3 = 6$ new possibilities. It must also be observed, however, that each source system may contain variables of different methodological types (in the sense of Figure 2.2a).

Let the total number of methodological distinctions recognized at the source system level be denoted by $\# S$. Then, under the realistic assumption that the number of supports does not exceed 9 ($m \leq 9$), we get

$$\# S = 6 \times \sum_{i=1}^{k} \binom{9}{i} \times \sum_{j=1}^{m} \binom{9}{j}, \tag{2.20}$$

where $k = \min(9, n)$.

If source systems with variables and supports of mixed methodological types were not allowed, the number of methodological distinctions would become $6 \times 9 \times 9 = 486$. If, in addition, only discrete variables were considered, the number would reduce to $6 \times 5 \times 5 = 150$. This is the range within which some implementation aspects of the GSPS are illustrated in this book.

Methodological distinctions recognized for source systems are important since they are applicable to all epistemologically higher types of systems. Let us illustrate them, together with other aspects of source systems, by the following two examples.

Example 2.3. Let the object of investigation be a stand of northern hardwood timber in western New York. Foresters usually mark trees for selective harvests at fairly regular intervals. The two main purposes of marking trees for cutting are to maintain or improve the overall quality of the woodlot and to remove timber of sufficient value to be economically attractive to the land owner and the timber processor. The aim of defining a source system on this object is to determine characteristics of trees that are marked for cutting, evaluate them critically, and develop more desirable and precise guidelines for marking trees in the future.

The backdrop in this example consists of a population of trees in the stand that are selected for the investigation. Assume that each tree investigated is labeled by an integer. Then, function ω is one-to-one and ε is the identity function.

Assume that seven attributes are selected on the object for this investigation. They are described and the corresponding variables defined for them in the following paragraphs.

Tree species: attribute a_1. Only four classes of all species recognized in the stand are distinguished in the investigation. Hence, a specific variable $\dot{v}_1$ with four states is needed to represent the attribute. Function o_1, through which the variable is related to the attribute, is defined in Figure 2.5a. Function e_1, defined in the same figure, is (as always) a simple relabeling scheme. No properties are recognized in sets A_1 and $\dot{V}_1$ and,

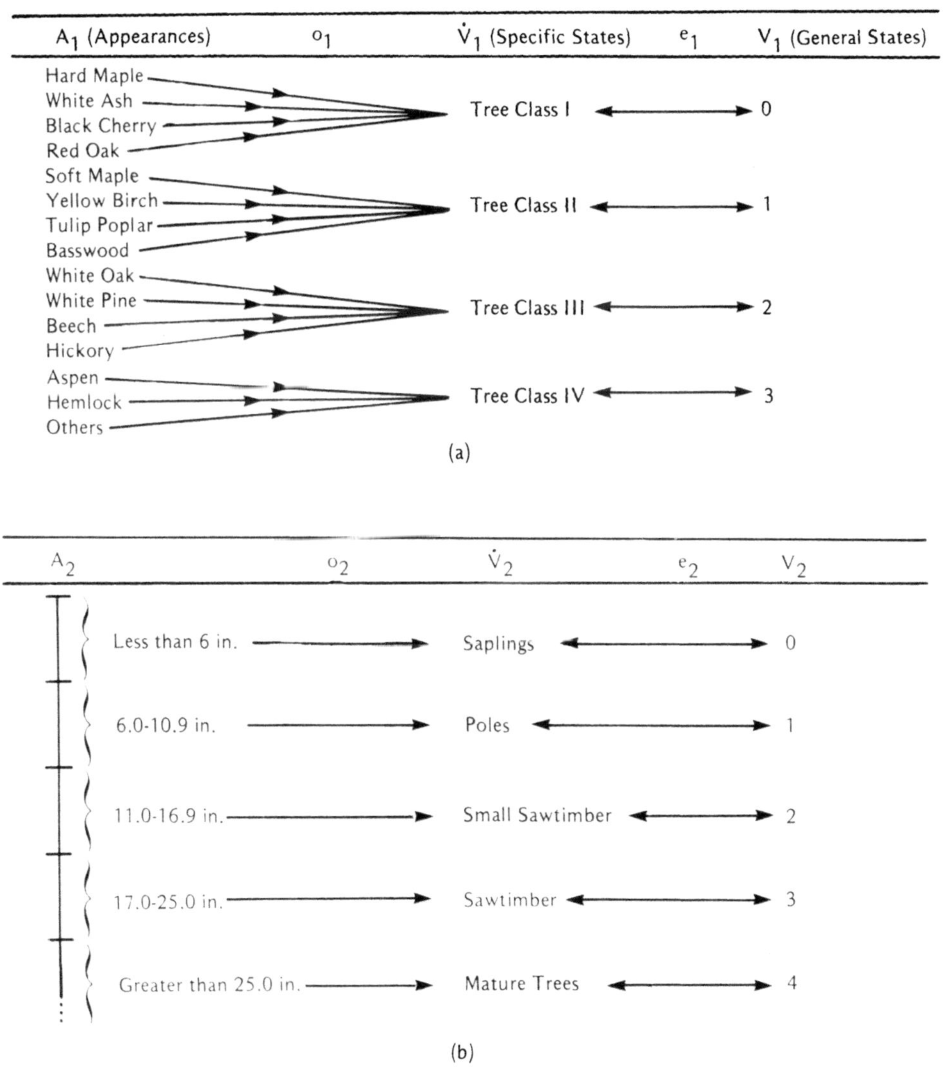

Figure 2.5. Definitions of variables for attributes a_1 and a_2 in Example 2.3.

consequently, the various properties of integers in set V_1 cannot be utilized in this case. The observation channel is crisp, i.e., it is directly represented by the function o_1.

DBH (diameter at breast height): attribute a_2. This attribute is more precisely defined as the diameter of the tree stem at $4\frac{1}{2}$ feet above ground level on its uphill side. Although DBH can be measured to the nearest 0.1 inch with a tape, it is customarily estimated or measured to the nearest inch or even-numbered inch when estimating timber volumes. However, for the purposes of selecting trees for cutting, 5 categories of diameter size are sufficient. They are defined in Figure 2.5b, together with functions o_2 and e_2. Although some measurement uncertainty may occur near the boundaries between blocks of the partition A_2/e_2, the observation channel o_2 can be viewed as crisp because this uncertainty would have little practical significance. Sets A_2, $\dot{V}_2$, V_2 can be viewed as linearly ordered with metric distance and, consequently, properties of the integers in set V_2 can be utilized if desirable.

Merchantable height: attribute a_3. Although this attribute can be measured quite precisely, it is sufficient to estimate its values and distinguish three ranges: less than 24 ft, 24 ft to 48 ft, greater than 48 ft. They can be mapped to states 0, 1, 2 of V_3, respectively. Order and distance of the integers in V_3 are meaningful.

Crown class: attribute a_4. Crown class refers to the size and position of a tree top relative to the tops of neighboring trees. Suppressed trees are overtopped by others, have relatively little influence on their neighbors, and respond poorly to release from competition. On the other hand, dominant trees have much influence on their neighbors. This attribute is considered very important as an indicator of a tree's ability to respond to release and develop into a desirable crop tree. The forestry profession has developed well-defined standards for classifying the actual appearances into four states of $\dot{V}_4$: dominant, codominant, intermediate, suppressed; they can be mapped, respectively, to states 0, 1, 2, 3 of V_4, to preserve the linear order in V_4.

U.S.F.S. grade: attribute a_5. The appearance of this attribute is dependent upon the number, size, and relative location of branches, branch scars, and other evidence which indicates the presence of knots in the wood. Four grades (states of $\dot{V}_5$) are recognized which are well defined and based on standards developed by the National Hardwood Lumber Manufacturers' Association and the U.S.D.A. Forest Service. They are called U.S.F.S. Grade 1, 2, or 3, and local use; when mapped to integers 0, 1, 2, 3 of V_5, their order is preserved.

Unsound defect: attribute a_6. The appearance of this attribute is an indicator of defects such as rot, crook, and sweep that reduce the amount of lumber from that which could be cut from a defect-free tree of the same gross dimensions. Three states are distinguished in $\dot{V}_6$: little or no defect, partly defective, and cull trees; they can be mapped to 0, 1, 2 in V_6 to preserve the order. Because this variable is an estimate of internal defect based on external evidence, it is subject to measurement and judgement error by even the most competent observers. It is thus desirable to use the fuzzy observation channel to allow the observer to express his uncertainty in each individual observation. The function o_6 is not defined explicitly in this case, but it is represented by the observer himself.

Tree marking: attribute a_7. The observed tree is either marked for cutting or not. Let these two appearances be mapped to 1 or 0 of V_7, respectively. No property is recognized in V_7.

We can see that the source system defined in this example is neutral. However, for the purpose of formulating rules for marking trees to be harvested, it would be redefined as a directed system with input variables v_1 through v_6 and output variable v_7. Since one of the observation channels is fuzzy while the others are crisp, the source system is a mixture of crisp and fuzzy variables. The support set has no recognized property and the state sets are of two or, possibly, three kinds: with no property, with linear ordering and, possibly, with both linear ordering and metric distance.

Example 2.4. Let the object of investigation be a particular female patient suffereing from anemia. The purpose of investigation is to monitor the so-called complete blood count (CBC) of the patient for some period of time to determine whether her anemic condition is improving on its own, or whether a regimen of treatment should be introduced.

The support is time. Measurements were taken once a day, at 7 a.m., for the entire month of September, 1982. The specific support set thus consists of dates 9/1/82, 9/2/82, . . . , 9/30/82, which can be mapped to integers 0, 1, 2, . . . , 29 of the corresponding general support set so that the linear ordering and distance are preserved.

Each measurement consists of determining states of the following four variables in a 10 cm^3 blood specimen taken from the patient. Each of the variables is metric and is defined by a crisp observation channel o_i such that the corresponding partition o_i/A_i contains blocks of equal size; states of the variable assigned to appearances in the individual blocks by o_i represent their midpoints. The four variables are a base from which several other variables included in the complete blood count are derived by specific calculations.

Red blood cell count: attribute a_1, defined as the number of red blood cells per cubic millimeter of whole blood and expected in the range of 4.2–5.4 million/mm^3 for normal females. It is measured to an accuracy of 10,000 cells. State set $\dot{V}_1$ consists of values 4.20, 4.21, . . . , 5.40 million/mm^3; according to the previous general remark regarding the nature of observation channels in this example, blocks $o_1^{-1}(x)$ of the partition A_1/o_1 that are represented by these values $x \in \dot{V}_1$ are thus, respectively,

$$4.105 \leq o_1^{-1}(4.20) < 4.205,$$
$$4.205 \leq o_1^{-1}(4.21) < 4.215,$$
$$\dots$$
$$5.395 \leq o_1^{-1}(5.40) < 5.405.$$

Specific states 4.20, 4.21, . . . , 5.40 can be then mapped by e_1^{-1} to integers 0, 1, . . . , 120, respectively, so that the ordering and distance properties of $\dot{V}_1$ are preserved in the set of integers V_1.

White blood cell count: attribute a_2, defined as the number of white blood cells per cubic millimeter of whole blood and expected in the range of 5–10 thousand/mm^3 for normal females. It is measured to an accuracy of 100 cells. State set V_2 consists of values 5.0, 5.1, . . . , 10.0 thousand/mm^3 that are represented, respectively, by the blocks

$$4.95 \leq o_2^{-1}(5.0) < 5.05,$$
$$5.05 \leq o_2^{-1}(5.1) < 5.15,$$
$$\ldots$$
$$9.95 \leq o_2^{-1}(10.0) < 10.05.$$

Function e_2 maps integers 0, 1, . . . , 50 to numbers 5.0, 5.1, . . . , 10.0, respectively.

Hematocrit: attribute a_3, defined as the volume of red blood cells expressed as a percentage of the volume of whole blood in a sample and expected in the range of 37–47 %. It is measured to an accuracy of 1 %. State set V_3 consists of values 37 %, 38 %, . . . , 47 % that represent, respectively, the following blocks:

$$36.5 \leq o_3^{-1}(37) < 37.5,$$
$$37.5 \leq o_3^{-1}(38) < 38.5,$$
$$\ldots$$
$$46.5 \leq o_3^{-1}(47) < 47.5.$$

Function e_3 maps integers 0, 1, . . . , 10 to numbers 37, 38, . . . , 47, respectively.

Hemoglobin: attribute a_4, measured by the amount of hemoglobin in grams per 100 milliliters of whole blood and expected in the range 12–16 g/ml. It is measured to an accuracy of 0.01 g. State set V_4 consists of values 12.00, 12.01, . . . , 16.00; functions o_4 and e_4 are determined in the same way as shown for attributes a_1, a_2, a_3.

Several other variables are included in the complete blood count set, each of which is defined by a specific formula on the basis of the four introduced variables. One such variable, referred to as *mean corpuscular volume* (MCV), is defined as the average volume of an individual red blood cell expressed in cubic microns and is expected in the range of 82–92 cubic microns. It is determined to an accuracy of 1 cubic micron by the formula

$$\text{MCV} = \frac{\text{hematocrit} \times 10}{\text{red blood cell count}}.$$

When a source system contains variables that are defined in terms of other of its variables, such as the variable MCV in the previous example, it contains artificial relationships among the variables introduced by the investigator. These artificial relationships must be clearly identified in the definition of the source system and must be separated from genuine relationships which emerge directly from the investigated phenomena.

2.6. DATA SYSTEMS

> *A system is a big black box*
> *Of which we can't unlock the locks*
> *And all we can find out about*
> *Is what goes in and what comes out.*
>
> KENNETH L. BOULDING

A source system is a frame within which observations of selected attributes can be made. If the observation channel is crisp, any actual observation is recorded in terms of an ordered pair that consists of an overall support instance at which the observation is made and the observed overall state of the variables involved. Since only one observation of the variables can be made at one support instance, the set of all these ordered pairs is a function from the overall support set into the overall state set. This function constitutes data or, more precisely, crisp data.

► Within the GSPS, data are always assumed to be expressed in terms of the general supports and variables (see Figure 2.3). To formalize the notion of data, we may thus consider only the general image system **I**, as defined by (2.12). Let

$$\mathbf{W} = W_1 \times W_2 \times \cdots \times W_m,$$
$$\mathbf{V} = V_1 \times V_2 \times \cdots \times V_n.$$

Then, *crisp data* are expressed by a function

$$d : \mathbf{W} \to \mathbf{V}. \qquad (2.21)$$

For each overall support instance, one overall state of the variables involved is assigned by function d.

While the image system **I** characterizes only potential states of the variables, function d provides information about their actual states within the delimited support set. When **I** is supplemented with d, it is thus reasonable to view this new ensemble (i.e., **I** and d) as a system at a higher epistemological level (level 1). Let such a system be called a *data system* and denote it by **D**. Then,

$$\mathbf{D} = (\mathbf{I}, d). \qquad (2.22)$$

Although this formulation lacks any semantic content, it is sufficient and convenient for developing and describing methodological features of the GSPS. For any particular application, however, the meaning of data d must be added to the formulation. This can be done by replacing the image system **I** in (2.22) by a relevant source system **S**. Let the resulting system be called a *data system with semantics* and denote it by ${}^s\mathbf{D}$. Then,

$${}^s\mathbf{D} = (\mathbf{S}, d), \qquad (2.23)$$

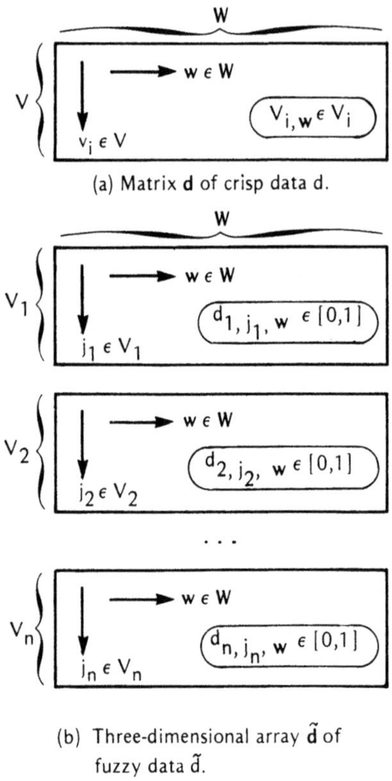

Figure 2.6. Standard forms of data representation for discrete variables. (a) Matrix **d** of crisp data d. (b) Three-dimensional array $\tilde{\mathbf{d}}$ of fuzzy data $\tilde{d}$.

where d is the same function as in (2.22). In this case, however, d is related to **S** as follows: when an observation characterized by

$$o_i \circ e^{-1}(x_i) = y_i$$

for all $i \in N_n$ (x_i denotes the presumed appearance of attribute a_i, and y_i denotes the corresponding state of variable v_i) is associated with the overall support instance $\mathbf{w} \in \mathbf{W}$, then

$$d(\mathbf{w}) = \mathbf{v},$$

where $\mathbf{v} = (y_1, y_2, \ldots, y_n) \in \mathbf{V}$. Depending on the problem of concern, d is actually determined in one of at least three different ways. First, it may result from observations or measurements, which is the case in all sorts of empirical investigations. Second, it may be derived from a higher-level system, as discussed in Chapters 3–5. Third, it may be defined for a specific purpose by the user, as in the problem of systems design.

Data systems **D** and ${}^{S}\mathbf{D}$ are neutral since they are defined in terms of a neutral image system **I** and a neutral source system **S**, respectively. A modification to their directed counterparts, say $\hat{\mathbf{D}}$ and ${}^S\hat{\mathbf{D}}$, is trivial. It amounts to replacing $\mathbf{I}$ by $\hat{\mathbf{I}}$ and $\mathbf{S}$ by $\hat{\mathbf{S}}$, respectively. Hence,

$$\hat{\mathbf{D}} = (\hat{\mathbf{I}}, d), \tag{2.24}$$

$${}^S\hat{\mathbf{D}} = (\hat{\mathbf{S}}, d) \tag{2.25}$$

are *directed data systems*, without semantics and with semantics, respectively.

If variables are defined in terms of fuzzy observation channels, then each actual observation is recorded as an ordered pair that consists of an overall support instance, with which the observation is associated, and an n-tuple $(h_1, h_2, \ldots, h_n)$ of particular functions

$$h_i: V_i \to [0, 1], \tag{2.26}$$

$i \in N_n$, where $h_i(y)$ expresses the degree of certainty that y is the observed state of variable v_i. To formalize the notion of fuzzy data, let

$$\tilde{\mathbf{V}} = \{V_1 \to [0, 1]\} \times \{V_2 \to [0, 1]\} \times \cdots \times \{V_n \to [0, 1]\}.$$

Then, *fuzzy data* are expressed by a function

$$\tilde{d}: \mathbf{W} \to \tilde{\mathbf{V}}. \tag{2.27}$$

For each overall support instance $\mathbf{w} \in \mathbf{W}$,

$$\tilde{d}(\mathbf{w}) = \mathbf{h},$$

where

$$\mathbf{h} = (h_1, h_2, \ldots, h_n) \in \tilde{\mathbf{V}}.$$

If data are fuzzy, then definitions of data systems must be modified by replacing function d in Eqs. (2.22)–(2.25) with function $\tilde{d}$. Since it always follows from the context which of the two cases actually occurs, it is not necessary to use different symbols for data systems with crisp and fuzzy data.

If a source system $\mathbf{S}$ is included in the definition of a data system with fuzzy data, $\tilde{d}$ and $\mathbf{S}$ are related as follows: when an observation associated with an overall support instance is characterized by

$$\tilde{o}_i(x_i, \dot{y}_{i,k}) = z_{i,k}$$

and

$$e^{-1}(\dot{y}_{i,k}) = y_{i,k}$$

for all $i \in N_n$, where x_i denotes the presumed appearance of attribute a_i, then

$$h_i(y_{i,k}) = z_{i,k}$$

for all $y_{i,k} \in V_i$ and all $i \in N_n$.

Crisp data can be represented in a number of alternative forms. Let a standard form of representation for discrete variables and supports be a matrix

$$\mathbf{d} = [v_{i,w}]$$

whose entries $v_{i,w}$ are states of variables v_i observed at overall support instances $\mathbf{w}$ (Figure 2.6a). Each column in $\mathbf{d}$ thus represents an overall state observed at $\mathbf{w}$, and each row represents all observations of one variable within the support set $\mathbf{W}$. If $\mathbf{W}$ is linearly ordered, columns in $\mathbf{d}$ should be ordered in the same way. When several supports are involved, such as population-time, space with more than one dimension or space-time, other forms of representation may be preferable. Some of them are introduced later for various examples.

For fuzzy data, a standard form of representation, which is similar to matrix $\mathbf{d}$, is a three-dimensional array

$$\tilde{\mathbf{d}} = [\tilde{d}_{i,\,j_i,\,w}]$$

whose entries are degrees of certainty that state j_i of variable v_i is observed at support instance $\mathbf{w}$ (Figure 2.6b). Clearly, $i \in N_n$, $j_i \in V_i$, $\mathbf{w} \in \mathbf{W}$, and $\tilde{d}_{i,\,j_i,\,w} \in [0, 1]$. Array $\mathbf{d}$ is represented by n matrices (pages, planes), one for each variable. Column $\mathbf{w}$ in the matrix for variable v_i represents a function h_i, given by Eq. (2.26), that is associated with the observation identified by $\mathbf{w}$. ◄

To illustrate various aspects of data systems and their representations, let several examples of specific data systems (i.e., data systems with semantics) be discussed in detail.

Example 2.5. In their study of animal behavior, ethologists use methods of observation that disturb as little as possible the natural habitat of the animals investigated. One of the methods used in studying groups of animals is to make motion-picture films and determine relevant behavior sequences from them. For each specific kind of animal, some significant postures and movements are usually recognized. Ethologists often specify them by characteristic pictures supplemented by verbal descriptions. For instance, principal postures of herring gulls are pictorially specified in Figure 2.7a and are given suggestive names such as "rest," "facing away," "forward," etc. Each of them is also described verbally; for example: "choking begins with bending down over the nest (or any depression in the ground similar to a nest, such as a human footprint), followed by a rhythmic up-and-down movement of the head."

The object of investigation in this example consists of two male herring gulls, identified as I and II. Variables are defined on the following attributes for both gulls:

a_1—type of action of gull I,

a_2—type of action of gull II.

The attributes are observed in time. The period of observation is 90 sec. It is divided into

(a)

$\dot{V}_1 = \dot{V}_2$	$e_1 = e_2$	$V_1 = V_2$
ATTACK	↔	0
UPRIGHT	↔	1
GRASS PULLING	↔	2
CHOKING	↔	3
RETREAT	↔	4

(b)

t =	1	2	3	4	5	6	7	8	9	10	11	12	13	14	15	16	17	18	19	20
v_1	1	1	0	3	3	3	3	3	4	3	3	0	2	1	1	1	1	4	4	4
v_2	4	3	4	3	3	3	3	3	4	4	3	3	4	2	1	1	1	4	4	4

t =	21	22	23	24	25	26	27	28	29	30	31	32	33	34	35	36	37	38	39	40
v_1	4	4	0	2	2	4	4	4	4	4	4	4	2	2	2	2	2	0	2	1
v_2	4	3	1	4	2	2	4	4	4	4	4	4	4	4	1	1	1	3	3	1

t =	41	42	43	44	45
v_1	0	2	1	1	4
v_2	3	3	3	3	4

(c)

Figure 2.7. Data regarding a boundary clash between two male gulls (Example 2.5).

intervals of 2 sec, each of which represents one observation. Hence, the specific time set (which is the support set in this example), say $\dot{T} = \{t_1, t_2, \ldots, t_{45}\}$, can be defined by the partition imposed on the time period of 90 sec by the observation channel:

$$0 \leq \omega^{-1}(t_1) < 2\,\text{sec},$$
$$2 \leq \omega^{-1}(t_2) < 4\,\text{sec},$$
$$\ldots$$
$$88 \leq \omega^{-1}(t_{45}) < 90\,\text{sec}.$$

$\dot{T}$ is a linearly ordered set with a metric distance; when mapped into a set of integers, say set $T = N_{45}$, by the function

$$\varepsilon(k) = t_k (k \in N_{45}),$$

the order and distance are preserved.

Each observation of the two attributes is represented by those pictures in the film that correspond to the respective period of 2 sec. They are analyzed by the ethologist and one of several previously defined types of action (a state from state sets $\dot{V}_1, \dot{V}_2$) is recognized for each gull. In this example, in which a boundary clash between two male gulls is investigated, the same five types of action are sufficient for each attribute (i.e., $\dot{V}_1 = \dot{V}_2$); their names are listed in Figure 2.7b, together with their integer labels (elements of general state sets V_1, V_2). The actions called "upright" and "choking" are two of the basic postures defined in Figure 2.7a. "Grass pulling" is defined as "pecking violently at the ground, uprooting plants and tossing them sideways with a flick of the head." The remaining two actions—"attack" and "retreat"—are well characterized by their names. A data matrix obtained from an actual motion-picture film is shown in Figure 2.7c, where $t \in T$ and $v_{1,t}, v_{2,t} \in V_1 (= V_2)$.

The data system defined in this example is a neutral system with semantics. Its variables are discrete and are supported by (observed in) time. The support set (time set) is linearly ordered and metric. The data are crisp.

Two remarks should be made about this example. First, the two identical observation channels

$$o_i : A_i \to \dot{V}_i \qquad (i = 1, 2),$$

where A_i denotes the presumed set of all possible appearances of a gull, are represented by the ethologist himself and cannot be mathematically defined. They are defined by the combination of pictorial and verbal characterizations.

Second, each observation (one column in the data matrix) is a summary (made by the ethologist) of what happened during the respective period of 2 sec. This is somewhat problematic, since actions of the gulls (in terms of the types recognized in the state set) do not necessarily begin and stop according to the predefined time scale. It seems more adequate in this kind of investigation to define the time set $\dot{T}$ implicitly, by changes in states of the variables, rather than explicitly. Time set $\dot{T}$ is defined implicitly by the following rule: the whole period of observation (90 sec in our example) is partitioned into time intervals during which none of the variables (two variables in our case) changes its state; when at least one variable changes its state, one time interval terminates and a new one begins. If it is important to keep information about durations of the individual actions, a new variable can be introduced by which the duration of each of the implicitly defined time intervals is measured (with a desirable precision) and recorded as part of the data. The *implicit definition of time sets* (with or without a supplementary variable that expresses durations of the implicitly defined time intervals) is often more adequate than any explicit definition. It is also meaningful and frequently desirable for variables whose support is space of some sort.

Example 2.6. Typical objects of investigations in musicology are musical compositions. Let the object in this example be a modern blues tune. Its score, which is given in Figure 2.8a, consists of two parts: a melody and a harmony. While the melody is defined in standard musical notation, the harmony is expressed in terms of the so-called "fake book" notation, frequently used by jazz musicians. To define a meaningful system on this blues tune (or any musical composition), three kinds of attributes have to be considered: pitch, rhythm, and harmony. All of them change in time. Relevant time intervals can be defined in terms of the duration of the shortest note in the composition, say duration Δt. In this example, Δt represents the duration of the one eighth measure. Time (as a support) can be defined explicitly or implicitly. When defined explicitly, elements of the corresponding time set $\dot{T}$ would be labels of time intervals $[0, \Delta t)$, $[\Delta t, 2\Delta t)$, etc., and would be mapped to an appropriate set of integers of the general time set T (e.g., 1, 2, . . ., respectively, to preserve the order and distance). When defined implicitly, the time intervals represented by $\dot{T}$ would be determined by durations of the individual notes in the melody. In this example, the implicit definition of time seems preferable and is adopted.

The three attributes—pitch, rhythm, and harmony—can be represented by variables in a number of different ways. Pitch of a single melody, for example, can be represented by one variable, say variable v_1, as shown in Figure 2.8b for the pitch range of our melody. However, it can also be represented by two variables, one of which would distinguish octaves, while the other would distinguish the 12 standard levels in a chromatic scale; or it can be represented by three variables, one for octaves, one for the seven basic levels (a, b, . . . , g) in each octave, and one for the modifiers b and # . Rhythm of a melody is a time-oriented attribute and, consequently, the definition of a variable by which it is represented depends on the definition of the support time. When the support time is defined implicitly (as in our example), the variable for rhythm, say variable v_2, must identify durations of the individual tones in the melody. This is done by multiples of Δt (or the 1/8 measure) as shown in Figure 2.8b; the observation channel o_2 by which time intervals $[0, \Delta t)$, $[\Delta t, 2\Delta t)$, . . ., $[4\Delta t, 5\Delta t)$ are introduced is obvious. As far as harmony is concerned, let a general variable for harmony, say variable v_3, be related to the corresponding specific variable $\dot{v}_3$ by the abstraction channel specified in

(C)
C
F7
C
C7
F7
F7
C
C
G7
F7
1.
2.
C
C

(a)

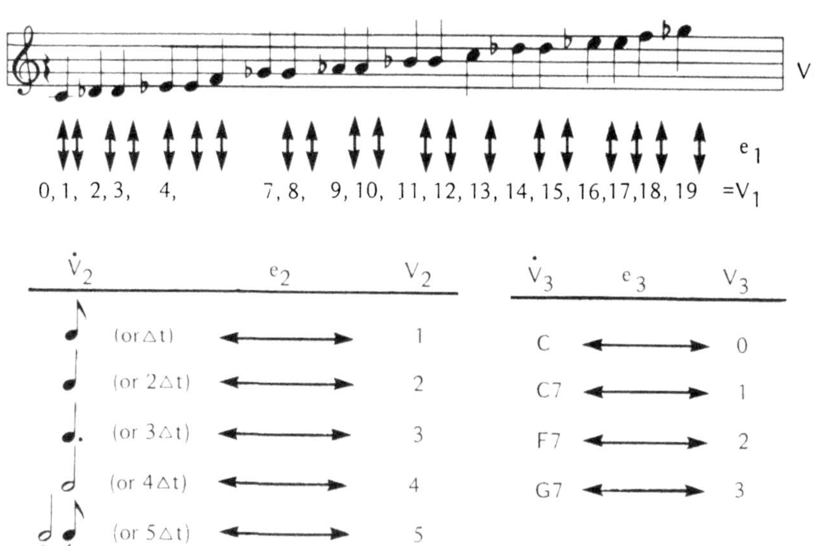
V_1
e_1
0, 1, 2, 3, 4, 7, 8, 9, 10, 11, 12, 13, 14, 15, 16, 17, 18, 19 $=V_1$
$\dot{V}_2$ e_2 V_2
(or Δt) 1
(or $2\Delta t$) 2
(or $3\Delta t$) 3
(or $4\Delta t$) 4
(or $5\Delta t$) 5
$\dot{V}_3$ e_3 V_3
C 0
C7 1
F7 2
G7 3

(b)

t =				68	69	70	71	72	73	74	75	76	77	78	79	80	81	82
	1	2	3	4	5	6	7	8	9	10	11	12	13	14	15	16	17	18
v_1	8	10	13	17	17	13	8	10	13	16	16	13	8	10	13	17	17	13
v_2	1	1	1	1	3	1	1	1	1	1	3	1	1	1	1	1	3	1
v_3	0	0	0	0	0	0	0	0	0	2	2	2	2	2	2	0	0	0

t =	83	84	85	86	87	88	89	90	91	92	93	94	95	96	97	98	99	100
	19	20	21	22	23	24	25	26	27	28	29	30	31	32	33	34	35	36
v_1	8	10	8	11	0	13	18	19	18	0	13	19	18	17	13	11	8	5
v_2	1	1	1	5	1	1	1	3	1	1	1	1	3	1	1	1	1	2
v_3	0	0	0	1	1	1	2	2	2	2	2	2	2	2	2	2	2	0

t =	101	102	103	104	105	106	107	108	109	110	111	112	113	114	115	116	117	118
	37	38	39	40	41	42	43	44	45	46	47	48	49	50	51	52	53	54
v_1	5	6	7	8	8	10	8	15	15	12	8	10	8	15	15	12	8	10
v_2	2	2	2	5	1	1	1	1	3	1	1	1	1	1	3	1	1	1
v_3	0	0	0	0	0	0	0	3	3	3	3	3	3	2	2	2	2	2

t =	119	120	121	122	123									124	125	126	127	128
	55	56	57	58	59	60	61	62	63	64	65	66	67					
v_1	8	13	11	10	9	8	6	5	1	0	8	10	13	8	10	12	13	0
v_2	1	1	3	1	3	1	1	1	1	1	1	1	1	1	1	1	1	4
v_3	2	0	0	0	0	0	0	0	0	0	0	0	0	0	0	0	0	0

(c)

Figure 2.8. Data system representing a blues tune (Example 2.6).

Figure 2.8b. The observation channel o_3 is in this case represented by the standard definitions of the "fake book" harmonic symbols (elements of V_3). Observe that harmony could also be represented by two variables, one for the base note (C, F, G) and one for the chord type (C versus C7).

If the variables representing pitch, rhythm, and harmony are defined as shown in Figure 2.8, the blues tune is fully described by the data matrix in Figure 2.8c. That part of the tune which is repeated is represented in the data matrix by columns with two time labels.

Example 2.7. For the purpose of designing a traffic light system, a periodic data matrix that specifies the required sequence of lights at an intersection is given in Figure 2.9a. Specific variables describing the lights for traffic bound north–south,

	1st Period						*2nd Period*						
t	t_1	t_2	t_3	t_4	t_5	t_6	t_7	t_8	t_9	t_{10}	t_{11}	t_{12}	...
NS	*g*	*g*	*g*	*y*	*r*	*r*	*g*	*g*	*g*	*y*	*r*	*r*	...
NE	*a*	*n*	*n*	*n*	*n*	*n*	*a*	*n*	*n*	*n*	*n*	*n*	...
SN	*r*	*r*	*g*	*y*	*r*	*r*	*r*	*r*	*g*	*y*	*r*	*r*	...
SE	*a*	*a*	*a*	*n*	*n*	*n*	*a*	*a*	*a*	*n*	*n*	*n*	...
WE = EW	*r*	*r*	*r*	*r*	*g*	*y*	*r*	*r*	*r*	*r*	*g*	*y*	...

(a)

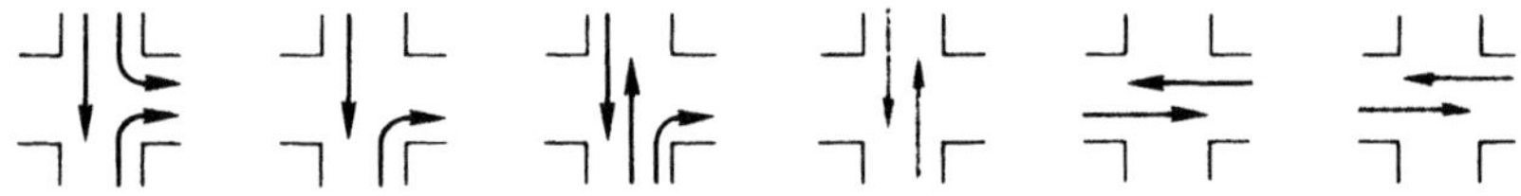

(b)

Figure 2.9. Defined activity of traffic lights at an intersection (Example 2.7).

south–north, west–east, and east–west are denoted by NS, SN, WE, and EW, respectively; each of these variables acquires three states: red, yellow, green, abbreviated by r, y, g, respectively. The left-turn arrow for traffic bound north–east is denoted by NE, and the right-turn arrow for traffic bound south–east is denoted by SE. These variables acquire two states: the arrow is either lighted or not, abbreviated as a and n, respectively. The support is time. Time set T consists of six labels $t_1, t_2, \ldots, t_6$, which represent the following partition of the time interval by 90 sec imposed by the observation channel:

$$\begin{aligned} 0 &\leq \omega^{-1}(t_1) < 15, & 50 &\leq \omega^{-1}(t_4) < 60, \\ 15 &\leq \omega^{-1}(t_2) < 25, & 60 &\leq \omega^{-1}(t_5) < 80, \\ 25 &\leq \omega^{-1}(t_3) < 50, & 80 &\leq \omega^{-1}(t_6) < 90. \end{aligned}$$

The actual situations at the intersection for the time intervals labeled by $t_1, t_2, \ldots, t_6$ are schematically illustrated in Figure 2.9b. The data system is completely determined by specifying one period of the data. It is trivial to replace the data matrix in Figure 2.9a by a data matrix based on general variables; since no properties are recognized in the state sets, their mapping to appropriate sets of integers is arbitrary.

Example 2.8. Sea-ice cover is one of the attributes that have been monitored in climatological studies. Such attributes are usually observed in space and time. Data in this example, derived from satellite imagery, regard percentage of ice cover in the Southern Ocean in the latitude range from 50° to 76° and for a 1 year period. It is assumed that six states representing the percentage of ice cover are used in the observation channel o: no cover, low, medium, high, or very high percentage (but less than 100%), and full cover (i.e., 100% cover); they are labeled by integers 0, 1, . . ., 5 so that the linear order is preserved. The states are defined in terms of the following partition of the interval [0, 100%]:

$$o^{-1}(\text{no cover}) = 0\%, \qquad 50 < o^{-1}(\text{high } \%) \leq 75\%,$$
$$0 < o^{-1}(\text{low } \%) \leq 25\%, \qquad 75 < o^{-1}(\text{very high } \%) < 100\%,$$
$$25 < o^{-1}(\text{medium } \%) \leq 50\%, \qquad o^{-1}(\text{full cover}) = 100\%.$$

SPACE → l

TIME ↓ t

	Degrees of latitude													
TIME	50	52	54	56	58	60	62	64	66	68	70	72	74	76
Jan.	0	0	0	1	1	1	1	1	2	3	3	4	4	4
Feb.	0	0	0	0	1	1	1	1	2	2	3	3	3	4
March	0	0	0	0	0	1	1	1	2	2	3	4	4	4
April	0	0	0	0	0	1	1	1	2	3	4	4	5	5
May	0	0	0	0	1	1	1	2	3	4	4	5	5	5
June	0	0	0	1	1	2	2	3	4	4	5	5	5	5
July	0	0	1	1	1	2	3	4	4	4	5	5	5	5
Aug.	0	1	1	1	2	3	4	4	4	5	5	5	5	5
Sept.	0	1	1	1	2	3	4	4	5	5	5	5	5	5
Oct.	0	1	1	1	2	3	3	4	4	5	5	5	5	5
Nov.	0	1	1	1	2	2	2	3	4	4	5	5	5	5
Dec.	0	0	1	1	1	2	2	2	3	4	4	4	5	5

(a) Pacific sector of the Southern Ocean.

	Degrees of latitude													
	50	52	54	56	58	60	62	64	66	68	70	72	74	76
Jan.	0	0	0	0	0	1	1	1	1	2	4	4	4	4
Feb.	0	0	0	0	0	0	0	1	1	1	3	4	4	3
March	0	0	0	0	0	0	0	0	1	1	3	4	4	4
April	0	0	0	0	0	0	0	1	1	2	4	5	5	5
May	0	0	0	0	0	1	1	1	2	4	4	5	5	5
June	0	0	0	0	0	1	1	2	3	4	5	5	5	5
July	0	0	0	0	0	1	1	3	4	5	5	5	5	5
Aug.	0	0	0	0	0	1	2	3	4	4	5	5	5	5
Sept.	0	0	0	0	1	1	1	3	4	5	5	5	5	5
Oct.	0	0	0	0	1	1	1	3	4	5	5	5	5	5
Nov.	0	0	0	0	0	1	1	3	4	4	5	5	5	5
Dec.	0	0	0	0	0	1	1	2	2	3	4	5	5	5

(b) Atlantic sector of the Southern Ocean.

Figure 2.10. Satellite-derived five-year monthly averages (1973–1977) of sea-ice cover in the Southern Ocean (Example 2.8). (a) Pacific sector of the Southern Ocean. (b) Atlantic sector of the Southern Ocean.

There are two supports in this case: time and one-dimensional space that has the meaning of latitude. Time is expressed in months of one year and is defined in the usual way of partitioning the whole period of one year into 12 time intervals. Latitude is measured to an accuracy of 2° and is also defined in the usual way.

Two data sets are given in Figure 2.10, one for the Pacific sector and one for the Atlantic sector of the Southern Ocean. This means that two different data systems are really defined in this example. The data entries are actually based on monthly averages over a period of five years (1973–1977). Each of the data sets is represented in a convenient matrix form. Each row in the matrix represents a "snapshot," i.e., a one-time observation of the ice cover at various latitudes; each column represents the development of the ice cover situation at one latitude over the whole year. If more than one variable were involved, each entry in the data matrix would contain a tuple consisting of a particular state for each variable (i.e., an element of **V**).

Example 2.9 Let the source system defined in Example 2.4 (complete blood count) be extended by one additional support—a population of patients suffering from anemia—everything else being the same. Then, data can be represented conveniently by the matrix form shown in Figure 2.11. Each entry of the matrix is a quadruple consisting of a particular overall state of the four variables, observed on a particular day for a particular patient. This example illustrates a systematic development of medical data for further processing.

Example 2.10 To illustrate fuzzy data, let two variables from Example 2.3 (a stand of hardwood timber) be considered. They are: merchantable height (v_3) and unsound defect (v_6). Assume that both of them are defined by fuzzy observation channels which, however, are not defined explicitly, but are represented by the observer himself. Everything else is the same as in Example 2.3. A fuzzy data array would then have the form illustrated in Figure 2.12a. Each entry in the array expresses the certainty (of the investigator) that a particular tree (labeled by an integer **w**) is characterized by a

	DAY 1	DAY 2	.....
PATIENT 1	4.28 6.2 41 14.05	4.25 5.8 44 13.95	.....
PATIENT 2	5.21 5.1 48 15.21	5.22 5.1 48 15.25	.....
⋮	⋮	⋮	
PATIENT n	4.36 7.8 32 14.98	4.36 7.9 35 14.95	.....

Figure 2.11. Medical data (Examples 2.4 and 2.9).

		w = 1	2	3	4	5	6	7	8	...
V_3	0	0.2	0.8	0.0	0.0	1.0	0.0	0.5	0.0	
	1	0.9	0.4	0.7	0.0	0.0	0.6	0.5	1.0	...
	2	0.0	0.0	0.3	1.0	0.0	0.5	0.0	0.0	
V_6	0	0.3	0.9	0.0	0.4	1.0	0.0	0.7	0.2	
	1	0.7	0.3	0.0	0.6	0.0	0.5	0.3	0.9	...
	2	0.2	0.0	1.0	0.4	0.0	0.5	0.1	0.0	

(a) Fuzzy data.

w =	1	2	3	4	5	6	7	8	...
v_3	1	0	1	2	0	1	1	1	...
v_6	1	0	2	1	0	2	0	1	...

(b) Comparable crisp data.

Figure 2.12. Illustration of fuzzy data (Example 2.10).

particular state of one of the variables. The three-dimensional array of fuzzy data can be compared with the data matrix in Figure 2.12b, which is based on the assumption that both of the variables are defined in terms of crisp observation channels.

Let two methodological distinctions be recognized for data systems, in addition to those introduced for source systems. The first one is a distinction between *completely specified data* and *incompletely specified data.* The data are called completely specified if and only if all entries in its data matrix or array are specified; otherwise it is called incompletely specified. Let two types of incompletely specified data be further distinguished:

i. all cases in which some data within the defined support set are *not available* (as in some experimental and historical investigations);
ii. all cases in which *it does not matter* what some of the data are (as in some problems of systems design, where such instances are usually referred to as *don't care conditions*).

When data are incompletely specified, individual state sets must be supplemented by some convenient (standard) symbols reserved for the identification of the "not available" or "don't care" entries in the data arrays. To deal with these entries, the GSPS must be equipped with appropriate methodological capabilities.

The second methodological distinction for data systems is applicable only to those systems whose overall support sets are linearly ordered. It permits us to speak of *periodic data*, i.e., data that repeat in the same order when the support set is extended. Instances of periodic data are in Examples 2.6 and 2.7

Source and data systems are epistemological types of systems that are predominantly of an empirical nature. As such, their relationship to the various traditional disciplines of science and other areas is much stronger than that of systems types at higher epistemological levels, which are predominantly of a theoretical nature. Indeed, to define a source system on an object, so that the purpose for which it is defined is well served, requires considerable knowledge and experience in a particular discipline. There are usually many ways in which a source system can be defined and it is the principal challenge of the investigator to select one within which relevant questions can be best formulated and dealt with. Once a source system is defined, special skills, knowledge, and instruments are again required to obtain meaningful data. The main purpose of the diverse examples discussed in this chapter is to illustrate some of the issues involved in the process of defining source systems and obtaining data for them.

NOTES

2.1. The distinction between an object and a system defined on the object is often not made in the literature. This lack of terminological precision has been the source of much confusion. In some instances, the terms "real system" and "model" are used for our concepts of object and system, respectively. This terminology is unfortunate since, according to it, systems science would actually deal with models rather than systems. I believe that it is more appropriate to use the term "system" exclusively for any operationally described abstract representations of manageable sets of attributes and backdrops. The term "model" can be then reserved for the various kinds of similarity relationships between pairs of comparable systems, i.e., systems of the same epistemological type (as discussed in Chapter 8).

2.2. In many instances, the problems of defining source systems on objects of interest and gathering data for them involve various issues of the theory and practice of measurement. Such issues cannot be separated from the individual traditional disciplines of science. As such, they are outside the scope of the GSPS and, consequently, are not covered in this book. As supplementary reading, several books devoted to general aspects of measurement are recommended [EL1, KR1, PF1, TO1].

2.3. The theory of fuzzy sets was introduced by Lotfi Zadeh in 1965 [ZA3]. Its development since the publication of Zadeh's seminal paper has been dramatic. A survey of the status of the theory and its applications in the late 1970s is well covered in a book by Dubois and Prade [DU1]. Current contributions to the theory of fuzzy sets are scattered in many journals, but the most important source is the specialized journal *Fuzzy Sets and Systems* (North-Holland).

2.4. The syntactic, semantic, and pragmatic aspects that are recognized in source systems (Figure 2.3), can be briefly characterized as follows. The *syntactic aspects* are those which involve the relationships among signs, such as rules of constructing sentences from words, but without any reference to the meaning and use of the signs (words, sentences). The *semantic aspects* involve the relationships of signs to things other than signs by which a meaning is given to the signs (word, sentences), but without any reference to their use. The *pragmatic aspects* involve the relationships of signs to things other than signs by which some use is ascribed to the signs. These three kinds of aspects involving signs are studied by three areas of semiotics (a general theory of signs introduced by Charles Morris in 1938): *syntactics* (or syntax), *semantics*, and *pragmatics*, respectively. The term "semiotics" is derived from the Greek word "sema," which means sign. *Semiotics* is defined by Morris as "a general theory of signs in all their forms and manifestations, whether in animals or men, whether personal or social" [MO1–3].

2.5. The best sources of information about the recent developments and trends in reformulating the various areas of theoretical and applied physics (as well as some other disciplines of natural sciences) in terms of discrete variables and supports are two books of Donald Greenspan [GR1, GR5], a Special Issue of the *International Journal of General Systems on Discrete Models* (Vol. 6, No. 1, 1980, pp. 1–45) and a book by Herbert S. Ingham [IN1].

2.6. Data used in Example 2.5 (Figure 2.7) were published in a paper "The evolution of behaviour in gulls" by N. Tinbergen (*Scientific American*, Dec. 1960). Data used in Example 2.8 (Figure 2.10) were published by Burckle, Robinson, and Cooke in *Nature* (September 30, 1982).

EXERCISES

2.1. Determine the total number of methodological distinctions for source systems that contain
   (a) two variables and one support;
   (b) two variables and two supports;
   (c) five variables and three supports.

2.2. Repeat Exercise 2.1 for data systems.

2.3. Repeat Exercises 2.1 and 2.2 under the conditions that
   (a) only discrete variables and supports are considered;
   (b) all variables are of the same methodological type;
   (c) all supports are of the same methodological type;
   (d) all three conditions (a), (b), (c) are satisfied, but variables may be of a different methodological type than supports.

2.4. Under the assumption that time is defined explicitly in Example 2.6
   (a) define a variable describing the rhythm;
   (b) determine an appropriate data matrix for the tune.

2.5. Suggest suitable forms of representing data in terms of arrays for the following source systems:
   (a) a source system similar to the one defined in Example 2.8, but with three supports: time, latitude, and longitude;
   (b) a source system similar to the one considered in (a), but with a fuzzy observation channel;
   (c) the source system defined in Example 2.3;
   (d) a source system with two discrete variables based on crisp observation channels and a three-dimensional discrete space represented by the Cartesian coordinates.

2.6. As in Example 2.7 (traffic lights), define an appropriate source system and data matrix for the following activities associated with man-made objects:
   (a) the activity of the intake valve, exhaust valve, piston, and spark plug of one cylinder in an internal combustion gasoline engine;

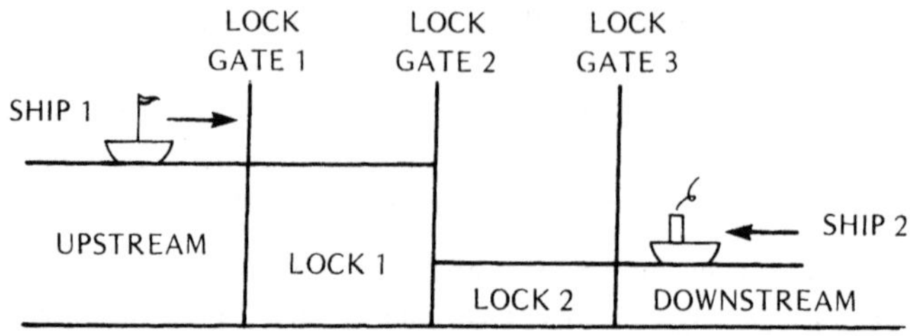

Figure 2.13. Illustration of Exercise 2.6b.

(b) the activity of two locks and three lock gates of a typical canal lock ensemble with a constant stream of ships moving both ways (Figure 2.13);

(c) a weekly flight schedule at a small airport (consider all features relevant to passengers, such as times of arrivals or departures, gates, carriers, etc., as variables or supports);

(d) the control activity of a music box to play the melody discussed in Example 2.6 (Figure 2.8a) and based on the assumption that each note is controlled in time by two-valued signals.

2.7. Define a source system for some purpose on an object with which you interact regularly in your daily life and collect appropriate data for the system. The object may include you yourself; such a self-investigation may be motivated by some purpose related to the quality of your life and may include a variety of diverse kinds of variables characterizing, e.g., various physiological attributes, diet, sleep, working conditions, exercise, weather conditions, intake of drugs, etc.; they may be recorded on a daily basis or at some other, more appropriate time scale. The data can then be analyzed by some of the methods described in Chapters 3–5.

2.8. Define an appropriate fuzzy observation channel for electric current measured in the range 0–10 milliamperes (mA), to an accuracy of 1 mA, and under the following assumptions: (i) the maximum measurement error is 0.1 mA; and (ii) the likelihood of error decreases linearly with the distance from each boundary between two blocks of the partition of the interval [0, 10] imposed by the underlying crisp observation channel (the blocks are appropriate intervals of 1 mA in this case).

2.9. Suppose that two-dimensional (or three-dimensional) space is represented by Cartesian coordinates. Clearly, state sets representing each coordinate are totally ordered and metric.

(a) Show that the space can be ordered only partially if the total orders of the coordinates are required to be preserved.

(b) Define the partial ordering of the space under which the coordinates remain totally ordered.

(c) Define a distance for the space based on distances defined for the coordinates.

2.10. Show for some attributes in Example 2.4 that the total ordering and distance presumed in the set of appearances of the attribute are preserved in the state set of the corresponding abstract variable.

2.11. Define mathematically the observation and abstraction channels for variable v_4 in Example 2.4 and supports (time, latitude) in Example 2.8

3

GENERATIVE SYSTEMS

A basic purpose of theorizing is to organize information in a way that will develop its nonobvious implications.

—David R. Heise

3.1. EMPIRICAL INVESTIGATION

> *Science must start with facts and end with facts, no matter what theoretical structures it builds in between.*
>
> —JOHN G. KEMENY

There are three prerequisites for every meaningful empirical investigation. First, an object of investigation must be identified; second, a purpose of investigating the object must be known; third, constraints imposed upon the investigation must be assessed.

The *object of investigation* is defined in Chapter 2 as a part of the world identifiable as a single entity for an appreciable length of time and desirable for a particular investigation.

The *purpose of investigation* can be viewed as a set of questions regarding the object which the investigator (or his client) wants to answer. For example, if the object of investigation is New York City, the purpose of the investigation might be represented by questions such as "How can crime be reduced in the city?" or "How can transportation be improved in the city?"; if the object of investigation is a computer installation, the purpose of investigation might be to answer questions "What are the bottlenecks in the installation?", "What can be done to improve performance?", and the like; if a hospital is investigated, the questions might be "How can the ability to give immediate care to all emergency cases be increased?", "How can the average time spent by a patient in the hospital be reduced?", or "What can be done to reduce the cost while preserving the quality of services?"; if the object of interest of a musicologist is a musical composer, say Igor Stravinsky, his question is likely to be "What are the basic characteristics of Stravinsky's compositions which distinguish him from other composers?"

Constraints associated with an empirical investigation consist of limitations in the availability of appropriate instruments, financial and time limitations, limited manpower or computer resources, and legal, ethical, or other restrictions imposed upon the investigators.

Basic stages involved in every empirical investigation are illustrated in Figure 3.1; it is used as a guide for further discussion in this section.

The first stage in each particular empirical investigation is to *define a source system* on the relevant object. This is described in sufficient detail in Chapter 2 and summarized

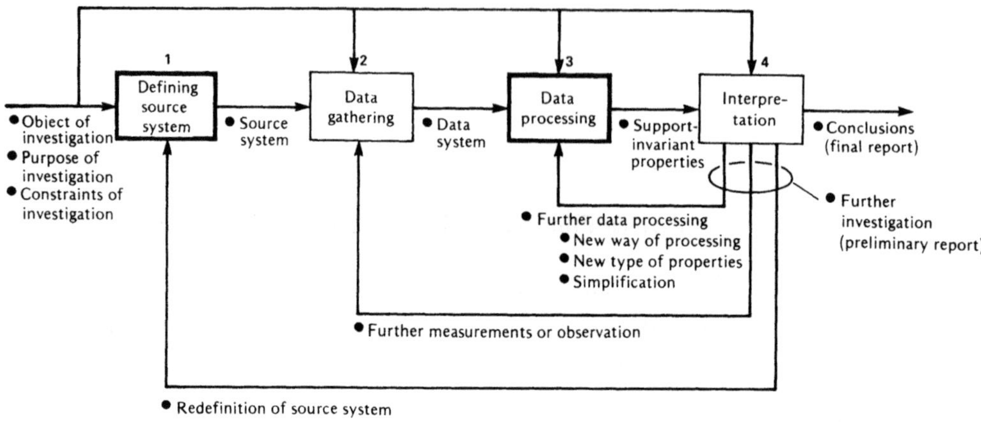

Figure 3.1. Basic stages in empirical systems investigations.

in Figure 2.3. The main issue at this stage of investigation is to select, from usually a large multitude of possibilities, a source system that is most appropriate for the purpose of investigation under the given constraints. This issue is clearly context dependent. It requires knowledge and experience in the specific area of inquiry as well as some ingenuity on the part of the investigator. He often examines some feasible hypothetical systems at some higher epistemological levels before choosing one particular source system for the empirical investigation.

Two issues are initially involved in the process of selecting an appropriate source system: (i) a selection of attributes and backdrops, and (ii) a selection of observation channels for them. Abstraction channels become involved later, when the need arises to translate the system into the GSPS language.

The selection of attributes and backdrops is perhaps the most important decision in the process of empirical investigation, since it affects all subsequent steps in the process. It is a difficult decision, which usually cannot be well characterized in terms of rational criteria. It is frequently based on some preconceived ideas which, when sufficiently crystallized in the mind of the investigator in potentially experimental terms, are called scientific theories. Some philosophers of science insist that any meaningful selection of attributes for an empirical study is always based on some underlying theory, be it one that is explicitly stated, one that emerges from the investigator's unconscious, or one that is a part of our genetically inherited innate knowledge.

It is important that in the process of defining a source system the investigator be aware of the full range of capabilities of modern systems methodology. Otherwise, he may restrict his choices unnecessarily. One such restriction is well expressed by Ashby [AS9]:

> The worker who has some training in mathematics can only too easily fall into the habit (or trap) of thinking that a "variable" must mean a numerical scale with an additive metric. This assumption is quite unnecessarily restrictive, sometimes fatally so. The meteorologist has long worked with his five "types of cloud," the veterinarian with the various "parasites of the pig," the hematologist with the four basic types of "blood-groups." Modern mathematics, using the method of set theory, is quite able to handle such variables, which are often unavoidable in behavioral sciences.

Another restriction, often encountered in empirical investigations, is to consider only crisp observation channels, even though their fuzzy counterparts would be far better for some attributes and situations. Such restriction is not necessary since methods for dealing with fuzzy data are now available.

After attributes and backdrops are selected, the investigator must define observation channels for them. As discussed in Chapter 2, observation channels impose partitions on a given set of appearances or backdrop instances. Let each of these partitions be called a *resolution form*. Although resolution forms sometimes cannot be defined mathematically (unless we accept some metaphysical assumptions), it is perfectly possible to determine whether one resolution form is a refinement or coarsening of another form (in terms of the standard refinement ordering defined on partitions of a given set). Such comparison of two resolution forms is not done mathematically, but by comparing the corresponding measurement procedures. In each case, the range of possible resolution forms has an upper bound represented by the resolution capability of the available measuring instruments. A lower bound is any resolution form that contains only two blocks. Which resolution form to select within this range depends on the purpose of investigation.

When the source system is defined, data gathering is then possible. This amounts to making observations or measurements of the chosen attributes at defined support instances and recording the observations in some convenient form, as discussed in Section 2.6. If the investigator can control some of the attributes, he may take advantage of this. If he does, the attributes he intends to control are viewed as input attributes. This results in a directed source system. The investigator then designs some experiments in which input attributes are manipulated, according to some experimentally feasible strategy related to the purpose of investigation, and output attributes are observed. The result is a data system.

After the data system is finalized, the next stage in empirical investigation is data processing. Its aim is to determine some *support-invariant properties* of the variables involved through which the data can be represented in a parsimonious fashion and, if desirable, generated. This is a stage at which the GSPS can be of great help to the investigator. Either all data are employed for deriving the required support-invariant properties or only some data are processed initially while the rest is reserved for subsequent testing of the derived properties.

There is a variety of support-invariant properties, but they all have a common denominator. Each of them characterizes a *constraint* among the variables of the source system that does not change within the support set. For instance, if the support is time, then any time-invariant property describes a constraint among the variables that does not change in time. Different support-invariant properties may characterize types of constraints that are associated with different epistemological levels or, alternatively, they may differ only in the manner in which the same constraint type associated with a particular epistemological level is represented. The former differences are a basis for distinguishing epistemological types of systems; the latter represent methodological distinctions recognized for each particular epistemological type.

After data have been processed in some fashion and appropriate support-invariant properties of the variables determined, they have to be given a proper *interpretation* with respect to the purpose of investigation. That is, their use in answering the various questions posed in the investigation have to be explored. If the questions can be answered adequately, the investigation is successfully concluded and the investigator is in a position to summarize his conclusions and prepare a final report. Otherwise, he may try to process the data again, in a different way. This may be repeated several times and may involve a search for the same type of support-invariant properties or a search for properties of different epistemological types. In the end, the investigator is provided with a set of generative systems or higher-level systems, each of which correctly represents the data from a particular point of view. Such a set of complementary systems, each reflecting certain aspects of the data, may frequently give the investigator much better insight than any of them could furnish alone.

The system (or systems) obtained by processing the data is sometimes too complex to be comprehensible to the human mind and, consequently, does not help the investigator to develop his insight. In such cases, a *reduction in complexity* of the system is necessary or, at least, desirable. The GSPS should thus have the capability of simplifying systems of the various types according to simplification criteria specified by the user.

After processing the data and interpreting the obtained properties, the investigator may also decide to gather some more data, in order to either increase his confidence in the derived properties or to revise them on the basis of the new data. This *renewed data gathering* changes the data system, but leaves the source system unchanged. However, the investigator may also decide to make a more drastic change—*to redefine the source system*. Then, of course, he has to repeat the entire process for the new source system.

Referring to Figure 3.1 as a guide, the overall procedure of empirical systems investigation can be now summarized as follows:

1. given an object, purpose, and constraints of an empirical investigation, a source system is defined on the object (details are shown in Figure 2.3);
2. data are gathered for the defined source system and organized in a suitable form, usually a data array;
3. the data are processed with the objective of determining some support-invariant properties representing them;
4. the support-invariant properties obtained are interpreted with respect to the

purpose of the investigation, and either some final conclusions are reached or further investigation is initiated, beginning at stage 3, 2, or 1.

The GSPS should be able to handle all problems associated with the data processing stage that is desirable to recognize. This requires that it possesses the capabilities of (i) deriving support-invariant properties of all desirable types from given data, (ii) comparing the derived properties and excluding those systems whose properties are inferior according to the user's criteria, and (iii) simplifying systems of the various types according to simplification criteria specified by the user. In this chapter, these three GSPS capabilities are discussed only with respect to the epistemological level immediately above that of data systems, referred to as level 2 (Figure 1.3). The support-invariant properties at this level are direct characterizations (of various kinds) of the overall constraint associated with the variables involved. Systems that contain such characterizations are called *generative systems*. The purpose of this chapter is to define generative systems and illustrate some of the problem types in which they are involved.

Generative systems (as well as systems of epistemologically higher types) are defined and discussed in this book in terms of general image systems (abstract variables and supports). This means that they are presented in terms of the GSPS language, i.e., without semantics. When an application is discussed, however, the generative system is supplemented with a source system, through which the relevant semantic aspects are introduced. The general image system, which is included in both of the systems, represents the interface between the GSPS language and some object-oriented language of a specific discipline. Consequently, it must be exactly the same in both of the systems.

3.2. BEHAVIOR SYSTEMS

> *Behavior system is in the eyes of the masker.*
>
> —Richard Karney

The term "behavior" is used in this book for a simple characterization of the overall support-invariant constraint among variables of a general image system and, possibly, some additional abstract variables. Each of the additional variables is defined in terms of a specific *translation rule* in the support set. The rule can be applied either to a variable in the given image system or to a hypothetical variable, introduced for various methodological reasons and usually referred to as an *internal variable*. Issues associated with internal variables are discussed in Section 3.10; the rest of this chapter is based on the assumption that no internal variables are involved. Since a description of the support-invariant constraint among the variables considered can be used for generating states of the variables within the support set, systems that contain such descriptions are called generative systems. Behavior is one of the forms in which the constraint can be expressed.

Given a general image system, the range of possible kinds of support-invariant constraints among its variables depends on the properties recognized in the support set. If no properties are recognized in it (as in most populations), then states of the variables can be constrained solely by each other. If, however, the support set is ordered, then they can be constrained not only by each other, but also by states in a chosen *neighborhood* of each particular support instance. Since the neighborhood is a basis in terms of which a support-invariant constraint is expressed, it must itself be support-invariant.

A neighborhood in an ordered support set, which is usually referred to as a *mask* (for reasons explained later), is defined in terms of the variables involved, the support set, and a set of translation rules in the support set. A *translation rule*, say r_j, is a one-to-one function

$$r_j\colon \mathbf{W} \to \mathbf{W}, \tag{3.1}$$

by which each element in $\mathbf{W}$ is assigned another (unique) element in $\mathbf{W}$. For instance, when the support set is totally ordered (as in the case of time or one-dimensional space) and represented by a set of consecutive positive integers, each translation rule can be expressed by a simple equation

$$r_j(\mathbf{w}) = \mathbf{w} + \rho, \tag{3.2}$$

where ρ is an integer constant (positive, negative, or zero). When $\rho = 0$, r_j is called an *identity translation rule*.

Assume that a general image system $\mathbf{I}$ specified by Eq. (2.12) is given. Let V denote the set of variables in $\mathbf{I}$ and let R denote a set of translation rules that are considered for the variables. Then a set of variables

$$S = \{s_1, s_2, \ldots\}$$

referred to as *sampling variables*, can be introduced by the equations

$$s_{k,\mathbf{w}} = v_{i,r_j(\mathbf{w})} \tag{3.3}$$

for some variables $v_i \in V$ and some translation rules $r_j \in R$; $s_{k,\mathbf{w}}$ denotes the state of sampling variable s_k at support instance $\mathbf{w}$ and $v_{i,r_j(\mathbf{w})}$ denotes the state of variable v_i at support instance $r_j(\mathbf{w})$, i.e., a support instance obtained for any given $\mathbf{w}$ by the translation rule r_j. For a totally ordered support set, whose translation rules are expressed by (3.2), Eq. (3.3) may be written in a more specific form

$$s_{k,\mathbf{w}} = v_{i,\mathbf{w}+\rho}. \tag{3.4}$$

Since any translation rule in R can be applied to any variable in V, the set of all possible sampling variables is represented by the Cartesian product $V \times R$. Sampling variables that are actually considered are characterized by the relation

$$M \subseteq V \times R \tag{3.5}$$

in such a way that each pair $(v_i, r_j) \in M$ corresponds to one equation in (3.3). Relation M represents a neighborhood pattern in the support set in terms of which the sampling variables are defined. As mentioned previously, it is usually called a *mask*. Clearly, some one-to-one function (labeling)

$$\lambda\colon M \to N_{|M|} \tag{3.6}$$

must be used to introduce identifiers k of the sampling variables: ($|M|$ denotes the cardinality of M).

When sampling variable s_k is defined in terms of variable v_i and some translation rule, according to Eq. (3.3), then the state set of s_k is obviously the same as the state set of v_i, i.e., the set V_i. For notational convenience, however, let the state set of sampling variable s_k be denoted by S_k; the meaning of each $S_k (k \in N_{|M|})$ in terms of one of the sets $V_i (i \in N_n)$ is uniquely determined by the mask. Then, the Cartesian product

$$\mathbf{C} = S_1 \times S_2 \times \cdots \times S_{|M|}$$

represents the set of all overall states of the sampling variables.

Let us develop the notion of a mask and the associated behavior for image systems with totally ordered support sets first, and extend it to partially ordered support sets later. Let totally ordered support sets be denoted by T and let t denote elements of $T (t \in T)$. Equation (3.4) thus becomes slightly modified:

$$s_{k,t} = v_{i,t+\rho}. \tag{3.7}$$

For totally ordered support sets, a mask can be depicted as a cut in the matrix representing the Cartesian product $V \times R$. This is illustrated in Figure 3.2a, where the rows are labeled by identifiers i of the variables in set V and columns are labeled by integer constants ρ associated with the translation rules of the form (3.2). Entries in the matrix are either empty or contain indentifiers k of sampling variables assigned to pairs (i, ρ) by Eqs. (3.7); the empty entries identify those elements of $V \times R$ which are not included in the mask. The reason for using the term "mask" becomes clear when considering this visual representation.

It is often convenient to partition a mask M into submasks M_i, each associated with one variable v_i of the image system. Formally,

$$M_i = \{(\alpha, \beta) \mid (\alpha, \beta) \in M, \alpha = v_i\}. \tag{3.8}$$

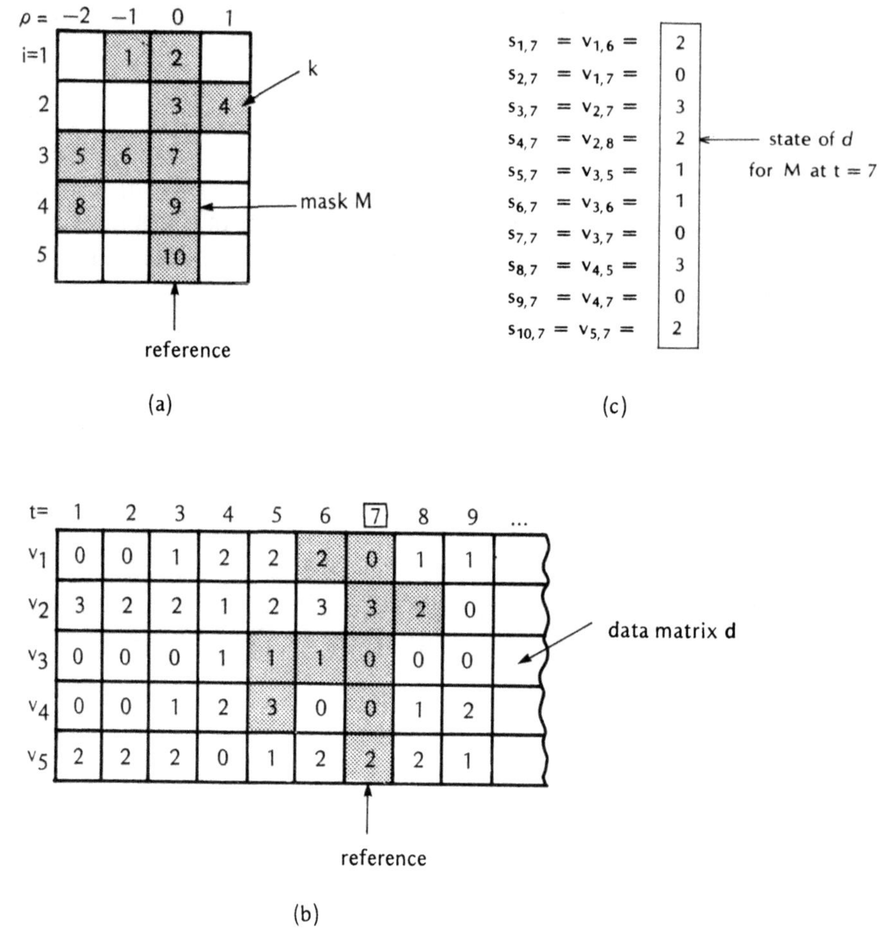

Figure 3.2. Illustration of the concept of mask for totally ordered support sets.

In the visual (matrix) representation of M, each of its submasks M_i is recognized as a row in M.

One of the columns in each mask is associated with the identity translation rule ($\rho = 0$). Such a column has a special significance since the sampling variables associated with it are identical with the basic variables of the given image system. For each mask, let this column be called its *reference*. When a mask is placed on a data matrix so that its reference coincides with a particular value of t, the mask makes only a subset of the matrix entries transparent—those specific entries which represent the overall state of the sampling variables at the support instance t. In Figure 3.2b, for example, the mask (defined in Figure 3.2a) is located on the data matrix d at $t = 7$ (its reference coincides with $t = 7$). The overall state of the sampling variables for this location is specified in

Figure 3.2c. Observe that states of the reference sampling variables $s_2, s_3, s_7, s_9, s_{10}$ are exactly the same (for any t) as states of the basic variables v_1, v_2, v_3, v_4, v_5, respectively. The remaining sampling variables represent states of the support neighborhood of t. For each mask, the neighborhood pattern remains the same for any t. If t were time, then variable s_4 would represent a future state of variable v_2 (with respect to each value of t that is considered), while variables s_5 and s_6, for example, would represent past states of variable v_3.

Each mask represents a specific point of view according to which the constraint among the basic variables is represented. The simplest way of expressing the constraint for a particular mask is to list all possible overall states of the associated sampling variables. Such a list is generally a subset of the Cartesian product **C**, i.e., a multidimensional relation defined on **C**. It can be defined by a function

$$f_B: \mathbf{C} \rightarrow \{0, 1\} \tag{3.9}$$

such that $f_B(\mathbf{c}) = 1$ if state $\mathbf{c}$ actually occurs and $f_B(\mathbf{c}) = 0$ if it does not. Function f_B is thus a typical selection function. It selects states that the sampling variables actually take on from the set of all their potential states (the Cartesian product **C**). Since such a selection provides at least some information about the behavior of the variables, function f_B is usually called a *behavior function*, which explains why the subscript B is used. The function defined by (3.9) is only one of several types of behavior functions, each of which characterizes the constraint among the variables in some particular fashion. Different behavior functions, which are viewed as methodological distinctions, are introduced in Section 3.3. In this section, the discussion is restricted to the selection behavior function defined by (3.9).

Observe that the behavior function f_B specifies states of **C** that actually occur, but it does not specify at which support instances they occur. Hence, it is support-invariant. Observe also that the domain of f_B, which is the same for all types of behavior functions, is defined in terms of a mask which, in turn, is defined in terms of variables and supports of an image system. This implies that a system, say system $\mathbf{F}_B$, which is supposed to characterize a support-invariant constraint of a set of variables in terms of a behavior function, is defined by the triple

$$\mathbf{F}_B = (\mathbf{I}, M, f_B), \tag{3.10}$$

where **I** is a general image system, M is a mask defined in terms of **I**, and f_B is a behavior function whose domain is defined by M and **I**. Let such a system be called a *behavior system*.

Although each behavior system defined by Eq. (3.10) characterizes, in some particular support-invariant manner, the constraint among variables of an image system, it does not include a description of how to utilize the constraint to generate data.

To develop such a description, the sampling variables must be partitioned into two subsets:

i. variables whose states are generated through the constraint—let them be called *generated variables*;
ii. variables whose states are employed as conditions in the generating process—let them be called *generating variables.*

► Given a behavior system, one way in which generated and generating variables can be defined is to associate with them two submasks, say M_g and $M_{\bar{g}}$, respectively, of the given mask M. Let

$$M_G = (M, M_g, M_{\bar{g}}), \tag{3.11}$$

where

$$\begin{aligned} M_g, M_{\bar{g}} &\subset M, \\ M_g \cup M_{\bar{g}} &= M, \\ M_g \cap M_{\bar{g}} &= \varnothing, \end{aligned}$$

be called a *generative mask*; it is a mask M with its partition into a generated submask M_g and generating submask $M_{\bar{g}}$.

In analogy with the partition of M into M_g and $M_{\bar{g}}$, the set $N_{|M|}$ of identifiers k of the sampling variables based on M can be partitioned into two subsets, say K_g and $K_{\bar{g}}$, which represent identifiers of the generated and generating variables, respectively. For notational convenience, the labeling function (3.6) may then be replaced with two functions

$$\begin{aligned} \lambda_g &: M_g \to K_g, \\ \lambda_{\bar{g}} &: M_{\bar{g}} \to K_{\bar{g}}, \end{aligned} \tag{3.12}$$

by which state sets $\mathbf{G}$ and $\mathbf{G}$ of the generated and generating variables, respectively, are defined by the Cartesian products

$$\begin{aligned} \mathbf{G} &= \bigtimes_{k \in K_g} S_k, \\ \overline{\mathbf{G}} &= \bigtimes_{k \in K_{\bar{g}}} S_k. \end{aligned} \tag{3.13}$$

The way in which a state of the generated variables (say $\mathbf{g} \in \mathbf{G}$) is determined on the basis of a state of the generating variables (say $\bar{\mathbf{g}} \in \overline{\mathbf{G}}$) can now be expressed by a function

$$f_{GB}: \overline{\mathbf{G}} \times \mathbf{G} \to \{0, 1\}, \tag{3.14}$$

where

$$f_{GB}(\bar{\mathbf{g}}, \mathbf{g}) = \begin{cases} 1, & \text{if } g \text{ can occur when } \bar{\mathbf{g}} \text{ occurs,} \\ 0, & \text{if } g \text{ cannot occur when } \bar{\mathbf{g}} \text{ occurs.} \end{cases}$$

Let this function be called a *generative behavior function.*

When M and f_B in Eq. (3.10) are replaced with M_G and f_{GB}, respectively, an alternative system

$$\mathbf{F}_{GB} = (\mathbf{I}, M_G, f_{GB}) \tag{3.15}$$

is obtained. Let this system be referred to as a *generative behavior system.*

The use of a generative behavior function for generating data involves basically the following two steps:

a. given a state $\bar{\mathbf{g}} \in \bar{\mathbf{G}}$ for some value of $t \in T$, function f_{GB} is used to determine state $\mathbf{g} \in \mathbf{G}$ for the same value of t;
b. the value of t is replaced with a new value and step (a) is repeated.

Several issues associated with this two-step generative procedure must be clarified. First, it is tacitly assumed in step (a) that the state $\bar{\mathbf{g}}$ for the given value of t is known. When the step is used for the first time, this state is specified by user as a desirable *initial condition.* After that, however, it must be fully determined by the generating process itself, i.e., by states $\bar{\mathbf{g}}$ and $\mathbf{g}$ associated with the previous value of t. This implies that values of t must be changed in step (b) according to the order of set T. Hence t can be changed either by replacing t with $t+1$ or, alternatively, by replacing t with $t-1$. If the former alternative is employed, the intial condition must be specified for the smallest possible value of t; for the latter alternative, it must be specified for the largest possible value of t.

Second, the necessity of generating data in one of the two orders implies that there are only two meaningful partitions of a mask M into M_g and $M_{\bar{g}}$, each corresponding to one of the two generating orders. If data are generated in increasing (decreasing) order of t, then M_g contains exactly one element from each of the submasks M_i $(i \in N_n)$ defined by Eq. (3.8), the one with the largest (smallest) value of ρ; all remaining elements of M are included in $M_{\bar{g}}$. In the visual representation, M_g is thus the set of all the right-most elements of M (the right edge of the mask) or, alternatively, the set of the left-most elements of M (the left edge of the mask).

Third, for each particular state $\bar{\mathbf{g}} \in \bar{\mathbf{G}}$, it is assumed that at least one state $\mathbf{g} \in \mathbf{G}$ is permitted by function f_{GB} [i.e., $f_{GB}(\bar{\mathbf{g}}, \mathbf{g}) = 1$]. If only one state is permitted, the data generation is unique for each initial condition; such systems are called *deterministic.* If, however, more than one state is permitted, the data generation is problematic because the generated state is not always determined uniquely. Generative systems with this undesirable property are called *nondeterministic systems.* Selection behavior functions are not suitable for representing such systems. They can be more meaningfully characterized by behavior functions of other types, as discussed in Section 3.3. For deterministic systems, form (3.14) of the generative behavior function f_{GB} can be replaced with the simpler form

$$f_{GB}: \bar{\mathbf{G}} \rightarrow \mathbf{G} \blacktriangleleft \tag{3.16}$$

Example 3.1. To illustrate the generation of data by a generative behavior system of the type defined by Eq. (3.15), let the image system consist of the totally ordered support set $T = N_{99}$ and five variables $v_1, \ldots, v_5$ whose state sets are defined later. Let us use the mask defined in Figure 3.2. Data can be generated either in increasing or decreasing order of t. The two alternatives are illustrated in Figures 3.3 and 3.4, respectively.

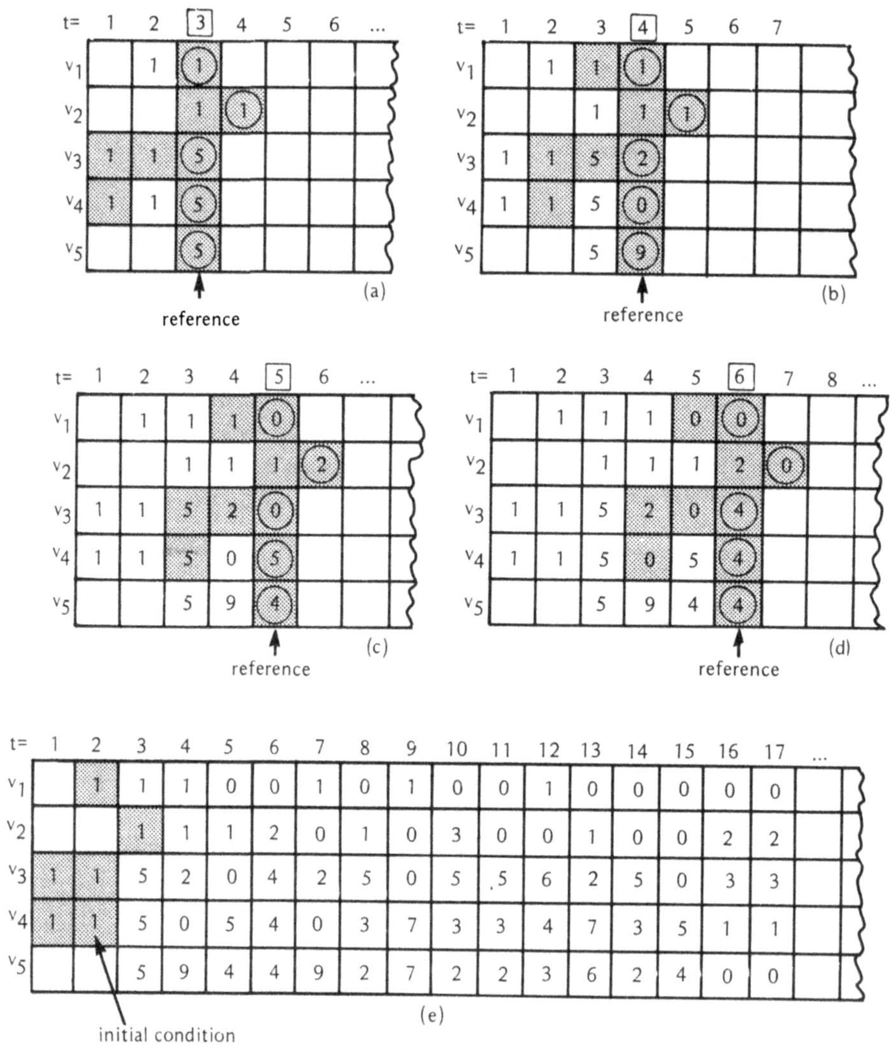

Figure 3.3. Data generated in the order of increasing values of support t (Example 3.1).

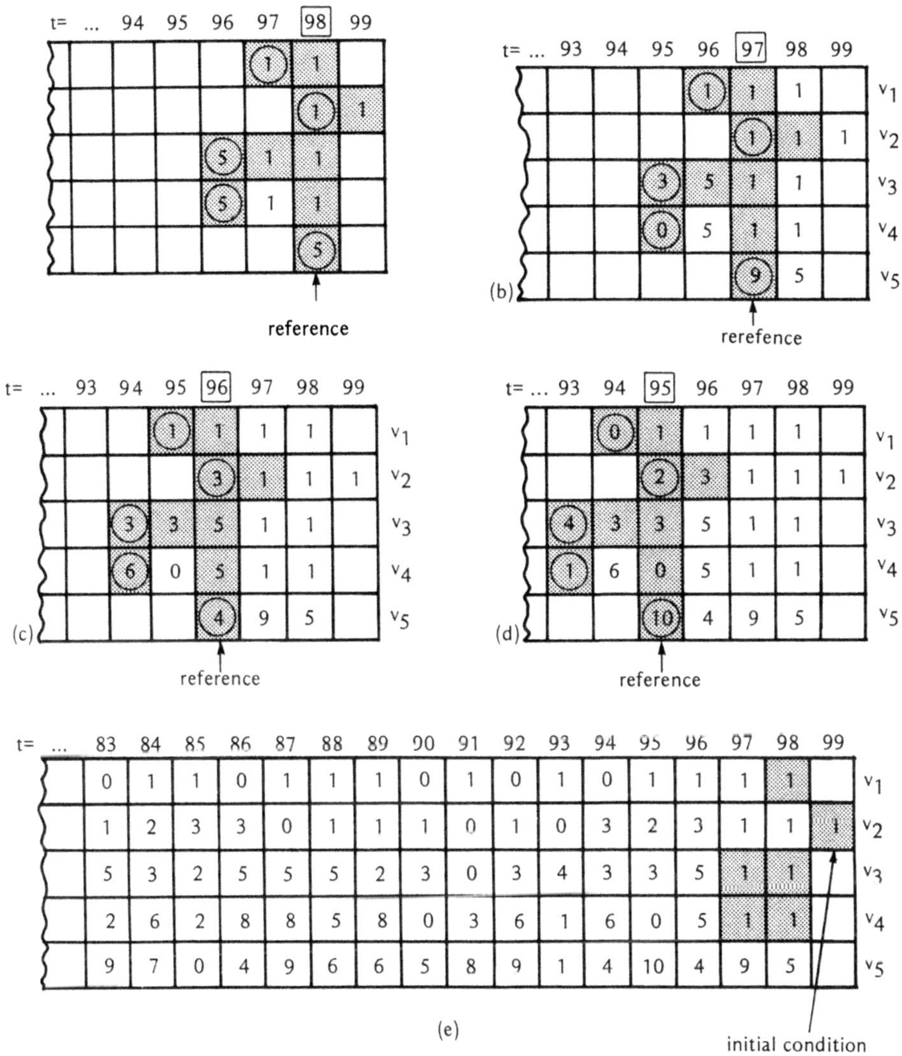

Figure 3.4. Data generated in the order of decreasing values of support t (Example 3.1).

In the case of the first alternative (Figure 3.3), the generated sampling variables are those associated with the right edge of the mask, i.e., variables s_2, s_4, s_7, s_9, s_{10}; the remaining sampling variables are generating. The generation of data proceeds from the left to the right in the data matrix. Let the generative behavior function f_{GB} in the form (3.16) be defined by the equations

$$s_{k,t} = s_{1,t} + s_{3,t} + s_{5,t} + s_{6,t} + s_{8,t} \text{ (modulo } k)$$

for $k = 2, 4, 7, 9, 10$. State sets of the generated variables are implied by the equations. State sets of the generating variables then follow from their meaning (location) in the mask. For instance, the state set of generated variable s_4 is 0, 1, 2, 3 since the equation for s_4 is taken modulo 4; generating variable s_3 has the same state set as s_4 since both of these variables are defined in terms of the same variable of the image system (i.e., $S_3 = S_4 = V_2$).

The first meaningful position of the mask on the data matrix (defined by the location of its reference) is for $t = 3$; the positions for $t = 1$ and $t = 2$ are not meaningful since some of the sampling variables are not defined for them ($t + \rho$ is not in set T). The initial condition consists of six entries in the data matrix: $v_{1,2}, v_{2,3}, v_{3,1}, v_{3,2}, v_{4,1}, v_{4,2}$; as an example, we assume that all these entries are equal to 1. Five additional entries of the data matrix—$v_{1,1}, v_{2,1}, v_{2,2}, v_{5,1}, v_{5,2}$—cannot be generated; they may be defined by the user, but they are not required for the data generation. The generation for $t = 3, 4, 5, 6$ is illustrated in detail in Figures 3.3a, b, c, d, respectively; symbols of the generated states in each of the four situations are circled. Figure 3.3e shows the initial condition and a larger segment of the generated data matrix.

If the data are generated in decreasing order of t (Figure 3.4), the generated variables are those represented by the left edge of the mask, i.e., variables $s_1, s_3, s_5, s_8, s_{10}$. The data are generated from the right to the left in the data matrix. Assume that f_{GB} is now defined by equations

$$s_{k,t} = s_{2,t} + s_{4,t} + s_{6,t} + s_{7,t} + s_{9,t} \text{ (modulo } k + 1)$$

for $k = 1, 3, 5, 8, 10$. Details of the data generation for $t = 98, 97, 96, 95$ are illustrated in Figures 3.4a, b, c, d, respectively. Figure 3.4e shows the initial condition and a larger segment of the generated data matrix.

3.3. METHODOLOGICAL DISTINCTIONS

> *With respect to methodology, at any rate, the pragmatists were surely right—there is certainly no better way of justifying a method than by establishing "it works" with respect to the specific tasks held in view.*
>
> —NICHOLAS RESCHER

The support-invariant constraint among a set of sampling variables can be characterized in various ways. A simple characterization, which is discussed in Section 3.2, can be accomplished by a *selection function* defined on the relevant set of states. Although the selection function is perhaps the most appropriate formal apparatus to characterize constraints of deterministic systems, whose data generation can conveniently be described by means of function (3.16), it is not adequate for dealing with nondeterministic systems.

Nondeterministic systems have been traditionally handled in terms of probability theory. The key concept for characterizing constraints among their variables has been the probability measure. Athough it remains the best developed and primary mathematical tool for dealing with nondeterministic systems, the probability measure is now viewed as a special case of a more general class of measures referred to as fuzzy measures.

Any measure assigns real numbers to the various subsets of a given set, by which degrees of some property associated with the individual subsets are characterized (measured). For our purpose, the set of concern is the set of all states of the sampling variables involved, and the property of interest is the degree of likelihood that any of the states in each particular subset can occur. The degree of likelihood is usually characterized by a real number in the unit interval; the greater the number, the higher is the degree of likelihood. Each class of measures is defined in terms of some mathematical properties; they are operationally expressed by a set of computational rules, referred to as the calculus of the respective class of measures. In an attempt to relate the mathematical properties to commonsense notions, different classes of measures have been given suggestive names such as probability, possibility, plausibility, or credibility measures. Although such names are useful for a quick orientation, they must not be taken literally. Whether a measure is suitable or not for a particular application (and similar questions) must be decided on the basis of its mathematical properties and not by resorting to the common-sense meaning of the name given to it.

► For our purpose, measures are defined on subsets of the Cartesian product **C**. A measure is thus defined by a function

$$\mu : \mathscr{P}(\mathbf{C}) \rightarrow [0, 1], \tag{3.17}$$

where $\mathscr{P}(\mathbf{C})$ denotes the power set of **C**. To qualify as a measure, function μ must satisfy at least the following requirements of *fuzzy measures*:

(μ1) $\mu(\varnothing) = 0; \quad \mu(\mathbf{C}) = 1;$

(μ2) if $X_1 \subseteq X_2$, then $\mu(X_1) \leq \mu(X_2)$;

(μ3) if $X_1 \subseteq X_2 \subseteq \cdots$, or $X_1 \supseteq X_2 \supseteq \cdots$, then $\lim_{i \to \infty} \mu(X_i) = \mu(\lim_{i \to \infty} X_i)$. ◄

Requirement (μ1) is obvious. Requirement (μ2), which is usually called the requirement of *monotonicity*, does not allow a subset of another subset of **C** to have a larger degree of the measured property than the latter subset has. According to requirement (μ3), which is called the requirement of *continuity*, the limit of the degrees of the measured property for any infinite monotonic sequence of subsets of C must be the same as the degree associated with the limit of the sequence. For discrete systems, in which **C** is always a finite set, the requirement of continuity is not applicable.

Various special classes of fuzzy measures, each with some additional properties, have been suggested in the literature. Names of some of these measures, together with a

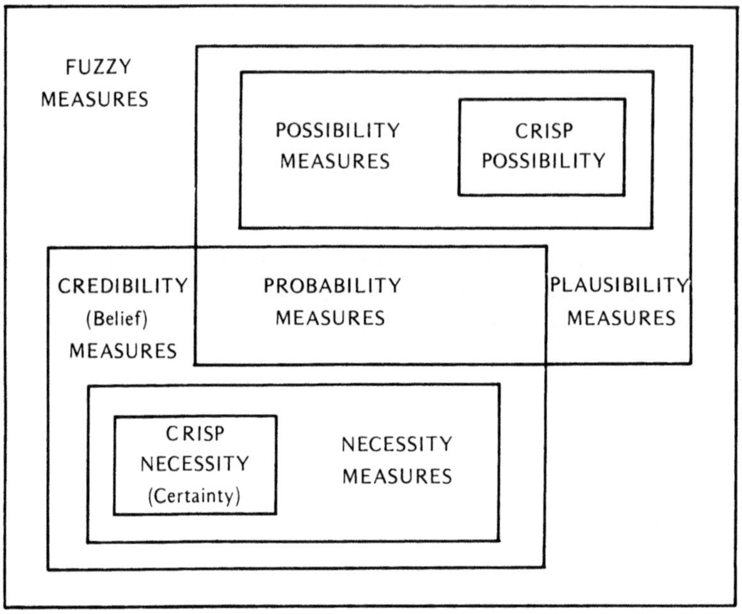

Figure 3.5. A summary of some classes of fuzzy measures.

diagram of the inclusion relationships among them, are given in Figure 3.5. For example, the class of probability measures is included in the class of plausibility measures as well as credibility measures, but it does not overlap with the class of possibility measures or necessity measures.

The individual classes of fuzzy measures are viewed as methodological distinctions. They are applicable to generative systems and all epistemologically higher types of systems. To see how the various systems problems are affected by the choice of a particular class of measure, two classes of fuzzy measures are used in this book in the context of the various systems problems. One is the classical and well-developed class of probability measures; the other is the class of possibility measures. It should be mentioned that the possibility measures are applicable only to finite sets and to some special cases of infinite sets; in general, they do not satisfy the continuity requirement. Hence, their applicability is guaranteed for discrete systems, but not for continuous systems. ◄

It is assumed in this book that the reader is familiar with the fundamentals of probability theory, to which the notion of the probability measure is central. It is well known in probability theory that any probability measure, say measure p, can be uniquely determined by a probability distribution function:

$$f_B: \mathbf{C} \to [0, 1], \tag{3.18}$$

which must satisfy the appropriate requirements, via the formula

$$p(X) = \sum_{\mathbf{c} \in X} f_B(\mathbf{c}), \tag{3.19}$$

where $X \in \mathscr{P}(\mathbf{C})$. The subscript B is used to indicate that the probability distribution function is employed as a behavior function in the same sense as the function defined by (3.9). Functions (3.9) and (3.18) play essentially the same role in defining a behavior system (3.10), even though they are methodologically distinct and neither is a special case of the other. Hence, the same symbol is used for both of them. Which of the two functions is actually used in each particular case must follow from the definition of the overall methodological distinction, which is necessary for each problem statement.

► A possibility measure is a function

$$\pi : \mathscr{P}(\mathbf{C}) \rightarrow [0, 1] \tag{3.20}$$

that satisfies the following requirements:

$$(\pi 1)\ \pi(\varnothing) = 0; \quad \pi(\mathbf{C}) = 1;$$

$$(\pi 2)\ \pi(\bigcup_i X_i) = \max_i \pi\,(X_i).$$

It is obvious that (π2) implies the monotonicity requirement of fuzzy measures. As mentioned previously, the continuity requirement is not always satisfied by π and, consequently, the possibility measure is not useful for systems with continuous variables. ◄

It is well known that any possibility measure π can be uniquely determined by a possibility distribution function f_B of the form (3.18) via the formula

$$\pi(X) = \max_{\mathbf{c} \in X} f_B(\mathbf{c}). \tag{3.21}$$

The same symbol f_B is used again for the reasons mentioned in the context of probability distribution functions.

Observe that the selection function (3.9) is a special case of the possibility distribution function, but it is not a special case of the probability distribution function. It is a possibility distribution function in which the degrees of possibility $f_B(\mathbf{c})$ are either 0 or 1 for each $\mathbf{c} \in \mathbf{C}$. This special case is usually called a *crisp possibility distribution function* (Figure 3.5).

The generative behavior function f_{GB} for the probabilistic or possibilistic methodological distinctions has the form

$$f_{GB} : \overline{\mathbf{G}} \times \mathbf{G} \rightarrow [0, 1], \tag{3.22}$$

where f_{GB} $(\bar{\mathbf{g}},\mathbf{g})$ is a conditional probability or possibility, respectively, based on condition $\bar{\mathbf{g}}$. To emphasize that f_{GB} represents conditional probabilities or possibilities, the standard notation $f_{GB}(\mathbf{g}|\bar{\mathbf{g}})$ is used instead of $f_{GB}(\bar{\mathbf{g}},\mathbf{g})$ for the probability (or possibility) of $\mathbf{g}$ given $\bar{\mathbf{g}}$.

Selection function (3.14) may be viewed as a special (crisp) case of the possibilistic interpretation of function (3.22), but not a special case of its probabilistic interpretation. For deterministic systems, however, form (3.16) of the generative behavior function is methodologically appealing and, consequently, it seems useful to view it as methodologically distinct from the possibilistic alternative. Whether function (3.16) is actually utilized or not is an issue associated with the GSPS implementation and not its architecture. From the standpoint of the GSPS user, it is sufficient to distinguish only the probabilistic and possibilistic alternatives and, perhaps, some other useful classes of fuzzy measures.

Thus far, only neutral behavior systems (basic and generative) have been considered. To describe their directed counterparts, the relevant set of sampling variables must be partitioned into two subsets:

i. sampling variables that are determined by the environment, i.e., those defined as input variables [variables v_i for which $u(i) = 0$];
ii. all remaining sampling variables associated with a mask under consideration.

These two subsets of sampling variables can be defined by partitioning the given mask, say M, into two submasks. Let submask M_e define the sampling variables determined by the environment and let $M_{\bar{e}}$ define the remaining ones. Then, the triple

$$\hat{M} = (M, M_e, M_{\bar{e}}), \tag{3.23}$$

where

$$M_e, M_{\bar{e}} \subset M$$
$$M_e \cup M_{\bar{e}} = M$$
$$M_e \cap M_{\bar{e}} = \varnothing,$$

characterizes a mask of a directed behavior system.

► According to the partition of M into M_e and $M_{\bar{e}}$, the set $N_{|M|}$ of identifiers of the sampling variables defined by M is now partitioned into two subsets, K_e and $K_{\bar{e}}$. The labeling function (3.6) is replaced by two functions,

$$\begin{aligned} \lambda_e &: M_e \to K_e, \\ \lambda_{\bar{e}} &: M_{\bar{e}} \to K_{\bar{e}}, \end{aligned} \tag{3.24}$$

and the following two sets of states, which are needed for directed systems, are defined:

$$\mathbf{E} = \bigtimes_{k \in K_e} S_k,$$
$$\overline{\mathbf{E}} = \bigtimes_{k \in K_{\bar{e}}} S_k. \tag{3.25}$$

The behavior function of directed systems has the form

$$\hat{f}_B: \mathbf{E} \times \overline{\mathbf{E}} \to [0, 1], \tag{3.26}$$

where $\hat{f}_B(\mathbf{e}, \bar{\mathbf{e}})$ has the meaning of a conditional probability or possibility (or some other measure) and, hence, the standard symbol $\hat{f}_B(\bar{\mathbf{e}} | \mathbf{e})$ is used instead of $\hat{f}_B(\mathbf{e}, \bar{\mathbf{e}})$. The *directed behavior system* can now be defined as the triple

$$\hat{\mathbf{F}}_B = (\hat{\mathbf{I}}, \hat{M}, \hat{f}_B). \tag{3.27}$$

A generative behavior function for directed systems can be introduced by partitioning $M_{\bar{e}}$ into two subsets, M_g and $M_{\bar{g}}$, associated with generated and generating variables, respectively. This is done in exactly the same way as previously described for M. The generative mask for directed systems is then defined by the quadruple

$$\hat{M}_G = (M, M_e, M_g, M_{\bar{g}}),$$

where $\{M_e, M_g, M_{\bar{g}}\}$ is a partition of M. Labeling functions (3.12) are again defined, where $\{M_g, M_{\bar{g}}\}$ is now viewed as a partition of $M_{\bar{e}}$, and two sets of states, $\mathbf{G}$ and $\overline{\mathbf{G}}$, are introduced by (3.13). Then,

$$\hat{f}_{GB}: \mathbf{E} \times \overline{\mathbf{G}} \times \mathbf{G} \to [0, 1], \tag{3.28}$$

where $\hat{f}_{GB}(\mathbf{e}, \bar{\mathbf{g}}, \mathbf{g})$ is a conditional probability or possibility (or another measure) and, hence, the symbol $\hat{f}_{GB}(\mathbf{g} | \mathbf{e}, \bar{\mathbf{g}})$ is used to conform to the literature. For deterministic systems, f_{GB} can also be expressed in a more convenient form

$$\hat{f}_{GB}: \mathbf{E} \times \overline{\mathbf{G}} \to \mathbf{G}, \tag{3.29}$$

which is a directed counterpart of the generative behavior function defined by (3.16). Assuming that the meaning (methodological distinction) of $\hat{f}_{GB}$ is specified, the *directed generative behavior system* is defined by the triple

$$\hat{\mathbf{F}}_{GB} = (\hat{\mathbf{I}}, \hat{M}_G, \hat{f}_{GB}). \blacktriangleleft \tag{3.30}$$

The partition of a mask into the three submasks M_e, M_g, $M_{\bar{g}}$ (and the

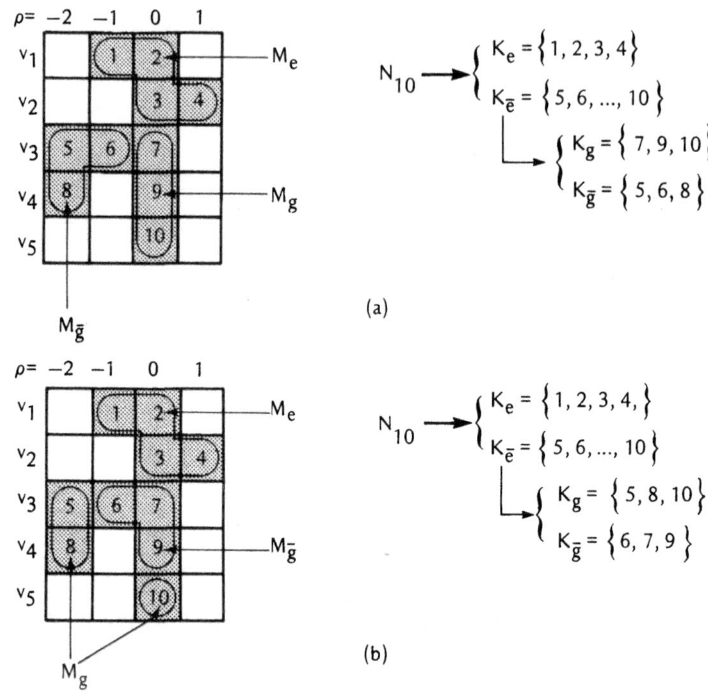

Figure 3.6. Partitions of a mask based on a directed image system with totally ordered support set and $\mathbf{u} = (0, 0, 1, 1, 1)$ for the two possible orders of generating data.

corresponding partition of the identifiers of sampling variables) is illustrated in Figure 3.6 under the assumption that v_1 and v_2 are input varIables. The two alternatives shown in Figure 3.6a and 3.6b correspond to the generation of data in increasing and decreasing order of the support values, respectively.

3.4. FROM DATA SYSTEMS TO BEHAVIOR SYSTEMS

> *Histories of science written in terms of processes that discover patterns in nature would seem closer to the mark than histories that emphasize the search for data to test hypotheses created out of whole cloth.*
>
> —Herbert A. Simon

One important class of systems problems, often referred to as *inductive systems modeling*, can be described loosely (in the context of the GSPS) as the set of problems associated with the process of climbing up the epistemological hierarchy of systems. All problems in this class are characterized by the following common description: given

- a particular system, say system x, of some epistemological type;
- the set of all particular systems of some higher epistemological type that are compatible with system x (i.e., are based on the same image system and methodological distinctions), say set Y; and
- a set of relevant requirements Q regarding some properties of systems in set Y, one of which is the requirement that the given system x be approximated by the higher-level system as closely as possible,

determine a subset Y_Q of Y such that each system in Y_Q satisfies all the requirements specified in Q.

To illustrate in this section the problem type of determining behavior systems that represent a given data system and has some additional desirable properties, let x be a data system with nominal variables, let Y be the set of all behavior systems with either probability or possibility behavior functions that are compatible with x, and let Q consist of

i. a restriction of the set Y to a subset Y_r defined either by the user or by the GSPS (as a default option);
ii. a requirement that the misfit (disagreement) between the constraint among relevant variables of the given data system and a behavior system in Y_Q be as small as possible;
iii. a requirement that the degree of nondeterminism in generating data by a behavior system in Y_Q be as small as possible;
iv. a requirement that the system in Y_Q be as simple as possible;
v. the precedence of requirement (ii) over requirements (iii) and iv.

In this general formulation, requirement (i) amounts to a specification of a set of acceptable masks. If the support set is not ordered, then the notion of support neighborhood is vacuous and, consequently, only one mask is meaningful. It is the mask based only on the identity translation rule, which is usually called a *memoryless mask*. Since there is only one acceptable mask in this case, the problem is rather trivial [requirements (iii), (iv) and (v) are not applicable]. It amounts to deriving either the probability or possibility distribution function from the given data to satisfy requirement (ii). This is done by an exhaustive sampling of the data in terms of the memoryless mask (order of sampling is irrelevant in this case) and determining for each state $\mathbf{c}$ of the sampling variables (which are in this case identical with the basic variables) the number of its occurrences in the data, say $N(\mathbf{c})$. Numbers $N(\mathbf{c})$ for all $\mathbf{c} \in \mathbf{C}$, often called *frequencies* of the individual states $\mathbf{c}$, are then employed for calculating the corresponding probabilities or possibilities $f_B(\mathbf{c})$ according to some rules.

The rules for calculating probabilities or possibilities from the frequencies are not unique. They depend on the meaning given to the probabilities or possibilities by the user. For example, if probabilities are viewed purely as means of characterizing the given data, they are normally calculated as *relative frequencies*, i.e., ratios of the individual frequencies $N(\mathbf{c})$ to the total number of available samples in the data for the mask used. Hence,

$$f_B(\mathbf{c}) = \frac{N(\mathbf{c})}{\sum_{\alpha \in \mathbf{C}} N(\alpha)}. \tag{3.31}$$

If, however, probabilities are viewed as estimators of frequencies of yet-to-be-observed outcomes, they are calculated by the formula

$$f_B(\mathbf{c}) = \frac{N(\mathbf{c}) + 1}{\sum_{\alpha \in \mathbf{C}} N(\alpha) + |\mathbf{C}|} \tag{3.32}$$

Since possibility distributions are less restrictive than their probabilistic counterparts (e.g., they are not required to add to 1), there is an even greater variety of rules for calculating them from the frequencies $N(\mathbf{c})$. A natural way of calculating possibilities, which may be viewed as a possibilistic analog of formula (3.31), is to consider them as ratios of the individual frequencies $N(\mathbf{c})$ to the maximum frequency observed in the data, i.e.,

$$f_B(\mathbf{c}) = \frac{N(\mathbf{c})}{\max_{\alpha \in \mathbf{C}} N(\alpha)} \tag{3.33}$$

Another formula is based on calculating possibilities from the corresponding probabilities. Let $f_B(\mathbf{c})$ and $f'_B(\mathbf{c})$ denote the possibility and probability of state $\mathbf{c}$ ($\mathbf{c} \in \mathbf{C}$), respectively. Then,

$$f_B(\mathbf{c}) = \sum_{\alpha \in \mathbf{C}} \min[f'_B(\mathbf{c}), f'_B(\alpha)]. \tag{3.34}$$

According to this formula, possibilities are expressed in terms of upper bounds of the probability values (Note 3.3).

As an example of the use of formulas (3.31)–(3.34), probability and possibility distributions calculated by these formulas for a specific frequency distribution are given in Table 3.1.

▶ For fuzzy data, $N(\mathbf{c})$ cannot be obtained by counting since each overall state $\mathbf{c}$ occurs at each particular support instance $\mathbf{w}$ with some degree of certainty, say $d_{c,w}$. This degree, which for crisp data is either 0 or 1, is determined by some function

$$a: [0, 1]^2 \to [0, 1],$$

through which the individual degrees $d_{i, j_i, w}$ associated with the components of $\mathbf{c}$ are

TABLE 3.1
A Comparison of Different Probability and Possibility Distributions for the Same Frequencies $N(\mathbf{c})$

					$f_B(\mathbf{c})$ calculated by formula			
v_1	v_2	v_3	$N(\mathbf{c})$	(3.31)	(3.32)	(3.33)	(3.34) and (3.31)	(3.34) and (3.32)
0	0	0	20	0.20	0.194	0.4	0.7	0.722
0	0	1	0	0.00	0.009	0.0	0.0	0.072
0	1	0	10	0.10	0.102	0.2	0.5	0.538
0	1	1	5	0.05	0.056	0.1	0.3	0.354
1	0	0	0	0.00	0.009	0.0	0.0	0.072
1	0	1	50	0.50	0.472	1.0	1.0	1.000
1	1	0	10	0.10	0.102	0.2	0.5	0.538
1	1	1	5	0.05	0.056	0.1	0.3	0.354

aggregated. Function a, which is called an *aggregation funtion*, must satisfy at least the following requirements:

(a1) a is a *continuous* function;
(a2) a is *symmetric*, i.e., $a(x, y) = a(y, x)$;
(a3) a is *associative* with respect to composition, i.e., $a(x, a(y, z)) = a(a(x, y), z)$;
(a4) a is *monotonic nondecreasing*, i.e., if $y > z$, then $a(x, y) \geq a(x, z)$.

It is clear that the class of functions that satisfy these requirements is infinite, but it is outside the scope of this book to study this class. Obvious examples of the aggregation function are functions

$$a(x, y) = xy \quad \text{and} \quad a(x, y) = \min(x, y).$$

After the aggregate degrees of certainty $d_{\mathbf{c},\mathbf{w}}$ are determined for all samples in the given data, each identified by a particular support instance w, and all overall states $\mathbf{c} \in \mathbf{C}$, their values for each particular $\mathbf{c}$ can be added. This results in numbers that are analogous with the frequencies $N(\mathbf{c})$. It is reasonable to call them *pseudofrequencies* (as they are not necessarily integers) and use for them the same symbol $N(\mathbf{c})$ that is used for frequencies. Then, $N(\mathbf{c})$ are defined for fuzzy data by the formula

$$N(\mathbf{c}) = \sum_{\mathbf{w}} d_{\mathbf{c},\mathbf{w}},$$

where the summation is taken over all meaningful samples in the data, each associated with a particular support value w that uniquely defines the location of the reference in the support set of the used mask. Once the pseudofrequencies $N(\mathbf{c})$ are determined, they can be employed for calculating the probabilities or possibilities in exactly the same way as frequencies, e.g., by using formulas (3.31) or (3.33), respectively, or by any other acceptable and desirable rules. ◀

It is obvious from this brief exposure that the variety of ways in which probabilities or possibilities of overall states ($\mathbf{c} \in \mathbf{C}$) may be determined from the frequencies or pseudofrequencies $N(\mathbf{c})$ is infinite, even though the values $N(\mathbf{c})$ are unique for any specific data and a particular mask. It is therefore important that the GSPS allow the user to specify his own choice of how to employ the frequencies or pseudofrequencies. If no choice is specified the GSPS should offer a "menu" of options, preferably those which have been frequently used. If he indicates no preference, then the GSPS should use a standard default option, say the one expressed by Eqs. (3.31) or (3.33) for probabilities or possibilities, respectively, and some standard aggregation operation for fuzzy data (e.g., min or product operations). Some typical options are illustrated in this book by various examples.

In addition to providing options for calculating probabilities or possibilities, it is also important to allow the user to incorporate in the calculation any additional information regarding the constraint among the variables. Let such additional information, which is not included in the data, be called *background information*. It can take many different forms, some of which are illustrated by examples.

Example 3.2. Data regarding 12 variables were collected in 1976 for a population of 200 government employees within a particular organizational unit. These data were used in a study whose purpose was to discover possible inequities. For this example, only five of the 12 variables are considered; they are defined in Table 3.2. The support set (the population of 200 employees) is not ordered and, hence, only the memoryless mask is applicable.

The user decided to characterize the constraint among the variables by a possibility distribution function and selected formula (3.33) for calculating degrees of possibilities from observed frequencies of the individual states. In addition, however, he wanted to utilize available background information regarding constraint among the variables due to law and other regulations. This was done by distinguishing three kinds of states:

a. states that occur in the data—their degrees of possibility $f_B(\mathbf{c})$ are calculated by formula (3.33);
b. states that are not possible according to the background information (are prohibited by law or other regulations)—clearly, $f_B(\mathbf{c}) = 0$ for these states;
c. states that are in principle possible even though they do not occur in the data—it is reasonable to define for these states a nonzero degree of possibility that is smaller than the minimal degree calculated for the observed states, i.e., to use some value in the interval

$$0 < f_B(\mathbf{c}) < \min_{\alpha} f_B(\alpha),$$

TABLE 3.2
Definitions of Variables in Example 3.2

Attribute	Variable	States of the variable
Date of birth	v_1	1: 1930 or earlier 2: 1931–1945 3: 1946 or later
Sex	v_2	1: male 2: female
Date of hiring	v_3	1: 1960 or earlier 2: 1961–1970 3: 1971 or later
Total performance score	v_4	1: 20% or less 2: 21%–40% 3: 41%–60% 4: 61%–80% 5: 81% or more
Average hourly earning	v_5	1: \$4.99 or less 2: \$5.00–\$9.99 3: \$10.00–\$14.99 4: \$15.00 or more

where α indexes the observed states; in this study it was decided to use the value

$$f_B(\mathbf{c}) = \tfrac{1}{2} \min_{\alpha} f_B(\alpha).$$

All states $\mathbf{c}$ that are possible according to the background information are listed in Table 3.3, together with their frequencies $N(\mathbf{c})$ and degrees of possibility $f_B(\mathbf{c})$. Clearly, only those states for which $N(\mathbf{c}) \neq 0$ occur in the data.

Suppose now that the support set is totally ordered. In that case, many different behavior systems can be obtained for the same data system, each one based on a particular mask. If properly derived from the given data, they all satisfy the misfit requirement equally well. More specifically, the term "properly derived" means that the behavior function is in perfect agreement with the data (and, possibly, some background information) in terms of the mask and the type of constraint characterization chosen.

As explained previously for memoryless masks, a behavior function that agrees perfectly with the given data and background information can be obtained from state frequencies (of the respective sampling variables) determined by an exhaustive sampling of the data in terms of the mask considered. Each mask can be viewed as a window through which a sample is seen in a data matrix (or a higher-order array). When this window is moved across the entire data matrix, state frequencies of the respective sampling variables are determined by observing the samples and counting how many times each state occurs. The order of moving the mask on the data matrix does not

TABLE 3.3
Possibility Behavior Function in Example 3.2

c	v_1	v_2	v_3	v_4	v_5	$f_B(\mathbf{c})$	N(c)
1	1	1	1	2	1	0.111	2
2	1	1	1	3	4	0.056	0
3	1	1	2	2	1	0.111	2
4	1	1	2	4	1	0.111	2
5	1	1	3	3	1	0.056	0
6	1	1	3	4	1	0.111	2
7	1	2	1	1	2	0.056	0
8	1	2	1	1	4	0.111	2
9	1	2	1	2	2	0.167	3
10	1	2	1	2	4	0.889	16
11	1	2	1	3	2	0.056	0
12	1	2	1	3	4	0.222	4
13	1	2	1	4	1	0.056	0
14	1	2	1	4	4	0.778	14
15	1	2	1	5	4	0.222	4
16	1	2	2	3	4	0.056	0
17	1	2	2	4	4	0.167	3
18	1	2	2	5	4	0.056	0
19	1	2	3	2	4	0.056	0
20	1	2	3	3	1	0.056	0
21	1	2	3	3	4	0.056	0
22	1	2	3	4	4	0.056	0
23	1	2	3	5	4	0.056	0
24	2	1	3	2	1	0.056	0
25	2	1	3	4	1	0.056	0
26	2	1	3	4	2	0.111	2
27	2	2	1	1	2	0.056	0
28	2	2	1	1	4	0.222	4
29	2	2	1	2	1	0.111	2
30	2	2	1	2	4	1.000	18
31	2	2	1	3	4	0.167	3
32	2	2	1	4	2	0.056	0
33	2	2	1	4	4	0.667	12
34	2	2	1	5	4	0.167	3
35	2	2	2	1	2	0.056	0
36	2	2	2	2	2	0.056	0
37	2	2	2	2	4	0.222	4
38	2	2	2	3	2	0.056	0
39	2	2	2	3	4	0.056	0
40	2	2	2	4	4	0.111	2
41	2	2	2	5	4	0.056	0
42	2	2	3	2	2	0.056	0
43	2	2	3	2	3	0.056	0
44	2	2	3	2	4	0.111	2
45	2	2	3	4	1	0.056	0
46	2	2	3	4	3	0.056	0
47	2	2	3	4	4	0.333	6
48	2	2	3	5	4	0.222	4
49	3	1	3	4	1	0.056	0
50	3	2	1	1	4	0.056	0
51	3	2	1	2	2	0.056	0
52	3	2	1	2	4	0.167	3
53	3	2	1	3	4	0.167	3
54	3	2	1	4	2	0.056	0
55	3	2	1	4	4	0.333	6
56	3	2	1	5	4	0.056	0
57	3	2	2	1	4	0.111	2
58	3	2	2	2	2	0.222	4
59	3	2	2	2	4	0.667	12
60	3	2	2	3	2	0.056	0
61	3	2	2	3	4	0.333	6
62	3	2	2	4	2	0.222	4
63	3	2	2	4	3	0.056	0
64	3	2	2	4	4	0.167	3
65	3	2	2	5	4	0.111	2
66	3	2	3	1	1	0.056	0
67	3	2	3	1	2	0.111	2
68	3	2	3	1	4	0.111	2
69	3	2	3	2	1	0.167	3
70	3	2	3	2	2	0.167	3
71	3	2	3	2	4	0.556	10
72	3	2	3	3	2	0.222	4
73	3	2	3	3	4	0.167	3
74	3	2	3	4	2	0.056	0
75	3	2	3	4	4	0.500	9
76	3	2	3	5	1	0.056	0
77	3	2	3	5	4	0.167	3

matter, provided that all possible sampling positions are covered, but it is convenient to proceed according to the order of the support set (from left to right or vice versa).

Some masks may be better than others for each specific purpose, but none of them is either right or wrong. This important point is well expressed by James Keys*:

* *Only Two Can Play This Game*, Bantam Books, New York, 1974, p. 99 (original edition published by Julian Press, New York, 1972). James Keys is a pseudonym of G. Spencer Brown.

> You may look at the world any way you please, through any window you choose. Nor does it always have to be the same window. Naturally how the world appears, what you see and what you miss, and the angle on what you see, depends on which window you are using, but can a window be right or wrong? A window is a window. . . . it is perfectly OK to try another window if what you see through yours seems meaningless and inadequate. Naturally if you enjoy the view, there is no need to change it. Alternatively, if you come to another window, it may take time to adjust to what you see.

If the mask considered consists of a single column (a memoryless mask), samples for all support values are complete. However, if it consists of more than one column, then some samples at the beginning and end of the support set (the left and right end of the data matrix) are not complete (see Figures 3.3 and 3.4). More specifically, the number of incomplete samples at each end of the data matrix is equal to the number of columns in the mask minus one. Let the number of columns in the mask be called the *depth of a mask* and let it be denoted by ΔM for mask M. Then,

$$\Delta M = 1 + \max \rho - \min \rho, \tag{3.35}$$

where the max and min operators are applied over all integers such that $(v_i, t + \rho) \in M$. For instance, $\Delta M = 4$ for the mask defined in Figure 3.2; $\Delta M = 1$ for any memoryless mask.

There are at least two reasons why masks with large depths are generally undesirable. First, if a mask is employed for generating data, as discussed in Section 3.2, then the larger its depth the larger is the required initial condition. This, generally, is not desirable. Second, if the mask is used for sampling data, the number of incomplete samples is equal to $2(\Delta M - 1)$. This means that as the depth of the mask increases, fewer of the available data are utilized for deriving a behavior function. Consequently, the empirical support of the derived behavior function weakens with increasing depth of the mask used. This, again, is clearly undesirable. For both of these reasons, as well as for practical reasons related to computational complexity, mask depths are usually restricted. This provides a rationale for requirement (i) in the problem type under discussion.

Given a data system, background information (if available), a mask, and a characterization of the constraint among sampling variables, the behavior function is uniquely determined by the sampling procedure explained previously. This uniqueness is a direct consequence of the misfit requirement. The data system and background information are fixed in each particular problem. Assume that the type of constraint, as a methodological distinction, is also fixed. Then, behavior systems that satisfy the misfit requirement (i.e., candidates for the solution set Y_Q) are uniquely identified (and distinguished from each other) by their masks. The restriction of the set Y to a subset Y_r, as required by (i), can thus be expressed as a restriction on the set of possible masks. According to our previous discussion, it is desirable to restrict the depth of the mask.

This can be accomplished by defining a *largest acceptable mask*, say mask **M**, by the Cartesian product

$$\mathbf{M} = V \times R,$$

where

$$R = \{(t+\rho) \mid \rho_1 \leq \rho \leq \rho_2\}.$$

Such a mask may be represented as a full matrix with n rows and $1+\rho_2-\rho_1 (= \Delta\mathbf{M})$ columns; let this matrix be referred to as an **M**-*matrix*. If the user specifies only $\Delta\mathbf{M}$, but not the actual values of ρ_1 and ρ_2, then some standard values are chosen by the GSPS, say $\rho_2 = 0$ and $\rho_1 = 1-\Delta\mathbf{M}$.

Given a largest acceptable mask **M**, all of its *meaningful submasks* represent the restricted set Y_r of behavior systems. The term "meaningful submask" is used here to characterize submasks of **M** that satisfy the following requirements:

(m1) at least one element in each submask $\mathbf{M}_i$, as defined by Eq. (3.8), is included (one element in each row of the **M**-matrix);

(m2) at least one element with the translation rule $t+\rho_2$ must be included (a right-most element in the **M**-mask).

Requirement (m1) is necessary for the sake of covering the given data system, i.e., to guarantee that each basic variable of the given data system is included in each of the behavior systems in the restricted set Y_r. Requirement (m2) is included to prevent duplicates of equivalent submasks, i.e., submasks that can be converted to each other solely by adding a constant to the translation rules $t+\rho$ (a column shift in the **M**-mask).

It is easy to derive the following formula for the number $N(n, \Delta\mathbf{M})$ of meaningful submasks of a largest acceptable mask **M** defined for n basic variables and with a depth of $\Delta\mathbf{M}$:

$$N(n, \Delta\mathbf{M}) = (2^{\Delta\mathbf{M}}-1)^n - (2^{\Delta\mathbf{M}-1}-1)^n. \tag{3.36}$$

The first term in (3.36) expresses the number of submasks of **M** that satisfy requirement (m1); the second term gives the number of masks that violate requirement (m2). Numbers of $N(n, \Delta\mathbf{M})$ for $n, \Delta\mathbf{M} \leq 10$ are shown in Table 3.4. The three areas indicated in the table identify such ranges of the largest acceptable mask that are (a) computationally tractable (the left top area), (b) potentially tractable, but would require heavy use of a very powerful computer (the middle area), and (c) considered as intractable (the right bottom area). These areas indicate, of course, only a typical situation. They depend, at least to some extent, on the available computing facilities. For instance, if specialized computer hardware with extensive parallel processing were available, it is likely that the range of tractable cases would be almost doubled.

If the number of meaningful masks is too large to be computationally tractable, the

TABLE 3.4
Numbers $N(n, \Delta M)$ of Meaningful Masks Given by Formula (3.36)

n \ ΔM	1	2	3	4	5	6	7	8	9	10
1	1	2	4	8	16	32	64	128	256	512
2	1	8	40	176	736	3,008	12,160	48,896	196,096	785,408
3	1	26	316	3,032	26,416	220,256	1.8×10^6	1.5×10^7	1.2×10^8	$\sim 10^9$
4	1	80	2,320	48,224	872,896	1.5×10^7	$\sim 2.4 \times 10^8$	$\sim 10^9$	$\sim 10^{11}$	$\sim 10^{12}$
5	1	242	16,564	742,568	$\sim 2.7 \times 10^7$	$\sim 9.6 \times 10^8$	$\sim 10^{10}$	$\sim 10^{12}$	$\sim 10^{13}$	$\sim 10^{15}$
6	1	728	116,920	$\sim 1.1 \times 10^7$	$\sim 8.8 \times 10^8$	$\sim 10^{11}$	$\sim 10^{12}$	$\sim 10^{14}$	$\sim 10^{16}$	$\sim 10^{18}$
7	1	2,186	821,356	$\sim 1.7 \times 10^8$	$\sim 10^{10}$	$\sim 10^{12}$	$\sim 10^{15}$	$\sim 10^{17}$	$\sim 10^{19}$	$\sim 10^{21}$
8	1	6,560	$\sim 5.8 \times 10^6$	$\sim 10^9$	$\sim 10^{12}$	$\sim 10^{14}$	$\sim 10^{17}$	$\sim 10^{19}$	$\sim 10^{21}$	$\sim 10^{24}$
9	1	19,682	$\sim 4 \times 10^7$	$\sim 10^{11}$	$\sim 10^{13}$	$\sim 10^{16}$	$\sim 10^{19}$	$\sim 10^{21}$	$\sim 10^{24}$	$\sim 10^{27}$
10	1	59,048	$\sim 2.8 \times 10^8$	$\sim 10^{12}$	$\sim 10^{15}$	$\sim 10^{18}$	$\sim 10^{21}$	$\sim 10^{24}$	$\sim 10^{27}$	$\sim 10^{30}$

GSPS should offer the user a "menu" of available additional restrictions within the largest acceptable mask. Such restrictions may include, for example,

- a fixed set of generated sampling variables;
- a fixed number of sampling variables;
- a fixed upper bound of the number of sampling variables;
- a restriction to masks without gaps (such as the element identified by $i = 4$, $\rho = -1$ of the mask defined in Figure 3.2a).

Restrictions such as these, some of which can be combined, reduce the set Y_r considerably and thus extend the size of computationally tractable largest acceptable masks.

Although these restrictions are important for reducing computational complexity, especially when the largest acceptable mask desired is intractable, each of them distorts the original problem. Unless justified by some context-dependent reasons, they should be used only as a last resort in dealing with computational complexity.

In the rest of this chapter, the focus is on the general problem type in which the restricted set Y_r consists of one behavior system for each meaningful submask of the chosen largest acceptable mask (assumed to be within the computationally tractable range). As explained previously, each of these systems is determined in such a manner that it is in perfect agreement with the given data system and background information with respect to its mask and the accepted type of constraint characterization. The misfit requirement was thus given a precedence over the other requirements, as demanded by (iv) in the problem statement. It now remains to employ requirements (iii) and (iv), which are usually called the *requirements of determinism and complexity*, respectively, for deriving the solution subset Y_Q of the restricted set Y_r.

Although various special requirements are sometimes added to the statement of any particular problem type, the requirements of determinism and complexity are of general significance. As such, neither of them is usually omitted. The solution set based on these two requirements (and, of course, the misfit requirement) is often determined first. Behavior systems included in it are then examined by the investigator. He may use all of them as complementary representations of the basic variables. If, however, further reduction is desirable, they are evaluated and compared by additional criteria, some of which may be context dependent or may express the investigator's preferences.

3.5. MEASURES OF UNCERTAINTY

> *As far as the laws of mathematics refer to reality, they are not certain; and as far as they are certain, they do not refer to reality.*
>
> —ALBERT EINSTEIN

Intuitively, the degree of nondeterminism should measure the average uncertainty associated with the generation of data. As such, it must be defined in terms of generative behavior functions f_{GB}, f_{GB}, given by (3.22) and (3.28) for neutral and directed behavior systems, respectively. When these functions are probability distribution functions, a measure of average uncertainty is well established—it is the *Shannon entropy*, introduced by Claude Shannon in 1948 [SH3].

Let P denote the set of all probability distributions that can be defined on finite sets of alternative (mutually exclusive) outcomes. Then, a *probabilistic measure of uncertainty* is a function

$$H: P \to [0, \infty)$$

that possesses some properties considered desirable for such a measure. The following properties, have been generally accepted as necessary properties of any meaningful measure of uncertainty (Note 3.6):

(H1) *symmetry*—uncertainty is invariant with respect to permutations of probabilities;
(H2) *expansibility*—uncertainty does not change when outcomes with zero probabilities are added to the set of outcomes considered;
(H3) *subadditivity*—the uncertainty of a joint probability distribution is not greater than the sum of the uncertainties of the corresponding marginal probability distributions;
(H4) *additivity*—for probability distributions of any two independent sets of outcomes, the uncertainty of the joint probability distribution is equal to the sum of the uncertainties of the individual probability distributions;
(H5) *continuity*—uncertainty is a continuous function in all its arguments.

It is well known that functions of the form

$$H(f(x) \mid x \in X) = -a \sum_{x \in X} f(x) \log_b f(x)$$

are the only functions that possess properties (H1)–(H5);

$$(f(x) \mid x \in X) \in P$$

denotes the probability distribution associated with a particular finite set X of alternative outcomes x, a is an arbitrary positive constant, and b is an arbitrary base of logarithms. When a reasonable *normalization property*

$$H(0.5, 0.5) = 1$$

is added to the properties (the uncertainty of two equally probable outcomes is equal to 1), the measure of uncertainty becomes unique:

$$H(f(x) \mid x \in X) = -\sum_{x \in X} f(x) \log_2 f(x) \tag{3.37}$$

Function (3.37) is usually referred to as the *Shannon entropy*. It measures uncertainty in units that are called *bits* (an abbreviation for *b*inary dig*its*). Such units are intuitively appealing since any integer value of uncertainity measured in them, say value u, is equivalent to the uncertainty in predicting the truth values of u propositions or values of u binary digits, under the assumption they are equally probable.

Assuming that each finite set X of alternative outcomes under consideration is characterized by a particular probability distribution, it is convenient to simplify the notation by using $H(X)$ instead of $H(f(x)|x \in X)$.

It is easy to show that

$$0 \leq H(X) \leq \log_2 |X|. \tag{3.38}$$

The lower bound, $H(X) = 0$, is obtained when probabilities of all outcomes except one are equal to 0; the upper bound is reached when probabilities of all events are the same, i.e., equal to $1/|X|$. The ratio

$$\mathbf{H}(X) = H(X)/\log_2 |X| \tag{3.39}$$

of the actual entropy to its upper bound is called a *normalized entropy*; clearly

$$0 \leq \mathbf{H}(X) \leq 1. \tag{3.40}$$

For our purpose, sets of outcomes are state sets **C**, **G**, $\bar{\mathbf{G}}$, **E**, and probability distributions are based on behavior functions f_B, f_{GB}, $\hat{f}_B$, $\hat{f}_{GB}$, defined by (3.18), (3.22), (3.26), (3.28), respectively. For the sake of notational simplicity, subscripts B and GB, as well as the caret, are omitted. Symbols

$$f(\mathbf{c}), \quad f(\mathbf{g}|\bar{\mathbf{g}}), \quad f(\bar{\mathbf{e}}|\mathbf{e}), \quad f(\mathbf{g}|\mathbf{e}, \bar{\mathbf{g}})$$

thus denote probabilities defined by (3.18), (3.22), (3.26), (3.28), respectively; the meaning of each symbol is uniquely determined by the argument shown in the parentheses. In addition, we define the marginal probabilities

$$f(\bar{\mathbf{g}}) = \sum_{\mathbf{c} \succ \bar{\mathbf{g}}} f(\mathbf{c}), \tag{3.41}$$

where $\mathbf{c} \succ \bar{\mathbf{g}}$ designates $\bar{\mathbf{g}}$ as a *substate* of **c**; formally, if

$$\mathbf{c} = (c_k | k \in N_{|\mathbf{C}|})$$

and

$$\bar{\mathbf{g}} = (\bar{\mathbf{g}}_j | j \in Z, Z \subset N_{|\mathbf{C}|}),$$

then $\bar{\mathbf{g}} \prec \mathbf{c}$ ($\bar{\mathbf{g}}$ is a substate of **c**) if and only if $\bar{g}_j = c_j$ for all $j \in Z$. For directed systems, the marginal probabilities ar calculated by a slightly modified formula

$$f(\bar{\mathbf{g}}|\mathbf{e}) = \sum_{\bar{\mathbf{e}} > \bar{\mathbf{g}}} f(\bar{\mathbf{e}}|\mathbf{e}). \tag{3.42}$$

The conditional probabilities, which characterize the process of generating data, are related to the basic (joint) and marginal probabilities by the formulas

$$f(\mathbf{g}|\bar{\mathbf{g}}) = \frac{f(\mathbf{c})}{f(\bar{\mathbf{g}})}, \tag{3.43}$$

$$f(\mathbf{g}|\mathbf{e}, \bar{\mathbf{g}}) = \frac{f(\bar{\mathbf{e}}|\mathbf{e})}{f(\bar{\mathbf{g}}|\mathbf{e})} \tag{3.44}$$

for neutral and directed systems, respectively.

Given a generative mask for a neutral system through which state sets $\mathbf{G}, \bar{\mathbf{G}}$ of some generated and generating sampling variables are defined, the *generative uncertainty* $H(\mathbf{G}|\bar{\mathbf{G}})$ is defined as the average uncertainty based on probabilities $f(\mathbf{g}|\bar{\mathbf{g}})$, weighted by the probabilities $f(\bar{\mathbf{g}})$ of the generating conditions:

$$H(\mathbf{G}|\bar{\mathbf{G}}) = -\sum_{\bar{\mathbf{g}} \in \bar{\mathbf{G}}} f(\bar{\mathbf{g}}) \sum_{\mathbf{g} \in \mathbf{G}} f(\mathbf{g}|\bar{\mathbf{g}}) \log_2 f(\mathbf{g}|\bar{\mathbf{g}}). \tag{3.45}$$

This value is taken as the *degree of nondeterminism* of the given neutral generative behavior system.

For directed systems, the generative uncertainty $H(\mathbf{G}|\mathbf{E} \times \bar{\mathbf{G}})$ is calculated by the formula

$$H(\mathbf{G}|\mathbf{E} \times \bar{\mathbf{G}}) = -\sum_{\mathbf{e} \in \mathbf{E}} \sum_{\bar{\mathbf{g}} \in \bar{\mathbf{G}}} f(\mathbf{e}, \bar{\mathbf{g}}) \sum_{\mathbf{g} \in \mathbf{G}} f(\mathbf{g}|\mathbf{e}, \bar{\mathbf{g}}) \log_2 f(\mathbf{g}|\mathbf{e}, \bar{\mathbf{g}}), \tag{3.46}$$

which is directly applicable only under the assumption that it is possible and meaningful to determine probabilities $f(\mathbf{e}, \bar{\mathbf{g}})$, e.g., when the directed system is derived from a neutral system. If probabilities of states in set $\mathbf{E}$ are not available or are viewed as irrelevant, then $f(\bar{\mathbf{e}}|\mathbf{e})$ are the basic probabilities [counterparts of probabilities f(c) for neutral systems] from which other desirable probabilities are calculated. The uncertainty $H(\mathbf{G}|\mathbf{E} \times \bar{\mathbf{G}})$ is then expressed by the formula

$$H(\mathbf{G}|\mathbf{E} \times \bar{\mathbf{G}}) = -\frac{1}{|\mathbf{E}|} \sum_{\mathbf{e} \in \mathbf{E}} \sum_{\bar{\mathbf{g}} \in \bar{\mathbf{G}}} f(\bar{\mathbf{g}}|\mathbf{e}) \sum_{\mathbf{g} \in \mathbf{G}} f(\mathbf{g}|\mathbf{e}, \bar{\mathbf{g}}) \log_2 f(\mathbf{g}|\mathbf{e}, \bar{\mathbf{g}}), \tag{3.47}$$

where probabilities $f(\bar{\mathbf{g}}|\mathbf{e})$ and $f(\mathbf{g}|\mathbf{e}, \bar{\mathbf{g}})$ are determined from the given probabilities $f(\bar{\mathbf{e}}|\mathbf{e})$ by Eqs. (3.42) and (3.44), respectively.

► Formulas (3.45), (3.46), (3.47) can be replaced by alternative formulas that are computationally more convenient. Equation (3.45), for instance, can be modified as follows:

$$\begin{aligned}
H(\mathbf{G}|\bar{\mathbf{G}}) &= -\sum_{\bar{\mathbf{g}}} f(\bar{\mathbf{g}}) \sum_{\mathbf{g}} f(\mathbf{g}|\bar{\mathbf{g}}) \log_2 f(\mathbf{g}|\bar{\mathbf{g}}) \\
&= -\sum_{\bar{\mathbf{g}}} \sum_{\mathbf{g}} f(\bar{\mathbf{g}}) f(\mathbf{g}|\bar{\mathbf{g}}) \log_2 f(\mathbf{g}|\bar{\mathbf{g}}) \\
&= -\sum_{\bar{\mathbf{g}}} \sum_{\mathbf{g}} f(\mathbf{c})[\log_2 f(\mathbf{c})/f(\bar{\mathbf{g}})] \\
&= H(\mathbf{C}) + \sum_{\bar{\mathbf{g}}} \sum_{\mathbf{g}} f(\mathbf{c}) \log_2 f(\bar{\mathbf{g}}) \\
&= H(\mathbf{C}) + \sum_{\bar{\mathbf{g}}} \log_2 f(\bar{\mathbf{g}}) \sum_{\mathbf{g}} f(\mathbf{c}) \\
&= H(\mathbf{C}) + \sum_{\bar{\mathbf{g}}} f(\bar{\mathbf{g}}) \log_2 f(\bar{\mathbf{g}}) \\
&= H(\mathbf{C}) - H(\bar{\mathbf{G}}). \blacktriangleleft
\end{aligned}$$

Hence, $H(\mathbf{G}|\bar{\mathbf{G}})$ can be calculated without the use of conditional probabilities by the formula

$$H(\mathbf{G}|\bar{\mathbf{G}}) = H(\mathbf{C}) - H(\bar{\mathbf{G}}). \tag{3.48}$$

By the same reasoning, Eqs. (3.46), (3.47) can be replaced by

$$H(\mathbf{G}|\mathbf{E} \times \bar{\mathbf{G}}) = H(\mathbf{C}) - H(\mathbf{E} \times \bar{\mathbf{G}}), \tag{3.49}$$

$$H(\mathbf{G}|\mathbf{E} \times \bar{\mathbf{G}}) = \frac{1}{|\mathbf{E}|}\left[\sum_{\mathbf{e} \in \mathbf{E}} H(\bar{\mathbf{E}}|\mathbf{e}) - \sum_{\mathbf{e} \in \mathbf{E}} H(\bar{\mathbf{G}}|\mathbf{e})\right]. \tag{3.50}$$

respectively.

The maximum value of generative uncertainty of any kind is $\log_2 |\mathbf{G}|$; hence, the normalized generative uncertainty is obtained from the given uncertainty by dividing it by this maximum value. For instance,

$$\mathbf{H}(\mathbf{G}|\bar{\mathbf{G}}) = H(\mathbf{G}|\bar{\mathbf{G}})/\log_2 |\mathbf{G}|.$$

Example 3.3. A probability behavior function $f(\mathbf{c})$ for four sampling variables s_1, s_2, s_3, s_4, each with two states 0, 1, is specified in Figure 3.7a; states with zero probabilities are not listed in the table. The sampling variables are defined in terms of two basic variables v_1, v_2 by the mask in Figure 3.7b. Since sampling variables s_2, s_3 are translations of the same basic variable v_1, probability distributions for their states must be equal; they, indeed, are equal: both have probabilities 0.7 and 0.3 for states 0 and 1, respectively. Similarly, variables s_1, s_4 (translations of v_2) have the same probability

	s_1	s_2	s_3	s_4	$f(\mathbf{c})$
$\mathbf{c} =$	0	0	0	0	0.2
	0	0	0	1	0.2
	0	0	1	0	0.1
	0	1	0	1	0.1
	1	0	0	0	0.1
	1	0	1	0	0.1
	1	1	0	1	0.1
	1	1	1	0	0.1

(a)

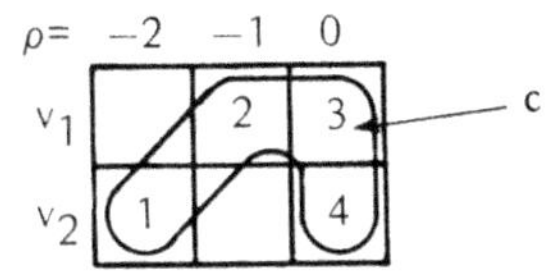

(b)

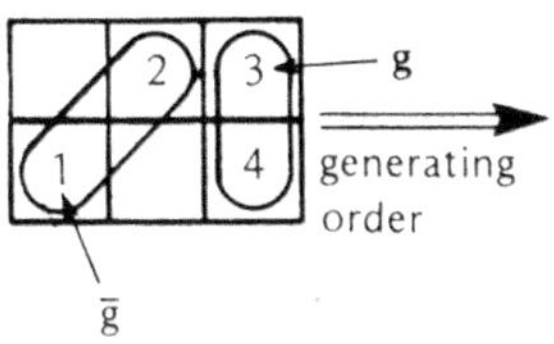

	s_1 s_2	$f(\bar{g})$
$\bar{g}=$	0 0	0.5
	0 1	0.1
	1 0	0.2
	1 1	0.2

(c)

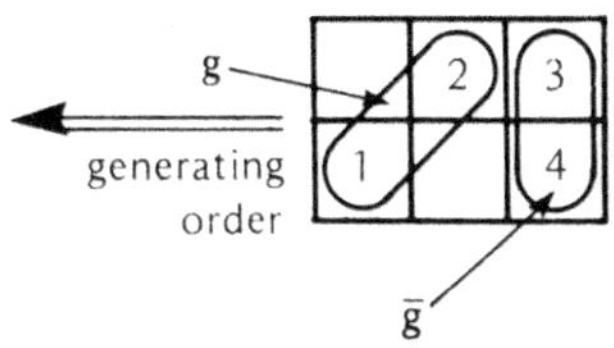

	s_3 s_4	$f(\bar{g})$
$\bar{g}=$	0 0	0.3
	0 1	0.4
	1 0	0.3

(d)

Figure 3.7. Illustration to Example 3.3: probabilistic neutral system.

distribution: 0.6 and 0.4 for states 0 and 1, respectively. Hence, the given probability distribution function is a legitimate behavior function for the specified mask.

If the system is viewed as a neutral system, the generative uncertainty $H(\mathbf{G}\,|\,\bar{\mathbf{G}})$ can be calculated by formula (3.48). For the first term, we get

$$H(\mathbf{C}) = -2 \times 0.2 \log_2 0.2 - 6 \times 0.1 \log_2 0.1 = 0.9288 + 1.9932 = 2.922.$$

The second term depends on the generative order and the corresponding generative mask. The two possible generative orders are illustrated in Figures 3.7c and 3.7d. For the generation from the left to the right, we obtain

$$H(\bar{\mathbf{G}}) = -0.5 \log_2 0.5 - 0.1 \log_2 0.1 - 2 \times 0.2 \log_2 0.2$$
$$= 0.5 + 0.3322 + 0.9288 = 1.761,$$
$$H(\mathbf{G}\,|\,\bar{\mathbf{G}}) = H(\mathbf{C}) - H(\bar{\mathbf{G}}) = 2.922 - 1.761 = 1.161.$$

For the other generating order (Figure 3.7d), we get

$$\begin{aligned} H(\bar{\mathbf{G}}) &= -2 \times 0.3 \log_2 0.3 - 0.4 \log_2 0.4 \\ &= 1.0422 + 0.5288 = 1.571, \\ H(\mathbf{G} \mid \bar{\mathbf{G}}) &= 2.922 - 1.571 = 1.351. \end{aligned}$$

Hence, if we are allowed to choose either of the two generating orders. the first one (Figure 3.7c) is preferable since it has a lower generative uncertainty. Since $\log_2 |\mathbf{G}| = 2$ in this example, we obtain normalized values of the calculated generative uncertainties by dividing each of them by two.

In some situations, only one of the generating orders is applicable. For instance, if the support is time, only one of the orders is meaningful in each case depending on the purpose for which the behavior system is employed. If it is employed for *prediction*, states must be generated in increasing order of time (from left to right); if it is employed for *retrodiction*, states must be generated in decreasing order of time. In this example, if the support is time, then it is easier to predict future states than to retrodict past states of the system.

Assume now that v_1 is viewed as an input variable and that the corresponding directed system is derived from the behavior function in Figure 3.7a. Then, formula (3.49) can be used for calculating the generative uncertainty. $H(\mathbf{C})$ was calculated before; $H(\mathbf{E} \times \bar{\mathbf{G}})$ depends on the generating order. In either case, $\mathbf{E}$ is represented by states of variables s_2, s_3; $\bar{\mathbf{G}}$ is represented by states of either s_1 (increasing order of support) or s_4 (decreasing order of support). In the former case, $H(\mathbf{E} \times \bar{\mathbf{G}})$ is the uncertainty associated with variables s_1, s_2, s_3:

$$\begin{aligned} H(\mathbf{E} \times \bar{\mathbf{G}}) &= -0.4 \log_2 0.4 - 6 \times 0.1 \log_2 0.1 \\ &= 0.5288 + 1.9932 = 2.522, \\ H(\mathbf{G} \mid \mathbf{E} \times \bar{\mathbf{G}}) &= H(\mathbf{C}) - H(\mathbf{E} \times \bar{\mathbf{G}}) = 2.922 - 2.522 = 0.4. \end{aligned}$$

In the latter case, it represents the uncertainty of variables s_2, s_3, s_4:

$$\begin{aligned} H(\mathbf{E} \times \bar{\mathbf{G}}) &= -0.3 \log_2 0.3 - 3 \times 0.2 \log_2 0.2 - 0.1 \log_2 0.1 \\ &= 0.5211 + 1.3932 + 0.3322 = 2.2465 \\ H(\mathbf{G} \mid \mathbf{E} \times \bar{\mathbf{G}}) &= 2.922 - 2.2465 = 0.6755. \end{aligned}$$

Hence, it is again easier to predict than to retrodict.

Assume now that no information about the input variable v_1 is available or, if it is available, that it is not relevant (e.g., when v_1 is controlled by the investigator). In this case, all calculations must be made in terms of the conditional probabilities $f(\bar{\mathbf{e}} \mid \mathbf{e})$ given in Figure 3.8a. As indicated in this figure, the probabilities as listed form four blocks, one for each state $\mathbf{e}$. Uncertainties for each of these blocks are given in Figure 3.8a. The partition of the mask into $\{M_e, M_{\bar{e}}\}$ is shown in Figure 3.8b.

e		ē			
s_2	s_3	s_1	s_4	$f(\bar{e} \mid e)$	$H(\bar{E} \mid e)$
0	0	0	0	0.4	
0	0	0	1	0.4	1.522
0	0	1	0	0.2	
0	1	0	0	0.5	1.0
0	1	1	0	0.5	
1	0	0	1	0.5	1.0
1	0	1	1	0.5	
1	1	1	0	1.0	0.0

(a)

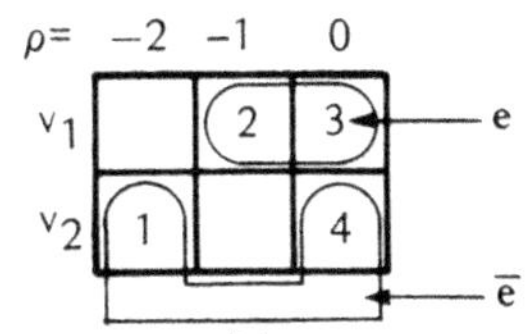

(b)

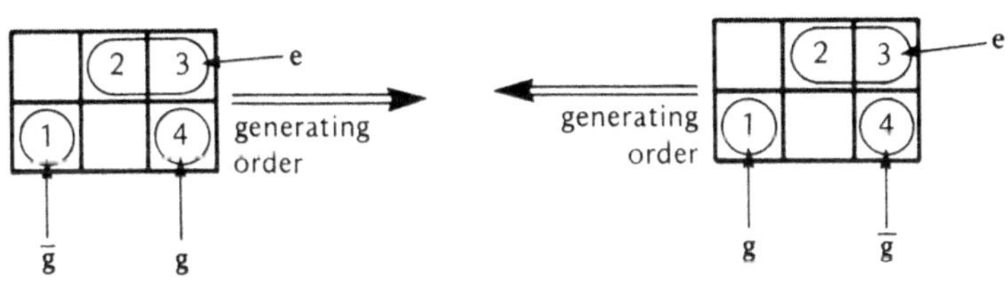

e		ḡ		
s_2	s_3	s_1	$f(\bar{g} \mid e)$	$H(\bar{G} \mid e)$
0	0	0	0.8	0.7219
0	0	1	0.2	
0	1	0	0.5	1.0
0	1	1	0.5	
1	0	0	0.5	1.0
1	0	1	0.5	
1	1	1	1.0	0.0

(c)

e		ḡ		
s_2	s_3	s_4	$f(\bar{g} \mid e)$	$H(\bar{G} \mid e)$
0	0	0	0.6	0.971
0	0	1	0.4	
0	1	0	1.0	0.0
1	0	1	1.0	0.0
1	1	0	1.0	0.0

(d)

Figure 3.8. Illustration to Example 3.3: probabilistic directed system.

The situation for generating states from the left to the right is illustrated in Figure 3.8c, including the values of $H(\bar{\mathbf{G}}|\mathbf{e})$ for each individual state $\mathbf{e}$. Applying formula (3.50),

$$H(\mathbf{G}|\mathbf{E}\times\bar{\mathbf{G}}) = \tfrac{1}{4}(3.522 - 2.7219) = 0.200025.$$

The other generating order is illustrated in Figure 3.8d, and

$$H(\mathbf{G}|\mathbf{E}\times\bar{\mathbf{G}}) = \tfrac{1}{4}(3.522 - 0.971) = 0.63775.$$

► Let us now discuss the generative uncertainty for systems characterized by possibility distribution functions. Let Π denote the set of all possibility distributions with at least one nonzero value that can be defined on finite sets of alternative outcomes (possibility distributions solely with zero values are not meaningful for our purpose). Then, a *possibilistic measure of uncertainty* is a function

$$U: \Pi \to [0, \infty) \tag{3.51}$$

that possesses appropriate properties. To be able to discuss these properties, several concepts associated with possibility distributions must be introduced first:

1. A possibility distribution, say

$$\mathbf{f} = (\phi_i | i \in N_{|X|}) \in \Pi, \tag{3.52}$$

on a finite set X of alternative outcomes x is called a *normalized possibility distribution* if and only if

$$\max_i \phi_i = 1;$$

clearly, $\phi_i = f(x)$ for some one-to-one correspondence between $N_{|X|}$ and X.

2. For each possibility distribution $\mathbf{f}$, say given by (3.52), and each real number $l \in [0, 1]$, let

$$c: \Pi \times [0, 1] \to \mathscr{P}(\mathbb{N}) \tag{3.53}$$

be a function such that

$$c(\mathbf{f}, l) = \{i \in N_{|X|} | \phi_i \geq l\}; \tag{3.54}$$

this function is called an *l-cut function* and the set $c(\mathbf{f}, l)$ is called an *l-cut of* $\mathbf{f}$.

3. Given a possibility distribution (3.52), let

$$L_f = \{l | (\exists\, i \in N_{|X|})(\phi_i = l) \text{ or } l = 0\} \tag{3.55}$$

be called a *level set of f.* Let

$$L_f = \{l_1, l_2, \ldots, l_q\}$$

denote the level set of **f**, where $l_1 = 0$, $q = |L_f|$, and $i < j$ implies $l_i < l_j$. For convenience, let

$$l_f = \max_i \phi_i.$$

Clearly, $l_f = l_q \in L_f$; further $l_f = 1$ if and only if **f** is a normalized possibility distribution.

4. For every $m \in \mathbb{N}$, let

$$^1\mathbf{f} = ({}^1\phi_i \mid i \in N_m) \in \Pi$$

$$^2\mathbf{f} = ({}^2\phi_i \mid i \in N_m) \in \Pi$$

be two possibility distributions. Then, $^1\mathbf{f}$ is called a *subdistribution* of $^2\mathbf{f}$ if and only if

$$\max_i {}^1\phi_i = \max_i {}^2\phi_i \quad \text{and} \quad {}^1\phi_i \leq {}^2\phi_i$$

for all $i \in N_m$.

Let $^1\mathbf{f} \leq {}^2\mathbf{f}$ be used to indicate that $^1\mathbf{f}$ is a subdistribution of $^2\mathbf{f}$. The relation "$^1\mathbf{f}$ is a subdistribution of $^2\mathbf{f}$" is clearly a partial ordering defined on each set of possibility distributions with some particular number m of elements, say set $^m\Pi$. Further, $(^m\Pi, \leq)$ is a lattice with join and meet defined, respectively, as

$$^1\mathbf{f} \vee {}^2\mathbf{f} = (\max[{}^1\phi_i, {}^2\phi_i] \mid i \in N_m)$$

$$^1\mathbf{f} \wedge {}^2\mathbf{f} = (\min[{}^1\phi_i, {}^2\phi_i] \mid i \in N_m)$$

for each pair $^1\mathbf{f}, {}^2\mathbf{f} \in {}^m\Pi$.

Equipped with the concepts necessary for a discussion of possibility distributions, we can now return to the main issue—the measure of possibilistic uncertainty. Intuitively, it is desirable that possibilistic counterparts of the properties (H1)–(H5), which are possessed by the Shannon entropy, be also possessed by the possibilistic uncertainty measure. The possibilistic conterparts of the properties can be formulated in the same way as (H1)–(H5), but the word "probability" must be replaced with the word "possibility." A function of the form (3.51) that satisfies all these properties is known and can be defined either in the form

$$U(\mathbf{f}) = \frac{1}{l_f} \sum_{k=1}^{q-1} (l_{k+1} - l_k) \log_2 |c(\mathbf{f}, l_{k+1})| \tag{3.56}$$

or in the simpler form

$$U(\mathbf{f}) = \frac{1}{l_f} \int_0^{l_f} \log_2 |c(\mathbf{f}, l)| \, dl. \tag{3.57}$$

This function is referred to as the *U-uncertainty*. Besides the possibilistic counterparts of properties (H1)–(H5), the U-uncertainty has some additional desirable properties. The most important of them is the property of *monotonicity*: for every pair ${}^1\mathbf{f}$, ${}^2\mathbf{f} \in {}^m\Pi\,(m \in N)$, if ${}^1\mathbf{f} \leq {}^2\mathbf{f}$, then $U({}^1\mathbf{f}) \leq U({}^2\mathbf{f})$,

Example 3.4. Calculate the U-uncertainties of the following possibility distributions:

$${}^1\mathbf{f} = (0.1, 0, 0.5, 0.8, 0.8, 0.8, 0.1, 0.7, 0.8),$$
$${}^2\mathbf{f} = (0.3, 0.2, 0.9, 1, 1, 1, 0.9, 0.8, 1).$$

The level sets are

$$L_{{}^1f} = \{0, 0.1, 0.5, 0.7, 0.8\},$$
$$L_{{}^2f} = \{0, 0.2, 0.3, 0.8, 0.9, 1\}.$$

Using either Eq. (3.56) or Eq. (3.57), we obtain

$$\begin{aligned} U({}^1\mathbf{f}) &= \frac{1}{0.8}(0.1 \log_2 8 + 0.4 \log_2 6 + 0.2 \log_2 5 + 0.1 \log_2 4) \\ &= 1.25(0.3 + 1.034 + 0.464 + 0.2) = 2.4975, \\ U({}^2\mathbf{f}) &= 0.2 \log_2 9 + 0.1 \log_2 8 + 0.5 \log_2 7 + 0.1 \log_2 6 + 0.1 \log_2 4 \\ &= 0.634 + 0.3 + 1.404 + 0.258 + 0.2 = 2.796. \end{aligned}$$

As in the case for probability distributions, it is reasonable to assume that each finite set X of alternative outcomes under consideration is characterized by a particular (unique) possibility distribution. Then, to emphasize the set for which the uncertainty is calculated, it is convenient to use $U(X)$ as a shorthand notation for $U(f(x)|x \in X)$.

It is easy to show that

$$0 \leq U(X) \leq \log_2 |X|, \tag{3.58}$$

which is analogous to (3.38). The lower bound is reached when possibilities of all outcomes except one are equal to 0; the upper bound is obtained when possibilities of all outcomes are equal (and, of course, nonzero). The ratio

$$\mathbf{U}(X) = U(X)/\log_2 |X| \tag{3.59}$$

is a *normalized U-uncertainty*, for which we get

$$0 \le \mathbf{U}(X) \le 1. \tag{3.60}$$

It is also known that a conditional U-uncertainty $U(Y|X)$ can be calculated without the use of conditional possibilities by the formula

$$U(X|Y) = U(X \times Y) - U(Y), \tag{3.61}$$

which has exactly the same form as its Shannon entropy counterpart. This conditional U-uncertainty becomes normalized when divided by $\log_2|X|$

In the same manner as for probability distributions, let the symbols

$$f(\mathbf{c}), \quad f(\mathbf{g}|\bar{\mathbf{g}}), \quad f(\bar{\mathbf{e}}|\mathbf{e}), \quad f(\mathbf{g}|\mathbf{e}, \bar{\mathbf{g}})$$

denote possibilities defined by (3.18), (3.22), (3.26), (3.28), respectively. In addition, the marginal possibilities are defined by

$$f(\bar{\mathbf{g}}) = \max_{\mathbf{c} > \bar{\mathbf{g}}} f(\mathbf{c}) \tag{3.62}$$

for neutral systems, and by

$$f(\bar{\mathbf{g}}|\mathbf{e}) = \max_{\bar{\mathbf{e}} > \bar{\mathbf{g}}} f(\bar{\mathbf{e}}|\mathbf{e}) \tag{3.63}$$

for directed systems.

The definition of conditional possibilities in terms of joint and marginal possibilities is a controversial issue in possibility theory. Fortunately, this controversy can be avoided by using formula (3.61). The possibilistic counterparts of the key formulas (3.48), (3.49), (3.50) are then, respectively,

$$U(\mathbf{G}|\overline{\mathbf{G}}) = U(\mathbf{C}) - U(\overline{\mathbf{G}}), \tag{3.64}$$

$$U(\mathbf{G}|\mathbf{E} \times \overline{\mathbf{G}}) = U(\mathbf{C}) - U(\mathbf{E} \times \overline{\mathbf{G}}), \tag{3.65}$$

$$U(\mathbf{G}|\mathbf{E} \times \overline{\mathbf{G}}) = \frac{1}{|\mathbf{E}|}\left[\sum_{\mathbf{e}\in\mathbf{E}} U(\overline{\mathbf{E}}|\mathbf{e}) - \sum_{\mathbf{e}\in\mathbf{E}} U(\overline{\mathbf{G}}|\mathbf{e})\right]. \tag{3.66}$$

No calculation of conditional possibilities is needed when these formulas are used. The conditional possibilities $f(\bar{\mathbf{e}}|\mathbf{e})$ that are needed for the calculation of $U(\overline{\mathbf{E}}|\mathbf{e})$ are given (or derived directly from data), but they are not calculated from the joint and marginal possibilities; possibilities $f(\bar{\mathbf{g}}|\mathbf{e})$ are marginals of $f(\bar{\mathbf{e}}|\mathbf{e})$ and are calculated by (3.63).

Example 3.5. Consider the situation characterized in Figure 3.9, where $f(\mathbf{c})$ is a possibility of state $\mathbf{c}$. We can see that $f(\bar{\mathbf{g}})$ is the same for both generating orders. Using formula (3.64), we get

$$
\begin{aligned}
U(\mathbf{C}) &= 0.25 \log_2 5 + 0.25 \log_2 3 + 0.5 \log_2 1 = 0.58 + 0.396 = 0.976,\\
U(\overline{\mathbf{G}}) &= 0.5 \log_2 2 + 0.5 \log_2 1 = 0.5,\\
U(G \mid \overline{\mathbf{G}}) &= U(\mathbf{C}) - U(\overline{\mathbf{G}}) = 0.976 - 0.5 = 0.476.
\end{aligned}
$$

Assume now that variable v_1 is viewed as an input variable and the following possibilities $f(\bar{\mathbf{e}} \mid \mathbf{e})$ are given:

s_1	s_2	s_3	$f(\bar{\mathbf{e}} \mid \mathbf{e})$
0	0	0	1.0
0	0	1	0.25
1	0	0	0.5
0	1	1	0.5
1	1	1	1.0

	s_1	s_2	s_3	$f(\mathbf{c})$
$\mathbf{c} =$	0	0	0	1.0
	0	0	1	0.25
	0	1	1	0.25
	1	0	0	0.5
	1	1	1	0.5

(a)

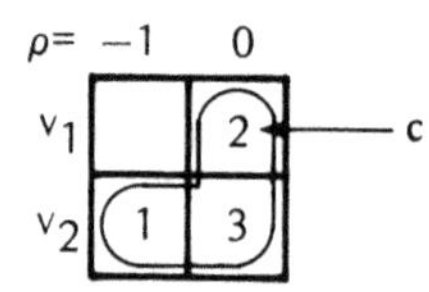

(b)

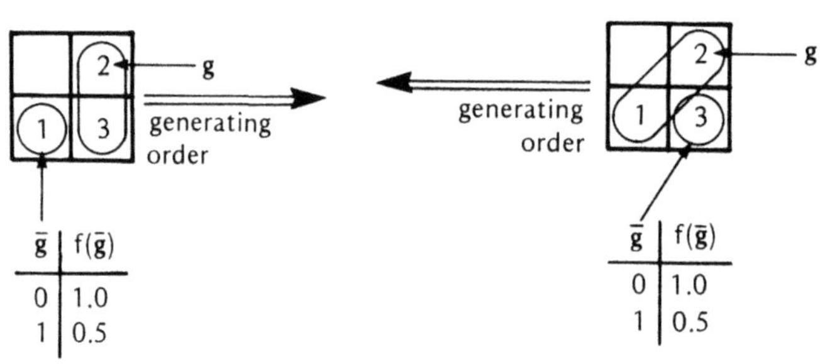

(c) (d)

Figure 3.9. Illustration to Example 3.5: possibilistic neutral system.

Then, $U(\mathbf{G}\,|\,\mathbf{E}\times\bar{\mathbf{G}})$ is calculated by formula (3.66). When the generating order is from the left to the right or vice versa, we obtain, respectively,

s_1	s_2	$f(\bar{\mathbf{g}}\,|\,\mathbf{e})$
0	0	1.0
1	0	0.5
0	1	0.5
1	1	1.0

s_2	s_3	$f(\bar{\mathbf{g}}\,|\,\mathbf{e})$
0	0	1.0
0	1	0.25
1	0	0.0
1	1	1.0

For the former generating order, we thus obtain

$$U(\mathbf{G}\,|\,\mathbf{E}\times\bar{\mathbf{G}}) = \tfrac{1}{2}(1.146-1) = 0.073;$$

for the latter generating order, the uncertainty is

$$U(\mathbf{G}\,|\,\mathbf{E}\times\bar{\mathbf{G}}) = \tfrac{1}{2}(1.146-0.25) = 0.448. \blacktriangleleft$$

It is well known that the average amount of uncertainty associated with a finite set of alternative outcomes, as measured by either the Shannon entropy or the U-uncertainty, can also be interpreted as the average amount of information expected from an experiment which involves the set. However, when a measure of uncertainty is adopted as an information measure, only the syntactic aspects of information, not its semantic or pragmatic aspects, are in fact measured.

3.6. SEARCH FOR ADMISSIBLE BEHAVIOR SYSTEMS

> *Facts do not arrange themselves.*
>
> —ROBERT M. HUTCHINS

Equipped now with the uncertainty measures, through which the degree of determinism is expressed, we return in this section to the problem type introduced in Section 3.4: given a data system **D** with a totally ordered parameter set and a largest acceptable mask **M** compatible with **D**, determine all behavior systems that satisfy the misfit, determinism, and complexity requirements, with the misfit requirement given a precedence over the other two.

As discussed previously, each largest acceptable mask **M** contains a set of meaningful masks, each of which is a subset of **M**. A behavior function (of the particular kind considered) that fits perfectly with the data can be obtained for each mask by an exhaustive sampling of the data. In practice, however, it is sufficient to perform the sampling only for **M**. Behavior functions of its submasks can then be determined by calculating the appropriate projections of the behavior function corresponding to **M**.

Given a behavior function f_B, defined in terms of overall states of some sampling variables, any of its *projections* is another behavior function that conforms to f_B in terms of substates based on a specific subset of the sampling variables. Let s_k ($k \in N_{|M|}$) be the sampling variables in terms of whose states f_B is defined; M indicates the mask through which the sampling variables acquire their meanings. Let $[f_B \downarrow Z]$ denote a projection of f_B, where Z denotes a subset of the set $N_{|M|}$ of identifiers of the sampling variables, i.e., $Z \subset N_{|M|}$. Then,

$$[f_B \downarrow Z]: \bigtimes_{k \in Z} S_k \rightarrow [0, 1] \tag{3.67}$$

such that

$$[f_B \downarrow Z](\mathbf{x}) = a(\{f(\mathbf{c}) \mid \mathbf{c} \succ \mathbf{x}\}) \tag{3.68}$$

where a is some aggregation function that is determined by the nature of the function f_B. For instance,

$$[f_B \downarrow Z](\mathbf{x}) = \sum_{\mathbf{c} \succ \mathbf{x}} f_B(\mathbf{c}) \tag{3.69}$$

when f_B is a probability distribution; ▶ for possibility distributions,

$$[f_B \downarrow Z](\mathbf{x}) = \max_{\mathbf{c} \succ \mathbf{x}} f_B(\mathbf{c}). \tag{3.70}$$

In the context of any particular problem, let ${}^1f_B$ denote the behavior function of the largest acceptable mask **M** and let ${}^if_B$ ($i = 2, 3, \ldots$) denote behavior functions of its various meaningful submasks ${}^iM$, each associated with some set ${}^iZ \subset N_{|M|}$ of identifiers of sampling variables. ◀

Except for very small data, it is computationally simper to determine behavior functions by projections rather than by sampling the data. The larger the data, the more computing time is saved when sampling is replaced by taking projections. It is thus desirable to perform the sampling only once, for the largest acceptable mask, and then determine the behavior functions of all its meaningful submasks as appropriate projections.

Example 3.6. Determine the projection of the probability behavior function defined in Figure 3.7a and the possibility behavior function defined in Figure 3.9a for

$Z = \{1, 2\}$ in both cases. Using formula (3.69) for the probability function, we obtain

	s_1	s_2	$[f\downarrow\{1,2\}](\mathbf{x})$
$\mathbf{x} =$	0	0	$0.5(=0.2+0.2+0.1)$
	0	1	0.1
	1	0	$0.2(=0.1+0.1)$
	1	1	$0.2(=0.1+0.1)$

For the possibility function, we use formula (3.70) to obtain

	s_1	s_2	$[f\downarrow\{1,2\}](\mathbf{x})$
$\mathbf{x} =$	0	0	1.0
	0	1	0.25
	1	0	0.5
	1	1	0.5

The given data system **D**, largest acceptable mask **M**, and the misfit requirement lead to the restricted set

$$Y_r = \{{}^i\mathbf{F}_B = (\mathbf{I}, {}^iM, {}^if) \mid i = 1, 2, \ldots, N(n, \Delta\mathbf{M})\},$$

which contains one behavior system for each meaningful mask, ${}^iM \subseteq \mathbf{M}$; for notational convenience, let ${}^1M = \mathbf{M}$. The next step in dealing with the problem under discussion must be the calculation of the degrees of nondeterminism and complexity for each system in the set Y_r.

As argued in Section 3.5, the *degree of nondeterminism* is defined by an appropriate measure of *generative uncertainty*, which for probabilistic or possibilistic systems is expressed by the Shannon entropy or U-uncertainty, respectively. The definition of generative uncertainty requires that a generating order (and the corresponding partition of each mask) be defined. If several generating orders are permitted, we accept for each mask only those with the smallest generative uncertainty.

As far as the *complexity measure* is concerned, many options are available (as discussed in Chapter 6). For the purpose of illustration, we adopt a measure that is simple yet meaningful and often employed by GSPS users—the size (cardinality) of the mask.

Let ${}^iq_u$ $(i = 1, 2, \ldots)$ denote values of the appropriate generative uncertainty for behavior systems ${}^i\mathbf{F}_B$ in the restricted set Y_r. Since each system ${}^i\mathbf{F}_B$ is uniquely identifiable by its mask ${}^iM$, whose cardinality $|{}^iM|$ represents its complexity, the status of ${}^i\mathbf{F}_B$ in terms of generative uncertainty and complexity can be conveniently described by the pair $(|{}^iM|, {}^iq_u)$. The problem under consideration can thus be discussed in terms of masks ${}^iM$ rather than the corresponding behavior systems ${}^i\mathbf{F}_B$.

The numerical ordering of the cardinalities of the masks ${}^iM$, which identify systems in Y_r, imposes a *complexity ordering* $\overset{c}{\leq}$ on set Y_r. The numerical ordering of values ${}^iq_u$ imposes an *uncertainty ordering* $\overset{u}{\leq}$ on Y_r. While the complexity ordering is fully determined by the masks themselves, the uncertainty ordering is determined only after the masks are evaluated. For any set of generative masks, we may define a partial ordering

$$ {}^iM_G \leq {}^jM_G \quad \text{iff } {}^i\mathbf{g} = {}^j\mathbf{g} \text{ and } {}^i\bar{\mathbf{g}} \prec {}^j\bar{\mathbf{g}} \tag{3.71} $$

(and ${}^i\mathbf{e} \prec {}^j\mathbf{e}$ for directed systems), for which the name "*submask ordering*" seems appropriate. This ordering is often useful in developing various heuristic procedures for searching through the systems in set Y_r.

An example of the complexity and submask ordering is shown in Figure 3.10 for the largest acceptable mask **M** with $n = 3$ and $\Delta\mathbf{M} = \mathbf{2}$. It is assumed that the generative order is from the left to the right. All meaningful submasks of **M** are specified by their matrices and are labeled by their identifiers i at the left top corner of each matrix. They are partitioned by their complexities into four blocks. Masks with the same complexity are placed at the same level in the diagram. For instance, masks identified by 2–7 form one block, associated with complexity 5, masks identified by 8–19 form another block with complexity 4, etc. From the standpoint of the complexity ordering, each mask at one level is an immediate successor of each mask at the next higher level and an immediate predecessor of each mask at the next lower level. The connections with arrows in Figure 3.10 indicate the submask ordering. It is obvious from this example that the complexity ordering is a connected quasiordering (reflexive and transitive relation in which each pair of systems is comparable).

The submask ordering is a partial ordering, but it does not form a lattice. However, it is a collection of lattices, one for each set of generated sampling variables (right-most elements of the masks in our example).

The uncertainty ordering is connected and, due to the fact that several different systems can have equal generative uncertainty, it is not antisymmetric. Hence, it is, generally, a connected quasiordering, which becomes a total ordering in some special cases.

Two connected quasiorderings are thus defined on the set Y_r—the complexity and uncertainty orderings. It is desirable to combine them in an appropriate manner. Since it is required in the problem type under discussion that both complexity *and* generative uncertainty of systems in the solution set Y_Q be minimized, the relevant *combined ordering* $\overset{*}{\leq}$ is defined as follows:

$$ {}^i\mathbf{F}_B \overset{*}{\leq} {}^j\mathbf{F}_B \quad \text{iff } |{}^iM| \overset{c}{\leq} |{}^jM| \text{ and } {}^iq_u \overset{u}{\leq} {}^jq_u, \tag{3.72} $$

where ${}^i\mathbf{F}_B, {}^j\mathbf{F}_B \in Y_r$. This ordering is not connected since pairs ${}^i\mathbf{F}_B, {}^j\mathbf{F}_B$ for which

$$ |{}^iM| < |{}^jM| \text{ and } {}^iq_u > {}^jq_u \quad \text{or} \quad |{}^iM| > |{}^jM| \text{ and } {}^iq_u < {}^jg_u $$

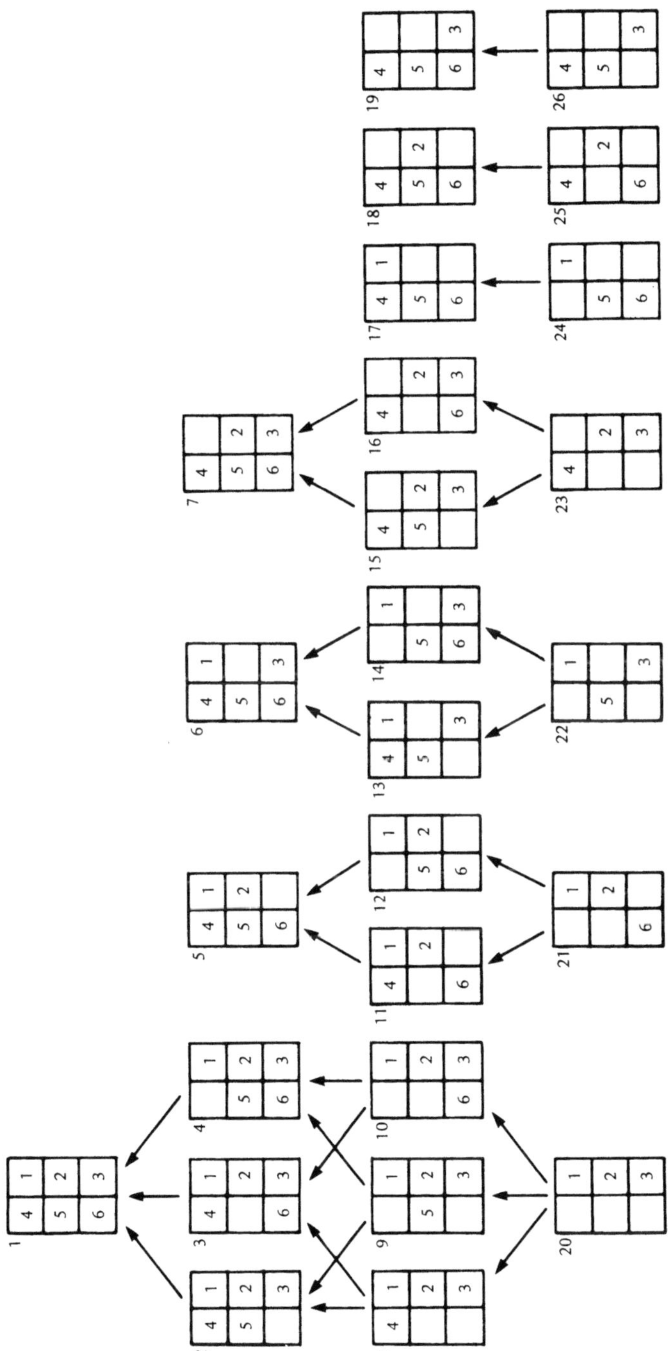

Figure 3.10. Meaningful masks for $n = 3$ and $\Delta\mathbf{M} = 2$, arranged according to the complexity and submask orderings.

(such pairs are certainly possible) are not comparable. It is not antisymmetric since the possibility of

$$|{}^iM| = |{}^jM| \quad \text{and} \quad {}^iq_u = {}^jq_u$$

for some $i \neq j$ is not excluded. Hence, the combined ordering (3.72) is a general quasiordering (reflexive and transitive relation) on Y_r.

The solution set Y_Q can now be defined as the set of all systems in Y_r that are either equivalent or noncomparable with respect to the combined ordering (3.72). Two systems in Y_r, say systems ${}^iF_B$ and ${}^jF_B$, are noncomparable in terms of the combined ordering if either of the following is satisfied:

(a) ${}^iF_B$ is more complex and more deterministic than ${}^jF_B$, or
(b) ${}^iF_B$ is less complex and less deterministic than ${}^jF_B$.

Formally,

$$Y_Q = \{{}^i\mathbf{F}_B \in Y_r \mid (\forall\, {}^j\mathbf{F}_B \in Y_r)({}^j\mathbf{F}_B \overset{*}{\leq} {}^i\mathbf{F}_B \Rightarrow {}^i\mathbf{F}_B \overset{*}{\leq} {}^j\mathbf{F}_B)\} \tag{3.73}$$

Let the systems in the solution set Y_Q be called *admissible behavior systems* of the problem type under discussion.

Example 3.7. To illustrate the various issues discussed in this section, let us consider the ethological data system described in Example 2.5 (and Figure 2.7). Let us determine all behavior systems that are admissible in the sense of (3.73) for this data system under the assumptions that the user wants to characterize the behavior systems in the probabilistic manner and to utilize them for prediction.

Assume first that $\Delta\mathbf{M} = 2$. Then, there are eight meaningful masks, which are shown in Figure 3.11a together with their submask ordering and the indication of the three levels of complexity. After exhaustive sampling is performed for the largest acceptable mask ${}^1M = \mathbf{M}$, probabilities $f_B(\mathbf{c})$ are calculated from the frequencies $N(\mathbf{c})$ by a formula specified by the user, and the generative uncertainty is calculated either by formula (3.45) or by the more convenient formula (3.48). If formula (3.31) is used for calculating the probabilities, the generative uncertainty becomes 1.11. Appropriate projections are then determined by formula (3.69) for the remaining seven meaningful masks and their generative uncertainties are calculated. The results are shown in Figure 3.11b (at the right bottom corner of each mask), together with the uncertainty ordering. We can see that it is a total ordering in this example since all of the masks have different uncertainty values. The combined complexity and uncertainty ordering (3.72) is shown for this example in Figure 3.11c. We can see that the minimal masks with respect to this combined ordering are those identified by 1, 2, 6. Hence, $Y_Q = \{{}^1\mathbf{F}_B, {}^2\mathbf{F}_B, {}^6\mathbf{F}_B\}$.

Assume now that $\Delta\mathbf{M} = 3$. Then, according to formula (3.36), there are 40 meaningful masks. After they are processed in the same way as described for $\Delta\mathbf{M} = 2$,

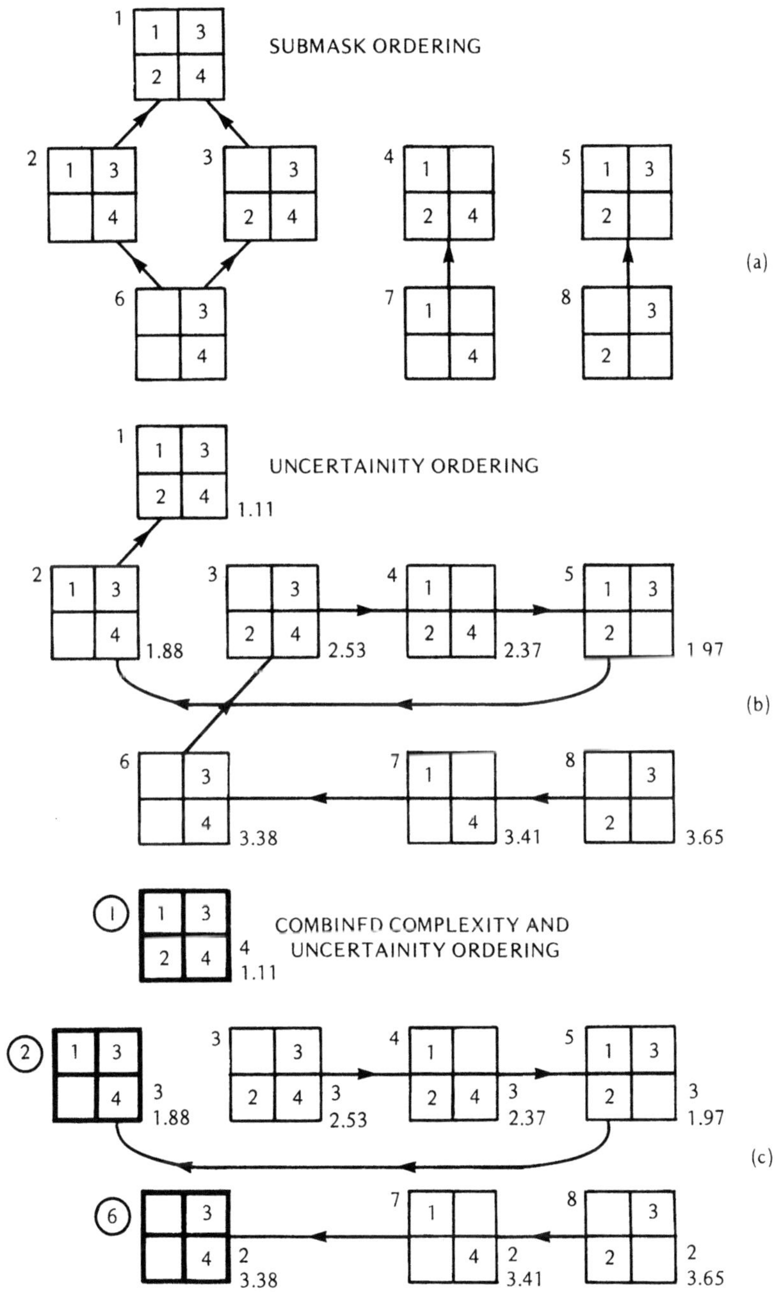

Figure 3.11. Illustration to Example 3.7.

we obtain five admissible behavior systems whose masks, complexities, and generative uncertainties are given in Figure 3.12a. All remaining 35 masks are inferior from both the complexity and uncertainty points of view and thus need not be considered at all. Figure 3.12a is a typical example of GSPS responses to users' requests. Various additional characteristics of the solution set can also be provided, if requested, such as the plot of uncertainty versus complexity shown in Figure 3.12b.

There are many different realizations of the search for admissible behavior systems outlined here. The basic principle is that the meaningful masks are derived by some algorithm from the largest acceptable mask in decreasing order of complexity. In each block of masks with equal complexity, only those with minimal generative uncertainty are accepted. If the value of this minimal uncertainty is smaller than or equal to that at the previous level of complexity, then all previously accepted systems are discarded. At the end of this procedure, we are left with only the admissible systems.

It is important to realize that this problem category is a theme on which there exist many variations. For instance, probabilities or possibilities can be calculated in a

i	MASK	$\lvert {}^{i}M \rvert$	${}^{i}H(G\lvert\bar{G})$
1	1 3 5 / 2 4 6	6	0.41
2	3 5 / 2 4 6	5	0.55
3	3 5 / 2 4	4	1.07
4	3 5 / 6	3	1.88
5	5 / 6	2	3.38

(a)

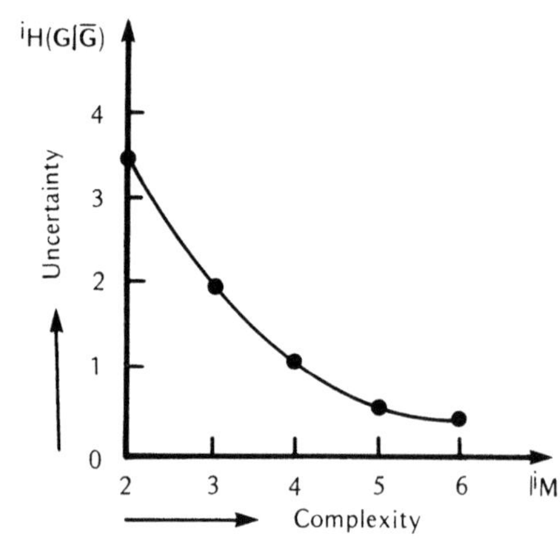

(b)

Figure 3.12. Admissible behavior systems in Example 3.7.

number of different ways, a variety of definitions of complexity can be used, additional requirements may be imposed such as a largest acceptable uncertainty or preference for masks with small depths, complexity and uncertainty (as well as additional requirements) may be weighted by the user when the combined ordering is defined, etc. These variations are basically minor methodological distinctions. The GSPS should be designed in such a way that it has a number of standard options for each key problem category, one of which is adopted as default option, but at the same time it should give the user as much freedom as possible to define his own problem variations.

3.7. STATE-TRANSITION SYSTEMS

> *The most fundamental concept in cybernetics is that of "difference," either that two things are recognisably different or that one thing has changed with time.*
>
> —W. Ross Ashby

Assume again that a data system is given whose support set is totally ordered. It is argued in Section 3.2–3.6 that the data system can be characterized in a support-invariant manner by a set of admissible behavior systems, which conform to the data system and satisfy requirements specified by the user. Although behavior systems are fully adequate to characterize the overall constraint among the variables investigated, there is an alternative form of expressing the constraint, which is often preferred by the user. This form, which is usually called a *state-transition relation* (or ST relation, in abbreviation) is defined in terms of pairs of successive states of sampling variables rather than single states; generative systems that use this form are called *state-transition systems* (or *ST-systems*).

► Masks, sampling variables, state sets of sampling variables, and their Cartesian product $\mathbf{C}$ are defined for ST-systems in exactly the same manner as for behavior systems, except for two differences: (1) the distinction of generated and generating sampling variables is not applicable to ST systems and, (2) meaningful masks for ST-systems are subject to an additional restriction (as explained later). The counterparts of behavior functions in ST-systems are *state-transition* functions (or ST-functions). Their domain is $\mathbf{C}^2 = \mathbf{C} \times \mathbf{C}$ rather than $\mathbf{C}$ for neutral systems, and $\mathbf{E}^2 \times \bar{\mathbf{E}}^2$ rather than $\mathbf{E} \times \bar{\mathbf{E}}$ for directed systems.

For neutral systems, the state-transition counterparts of the behavior functions defined by (3.18), (3.22), (3.16) are, respectively, the following ST-functions:

$$f_S: \mathbf{C}^2 \to [0, 1], \tag{3.74}$$

where $f(\mathbf{c}, \mathbf{c}')$ is the probability or possibility (or some other characterization) that state $\mathbf{c}'$ follows immediately after state $\mathbf{c}$ (according to the chosen generating order);

$$f_{GS}: \mathbf{C}^2 \to [0, 1], \tag{3.75}$$

where $f_{GS}(\mathbf{c}, \mathbf{c}')$ is the conditional probability or possibility that the next state is $\mathbf{c}'$ given that the present state is $\mathbf{c}$ and, consequently, the conventional symbol $f_{GS}(\mathbf{c}'|\mathbf{c})$ will be used;

$$f_{GS}: \mathbf{C} \rightarrow \mathbf{C}, \tag{3.76}$$

where $f_{GS}(\mathbf{c}) = \mathbf{c}'$, i.e., the next state $\mathbf{c}'$ is uniquely determined by the present state $\mathbf{c}$; the special form (3.76) of f_{GS} is, of course, applicable only to deterministic systems. Let functions f_{GS} be called *generative ST-functions*.

The state-transition counterparts of the neutral behavior systems (3.10) and (3.15) are, respectively, the *ST-system*:

$$\mathbf{F}_S = (\mathbf{I}, M, f_s) \tag{3.77}$$

and the *generative ST-system*

$$\mathbf{F}_{GS} = (\mathbf{I}, M_G, f_{GS}), \tag{3.78}$$

where $\mathbf{I}$, M, M_G have the same meaning for both behavior and ST-systems.

Given a data system and a mask, the ST-function f_s that fits perfectly with the data system for the mask can be determined by an exhaustive sampling of the data in a similar way as explained for the behavior function f_B. The only difference is that frequencies $N(\mathbf{c}, \mathbf{c}')$ of pairs of successive states result from the sampling rather than frequencies $N(\mathbf{c})$ of the individual states.

A pair $(\mathbf{c}, \mathbf{c}') \in \mathbf{C}^2$ is called a *transition* from state $\mathbf{c}$ to another state $\mathbf{c}'$ according to the declared generating order in the support set. One of the basic properties of ST-functions is that transitions to a state must balance with those from the same state. When probabilities are used, we have for each state $\mathbf{x} \in \mathbf{C}$

$$\sum_{\mathbf{c} \in \mathbf{C}} f_S(\mathbf{c}, \mathbf{x}) = f_B(\mathbf{x}), \tag{3.79}$$

$$\sum_{\mathbf{c}' \in \mathbf{C}} f_S(\mathbf{x}, \mathbf{c}') = f_B(\mathbf{x})$$

and, hence,

$$\sum_{\mathbf{c} \in \mathbf{C}} f_S(\mathbf{c}, \mathbf{x}) = \sum_{\mathbf{c}' \in \mathbf{C}} f_S(\mathbf{x}, \mathbf{c}'), \tag{3.80}$$

which expresses the *transition balance*. If possibilities are used, Eq. (3.80) becomes

$$\max_{\mathbf{c} \in \mathbf{C}} f_S(\mathbf{c}, \mathbf{x}) = \max_{\mathbf{c}' \in \mathbf{C}} f_S(\mathbf{x}, \mathbf{c}'). \tag{3.81}$$

States $\mathbf{c}$, $\mathbf{c}'$ may be viewed as states defined by two related masks M, M', respectively. The masks are related to each other by a simple translation of the translation rules involved:

$$(v_i, \rho) \in M \qquad \text{iff } (v_i, \rho + 1) \in M', \tag{3.82}$$

when the data generation proceeds in increasing order of the support, or

$$(v_i, \rho) \in M \qquad \text{iff } (v_i, \rho - 1) \in M', \tag{3.83}$$

when the data generation proceeds in the opposite order. The two masks M, M' are used simultaneously to characterize pairs of states **c**, **c**′.

In order to avoid either inconsistency or incompleteness in generating data, meaningful masks of ST-systems must satisfy the following requirement (in addition to the mask requirements for behavior systems):

- given a mask M, if $(v_i, \rho_1) \in M$ and $(v_i, \rho_2) \in M$, where $\rho_1 < \rho_2$, then $(v_i, \rho) \in M$ for all integers ρ such that $\rho_1 \leq \rho \leq \rho_2$.

This means that masks of ST-systems must not involve "gaps" such as the element $(v_4, -1)$ in Figure 3.2. Let any mask that satisfies this additional requirement be called a *compact mask.*

To justify this requirement assume that mask M of an ST-system is not compact. Then, there is at least one pair of elements in M, say pair

$$(v_i, \rho_1) \in M, \qquad (v_i, \rho_2) \in M,$$

such that $\rho_1 < \rho_2$,

$$(v_i, \rho_1 + 1) \notin M, \qquad (v_i, \rho_2 + 1) \notin M, \tag{3.84}$$

and $\rho_2 \geq \rho$ for all $(v_i, \rho) \in M$. By (3.82), we get

$$(v_i, \rho_1 + 1) \in M' \quad \text{and} \quad (v_i, \rho_2 + 1) \in M'. \tag{3.85}$$

Let sampling variables based on these elements of M' be denoted by s_1 and s_2, respectively. States of s_1, s_2 are components of **c**′. As such, they must be either determined for each support instance by state **c** or generated according to the probability or possibility distribution $f_{GS}(\mathbf{c}' \mid \mathbf{c})$ for each particular **c**. However, neither of these alternatives is possible for s_1. Due to (3.84), it cannot be determined by state **c**. It cannot be generated at any support instance t in a consistent manner since

$$s_{1,t} = s_{2,t+\rho_1-\rho_2}$$

and s_1 is thus committed to the state of variable s_2 at support instance $t - (\rho_2 - \rho_1)$. There is no guarantee that a generated state of s_1 would be consistent with this predetermined state. If, on the other hand, no state of s_1 is generated, then $\mathbf{c}'$ becomes incomplete since the predetermined state is not known (i.e., it is not a component of $\mathbf{c}$). Consequently, the state of s_1 at any value of t can neither be determined from c nor generated according to the generative ST-function.

This justification that masks with "gaps" are not acceptable for ST-systems is illustrated by an example in Figure 3.13: component y_4 of the next state $\mathbf{c}'$ for support instance t (Fig. d) can be neither determined from $\mathbf{c}$ at time t (Fig. c) nor generated by an ST-function since it was generated at support instance $t-3$ (Fig. a).

A convenient representation of the ST-functions (3.74) or (3.75) are square matrics whose rows and columns are associated, respectively, with $\mathbf{c}$ and $\mathbf{c}'$. Their entries are values $f_S(\mathbf{c}, \mathbf{c}')$ or $f_{GS}(\mathbf{c}' | \mathbf{c})$, respectively. ◄

Example 3.8. Hardware monitoring is one approach to computer performance evaluation whose significance seems to grow with the increasing complexity of the computer systems evaluated. In hardware monitoring, certain key variables, each usually describing the status of a particular unit of the computer system, are observed (by instruments called hardware monitors), within a specified period of time, while the computer system is serving its users. The data are processed by the hardware monitor

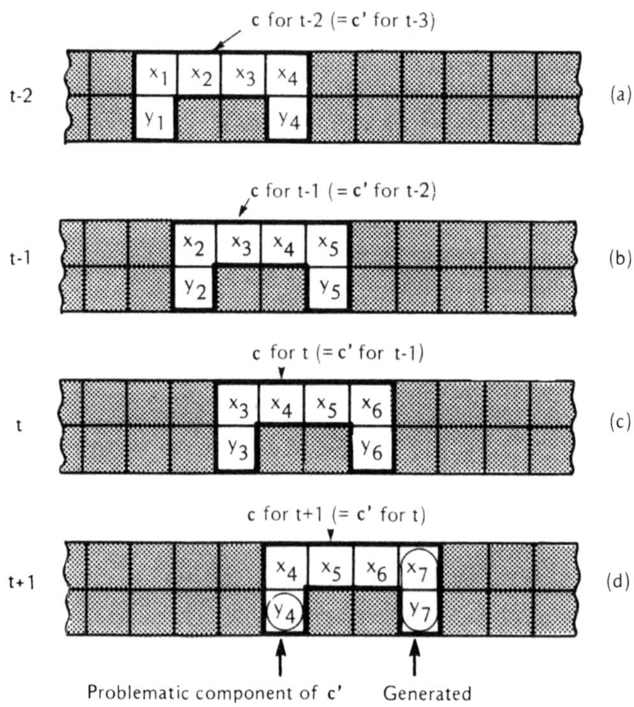

Figure 3.13. Illustration of the inconsistency or incompleteness problem for ST-systems with noncompact masks.

and analyzed with the aim of finding bottlenecks in the system and ways of increasing its performance, defined in some manner.

Hardware monitors are usually equipped with counters which, during the data gathering process, either count event occurrences (count mode) or measure time durations of events (time mode). This means that the hardware monitor normally provides the investigator with a summary of the data rather than with the actual data. For instance, the monitor specifies that the CPU (central processing unit) of the computer system was active 43% of the time during the observation period, that channel x was busy in 15% of all observations, etc., but does not make the actual sequence of observed events available for further processing and analysis.

Valuable information, which may help to better understand the computer performance issue at hand, is frequently lost in this process. In particular, dynamic aspects of the situation are completely lost.

In the GSPS spirit, all observations should be recorded and then processed in any desirable way (see Figure 3.1). In this example, 409, 610 observations made in time were recorded for four variables v_1, v_2, v_3, v_4. Each of the variables has two states, 0 and 1, which characterize the status of a particular hardware unit in the computer system: 0 means that the unit is not active at the time of observation while 1 means that it is active. Variable v_1 represents the activity of the CPU, the other variables represent activities of three communication channels of particular interest in the computer system. The huge data set of over 1.6 million bits was sampled for two successive states by the memoryless mask with the objective to determine a probabilistic ST-function. This led to the 15 states specified in Table 3.5a and 113 transitions. States 7–15 appeared with very low frequencies: the probability that the system was in any of these states is only 0.009. If we aggregate these states into one state (as discussed in Section 3.9), to simplify the ST-function (as demanded by the investigator), we obtain a matrix representation of the generative ST-function f_{GS} in Table 3.5b. Entries in the matrix are conditional probabilities $f_{GS}(\mathbf{c}'|\mathbf{c})$. Symbol ~ 0 is used in the matrix for probabilities that are negligible but not zero; 0 stands for transitions that were not observed at all. Underlined in the matrix are probabilities of each transition from a state to itself. Also shown in Table 3.5b is a column vector of values $f_B(\mathbf{c})$ of the behavior function f_B for the same (memoryless) mask, which would normally be the result of hardware monitoring. Clearly,

$$f_S(\mathbf{c}, \mathbf{c}') = f_{GS}(\mathbf{c}'|\mathbf{c}) \cdot f_B(\mathbf{c}). \tag{3.86}$$

so that values $f_s(\mathbf{c}, \mathbf{c}')$ can be calculated if necessary for some further processing. While the determination of the illustrated function f_{GS} or any other desirable representation of data collected for appropriate source systems defined on the computer complexes evaluated is within the domain of the GSPS, the interpretation of these results must be made by specialists in computer performance evaluation.

In some cases, a visual representation of the ST-function in the form of a diagram is preferable. Such a diagram consists of a set of nodes, one for each state of the involved sampling variables that actually occurs, and oriented connections of the nodes, which

TABLE 3.5
ST Function for a Computer Performance Evaluation Inquiry
(Example 3.8)

(a)

c	v_1	v_2	v_3	v_4	**c**	v_1	v_2	v_3	v_4
1	1	0	0	0	9	0	1	0	1
2	1	0	0	1	10	1	0	1	1
3	1	1	0	1	11	1	1	1	0
4	1	1	0	0	12	1	1	1	1
5	0	0	0	0	13	0	1	1	0
6	1	0	1	0	14	0	0	1	0
7	0	1	0	0	15	0	0	1	1
8	0	0	0	1					

(b)

	c′ =	1	2	3	4	5	6	7–15	
c =	1	<u>0.844</u>	0.064	0.004	0.057	0.028	0.002	0.001	0.458
	2	0.173	<u>0.757</u>	0.049	0.011	0.003	~0	0.007	0.175
	3	0.022	0.093	<u>0.725</u>	0.155	~0	0	0.005	0.092
	4	0.109	0.008	0.059	<u>0.816</u>	0.001	~0	0.007	0.242
	5	0.755	0.056	0.002	0.036	<u>0.146</u>	0.001	0.004	0.016
	6	0.103	0.007	0	~0	~0	<u>0.811</u>	0.079	0.008
	7–15	0.050	0.141	0.054	0.170	0.002	0.063	0.520	0.009
					$f_{GS}(\mathbf{c}'\mid\mathbf{c})$				$f_B(\mathbf{c})$

represent actual transitions. Nodes in the diagram must be labeled by the respective state identifiers **c** and connections marked with values of $f_S(\mathbf{c}, \mathbf{c}')$ or $f_{GS}(\mathbf{c}'\mid\mathbf{c})$; in the latter case, it is desirable to also mark the nodes with values $f_B(\mathbf{c})$ so that values $f_S(\mathbf{c}, \mathbf{c}')$ may be calculated, if necessary, by Eq. (3.86).

Example 3.9. ST-systems are often convenient for a concise description of legal constraints associated with various law-making bodies such as local, state, or federal government. The diagram in Figure 3.14 represents a crisp possibility ST-function that describes the constraints of the U.S. legal system on making the needs and desires of citizens (abbreviated as D) into a new law. The ST-system is based on the memoryless mask and its image system consists of the following seven basic variables and their state sets:

- v_1—political attractiveness (0, low; 1, high);
- v_2—*D* possesses congressional sponsor (0, no; 1, yes);
- v_3—house of representatives status of *D* (0, failed or not considered; 1, passed with simple majority; 2, passed with two-thirds majority);

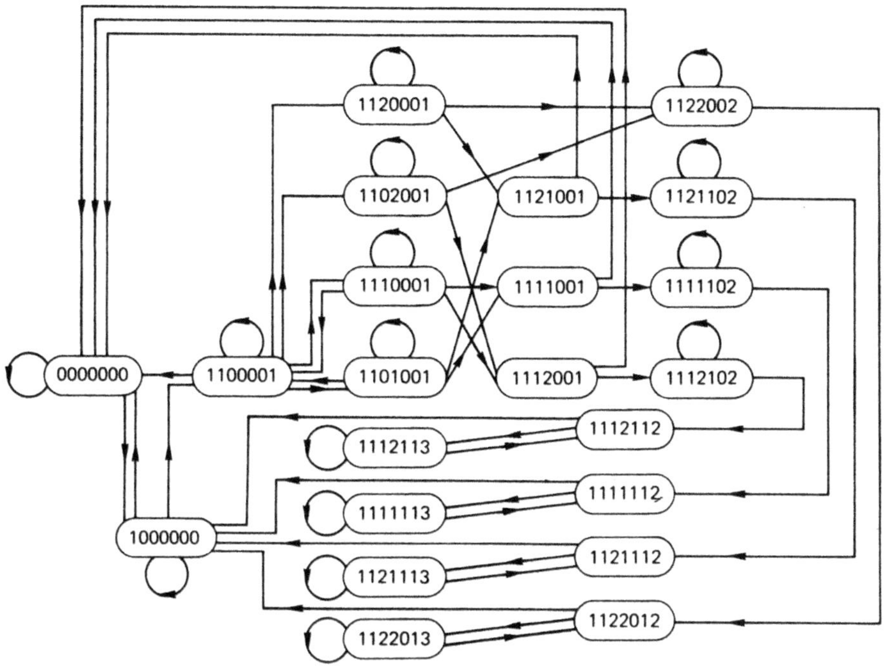

Figure 3.14. ST-diagram of the U.S. legal system (Example 3.9).

v_4—senate status of D (0, failed or not considered; 1, passed with simple majority; 2, passed with two-thirds majority);
v_5—president's approval of D (0, no; 1, yes);
v_6—D is tested in court (0, no; 1, yes);
v_7—legal status of D (0, none; 1, bill; 2, statute; 3, law).

The support is time and it is defined implicitly by changes in states of the variables.

Nodes in the diagram are labeled by the overall states of the variables in the order $v_1, v_2, \ldots, v_7$. Neither nodes nor connections are marked since it is natural in this case to dichotomize all states and transitions into possible (i.e., allowed by the current law) and impossible (i.e., prohibited by the law) and to represent only the former by the nodes and connections in the diagram. Hence, states and transitions that are shown in the diagram have a possibility degree 1, while all the other states and transitions have a possibility degree 0.

The diagram in Figure 3.14 describes only the legal constraints. It does not indicate the difficulty that is likely in making each individual transition. Data regarding previous instances of D can be employed for expressing the difficulty in terms of degrees of possibility in the range [0, 1]. For a particular D and specific political and other relevant circumstances, degrees of possibility can also be determined for the individual transitions subjectively, by the opinion of an expert (or a group of experts).

► Every ST-system can be converted easily into an isomorphic behavior system. To demonstrate how this conversion can be made, let an arbitrary ST-system.

$$\mathbf{F}_S = (\mathbf{I}, M, f_S)$$

be given, where M is, of course, a compact mask. Assume that each next state corresponds to a larger value of the support than the corresponding present state. Consider now a behavior system

$$\mathbf{F}_B = (\mathbf{I}, M^+, f_B),$$

where M^+ is defined in terms of M as follows:

$$(v_i, \rho) \in M^+ \quad \text{when } (v_i, \rho) \in M \quad \text{or} \quad (v_i, \rho - 1) \in M \tag{3.87}$$

Then, for any set of data regarding the common image system **I**, all samples of the data that yield the same pair of states for mask M, say pair $(\mathbf{c}, \mathbf{c}')$, yield also the same state for mask M^+, say state $\mathbf{c}^+$. If the data are exhaustively sampled by both of the masks, the frequencies of $\mathbf{c}, \mathbf{c}'$ and $\mathbf{c}^+$ must be exactly the same. Hence,

$$f_S(\mathbf{c}, \mathbf{c}') = f_B(\mathbf{c}^+),$$

where state $\mathbf{c}^+ \in \mathbf{C}^+$ consists of $\mathbf{c}$ and the generated part of $\mathbf{c}'$, say $\mathbf{c}'_g$. Behavior function f_B is thus equivalent to the given ST-function f_S under the one-to-one correspondence

$$\gamma: \mathbf{C}^2 \rightarrow \mathbf{C}^+, \tag{3.88}$$

where $\gamma(\mathbf{c}, \mathbf{c}') = \mathbf{c}^+$ if and only if $\mathbf{c}^+ = \mathbf{c}, \mathbf{c}'_g$.

Let mask M^+ defined by (3.87) be called an *extended mask* of M. It is based on the assumption that states are generated in increasing order of the support. If the generating order is inverted, an alternative extended mask, say mask ${}^+M$, is defined in a modified manner:

$$(v_i, \rho) \in {}^+M \quad \text{when } (v_i, \rho) \in M \quad \text{or} \quad (v_i, \rho + 1) \in M \tag{3.89}$$

It can be shown, by arguments analogous to those used for M^+, that behavior system

$$\mathbf{F}_B = (\mathbf{I}, {}^+M, f_B)$$

is isomorphic to a ST-system defined in terms of the same image system **I** and mask M.

The correspondence between masks M, M^+, as well as M, ${}^+M$, is illustrated in Figures 3.15a and 3.15b, respectively. Also illustrated in the figures are the one-to-one correspondence (3.88) and its counterpart for the mask ${}^+M$ and ${}^+\mathbf{c} \in {}^+\mathbf{C}$.

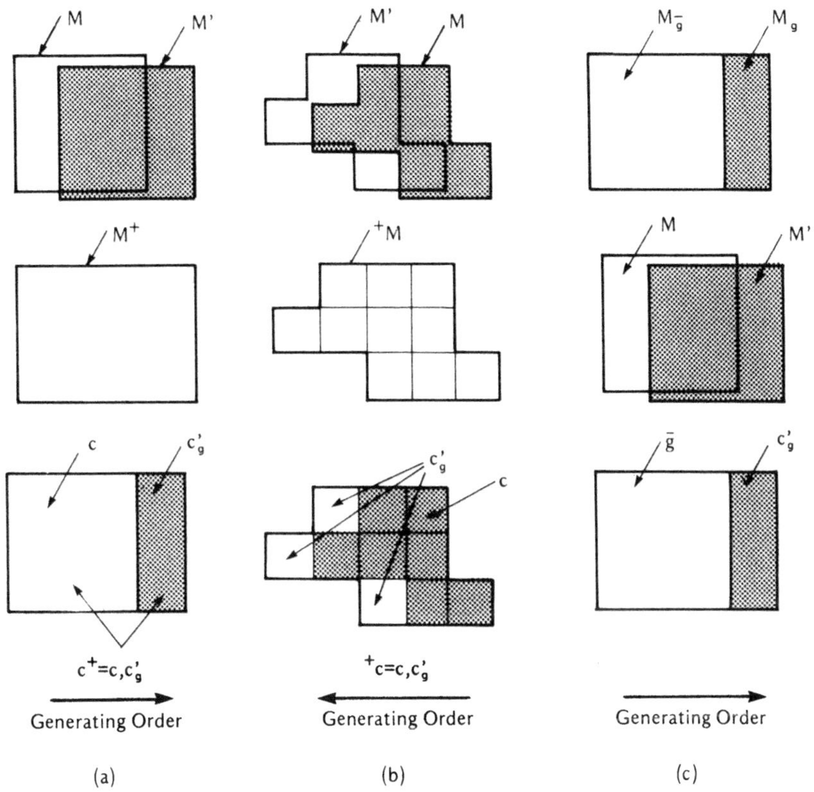

Figure 3.15. Isomorphism between behavior systems and ST-systems.

Given a behavior system

$$\mathbf{F}_B = (\mathbf{I}, M, f_B),$$

an isomorphic ST-system based on the same image system **I** exists only under the conditions that M is a compact mask and $|M_i| \geq 2$ for each of its submasks M_i. If these conditions are satisfied, then it is obvious that the ST-system

$$\mathbf{F}_S = (\mathbf{I}, M_{\bar{g}}, f_s),$$

where $M_{\bar{g}}$ is the generating part of M (according to the specified generating order), is isomorphic under an appropriate one-to-one correspondence between state sets **C** (based on M) and $\overline{\mathbf{G}} \times \overline{\mathbf{G}}$ (based on $M_{\bar{g}}$). This conversion from a behavior system to the isomorphic ST-system is illustrated in Figure 3.15c for one of the generating orders.

For directed systems, the state-transition counterparts of the behavior functions defined by (3.26), (3.28), 3.29) are, respectively, the following ST-functions:

$$\hat{f}_S: \mathbf{E}^2 \times \bar{\mathbf{E}}^2 \rightarrow [0, 1], \tag{3.90}$$

where $\mathbf{E}$ has the same meaning as for behavior systems and $\hat{f}_S(\bar{\mathbf{e}}, \bar{\mathbf{e}}' \mid \mathbf{e}, \mathbf{e}')$ is the conditional probability or possibility whose meaning follows uniquely from the standard notation;

$$\hat{f}_{GS}: \mathbf{E}^2 \times \bar{\mathbf{E}}^2 \rightarrow [0, 1], \tag{3.91}$$

where $\hat{f}_{GS}(\bar{\mathbf{e}}' \mid \mathbf{e}, \mathbf{e}', \bar{\mathbf{e}})$ are generative conditional probabilities or possibilities;

$$\hat{f}_{GS}: \mathbf{E}^2 \times \bar{\mathbf{E}} \rightarrow \bar{\mathbf{E}}, \tag{3.92}$$

where $\hat{f}_{GS}(\mathbf{e}, \mathbf{e}', \bar{\mathbf{e}}) = \bar{\mathbf{e}}'$.

The state-transition counterparts of the directed behavior systems (3.27) and (3.30) are, respectively, the *directed S T-system*

$$\hat{\mathbf{F}}_S = (\hat{\mathbf{I}}, \hat{M}, \hat{f}_S) \tag{3.93}$$

and the *directed generative S T-system*

$$\hat{\mathbf{F}}_{GS} = (\hat{\mathbf{I}}, \hat{M}_G, \hat{f}_{GS}), \tag{3.94}$$

where $\hat{\mathbf{I}}$ and $\hat{M}_G$ have the same meaning as defined for the behavior systems.

Functions (3.90) or (3.91) are conveniently represented by arrays of square matrices (three-dimensional arrays), one matrix for each condition, e, e'. Diagrams similar to those for neutral ST-systems are also convenient. For directed systems, connections in the diagrams are marked not only with values of the relevant ST-function, but also with the conditions $\mathbf{e}, \mathbf{e}'$. Functions (3.92) can be represented by matrices in which rows and columns characterize states $\mathbf{e}$ and $\mathbf{e}, \mathbf{e}'$, respectively, and entries are appropriate states $\mathbf{e}'$. They can also be represented by diagrams, tables and, in some cases, algebraic formulas. ◄

Example 3.10. A simple directed ST-system (with no interpretation) is defined in Figure 3.16. Its image system consists of input variable v_1 and output variable v_2, each with two states, 0 and 1. Mask M of the system is defined in Figure 3.16a and the relevant state components created by its two successive positions on the data matrix are shown in Figure 3.16b. Functions $\hat{f}_S$ and $\hat{f}_{GS}$ are defined in Figures 3.16c, and 3.16d, respectively, in the three-dimensional array form. The array consists of two matrices in this example.

Example 3.11. A deterministic directed ST-system by which the metabolism of bacteria of a certain class is characterized from the standpoint of biochemistry was

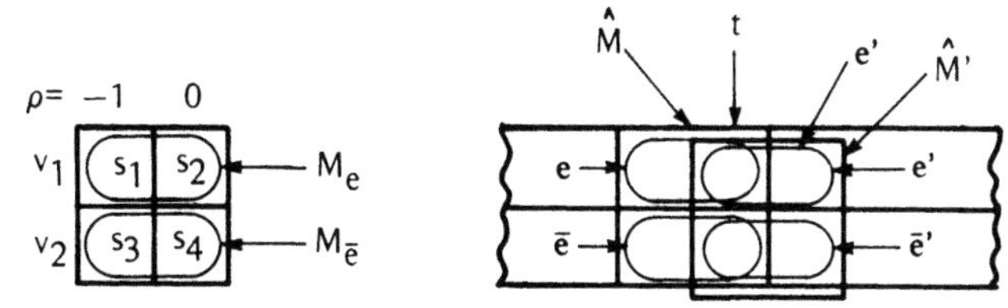

(a) (b)

$\bar{\mathbf{e}}' =$	00	01	10	11	
$\bar{\mathbf{e}} = 00$	0.1	0	0	0	
01	0	0	0.15	0.25	$\mathbf{e} = 01$
10	0.2	0.2	0	0	$\mathbf{e}' = 10$
11	0	0	0.05	0.05	

$\bar{\mathbf{e}}' =$	00	01	10	11	
$\bar{\mathbf{e}} = 00$	0	0.3	0	0	
01	0	0	0.2	0	$\mathbf{e} = 10$
10	0.1	0.1	0	0	$\mathbf{e}' = 01$
11	0	0	0.2	0.1	

(c)

$\bar{\mathbf{e}}' =$	00	01	10	11	
$\bar{\mathbf{e}} = 00$	1	0	0	0	
01	0	0	0.375	0.625	$\mathbf{e} = 01$
10	0.5	0.5	0	0	$\mathbf{e}' = 10$
11	0	0	0.5	0.5	

$\bar{\mathbf{e}}' =$	00	01	10	11	
$\bar{\mathbf{e}} = 00$	0	1	0	0	
01	0	0	1	0	$\mathbf{e} = 10$
10	0.5	0.5	0	0	$\mathbf{e}' = 01$
11	0	0	$2/3$	$1/3$	

(d)

Figure 3.16. Three-dimensional array representation of a directed ST-system (Example 3.10).

developed by Krohn *et al.* [KR3]. The bacteria are considered as composite multienzyme collections and the metabolism is viewed as the set of all possible sequences of biochemical reactions within the bacteria.

The mask in this example is memoryless. A diagram of the ST-function in the form (3.92) is given in Figure 3.17. States of output variables (denoted by numbers in Figure 3.17) represent a set of substrates produced by the corresponding chemical reactions. States of input variables (denoted by letters) represent coenzymes involved in the chemical reactions. A complete list of the transition-producing coenzymes and the produced substrates may be found in the paper by Krohn *et al.* Coenzymes denoted by S1, S2, S8, S36, S49 are equal to the substrates produced in states 1, 2, 8, 36, 49, respectively.

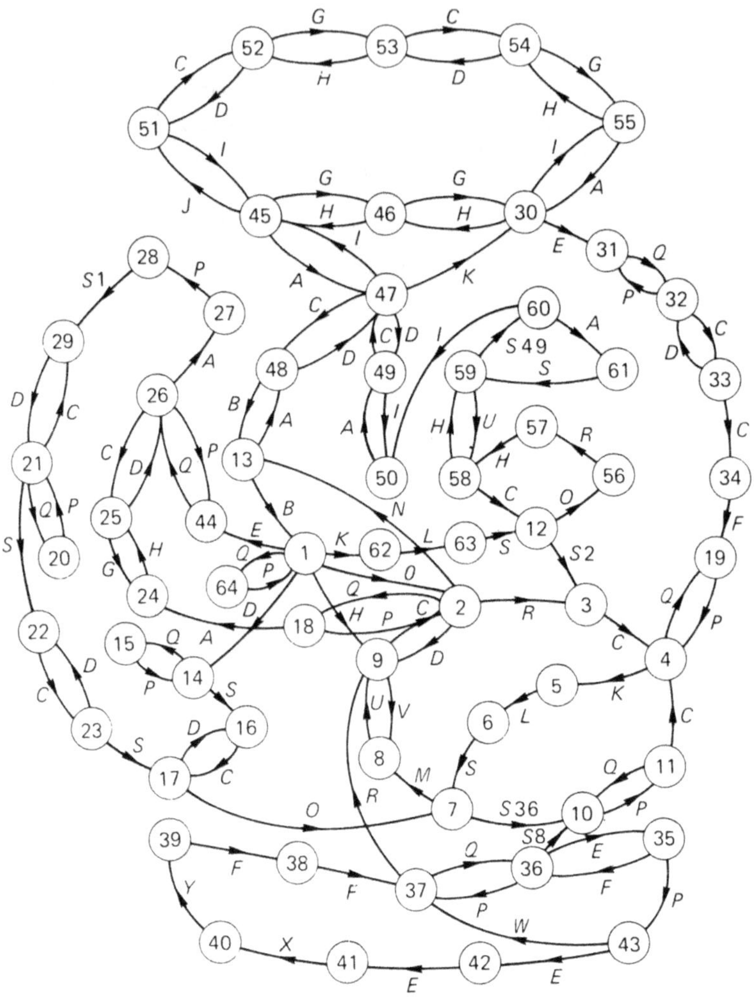

Figure 3.17. ST-function characterizing the metabolism of a class of bacteria.

Example 3.12. Four hormones are involved in the method of oral contraception: follicle stimulating hormone (usually abbreviated F.S.H.), luteinizing hormone (usually abbreviated L.H.), estrogen (E), and progesterone (P). Although levels of these hormones in the blood of a woman can be measured with high precision, it is not necessary to use highly refined data when the subject of interest is birth control based on external supply of estrogen by contraceptive pills. A critical threshold level can be defined for each of the hormones and it only matters whether the actual level of the hormone is below or above the threshold. A variable with two states is thus sufficient to characterize each of the hormones. Assume that the variable is in state 0 if the hormone level is below the threshold and it is in state 1 otherwise. The support is time and it is defined implicitly by changes in states of the variables. Besides these four variables, viewed as output variables, an input variable is included in the system to characterize the influence of the contraceptive pills. Assume that this variable is in state 0 if the pills are not used and in state 1 if they are used.

Data matrices for a normal menstrual period with and without the use of the contraceptive pills are shown, in Figures 3.18a and 3.18b, respectively. Variables v_1–v_5 have the following meaning: v_1, F.S.H.; v_2, L.H.; v_3, E, v_4, P; v_5, contraceptive pills. Assume that the mask defined in Figure 3.18 c is used for sampling the data matrices, which in this example are periodic (the last column in each of the matrices is followed by its first column). When sampling the data matrices for the mask and calculating probabilities, we obtain the diagram in Figure 3.18d. It represents the generative ST-function of the form (3.91). Nodes are labeled with states of the sampling variables s_1–s_5 in their natural order. Connections are marked with states of e' (of input variable

t =	1	2	3	4	5	6
v_1	0	0	0	1	0	0
v_2	0	0	0	0	0	0
v_3	0	0	1	1	1	1
v_4	0	1	1	1	1	0
v_5	1	1	1	1	1	1

(a)

t =	1	2	3	4	5	6	7	8	9	10	11	12
v_1	0	1	0	0	0	1	1	0	0	0	0	0
v_2	0	0	0	0	0	0	1	1	0	0	0	0
v_3	0	0	0	1	0	0	0	0	0	1	1	0
v_4	0	0	0	0	0	0	0	0	0	0	1	1
v_5	0	0	0	0	0	0	0	0	0	0	0	0

(b)

ρ =	−1	0
v_1	1	2
v_2		3
v_3		4
v_4		5
v_5		6

(c)

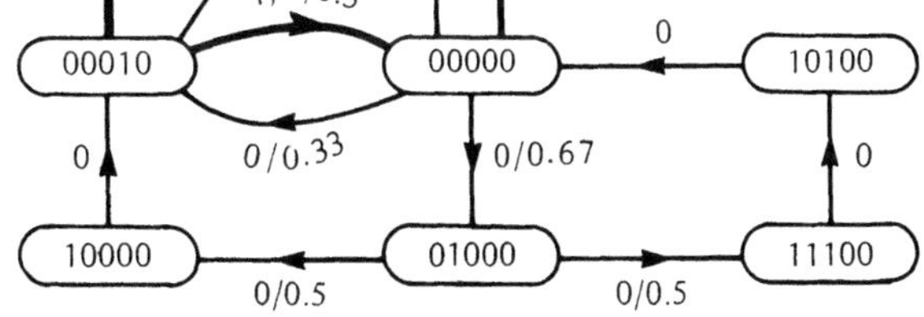

(d)

Figure 3.18. Birth-control system (Example 3.12).

$v_5 = s_6$ pertaining to the next state) and, if not unique, also by the probability of the corresponding transitions. It is assumed that $e = e'$ since data for the other possibilities are not available. We can see that the system is deterministic under the condition that the contraceptive pills are used as indicated by the boldface connections in the diagram. This unique sequence of transitions avoids states at which fertilization may occur.

Example 3.13. To illustrate the meaning of space-invariance, let us consider two examples of mosaic patterns. The first is the usual black and white chessboard pattern (Figure 3.19a). Its support set consists of 64 squares of the spatial grid. It is described by coordinates x and y. Although each of the coordinates is totally ordered, the support set is ordered only partially. One variable, say v, is defined on the support set. Its states are B (for black) and W (for white). Let $v_{x,y}$ denote the state of variable v for the square with coordinate values x, y. Then, sampling variables are defined by the equation

$$s_{k,x,y} = v_{x+\alpha,\,y+\beta},$$

where $s_{k,x,y}$ denotes the state of sampling variable s_k for the square with coordinate values x, y. Two sampling variables are sufficient to generate states of v in the whole support set, e.g.,

$$s_{1,x,y} = v_{x,y},$$
$$s_{2,x,y} = v_{x+1,\,y}.$$

Visual representation of the corresponding mask is shown in Figure 3.19b, together with the indication of possible generating orders (basef on the assumption that s_2 is the generated variable) and the behavior function of the form (3.16), derived from the data by sampling. Although s_2 is uniquely determined by s_1, the entire column corresponding to $x = 1$ is required as an initial condition. To reduce the initial condition to one square ($x = y = 1$), we employ one additional sampling variable defined by the equation

$$s_{3,x,y} = v_{x,\,y+1}.$$

The corresponding mask, generating orders, and behavior function are shown in Figure 3.19c. The generation must start in this case from the top left corner of the spatial grid. There are four similar masks with three elements shown in Figure 3.19d. In each of them we assume that variable s_1 is the reference variable and at the same time the generating variable. Each mask is associated with particular generating orders and requires a different initial condition (one of the four corners of the chess board).

As a second example, assume the same support set and variable, and let the mask and generative behavior function [again in the form (3.16)] be defined as shown in Figure 3.19e. Then, the pattern shown in Figure 3.19f is generated by the behavior function, provided that patterns in the first row and first column are given as initial conditions. In this way, we can generate a two-dimensional spatial pattern from two one-dimensional patterns, which in our example happen to be equivalent.

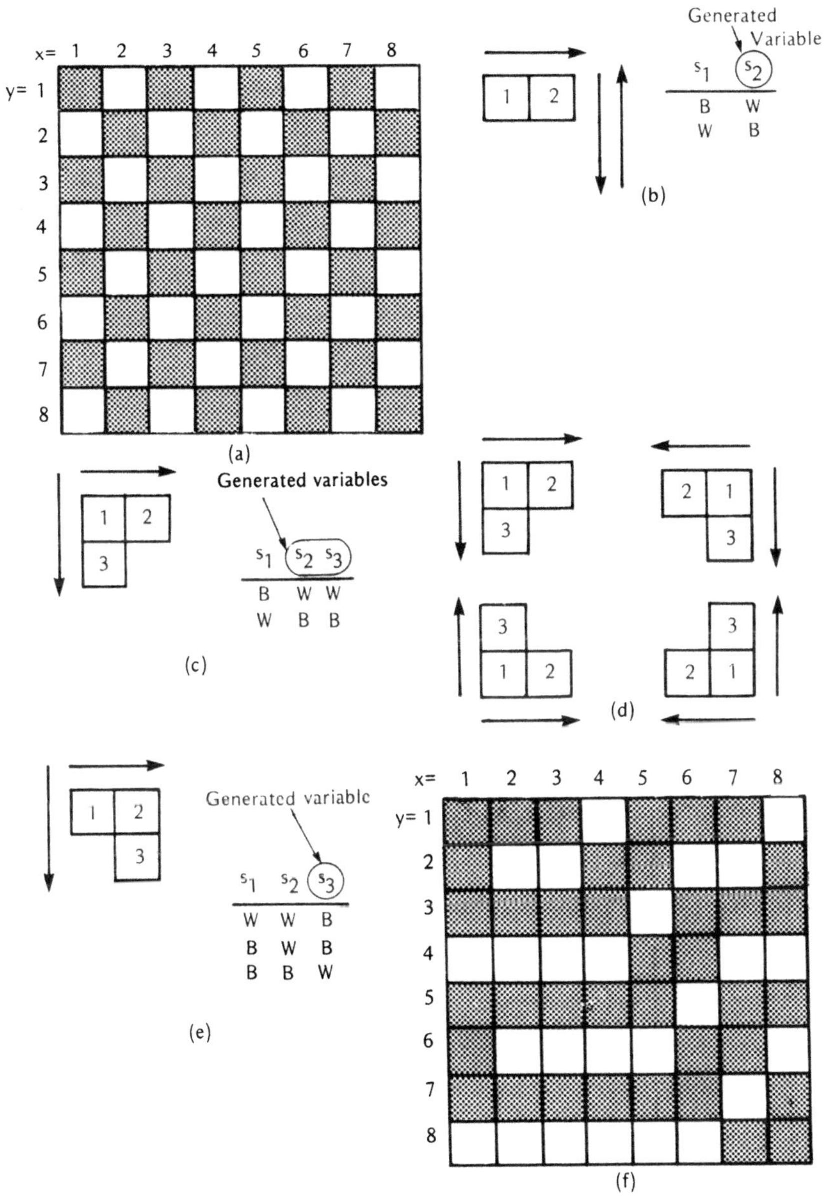

Figure 3.19. Illustration of spatial invariance (Example 3.13).

3.8. GENERATIVE SYSTEMS

> *. . . relations between percepts are, at least to some extent, creations of the mind which are then* imputed *to the external world. As such, they may be regarded as "working hypotheses," or, to use a more direct word,* models *of how the external world is organized.*
>
> —Robert Rosen

The term "generative system" is used in this book as a common name for any system associated with level 2 of the epistemological systems hierarchy. These are systems by which the overall support-invariant constraint among variables of interest is characterized from different points of view. Four types of *generative systems* are introduced in previous sections of this chapter, each of which can be either neutral or directed:

- *behavior systems*
  basic: Eq. (3.10)—neutral; Eq. (3.27)—directed;
  generative: Eq. (3.15)—neutral; Eq. (3.30)—directed;
- *ST-systems*
  basic: Eq. (3.77)—neutral; Eq. (3.93)—directed;
  generative: Eq. (3.78)—neutral; Eq. (3.94)—directed.

If, in some context, the differences between these four types are of no significance, it is convenient to refer to any of them as a neutral or directed generative system and use simpler symbols $\mathbf{F}$ or $\hat{\mathbf{F}}$, respectively.

As argued in Section 3.7, every ST-system can be converted to an isomorphic behavior system, while the inverse conversion is possible only for certain masks. Hence, behavior systems are more general than ST-systems.

Two main disadvantages of ST-systems are obvious: their restriction to compact masks and their inherent redundancy, which is a result of the overlap of the present and next states. This is a property of all masks except those in which only one translation rule is defined for each basic variable. As a consequence of these disadvantages, GSPS methodological tools for problems associated with generative systems are implemented solely in terms of behavior systems. If ST-systems are required by the user, his problem is solved in terms of the isomorphic behavior systems and within the additional restrictions imposed upon masks by ST-systems. When the problem is solved, the resulting behavior systems are converted to the isomorphic ST-systems and presented to the user in that form.

When applicable, ST-systems have certain advantages for users. It seems that generative ST-functions are easier to comprehend by the human mind than their behavior counterparts. This is probably because the generating and generated states in the ST-functions are drawn from the same state set, while they are drawn from two different state sets for behavior functions. The equality of the generating and generated state sets also makes it possible to use convenient diagrams, as illustrated in Figures 3.14, 3.17, 3.18.

Various methodological distinctions are recognized for generative systems. They include distinctions made at epistemologically lower types of systems as well as some new distinctions. The most significant among the former are:

- ordering of the support set, which makes it possible to introduce the important concept of a mask;
- ordering of state sets, which plays an important role in simplification procedures for generative systems (Section 3.9) and in dealing with incompletely specified data sets;
- the distinction between crisp and fuzzy observation channels, which leads to crisp or fuzzy data, respectively, and requires different methods for data processing;
- the distinction between neutral and directed systems, which must be handled in different ways.

Methodological distinctions that are applicable to generative systems, but not to data or source systems, are

- deterministic versus nondeterministic systems;
- for nondeterministic systems, various types of fuzzy measure are distinguished by which the support-invariant constraint of the variables involved is characterized, in particular probability and possibility measures;
- memoryless and memory-dependent generative systems are distinguished on the basis of the mask used.

These methodological distinctions are, of course, applicable to epistemologically higher types of systems as well.

3.9. SIMPLIFICATION OF GENERATIVE SYSTEMS

> *The only way to achieve any accuracy is to ignore most of the information available.*
>
> —Preston C. Hammer

At some stage in the processing of a given data system, it is often desirable to simplify the generative systems that are associated with it at that point. In some instances, a simplification is demanded by the user, for whom the existing generative systems are too complex to be comprehended. In other instances, it may be demanded by the intended utilization of the generative systems, or it may stem from various methodological considerations.

There are two basic methods of simultaneously simplifying data systems and their associated generative systems:

i. simplification by excluding some variables from the corresponding image system;
ii. simplification by defining equivalence classes of states of some variables.

► Assuming that the set of variables of a generative system, say set V, contains n variables, each proper subset of V except the empty set represents a meaningful simplification of the first kind. Hence, there are $2^n - 2$ nontrivial simplifications of this kind. They are partially ordered by the subset relationship. If the original set V and the empty set are included, for convenience, the simplification set with the partial ordering form a Boolean lattice. Let us call this lattice a *lattice of variables* or *V-lattice* and denote it by $\mathscr{L}_V$. Clearly, the V-lattice can be formulated either as

$$\mathscr{L}_V = (\mathscr{P}(V), \subseteq)$$

or as

$$\mathscr{L}_V = (\mathscr{P}(V), \cap, \cup).$$

Let f_B denote the behavior function of a given behavior system with variables in set V. When the system is simplified by reducing set V to a subset V', the new (simplified) behavior function f_B is determined by the projection

$$f_B'(\beta) = [f_B \downarrow V'](\beta), \tag{3.95}$$

defined by Eq. (3.68).

The second kind of simplification amounts to a reduction of the number of states that are recognized for the individual variables. One way of characterizing it is to define a function

$$\sigma_{i,j}: V_i \rightarrow V_i', \tag{3.96}$$

where V_i is a given state set (of variable v_i), V_i' is a simplified (reduced) state set for the same variable, $\sigma_{i,j}(\mathbf{x})$ is a new state assigned to the original state $\mathbf{x}$, and j is an identifier by which different functions of the form (3.96) are distinguished when applied to the state set of the same variable. If $\sigma_{i,j}(\mathbf{x}) = \sigma_{i,j}(\mathbf{y})$, then states $\mathbf{x}$ and $\mathbf{y}$ of V_i are not distinguished under the simplification. To be acceptable, function (3.96) must be homomorphic with respect to all mathematical properties in the original set V_i that are recognized as relevant to the problem at hand. Let any function of the form (3.96) that is homomorphic in this sense be called a *simplifying function*.

Each simplifying function induces a partition on the set V_i. Using the standard notation, let this partition be denoted by $V_i/\sigma_{i,j}$. Each partition $V_i/\sigma_{i,j}$ consists of blocks of states of V_i that are not distingushed under the simplification. Let such a partition (which preserves relevant properties in V_i) be called a *resolution form.*

Resolution forms defined on a particular state set V_i can be ordered by the usual refinement relation defined on partitions of a given set. It is well known that this refinement relation is a partial ordering and forms a lattice. Given two partitions, say X and Y, defined on the same set, we say that X is a *partition refinement* of Y if and only if for each block x in X there is a block y in Y such that $x \subseteq y$. If X is a partition refinement of Y, then Y is called a *partition coarsening* of X. Let the lattice of resolution forms defined on a state set V_i be called a *resolution lattice* of V_i and let us denote it by $\mathscr{L}_{V_i}$. Each resolution lattice on state set V_i can be defined either in the form

$$\mathscr{L}_{V_i} = (\{V_i/\sigma_{i,j}\}, \leq)$$

or in the form

$$\mathscr{L}_{V_i} = [\{V_i/\sigma_{i,j}\}, \times, +),$$

where $\times$ and $+$ denote the partition product and sum, respectively.

If the state set under consideration has no mathematical property to be preserved, then each of its partitions is acceptable as a resolution form. The resolution lattice contains in this case all partitions that can be defined on the state set. If the state set has m states, then the number of resolution forms in the lattice, say Λ_m, is given by the formula

$$\Lambda_m = \sum_{i=0}^{m-1} \binom{m-1}{i} \Lambda_i, \ \Lambda_o = 1. \tag{3.97}$$

The tremendous number of resolution forms, even for a small number of states, is indicated by the following table:

m	2	3	4	5	6	7	8	9	10
Λ_m	2	5	15	52	203	877	4,140	21,147	115,975

Since the least refined resolution form (all states in one block) is not meaningful and the most refined one does not represent a simplification, the number of meaningful simplifications is Λ_m-2.

When the state set is totally ordered and it is desirable to preserve the ordering in its simplifications, the number of resolution forms is considerably smaller than the number given by formula (3.97). Let $x_1, x_2, \ldots, x_m$ denote the states and let $x_k < x_{k+1}$ ($k = 1, \ldots, m-1$). Then for each $k \leq m-1$, x_k and x_{k+1} are either combined or not in a block. These decisions alone determine the partition. For m states, there are thus exactly $m-1$ binary decisions. Hence,

$$\Lambda_m = 2^{m-1} \tag{3.98}$$

for totally ordered state sets. It is obvious that this lattice for n states is isomorphic to the Boolean lattice of the subset ordering on the power set of any set with $m-1$ elements. To see the drastic reduction in the number of resolution forms for totally ordered state sets, when compared with unordered state sets, the following table lists values of Λ_m based on formula (3.98):

m	2	3	4	5	6	7	8	9	10
Λ_m	2	4	8	16	32	64	128	256	512

The number of meaningful simplifications is again Λ_m-2. ◀

Example 3.14. Let the states of a variable that characterizes the education of a person be

e—elementary education;
h—completed high school;
c—college degree;
g—graduate degree.

Clearly, $e < h < c < g$ is a natural ordering of the states and, consequently, there are eight resolution forms whose lattice is shown in terms of the Hasse diagram in Figure 3.20a. Blocks in the individual resolution forms, which are indicated by the bars over the respective letters, can be given appropriate names such as

cg—college or graduate degree;
hc—completed high school or college;
eh—no more than high school education;
ehc—any education except a graduate degree;
hcg—higher than elementary education.

The arrows in the diagram indicate the direction of increasing partition refinement. To simplify the original system, we have to proceed in the opposite direction.

For comparison, let states of the variable "the color of a traffic light" be the usual colors, i.e., red, yellow, green. Since they are not ordered, all partitions of the state set are acceptable as resolution forms. The Hasse diagram of their lattice is given in Figure 3.20b, where letters r, y, g stand for red, yellow, green, respectively.

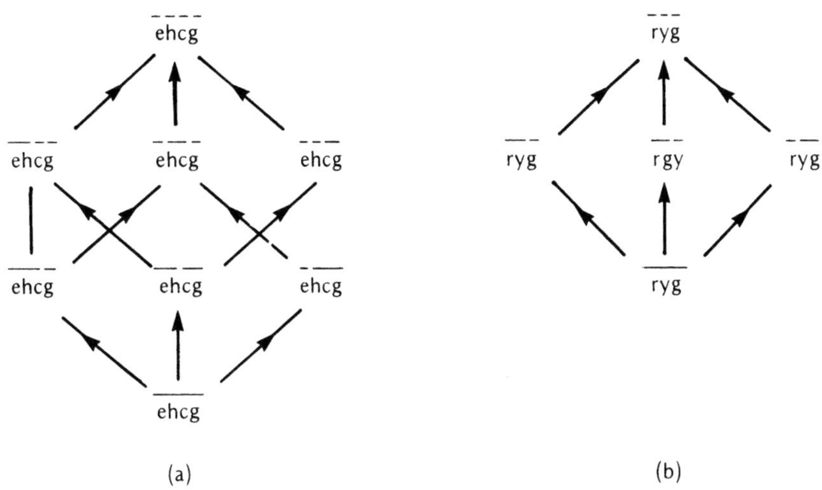

Figure 3.20. Lattices of resolution forms for: (a) a totally ordered set, (b) an unordered set (Example 3.14).

► Each element of the V-lattice represents a particular selection of variables on the original image system. For each variable selected, its resolution lattice contains all its possible resolution forms. If several variables are selected, any resolution form for one variable can be combined with any of the resolution forms for the other variables. All of these combinations can be incorporated in a single lattice representing the chosen set of variables. Let us call it a *joint resolution lattice*. Mathematically, it is a product of the individual resolution lattices. It is defined as follows:

Let $X_1, X_2, \ldots, X_n$ denote the sets of elements of the individual resolution lattices of the variables selected and let $\mathbf{X}$ denote the set of elements of the corresponding joint resolution lattice. Then,

$$\mathbf{X} = X_1 \times X_2 \times \cdots \times X_n$$

and, given two n-tuples

$$(x_1, x_2, \ldots, x_n), (y_1, y_2, \ldots, y_n) \in \mathbf{X},$$

we define

$$(x_1, x_2, \ldots, x_n) \le (y_1, y_2, \ldots, y_n)$$

if and only if $x_j \le y_j$ is satisfied for all individual resolution lattices ($j = 1, 2, \ldots, n$).

The total number of elements in the joint resolution lattice is clearly the product of the number of elements in the individual resolution lattices, i.e.,

$$|\mathbf{X}| = \prod_{j=1}^{n} |\mathbf{X}_j|,$$

but only some of them are meaningful simplifications. In particular, any combination that contains the least refined resolution form (a one-block partition) of any of the individual lattices is not meaningful. Also, the combination of all the most refined resolution forms does not represent a simplification. Hence, the total number of elements in the joint lattice that represent meaningful simplifications, say number $|X_s|$, is given by the formula

$$|X_s| = \prod_{j=1}^{n} (|X_j| - 1) - 1. \tag{3.99}$$

For the special case in which all the individual lattices are equal and each contains Λ_m resolution forms, we get

$$|X_s| = (\Lambda_m - 1)^n - 1. \tag{3.100}$$

Furthermore, if all the individual resolution lattices are based on totally ordered state sets with m states, we get

$$|X_s| = (2^{m-1} - 1)^n - 1. \blacktriangleleft \tag{3.101}$$

Example 3.15. Assume that two variables are selected for a simplification, each with three states 0, 1, 2. Assume further that the states are totally ordered: $0 < 1 < 2$. Then, the individual resolution lattices $(X_j, \leq)$ are equal and consist of four resolution forms ordered as shown in Figure 3.21a. Resolution forms of meaningful simplifications are indicated by the encircled symbols in the Hasse diagram. The joint lattice $(\mathbf{X}^2, \leq)$ is specified in Figure 3.21b. It consists of 16 resolution forms, but only those indicated by the encircled symbols in the Hasse diagram are meaningful simplifications.

► Suppose now that the original system, which is the subject of simplification, consists of n variables $v_1, v_2, \ldots, v_n$ that are associated with sets $X_1, X_2, \ldots, X_n$ of resolution forms, respectively. Then the total number of meaningful simplifications (including elimination of variables), denoted by $N(X_1, X_2, \ldots, X_n)$, is given by the formula

$$N(X_1, X_2, \ldots, X_n) = \prod_{j=1}^{n} |X_j| - 2. \tag{3.102}$$

This formula is based on the observation that a one block partition of a state set (such as partition d in Figure 3.21) can be viewed as an elimination of the corresponding

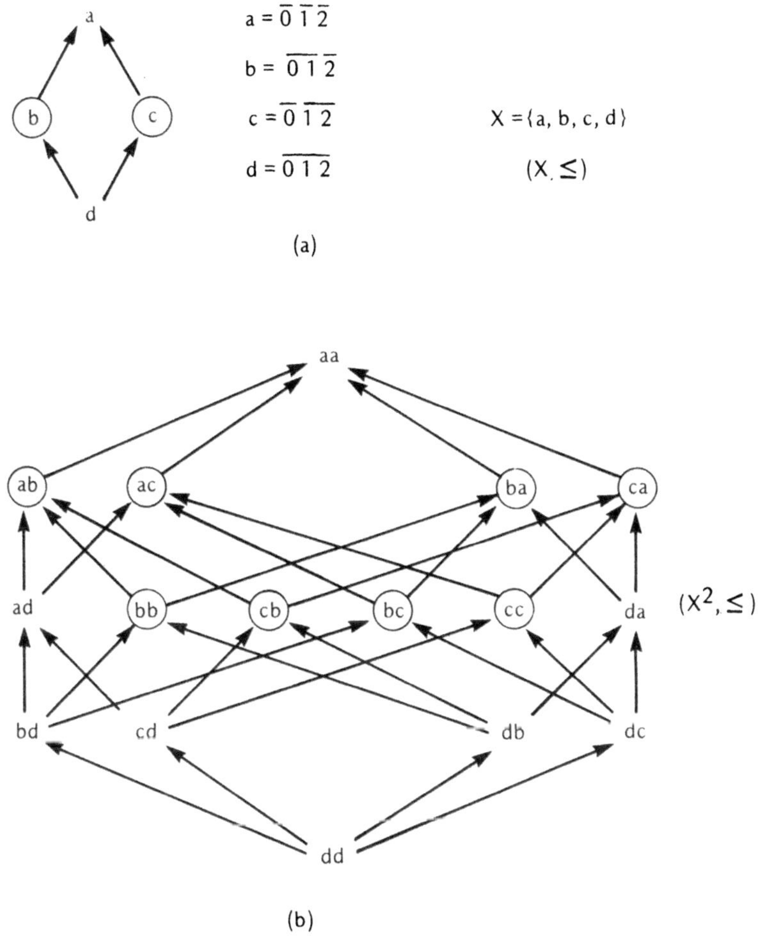

Figure 3.21. Resolution lattice $(x, \leqslant)$ and joint resolution lattice $(x^2, \leqslant)$ in Example 3.15.

variable. Then, each element in the joint resolution lattice is a meaningful simplification except the least refined and most refined joint resolution forms (*aa* and *dd* in Figure 3.21b). If all of the variables have the same set of resolution forms, say set **X**, then formula (3.102) becomes considerably simplified, namely,

$$N(X_1, X_2, \ldots, X_n) = |\mathbf{X}|^n - 2. \tag{3.103}$$

If, in addition, state sets of the variables are totally ordered and each of them contains m states, then

$$N(X_1, X_2, \ldots, X_n) = 2^{n(m-1)} - 2. \blacktriangleleft \tag{3.104}$$

Once it is decided which of the simplifications of a given behavior system to pursue, the behavior function of the simplified system must be determined on the basis of the behavior function of the given behavior system. If some of the variables of the given system are excluded, the projection (3.95) is calculated first. If further simplification by coarsening resolution forms of some of the remaining variables is required, further modifications similar to the projection must be made. First, the coarsenings of resolution forms are made as required. This creates blocks of states that are not distinguishable under the new resolution forms. Each block is then replaced by one state. Its probability (or possibility) is equal to the sum of probabilities (or the largest possibility) of all states in the block.

Example 3.16. Assume that a simplification of the probabilistic behavior function in Table 3.6a is required by excluding variable v_1 and applying the following simplifying function to state sets $V_2 = V_3$ of variables v_2 and v_3:

$V_2 = V_3$	$V'_2 = V'_3$
0	1
1	1
2	2

First, we use Eq. (3.69) to calculate the projection $[f_B \downarrow \{v_2, v_3\}]$ in Table 3.6b. Then, we identify blocks of states that are not distinguishable after the simplifying function is

TABLE 3.6
Simplification of a Behavior Function (Example 3.16)

(a)

	v_1	v_2	v_3	$f_B(\mathbf{c})$
$\mathbf{c} =$	0	0	0	0.20
	0	1	1	0.05
	0	2	2	0.04
	1	1	1	0.09
	1	1	2	0.06
	2	1	0	0.12
	2	1	1	0.10
	2	1	2	0.07
	2	2	0	0.15
	2	2	1	0.04
	2	2	2	0.08

(b)

	v_2	v_3	$[f_B \downarrow \{v_2, v_3\}](\mathbf{x})$	
$\mathbf{x} =$	0	0	0.20	block
	1	0	0.12	block
	1	1	0.24	block
	1	2	0.13	
	2	0	0.15	block
	2	1	0.04	block
	2	2	0.12	

(c)

	v_2	v_3	$f'_B(y)$
$\mathbf{y} =$	1	1	0.56
	1	2	0.13
	2	1	0.19
	2	2	0.12

applied (they are indicated in Table 3.6b). Finally, we add probabilities of states in each block and thus obtain the simplified behavior function in Table 3.6c.

Systems simplification is an important type of systems problem. It can be characterized loosely as the process of reducing the complexity (defined in some manner) of a system given at some epistemological level while, at the same time, preserving as much as possible of the information contained in the system. All problems in this class are subsumed under the following general description: given

- a particular system, say system **x**, of some epistemological type,
- a set of systems of the same type that are declared as meaningful simplifications of **x**, say set Y_x, and
- a set of requirements Q regarding some properties of systems in set Y_x,

determine a subset Y_Q of Y_x such that each system in Y_Q satisfies all requirements in Q.

To illustrate the class of simplification problems for generative systems, let **x** be a behavior system, let Y_x be the set of all meaningful simplifications of **x** based on the same set of variables as **x** (i.e., behavior systems based on all meaningful joint resolution forms derivable from x without excluding any of its variables), and let Q consist of

i. a requirement that the systems in Y_Q be as simple as possible;
ii. a requirement that the degree of generative uncertainty of the systems in Y_Q be as small as possible.

Let requirements (i) and (ii) be called a *complexity requirement* and an *uncertainty requirement*, respectively.

In order to make the complexity requirement specific, a particular measure of complexity must be defined for behavior systems. The GSPS should allow the user to specify his own complexity measure, but it should also be able to offer the user some common options, and should employ one of them as a default option. For the purpose of illustration, let the complexity of a behavior system be measured by the number of actual overall states of the system, i.e., the number of states with nonzero probabilities or possibilities. This is a simple measure yet it is perhaps the most meaningful. As such, it is a likely candidate for the default option. Let the symbol

$$|f_B| = |\{\mathbf{c} \mid f_B(\mathbf{c}) > 0\}| \tag{3.105}$$

be used for this measure of complexity of a behavior system $\mathbf{F}_B$, where f_B denotes the behavior function of $\mathbf{F}_B$.

As far as the uncertainty requirement is concerned, it is expressed in terms of the probabilistic uncertainty $H(\mathbf{G} \mid \overline{\mathbf{G}})$, defined by Eq. (3.45), or the possibilistic uncertainty $U(\mathbf{G} \mid \overline{\mathbf{G}})$, defined by Eq. (3.64). Let ${}^kq_u$ and $|{}^kf_B|$ ($k = 1, 2, \ldots$) denote, respectively, values of the appropriate generative uncertainty and the complexity for behavior systems ${}^k\mathbf{F}_B$ in the set Y_x. Superscripts k may be viewed as identifiers of the underlying joint resolution forms of the individual systems ${}^k\mathbf{F}_B$.

The numerical orderings of values ${}^kq_u$ and $|{}^kf_B|$ impose an *uncertainty ordering* $\overset{u}{\leq}$ and *complexity ordering* $\overset{c}{\leq}$, respectively, on the set Y_x. In general, these two *preference orderings* conflict with each other.

Both the uncertainty and complexity orderings are obviously reflexive, transitive and connected, but they are not antisymmetric. Hence, they are connected quasiorderings. The *combined preference ordering* $\overset{*}{\leq}$ is defined as follows:

$$ {}^j\mathbf{F}_B \overset{*}{\leq} {}^k\mathbf{F}_B \quad \text{iff} \quad {}^jq_u \overset{u}{\leq} {}^kq_u \quad \text{and} \quad |{}^jf_B| \overset{c}{\leq} |{}^kf_B|, \tag{3.106}$$

where ${}^j\mathbf{F}_B, {}^k\mathbf{F}_B \in Y_x$; it is a general quasiordering (reflexive and transitive relation) on Y_x.

The solution set $Y_Q \subset Y_x$, which is called the set of *admissible simplifications* of $\mathbf{x}$, can be now defined as the set of all systems in Y_x that are either equivalent or noncomparable in terms of the combined ordering (3.106). Formally,

$$ Y_Q = \{{}^jF_B \in Y_x \,|\, (\forall\, {}^k\mathbf{F}_B \in Y_x)\ ({}^k\mathbf{F}_B \overset{*}{\leq} {}^j\mathbf{F}_B \Rightarrow {}^j\mathbf{F}_B \leq {}^k\mathbf{F}_B)\} \tag{3.107}$$

After the set Y_Q of admissible simplifications of a given generative system is determined, the user may employ all systems in Y_Q as complementary simplifications of the original system, may choose one that is most appealing to him, or may use some additional criteria to reduce the set.

The simplification problem type, as formulated in this section, is, of course, subject to many variations. These are primarily due to alternative definitions of complexity or additional requirements. It is interesting to compare this problem type with the one of deriving admissible behavior systems from a given data system, which is discussed in detail in Sections 3.4 and 3.6. The two problem types certainly have some similarities, which can be utilized in the GSPS implementation. That is, some of the procedures developed for one of them can easily be adapted to the other. When simplification of a given behavior system $\mathbf{x}$ by the exclusion of variables is permitted, set Y_x becomes larger and some of the concepts introduced must be properly generalized, but the basic issues remain the same.

Example 3.17. Consider a probabilistic behavior system whose behavior function and mask are defined in Figures 3.22a and 3.22b. The image system is obvious. It is assumed that the support set and all state sets are totally ordered. Let us use the system to illustrate the problem of determining all its admissible simplifications without excluding any of its sampling variables.

First, we have to determine the set Y_x of all meaningful simplifications. Since there are two basic variables, each with a totally ordered set of three states, there are eight meaningful joint resolution forms. Each of these forms, which are specified in Figure 3.21, represents one meaningful simplification. When the simplifications are

	s_1	s_2	s_3	s_4	$f_B(\mathbf{c})$
$\mathbf{c} =$	0	0	0	1	0.2
	0	1	1	1	0.1
	0	1	1	2	0.1
	1	1	2	2	0.1
	1	2	1	2	0.2
	1	2	2	2	0.1
	2	2	0	0	0.2

(a)

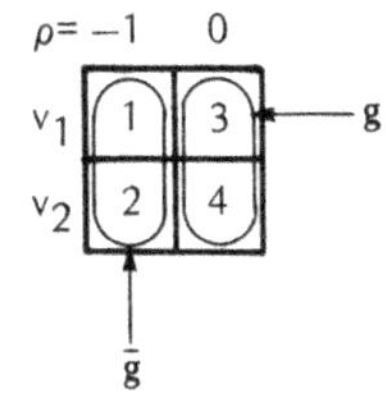

(b)

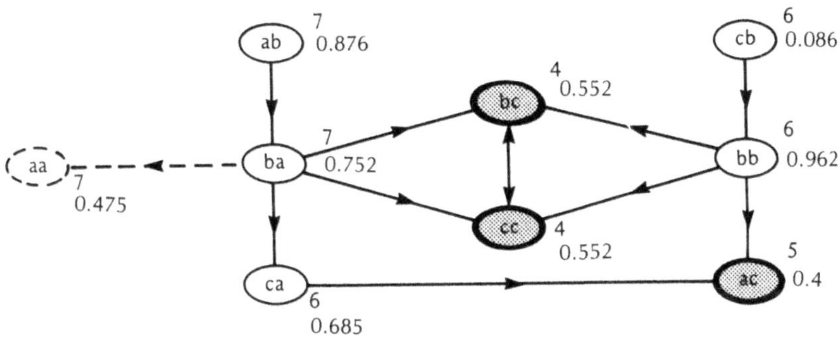

(c)

	s_1	s_2	s_3	s_4	${}^{ac}f_B(\alpha)$
$\alpha =$	0	0	0	1	0.2
	0	1	1	1	0.2
	1	1	2	1	0.2
	1	1	1	1	0.2
	2	1	0	0	0.2

(d)

	s_1	s_2	s_3	s_4	${}^{cc}f_B(\beta)$
$\beta =$	0	0	0	1	0.2
	0	1	1	1	0.2
	1	1	1	1	0.4
	1	1	0	0	0.2

(e)

Figure 3.22. Illustration of the simplification problem for behavior systems (Example 3.17).

determined, each of them is characterized by values of complexity and generative uncertainty. These values are given in Figure 3.22c, which specifies the resulting combined complexity/uncertainty ordering; the same labels are used for the resolution forms as those in Figure 3.21b. Values for the original system, labeled by aa, are included for comparison.

When inspecting the diagram in Figure 3.22c, we recognize three admissible simplifications of the given system, those based on resolution forms ac, bc, cc. The last two are equal in both complexity and uncertainty and, consequently, the ordering is not antisymmetric in this example. It is interesting to observe that the simplifications for ab and ba increase the generative uncertainty, but they do not reduce the complexity of the given system at all.

Behavior functions of two of the three admissible simplifications, those for resolution forms ac and cc, are specified in Figures 3.22d and 3.22e. Although details of determining these functions and calculating the numbers in Figure 3.22 are not presented here, the reader is encouraged to do at least some of the calculations.

3.10. SYSTEMS INQUIRY AND SYSTEMS DESIGN

> *Most of the problems associated with system design and analysis can be characterized as involving the creation and manipulation of models of real and conceptual phenomena. . . . The most difficult task is to create a framework, a context, within which any two solutions to the problem, no matter how different they may appear to be, may be compared, precisely, objectively, comprehensively.*
>
> —A. WAYNE WYMORE

Systems problems may arise in either of two principal contexts: systems inquiry or systems design. The aim of *systems inquiry* is to acquire knowledge regarding various sets of variables and supports defined for specific purposes on existing objects. The aim of *systems design*, on the other hand, is to utilize the acquired knowledge to construct new objects in which specified variables are constrained as required. Although systems problems of both systems inquiry and systems design exist at each level of the epistemological systems hierarchy, this section is restricted to a general discussion of problems that involve only source, data, and generative systems. Problems at higher levels are discussed in Chapter 4 and 5.

First, let us discuss some issues involved in systems design. The most fundamental feature of systems design is that a required support-invariant constraint among some specific variables is defined by the user. This is in sharp contrast with systems inquiry, where the constraint is not known and the task is to characterize it adequately for some specific purpose.

The constraint in systems design is defined either explicitly in terms of a specific generative system, usually a directed system, or it is defined implicitly in terms of a data system. In the former case, the design problem amounts to the determination of a set of structure systems that are admissible according to given requirements; this is discussed in Chapter 4. In the latter case, some generative systems that adequately capture the constraint embedded in the defined data must first be determined. This problem fits into the class of problems discussed in Sections 3.4 and 3.6 in the context of systems inquiry, but in the case of systems design the given data system contains (by definition) complete information about the manner in which the variables are constrained.

In the context of systems design, the data function is often defined implicitly by describing its properties rather than explicitly, in the form of a data matrix or array. For example, assume a simple directed system with one input variable whose state set consists of 26 Roman characters and empty space, and one output variable with two states, 0 and 1. The input variable is determined by a sequence of characters and spaces of an English text that is scanned. The output variable is required to be in state 1 under certain conditions, say if the last word of the scanned text ends with ING, and to assume state 0 otherwise. The problem is to convert this implicit definition of a data system into a generative system that for any English text would generate (in a deterministic manner) the required states of the output variable. Methods for dealing with problems of this kind have been well developed within the theory of finite state machines (or automata). A coherent collection of some of these methods should certainly be incorporated in the GSPS. Since abundant literature is available in this area (see Comment 3.8), there is no need to describe these methods in this book.

To compare systems inquiry with systems design at the levels of data and generative systems, two classes of data systems occurring in systems inquiries have to be distinguished. In one class are those data systems whose variables have no meaning outside the support sets for which they are defined. Examples of such data systems are

- a particular musical composition viewed as a data system (Example 2.6), where the variables obviously have no meaning beyond the time set covering the whole composition;
- any data system with spatial support, in which the space set cannot be extended, such as a system with spatial acoustic data for a particular concert hall or a system defined on the globe, where the support set consists of latitude and longitude values that cover the whole globe;
- any data system defined for an entire population of some sort, e.g., all compositions of a particular composer, all employees of a particular employer, and the like.

Data systems of this kind thus contain complete information about the constraints of their variables. As such, they are methodologically similar to the data systems defined in systems design. Let all such data systems be called *complete data systems*.

The second class of data systems in systems inquiries, which seem to occur considerably more frequently, are those systems whose variables are not restricted to the support set for which data are available. It is fair to state that virtually all systems whose support is time belong to this class (the example of a musical composition is a rare exception), and instances of complete data systems for other kinds of supports, while more frequent, are certainly not typical.

There are two fundamental approaches to systems inquiry. In one of them, admissible generative systems (or higher-level systems) based on certain requirements are derived from a given data system, as discussed for the most typical requirements in Sections 3.4 and 3.6. This approach is usually referred to as a *discovery approach*. In the other approach, a hypothetical generative system (or a higher level system) is postulated and its validity is then tested by comparing data it generates for appropriate initial conditions with empirical data. If the system fails the validation test, based on some specific validation (misfit) criteria, it is rejected and a new system is postulated. This approach to systems inquiry is usually called a *postulational approach.*

When using the discovery approach, it is obvious that each generative system derived properly from a data system is a genuine representation, in a parsimonious fashion, of some aspects of the data system. Which aspects are represented by the generative system depends on its mask and the nature of its behavior or ST-function. If the generative system is deterministic, then it is a parsimonious description of the entire data system, a sort of "shorthand" description.

If the data system is complete, the discovery approach amounts to finding patterns in its data. The discovered patterns can be then utilized for various purposes. If the data system is not complete, two issues regarding the discovered patterns (i.e., the derived admissible generative systems) must be distinguished:

- data *explanation* within the range of given support set, and
- data inference beyond the support set range, i.e., data *prediction, retrodiction,* or *generalization.*

The difference between these two issues, which are often confused in the literature, is well characterized in one of Herbert Simon's articles [SI3]:

> Law discovery means only finding pattern in the data; whether the pattern will continue to hold for new data that are observed subsequently will be decided in the course of testing the law, not discovering it .... The discovery process runs from particular facts to general laws that are somehow induced from them; the process of testing discoveries runs from the laws to predictions of particular facts from them ... The fact that a process can extract pattern from finite data sets says nothing about the predictive power of patterns so extracted for new observations. As we move from patterns detection to prediction, we move from the theory of discovery processes to the theory of processes for testing laws. To explain why the patterns we extract from observations frequently lead to correct predictions (when they do) requires us to face again the problem of induction, and perhaps to make some hypothesis about the uniformity of nature. But that hypothesis is neither required for, nor relevant to, the theory of discovery processes. The latter theory does not assert that data are patterned. Rather, it shows how pattern is to be detected if it is there. This is not a descriptive or psychological matter, it is normative and logical. By separating the question of pattern detection from the question of prediction, we can construct a true normative theory of discovery—a logic of discovery.

When the discovery approach is employed for data systems that are not complete, generative systems (or higher-level systems) are derived not only for the purpose of explaining the given data, but mainly for extending the data beyond the given support set, thus allowing prediction, retrodiction, or generalization. This requires, of course, that some sort of *inductive reasoning* be used. This means that the GSPS must be equipped with a package of well-founded methodological tools for inductive reasoning. Relevant issues of inductive reasoning are discussed in Chapter 4.

The problems of determining admissible generative systems are discussed in Sections 3.4 and 3.6 under the tacit assumption that sampling variables are defined solely in terms of the variables included in the given data system, i.e. *observed variables*. This is unnecessarily restrictive and may make it difficult in some cases to obtain reasonably simple generative systems with small or no generative uncertainty. The problems can be generalized by allowing the user to postulate hypothetical states of some additional variables, which are not among the observed variables. Such variables are usually called *internal variables* and their states are called *internal states*.

Although the hypothetical internal states may be introduced for various reasons, they are usually introduced for the purpose of improving the trade-off between generative uncertainties and complexities of admissible generative systems. The introduction of internal states requires that some pattern in the given data be recognized through which they can be generated while, at the same time, they help to reduce the overall generative uncertainty. Such pattern recognition is feasible only for complete data systems and has been investigated within the theories of deterministic and probabilistic finite state machines.

The concept of internal variables and states is important in systems design. The introduction of internal states in the process of systems design amounts to a convenient redefinition of the required constraint. After they are introduced at the abstract level, internal variables and their states can be exemplified in any way desired. In systems inquiry, on the other hand, the use of internal variables is problematic since they do not have any semantic content and it is not acceptable to exemplify them in an opportunistic fashion, as is done in systems design.

In summary, systems design is always a process of climbing up the epistemological hierarchy of systems. It starts with the definition of either a generative system or a data system and a set of requirements regarding structure systems. The problem of deriving admissible generative systems from the defined data system belongs to the same class of problems as those discussed in Sections 3.4 and 3.6, but internal variables can also be employed when convenient. Systems inquiry can be performed by

- climbing up the hierarchy by discovering higher-level systems whose lower-level systems have certain given properties and, if data are not complete, making appropriate inductive inferences (discovery approach);
- postulating generative systems or higher-level systems and rejecting those which fail a misfit test between empirical and generated data (postulational approach);
- any combination of the discovery and postulational approaches, e.g., climbing up to some level while postulating systems at a higher level.

Problems associated with the discovery approach are given more attention in this book than are other classes of systems problems. This is motivated by two reasons. One of them is of a pedagogic nature. The discovery approach, in which systems enter in increasing order of their conceptual complexity, is perfect for motivating, explaining, and formulating the whole GSPS conceptual framework. The second reason is that the discovery approach is not well covered in the literature, while both the postulational approach and systems design are covered quite well.

NOTES

3.1. The concepts of mask and sampling variables, which are central to generative systems, are due to Antonin Svoboda. He introduced them in the early 1960s for the purpose of designing and classifying switching circuits [SV1, SV2] and employed them later for developing a sophisticated and unorthodox methodology for dealing with switching circuits [SV4].

3.2. Literature on probability theory is abundant. Kolmogorov's classic book, first published in 1933, seems still to be the best choice for a standard axiomatic treatment of probability theory [KO3]. For a comprehensive survey and comparison of various axiomatic frameworks and interpretations of probability theory, a book by T.L. Fine is recommended [FI2].

3.3. Formula (3.32) was derived by R. Christensen [CH5] by using the principle of maximum entropy, which is one of the principles of inductive reasoning discussed in Chapter 4. He also derived a generalized formula for estimating probabilities, which accounts for any relevant background information available. Formula (3.34) was derived by D. Dubois and H. Prade [DU3]. The study of all acceptable functions for converting frequency distributions into possibility distributions is a subject of current research.

3.4. The concept of possibility measures, which is the basis for developing possibility theory, was suggested by Lotfi Zadeh in 1978 [ZA5]. Possibility measures form a small subset of the set of fuzzy measures, which were introduced by M. Sugeno in 1977 [SU1]. They do not overlap with probability measures, which also form a class of fuzzy measures. It was shown by Puri and Ralescu [PU1] that possibility measures can be defined only on finite sets and some special classes of infinite sets. The relationship among the various subsets of fuzzy measures, as summarized in Figure 3.5, is well described in Part II, Chapter 5, of a survey book by Dubois and Prade [DU1].

► **3.5.** As mentioned in Section 3.5, the notion of conditional possibilities is a controversial issue in possibility theory. The controversy emerges from the relationship between the concepts of noninteraction and independence. It is clear that these two concepts are equivalent within probability theory. Let $x \in X$ and $y \in Y$, where X and Y are some finite sets of events. Probabilistic noninteraction is defined by $p(x, y) = p(x) \cdot p(y)$ for all $x \in X$ and all $y \in Y$, where $p(x, y)$ denotes the joint probabilities. Probabilistic independence is defined by $p(x \mid y) = p(x)$ and $p(y \mid x) = p(y)$, where $p(x \mid y)$ denotes the conditional probability of x given y and $p(y \mid x)$ denotes the other conditional probability. Since joint probability is defined as $p(x, y) = p(x \mid y) \cdot p(y) = p(y \mid x) \cdot p(x)$, the sets X and Y are independent if and only if they do not interact.

It is not obvious that noninteraction and independence are equivalent concepts when defined in terms of possibility theory. Two views on this issue are expressed in the literature by Hisdal [HI6] and Nguyen [NG1]. Given two finite sets of events X and Y with possibility distributions

$f(x)$ and $f(y)$ $(x \in X,\ y \in Y)$, respectively, they are called noninteractive if and only if

$$f(x, y) = \min[f(x), f(y)] \tag{3.108}$$

for all $x \in X$ and $y \in Y$, where $f(x, y)$ denotes the joint possibilities of x and y. The sets are called independent if and only if

$$f(x \mid y) = f(x) \tag{3.109}$$

and

$$f(y \mid x) = f(y) \tag{3.110}$$

for all $x \in X$ and $y \in Y$, where $f(x \mid y)$ and $f(y \mid x)$ denote the conditional possibilities of x given y and y given x, respectively. Hisdal argues that the equations

$$f(x, y) = \min[f(y), f(x \mid y)] \tag{3.111}$$

$$f(x, y) = \min[f(x), f(y \mid x)] \tag{3.112}$$

must be satisfied for any two sets of events characterized by some possibility distributions $f(x)$, $f(y)$. Then, it is clear that (3.109) and (3.111) as well as (3.110) and (3.112) imply (3.108). Hence, the independence of possibilistic events implies their noninteraction. However, the converse is not true. Indeed, from (3.108) and (3.111) we obtain

$$f(x \mid y) = \begin{cases} f(x) & \text{if } f(x) < f(y) \\ [f(x), 1] & \text{if } f(x) \geq f(y) \end{cases} \tag{3.113}$$

(and similarly for the other conditional possibility). Hence, noninteraction does not imply independence. Nguyen takes a radically different approach to the meaning of the conditional possibilities. He defines "normalized" conditional possibilities in such a way that, by analogy with probability theory, possibilistic noninteraction is required to be equivalent to possibilistic independence. This requirement leads to the formula

$$f(x \mid y) = \begin{cases} f(x, y) & \text{if } f(x) \leq f(y) \\ f(x, y) \cdot \dfrac{f(x)}{f(y)} & \text{if } f(x) > f(y) \end{cases} \tag{3.114}$$

where $f(x)/f(y)$ is a normalization factor. ◄

3.6. The Shannon entropy, which is a natural measure of uncertainty and information for events characterized by probability distributions, has dominated the literature on information theory since it was proposed by Shannon in 1948 [SH3]. It was originally introduced for the purpose of analyzing and designing telecommunication systems, but its significance and applicability reaches far beyond this original purpose. The usefulness of the Shannon entropy in general systems methodology, perhaps one of its most significant roles, was recognized considerably later than many of its other applications. To the best of my knowledge, the only early proponent of this use of the Shannon entropy was the late Ross Ashby. Although his initial ideas along these lines were presented as early as 1956 in his classic book [AS2], he returned to this subject more forcefully one decade later [AS5] and continued to be strongly interested in it for the rest of his life [AS8, AS10, AS11].

The literature on information theory based on the Shannon entropy is plentiful. Among the many available books on information theory, a book by Aczél and Daróczy [AC4] seems to be best survey of possible axiomatic characterizations of the Shannon entropy as well as its various generalizations, a book by Guiasu [GU1] is a good overview of applications of information theory, and a book by Watanabe [WA6] is an excellent conceptual and mathematical treatment of the role of information in scientific inference.

3.7. It is generally recognized that the first measure of information and uncertainty was introduced by R. Hartley in 1928 [HA9]. He defined the information necessary to characterize an element of a finite set with n elements as the binary logarithm of n. The measure is frequently given one of two probabilistic interpretations. In the first, it is viewed as a special measure which distinguishes only between zero and nonzero probabilities and which, otherwise, is totally insensitive to the actual values of the probabilities. In the second interpretation, it is viewed as the Shannon entropy under the assumption that all elements of the set are equally probable. Such attempts to subsume the Hartley measure under the Shannon entropy are ill conceived since it is logically independent of probabilistic assumptions. In fact, it is incompatible with the Shannon entropy since it possesses a property of monotonicity (the larger the set, the larger its Hartley information), which is not applicable to probability distributions (they cannot be ordered in a comparable manner).

It is shown in one of my papers, which I coauthored with Masahiko Higashi [H12], that the Hartley information is a special case of possibilistic information, expressed by the U-uncertainty given by Eqs. (3.56) or (3.57), for crisp possibility distributions (with possibilities 0 or 1 only). It is also proven in the paper that the U-uncertainty satisfies possibilistic counterparts of all the axioms required for the Shannon entropy (listed in Section 3.5) and, in addition, that it satisfies a general requirement of monotonicity (the larger the possibility distribution, the larger its U-uncertainty). The controversy associated with conditional possibilities, as mentioned in Note 3.5, is avoided by requiring that the additivity requirement be satisfied by the class of noninteractive sets of outcomes, which subsumes the class of independent sets of outcomes. The conditional U-uncertainty, given by Eq. (3.64), is also derived in the paper without any use of the controversial concept of conditional possibilities.

3.8. Some methods emerging from the theory of finite state automata or machines (deterministic and probabilistic) were developed to deal with certain specific variations of the problem of deriving admissible generative systems from complete data systems. The methods are usually applicable only to directed systems and are based on the use of internal states. As such, although they are relevant to systems design, their utilization in systems inquiries is considerably limited. Among the many books available in the area of automata theory, a book by Taylor Booth [BO1] seems to offer the best coverage of both deterministic and probabilistic automata. A more methodologically oriented treatment can be found in one of my own books [KL5].

One particular contribution of automata theory deserves a special attention as an example of a perfect interface between the user and GSPS in dealing with a problem. It is a method, developed by Tal [TA2], by which a state-transition description of a deterministic finite-state machine (based on internal states) is constructed by asking the user, in an algorithmic fashion, simple questions of a yes/no type regarding the data system, e.g., whether a certain input/output sequence is possible or not. When the questioning is completed (in a finite number of steps if the data is complete), the ST-system is determined algorithmically from the solicited answers. This method thus does not require the user to define his problem completely, but helps him to define it. Two additional papers supplement the original paper by Tal [GU4, TA3]; they include a proof of the correctness of algorithm and discuss some subtle issues associated with it.

3.9. The possibility of using postulated internal states in systems inquiries was studied by Gerardy [GE3–4]. He demonstrates that the use of internal states in the process of deriving generative systems from empirical data is computationally intractable even for very small data. This result reinforces the conclusion, argued in Section 3.10, that the concept of internal states is of little significance for systems inquiries, while it plays an important role in problems of systems design.

3.10. Among the many available methodologies for systems design, the one developed by Wayne Wymore [WY2] seems currently to be the best candidate for being integrated into the GSPS. It is sufficiently general, well formulated, and conceptually sufficiently close to the GSPS conceptual framework to make the integration manageable.

3.11. The notions of explanation, prediction, and retrodiction are mentioned a number of times in this chapter. These are notions of great philosophical significance. Since it is beyond the scope of this book to cover them properly, I recommend two books as supplementary readings: a book by Satosi Watanabe [WA6], which deals with these notions in terms of information theory, and a book by Nicholas Rescher [RE2], which discusses fundamental philosophical issues associated with them (in terms of probabilistic ST-systems) and contains an extensive bibliography.

► **3.12.** For systems with continuous variables and supports, support-invariant constraints among variables are characterized, in general, by differential equations with constant coefficients. Derivatives, which are defined in terms of the basic variables and specific translation rules in the support set, are clearly sampling variables in this case. The highest-order derivative in a differential equation is an analog of the mask depth. Solutions of differential equations for various initial conditions have the meaning of generated data. Empirical data are any continuous functions of the support set.

As an example, let function $v(t) = \sin t$ represent empirical data for some domain of t, where v is a variable and t is time. Then, $\dot{v}(t) = \cos t$ and $\ddot{v}(t) = -\sin t$. Hence, the differential equation

$$\ddot{v}(t) + v(t) = 0$$

is a time-invariant characterization of variable v since it is the same for every value of t. When the differential equation is solved for appropriate initial conditions, the original function is obtained ◄.

3.13. Interesting discussions of the concept of environment are in the papers by G. C. Gallopin [GA5] and B. C. Patten [PA4].

3.14. A general formulation of the problem of determining admissible behavior systems, as exemplified by Eqs. (3.73) and (3.107) in two different contexts, was proposed by Brian Gaines [GA2].

► **3.15.** The concept of entropy was originally proposed by Boltzmann in 1896 in the form

$$H(f(x)|x \in [a, b]) = -\int_a^b f(x) \log f(x)\, dx,$$

where f is a probability density function defined for a continuous variable $x \in [a, b]$. Although similar by its form to the Shannon entropy, the Boltzmann entropy is not a counterpart of the Shannon entropy for continuous variables, as one might expect. In fact, the Boltzmann entropy is not a limit of the Shannon entropy and, consequently, it does not measure uncertainty and information. However, when modified to the form

$$H_B\left(\frac{f(x)}{g(x)|x \in [a, b]}\right) = \int_a^b f(x) \frac{\log f(x)}{\log g(x)},$$

it becomes a counterpart of the Shannon cross entropy

$$H_S\left(\frac{f(x)}{g(x)|x \in X}\right) = \sum_{x \in X} f(x) \frac{\log f(x)}{\log g(x)},$$

where X is a finite set of states of variable x, and f, g are two probability distribution functions defined on X. For further details, some books on information theory should be consulted [GU1, KU1, RE12]. ◄

EXERCISES

3.1. Let the following periodical time sequences of states 0, 1 of a single variable v be given, where one period is underlined in each case:

$$t = 1\ 2\ 3\ 4\ 5\ 6\ 7\ 8\ 9\ 10\ 11\ 12\ 13\ 14\ 15\ 16 \ldots$$

$$v_t = \underline{0\ 1}\ 0\ 1\ 0\ 1\ 0\ 1\ 0\ 1\ 0\ 1\ 0\ 1\ 0\ 1 \ldots$$

$$v_t = \underline{0\ 0\ 1\ 1}\ 0\ 0\ 1\ 1\ 0\ 0\ 1\ 1\ 0\ 0\ 1\ 1 \ldots$$

$$v_t = \underline{0\ 0\ 0\ 1\ 1\ 1\ 0\ 1}\ 0\ 0\ 0\ 1\ 1\ 1\ 0\ 1 \ldots$$

$$v_t = \underline{0\ 0\ 0\ 0\ 1\ 0\ 1\ 1\ 1\ 1\ 0\ 1\ 0\ 0\ 1\ 1} \ldots$$

For each of the sequences, find

(a) deterministic systems by which the sequence can be generated from the left to the right and vice versa, respectively;

(b) the ST-systems that are isomorphic to the behavior systems determined in (a).

3.2. Consider the two-dimensional 8 × 8 "chessboard" spatial grid (Example 3.13) on which a single variable v with two states, black and white, is defined. Assume a behavior system in which four sampling variables,

$$s_{1,x,y} = v_{x,y}, \qquad s_{3,x,y} = v_{x+1,y},$$

$$s_{2,x,y} = v_{x,y+1}, \qquad s_{4,x,y} = v_{x+1,y+1},$$

are involved, and whose generative behavior is defined as follows: s_4 is black if s_2 is white and $s_1 = s_3$ or if s_2 is black and $s_1 \neq s_3$; otherwise, s_4 is white.

(a) Determine the mosaic pattern generated by the behavior system under the assumption that the top row and left column (initial conditions) consist of sequences **W W B W W B W W** and **W B B W W B B W**, respectively, where **W** stands for "white" and **B** stands for "black."

(b) Define the ST-system isomorphic to the behavior system.

(c) Explore the effect of different initial conditions on the generated mosaic.

3.3. Define appropriate generative systems on some man-made objects with which you are familiar. The following are hints of some possible objects: a combination lock, an elevator, a vending machine, an embroidery machine, internal combustion gasoline engine, decimal adder.

3.4. Develop the following generative systems:

(a) A deterministic behavior system describing a mortgage on a house. Assume a 20-year mortgage of $50,000 at the interest rate of 10% or, alternatively, consider your own mortgage. The system should generate amounts (in dollars and cents) of the individual monthly payments of the principal and interest.

(b) A crisp possibilistic ST-system that characterizes (in a similar manner as the system described in Example 3.9) possible career paths of an employee of a large industrial company, say a computer firm. Let variables by which job categories are defined describe the levels of technical, managerial, customer interface, and selling aspects involved in each job category. Assume four states for each variable: none, low, medium, high. If you are not familiar with a particular company, use common sense to define a feasible system.

(c) A behavior system that fully describes an iterative numerical algorithm of some kind, say the Newton–Raphson algorithm for calculating the square root of a positive rational number. Assume some specific precision required in the final result.

(d) A crisp possibilistic ST-system that describes the constraints of a game such as chess or checkers. Do not attempt to specify the ST-function explicitly, but describe it by a set of propositions. For any given state, the propositions must be sufficient to determine all possible (legal) next states according to the rules of the game.

3.5. As a continuation of Exercise 2.7, determine all admissible behavior systems for the data system resulting from that exercise and for some manageable largest acceptable mask. Perform this exercise for both probabilistic and possibilistic methodological distinctions.

3.6. Consider a probabilistic ST-system based on one observed variable with three states (0, 1, 2) whose ST-function is defined in Figure 3.23a for the mask M_1 and for the purpose of prediction. Derive from this system

(a) ST-systems for masks M_2 and M_3;

(b) the isomorphic behavior system for the extended mask M_1^+;

(c) behavior systems for all meaningful submasks of M_1^+;

(d) admissible behavior systems among those determined in (c), provided that the generative uncertainty and mask size are to be minimized;

(e) isomorphic ST-systems for those of the behavior systems determined in (d) for which they exist.

3.7. Assume that frequencies $N(\mathbf{c})$ specified in Figure 3.23b were determined from empirical data for states $\mathbf{c}$ of four sampling variables based on two observed variables v_1, v_2, a totally ordered support set of 1,500 observations, and the mask M. Determine, for both the

c′ =	000	001	002	012	111	221	222
c = 000	.1	0	0.02	0	0	.01	0
001	0	.03	0	0	.05	0	0
002	0	0	.1	0	0	0	.03
012	.03	0	0	.2	0	0	0
111	0	0	.01	.01	.1	.03	0
221	0	.05	0	0	0	.2	0
222	0	0	0	.02	0	.01	0

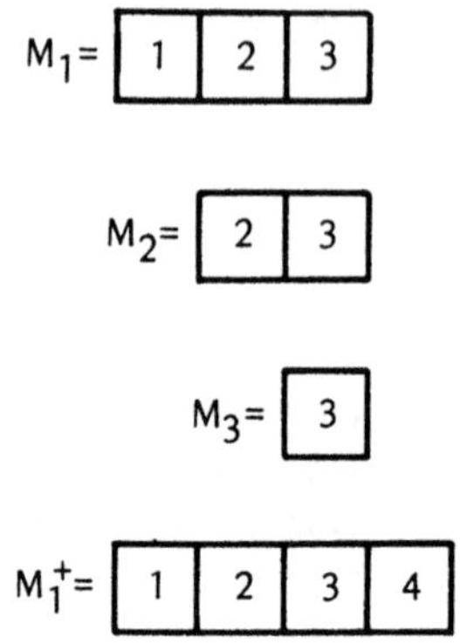

(a) Excercise 3.6

	s_1	s_2	s_3	s_4	N(c)
c =	0	0	0	0	525
	0	1	0	1	150
	1	1	0	0	300
	1	1	0	1	150
	1	0	1	0	150
	1	1	1	0	150
	1	1	1	1	75

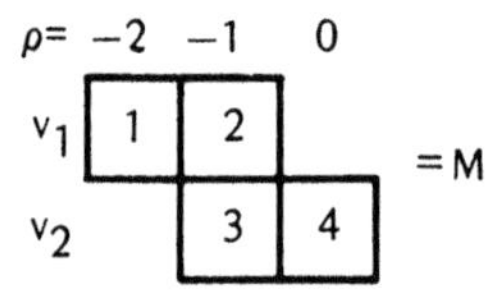

(b) Excercise 3.7

t =		0	1	2	3	4
v_1 =	0	1	.8	.5	0	0
	1	0	.2	.5	1	1
v_2 =	0	0	1	.4	.75	0
	1	1	0	.4	.25	0
	2	0	0	.2	0	1

1 period (columns t = 0 to 4)

(c) Exercise 3.8

Figure 3.23. Illustrations for (a) Exercise 3.6; (b) Exercise 3.7; (c) Exercise 3.8.

probabilistic and possibilistic methodological distinctions, and for the purpose of prediction the following:

(a) behavior systems for M;

(b) isomorphic ST-systems for the behavior systems determined in (a); check for each of them if it satisfies the transition balance expressed by Eqs. (3.80) and (3.81);

(c) behavior systems for all meaningful submasks of M; identify those which are admissible (in the usual sense);

(d) the directed behavior system for M under the assumptions that v_2 is an input variable;

also calculate its generative uncertainty, under the assumption that the available information about v_2 is utilized;

(e) the same as in (d), but under the assumption that no information about v_2 is available.

3.8. Determine, for the purpose of prediction, possibilistic and probabilistic behavior functions based on mask $M = \{(v_1, -1), (v_1, 0), (v_2, 0)\}$ for the periodic fuzzy data array (with totally ordered support set) specified in Figure 3.23c.

(a) Use both the product and min aggregation functions.

(b) Propose and use in this exercise some other aggregation functions, which satisfy the specified requirements for acceptable aggregation functions.

3.9. Repeat Example 3.7 for $\Delta M = 2$ and the possibilistic methodological distinction.

3.10. Consider a neutral data system with two variables, totally ordered support set, and data specified by the matrix

$$\mathbf{d} = \left[\begin{matrix} 0&0&1&1&1&2&0&2&1&2&1&1&0&2&2&2&2&1&0&1&2&1&2&2&2&0&2&2&0 \\ 0&0&1&1&2&2&1&2&0&2&2&0&0&0&2&2&1&1&2&1&0&0&1&2&0&1&0&0&2 \end{matrix} \right.$$

$$\begin{matrix} 2&2&2&1&0&2&0&0&1&0&1&0&2&1&1&0&2&2&2&2&1&0&1&2&1&2&2&2&0 \\ 2&2&2&1&0&1&2&1&2&2&2&0&2&2&0&0&0&2&2&1&1&2&1&0&0&1&2&0&1 \end{matrix}$$

$$\begin{matrix} 1&1&1&2&0&1&0&0&2&0&2&0&1&2&2&0&1&1&1&1&2&0&2&1&2&1&1&1&0 \\ 1&1&1&2&0&2&1&2&1&1&1&0&1&1&0&0&0&1&1&2&2&1&2&0&0&2&1&0&2 \end{matrix}$$

$$\begin{matrix} 0&0&0&0&0&1&1&2&2&1&2&0&0&2&1&0&2&0&0&1&0&1&0&2&1&1&0&2&2 \\ 0&0&1&0&1&0&2&1&1&0&2&2&2&2&1&0&1&2&1&2&2&2&0&2&2&0&0&0&2 \end{matrix}$$

$$\left. \begin{matrix} 0&1&2&0&1&2&0&1&2&0&1&2&2&0&2&2&0&0&0&2&2&1&1&2&1&0&0&1&2 \\ 0&1&2&0&1&2&0&1&2&0&1&2&0&1&0&0&2&0&2&0&1&2&2&0&1&1&1&1&2. \end{matrix} \right].$$

Assume that the state sets of the variables are totally ordered (i.e., $0 < 1 < 2$). Derive from the data system

(a) a probabilistic or possibilistic behavior system for a two-column mask, which is to be used either for prediction or retrodiction;

(b) all meaningful simplifications of the behavior system determined in (a) and based only on resolution form coarsening;

(c) for each simplification obtained in (b), determine all admissible behavior systems for all meaningful submasks of the two-column mask and the usual requirements.

3.11. Repeat Exercise 3.10 under the assumption that the data system is directed. Assume that the first variable is viewed as an input variable and that available information about this variable is not relevant (e.g., it is controlled by the user).

3.12. Suppose that for the purpose of testing a population of manufactured electronic chips of the same kind, you define a variable by which you characterize which of two types of defect occurred. The variable has four states:

0: neither of the two types of defect occurred
1: the first type of defect occurred, but not the second
2: the second type of defect occurred, but not the first
3: both occurred

This set of states is partially ordered since states 1 and 2 are not comparable.

(a) Determine all partitions of the state set that preserve the partial ordering.

(b) Describe (preferably in the form of a Hasse diagram) the resulting resolution lattice for this variable.

3.13. Let another variable be added to the one described in Exercise 3.12. It is introduced to characterize the total number of observed defects on each chip. Suppose the variable has ten states:

0: no defect

1: one defect

. . .

9: nine defects.

This state set can clearly be viewed as totally ordered.

(a) Determine the total number of elements in the joint resolution lattice and the number of meaningful resolution forms in the lattice.

(b) Give examples of at least three complete paths in the lattice, from the most refined to the least refined resolution form.

(c) Calculate the total number of meaningful simplifications that exist for generative or data systems based on the two variables.

3.14. Let the following three variables and state sets be defined for each member of a population:

- educational background (variable E):
  0—less than high school,
  1—high school,
  2—college degree,
  3—graduate degree;
- political affiliation (variable P):
  d—democrat,
  r—republican,
  i—independent;
- sex (variable S):
  f—female,
  m—male.

(a) Determine the lattice of variables (V-lattice) for this example.

(b) Decide which mathematical properties (if any) should be recognized in each of the state sets.

(c) Determine the resolution lattice and its subset of meaningful resolution forms for each of the state sets. This result depends, of course, on the decisions made in (b).

(d) Calculate the total number of meaningful simplifications that exist for any generative or data system based on the three variables.

3.15. Repeat Exercise 3.14 for the systems defined in Tables 3.1 and 3.2.

3.16. Assume that a generative system consists of three observed variables. One of them has six states that are partially ordered. Let the ordering be defined by the following list of immediate refinements: $0 \leqslant 1$, $0 \leqslant 2$, $1 \leqslant 3$, $1 \leqslant 4$, $2 \leqslant 5$, $3 \leqslant 5$, $4 \leqslant 5$. State sets of the other two variables are totally ordered; they have five states and seven states, respectively. Determine

(a) the total number of meaningful simplifications based only on resolution form coarsening;

(b) the total number of all meaningful simplifications, including those obtained by excluding some variables.

3.17. Derive formula (3.36) for the number of meaningful submasks of a largest acceptable mask.

3.18. Given a largest acceptable mask **M** with n rows and $\Delta\mathbf{M}$ columns, derive a formula for

(a) the number of meaningful submasks of **M** whose generated part is represented by the rightmost column of **M**;

(b) the number of meaningful submasks of **M** which contain $n+k$ elements, where $0 \leqslant k \leqslant (\Delta\mathbf{M}-1)n$;

(c) the number of meaningful compact submasks of **M**;

(d) the number of all meaningful submasks of **M** such that an isomorphic ST-system exists for a behavior system defined for any one of them.

3.19. Prove inequalities (3.38) for the Shannon entropy and inequalities (3.58) for the possibilistic uncertainty.

3.20. Derive formulas (3.102)–(3.104).

3.21. Calculate the numbers given in Figure 3.22 for Example 3.17.

3.22. Prove the following proposition: if $f_B(\mathbf{c})$ is calculated by formula (3.32), then

$$f_B(\mathbf{g}\,|\,\bar{\mathbf{g}}) = \frac{N(\bar{\mathbf{g}}, \mathbf{g})+1}{\Sigma_{\mathbf{g}\in G} N(\bar{\mathbf{g}}, \mathbf{g}) + |G|},$$

where $N(\bar{\mathbf{g}}, \mathbf{g}) = N(\mathbf{c})$.

3.23. Consider a generative system based on one basic variable v which is of ordinal scale and has three states labelled by integers 0, 1, 2. Assume further that the state set $\{0, 1, 2\}$ is linearly ordered in the same way as the integers. The support is time (linearly ordered). The system is characterized by the probabilistic behavior function in Table 3.7a, where the sampling variables s_1, s_2 have the following meaning:

$$s_{1,t} = v_{t-1}$$

$$s_{2,t} = v_t$$

TABLE 3.7
Illustration for Exercises 3.24–3.27.

(a)

s_1	s_2	$f_B(\mathbf{c})$
c = 0	0	0.05
0	1	0.20
0	2	0.15
1	0	0.10
1	1	0.05
1	2	0.10
2	0	0.25
2	2	0.10

(b)

s_1	s_2	s_3	$f_B(\mathbf{c})$
c = 0	0	1	0.20
0	1	0	0.15
0	1	1	0.10
1	0	0	0.20
1	0	1	0.05
1	1	0	0.10
1	1	1	0.20

Determine all meaningful simplifications based on coarsening the resolution from that are based upon the assumption that you use the behavior for the purpose of prediction. Then, determine which of them are members of the solution set.

3.24. A generative system, based on one basic variable v and linearly ordered support t, is characterized by the probabilistic behavior function in Table 3.7b, where the sampling variables s_1, s_2, s_3 have the following meaning:

$$s_{1,t} = v_t$$

$$s_{2,t} = v_{t+1}$$

$$s_{3,t} = v_{t+2}$$

Determine:

(a) the retrodictive generative uncertainty of the system;

(b) a matrix representation of the retrodictive basic ST-form corresponding to the given behavior;

(c) a matrix representation of the generative version of the ST-form obtained in (b);

(d) a diagram representation of the generative ST-form obtained in (c).

3.25. Repeat Exercise 2.24 for the purpose of prediction.

3.26. Evaluate the generative system given in Exercise 3.24 for all meaningful submasks under the assumption that it is used for the purpose of prediction and determine the solution set based on two preference orderings: complexity, expressed by the size of the mask, and generative uncertainty. Plot the dependence of the generative uncertainty on the size of the mask.

3.27. Repeat Exercise 3.26 for the purpose of retrodiction.

4

STRUCTURE SYSTEMS

The world does not present itself to us mostly divided into systems, subsystems, environments, and so on. These are divisions which we make ourselves, for various purposes, often subsumed under the general purpose evoked by saying "for convenience."

—Joseph A. Goguen and Francisco J. Varela

4.1. WHOLES AND PARTS

> *Once the whole is divided, the parts need names. There are already enough names. One must know when to stop. Knowing when to stop averts trouble.*
>
> —LAO TSU

The determination of a generative system (or a set of admissible generative systems), as discussed in Chapter 3, is only the first theoretical stage in systems inquiries. New challenges arise when higher epistemological types of systems become involved. This chapter is devoted to problems that arise in connection with structure systems.

A *structure system*, loosely speaking, is a set of source, data, or generative systems that are based on the same support set. The systems that form a structure system are usually referred to as its *elements*. They may share some variables. The shared variables, which are usually called *coupling variables*, represent interactions among the elements.

Three epistemological types of structure systems are distinguished, depending on whether the elements are source, data, or generative systems. It is natural to refer to these three systems types as *structure source systems*, *structure data systems*, and *structure generative systems*, respectively. More specific types of structure systems, such as *structure image systems*, *structure behavior systems*, or *structure ST-systems*, may also be recognized for some purposes.

Given a structure system of one of these types, there is one system associated with it that is defined in terms of *all* variables included in its elements. Such a system, which is assumed to be of the same type as the elements of the structure system, is viewed as an *overall system*, i.e., a system which represents, as a whole, all the variables involved. According to this view, elements of any structure system are interpreted as *subsystems* of the associated overall system and, similarly, the overall system is interpreted as a *supersystem* of the elements. Then structure systems become basically representations of overall systems in terms of their various subsystems.

The status of a system as either an overall system or a subsystem is, of course, not absolute. A behavior system, for example, may be viewed in one context as an element of a structure system (and, consequently, a subsystem of an overall behavior system), while in another context it may be viewed as an overall system whose subsystems form a structure system. Each source, data or generative system thus plays a dual role. It assumes the status of a subsystem in one context and the status of a supersystem in another context. We may thus say not only that "a part is a whole in a role" (as suggested by Ranulph Glanville), but also that a whole is a part in a role. This duality makes it possible to represent each overall system by a *hierarchy of structure systems*, i.e., by a structure system whose elements are also represented by structure systems, whose elements are also . . . , etc., up to elements consisting of single variables. The nature of this hierarchy is concisely captured by Goguen and Varela[GO1]:

> At a given level of the hierarchy, a particular system can be seen as an outside to systems below it, and as an inside to systems above it; thus, the status (i.e., the mark of distinction) of a given system changes as one passes through its level, in either the upward or the downward direction. The choice of considering the level above or below corresponds to a choice of treating the given system as autonomous or controlled (constrained).

Why is it desirable to represent overall systems by collections of their subsystems? There are several reasons. One of them is concerned with observation or measurement. When the support involves time, it is often technically impossible or, at least, impractical to observe (measure) simultaneously all variables that are considered relevant to some purpose of investigation. There is no choice in such cases but to compromise and collect data piecemeal, for the largest possible subsets of the variables. In other cases, the investigator must rely on second-hand data, often gathered for different purposes by various agencies or individual researchers, each covering only a subset of the variables of concern to him.

Another reason why structure systems are desirable is connected with complexity and that, in turn, is connected with manageability of the system under consideration. One aspect of systems manageability is expressed in terms of the size of computer memory required to store the system. Consider, for example, n variables, each of which has k states. When dealing with the overall system of these variables, nk^n memory cells, each of which can store any one of k states, must be made available for storing states of the system. On the other hand, when a structure system consisting of all subsystems with two variables is used, the number of memory cells that are needed for the same purpose is $k^2n(n-1)$. This number grows with increasing values of k and n at a considerably lower rate than that of nk^n, as illustrated in Figure 4.1 for $k = 10$. If the structure system contained only some of the two-variable subsystems, the comparison would be even more favorable. Although for some small values of n and k, structure systems may require more memory space than the corresponding overall systems, it is clear that their memory requirements are far less demanding in most cases of practical significance, especially for large values of k and n.

Another aspect of systems manageability is connected with the number of possible systems that must be considered in some problems. To make a comparison between the number of overall systems and the number of structure systems of some kind, let us again consider n variables, each with k states. In addition, let us distinguish for each state of the system considered only whether or not it is possible. Then, there are

$$2^{k^n}$$

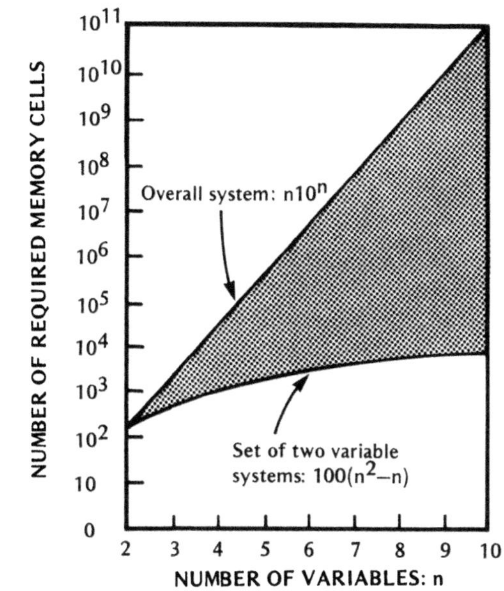

Figure 4.1. Comparison of memory requirements for an overall system and an associated structure system ($k = 10$).

possible overall systems,

$$n(n-1)2^{k^2-1}$$

possible structure systems consisting of all binary (two-variable) subsystems, and

$$n2^{k^2}$$

possible structure systems with only n binary subsystems. Although all these numbers are too large to allow an exhaustive search even for small values of n and k, the number of structure systems (in either of the two alternatives) grows at a considerably lower rate than the number of possible overall systems. For example, when $n = 10$ and $k = 2$, the number of structure systems with all binary subsystems is 720 while the number of all possible overall systems is 10^{308} (i.e., beyond the Bremermann limit discussed in Section 6.4). Hence, it is generally easier to search through the set of all structure systems of some kind rather than through the set of all possible overall systems, even though some restrictions are often unavoidable in either case.

There are numerous reasons why structure systems are desirable in engineering. Some of them are related to the manageability of the design process. These are basically variants of the reasons just discussed. Other reasons are associated with the availability of a limited inventory of prefabricated elements (modules) suitable for the given purpose, implementation efficiency, as well as various issues connected with reliability, testability, and maintainability of the system under design.

Practical reasons related to problems of manageability, efficiency, maintainability, and the like are not the only reasons why structure systems are desirable. In systems inquiries, structure systems have a more fundamental role. When properly justified, a structure system provides the investigators with some knowledge that is not available, at least explicitly, in the corresponding overall system. This additional knowledge may help him to answer certain questions associated with the investigation or, more generally, may help him to develop better insight into the problem.

Structure systems are at the center of one of the most controversial issues in philosophy—the issue of the relationship between *wholes* and *parts*. This issue can be traced not only to ancient Greek philosophy, but also to the much older Chinese philosophy of the *I Ching* and its various successors. It is well characterized in a paper by Archie Bahm[BA1]:

> No problem is more central to understanding the nature of existence, or knowledge, or values, or logic, than the problem of the nature and relations of a whole and its parts, and of wholeness and partiality.
>
> On the one hand, it is clear immediately that what we mean by "a part" is that it is "a part of a whole," and that what we mean by "a whole" is that it is "a whole of parts." Given these meanings, there are no parts which are not parts of a whole and no wholes which are not wholes of parts. Wholes and parts involve each other; each depends upon the other for being what it is, even though each is not the other. A part of a whole is not that whole, and a whole of parts is not one of its parts.
>
> However, difficulties in conceiving just how a whole and its parts are related to each other have given rise to theories which seem to deny, or at least modify, what is initially obvious. Some of these difficulties arise because there are different kinds of wholes and whole–part relations.

Goguen and Varela suggest four alternative criteria for the degree of wholeness of a system [GO1]:

> It is interesting to consider whether one can have a measure for the degree of wholeness of a system. One can, of course, always draw a distinction, make a mark, and get a "system" but the result does not always seem to be equally a "whole system", a "natural entity", or a "coherent object" or "concept". What is it that makes some systems more coherent, more natural, more whole, than others? . . . One point of view toward wholeness is that it co-occurs with interesting *emergent properties*. . . . Another point of view toward wholeness is that it can be measured by the difficulty of reduction . . . A third point of view is that a system is whole to the extent that its parts are *highly interconnected*, that is, to the degree that it is difficult to find relatively independent subsystems. . . . A fourth point of view is that a system seems more whole if it is more *complex*, that is, more difficult to reduce to descriptions as interconnections of lower level components.

The wholes–parts controversy in philosophy has been echoed by a controversy between two opposing methodological views in science—*reductionism* and *holism* (from Greek "holos" which means the adjective "whole"). Reductionism is based on the thesis that properties of a whole are explicable in terms of properties of constituent elements. Holism, on the other hand, rejects this thesis and claims that a whole cannot be analyzed without residue in terms of its parts. This claim is often expressed by the familiar statement "the whole is more than the sum of its parts," whose true author is likely to remain unknown.

Within the GSPS framework, the dichotomy of wholes and parts is manifested by the dual role of source, data, or generative systems as either supersystems or subsystems. The various issues regarding the relationship between wholes and parts, often surrounded by mysticism, can be clearly formulated as systems problems and studied accordingly. The two methodological doctrines—reductionism and holism—then emerge as complementary in a sense that is well described by Goguen and Varela [GO1]:

> Most discussions place holism/reductionism in polar opposition. This seems to stem from the historical split between empirical sciences, viewed as mainly reductionist or analytic, and the (European) schools of philosophy and social science that grope toward a dynamics of totalities.
>
> Both attitudes are possible for a given descriptive level, and in fact they are complementary. On the one hand, one can move down a level and study the properties of the components, disregarding their mutual interconnection as a system. On the other hand, one can disregard the detailed structure of the components, treating their behavior only as contributing to that of a larger unit. It seems that both these directions of analysis always coexist, either implicitly or explicitly, because these descriptive levels are mutually interdependent for the observer. We cannot conceive of components if there is no system from which they are abstracted; and there cannot be a whole unless there are constitutive elements . . .
>
> These descriptive levels haven't been generally realized as complementary largely because there is a difference between publicly announced methodology and actual practice, in most fields of research in modern science. A reductionist attitude is strongly promoted, yet the analysis of a system cannot begin without acknowledging a degree of coherence in the system to be investigated; the analyst has to have an intuition that he is actually dealing with a coherent phenomenon. Although science has publicly taken a reductionist attitude, in *practice* both approaches have always been active. It is not that one has to have a holistic view as opposed to a reductionist view, or *vice versa*, but rather that the two views of systems are complementary . . . . Reductionism implies attention to a lower level, while holism implies attention to a higher level. These are intertwined in any satisfactory description; and each entails some loss relative to our cognitive preferences, as well as some gain.

The same argument, that we need to be able to deal with both the parts and wholes of systems as our motivations dictate, is more concisely and poetically expressed by P. Suppes [SU4]:

> I am for the delicate dance from parts to wholes and back again. We should not be captured at either end. The dance should go forever.

Structure systems are associated with some of the most fundamental types of systems problems. These problem types are basically operational formulations, in the GSPS language, of the various issues regarding the relationship between wholes and parts. Some of them are connected with systems inquiry, others are involved in systems design; some of them are of a practical nature, others have theoretical significance or touch upon certain philosophical questions. The aim of this chapter is to define structure systems of various types and to discuss some of the key problems associated with them.

4.2. SYSTEMS, SUBSYSTEMS, SUPERSYSTEMS

> *The meaning of a whole and a part will concurrently exist in our mind only when we think about the relationship between them instead of about the things themselves.*
>
> —Amos Ih Tiao Chang

Systems of various types are introduced for three epistemological levels in Chapters 2 and 3. Given two systems of any one of these types, it is often desirable to determine whether they are related as whole and part. To make this possible, however, one must commit to some specific meaning of the whole–part relationship within the GSPS framework. Its choice is not arbitrary. It should reflect, in a satisfactory manner, our common-sense comprehension of such a relationship. This means, in turn, that the GSPS formulation of the whole–part relationship should be subject to some requirements through which its common sense meaning is adequately captured in terms of the GSPS language.

One of the obvious characteristics of the whole–part relationship is that the wholes and parts under consideration are compatible, i.e., they are things of the same sort. This leads to the requirement that systems involved in a whole–part relationship also be compatible in the same sense. To be compatible, systems must clearly be of the same type. In addition, our common sense requires that they be defined in terms of the same overall support set.

Systems compatibility is a necessary condition for the whole–part relationship among systems, but it is not a sufficient condition. Given two compatible systems, say **x** and **y**, our common sense would obviously accept **x** as a part of **y** only if **x** were totally included in **y** in some appropriate manner depending on the nature of the systems.

The requirements of *compatibility* and *inclusion* seem to capture the essence of the whole–part relationship of systems adequately. To keep the meaning of the relationship as general as possible, no additional requirements are desirable. It remains, of course, to define the whole–part relationship for source, data, and generative systems so that both of these requirements are satisfied.

First, let us introduce some convenient terminology and notation. Assume that system **x** is recognized as a part of system **y**. Then, let **x** be called a *subsystem* of **y**, or, alternatively, let **y** be called a *supersystem of* **x**. For formal purposes, let $\mathbf{x} \prec \mathbf{y}$ denote that **x** is a subsystem of **y** (and **y** is a supersystem of **x**).

Assume now that systems ${}^{x}\mathbf{S}$ and ${}^{y}\mathbf{S}$ are source systems. To qualify for the subsystem relationship (and its inverse—the supersystem relationship), they must be compatible. For source systems, it means that they must be of the same methodological type (i.e., based on the same methodological distinctions) and must be defined in terms of the same supports as well as the associated backdrops.

The inclusion requirement is expressed for source systems by several subset relationships: ${}^{x}\mathbf{S}$ is viewed as a *source subsystem* of ${}^{y}\mathbf{S}$ (assuming ${}^{x}\mathbf{S}$, ${}^{y}\mathbf{S}$ are compatible source systems) if and only if the set of variables (general as well as specific) and attributes of system ${}^{x}\mathbf{S}$ are subsets of the corresponding sets of system ${}^{y}\mathbf{S}$ and, accordingly, the families of state sets and appearance sets as well as the sets of observation and exemplification channels of system ${}^{x}\mathbf{S}$ are subsets of the corresponding families and sets of system ${}^{y}\mathbf{S}$. This collection of subset relationships, all of which must be satisfied to achieve a subsystem relationship, can be conveniently expressed in terms of a single index set. Its elements identify individual entities (variables, attributes, channels) in the various sets, and it is assumed that general variables, specific variables, and attributes corresponding to each other are labeled by the same element of the index set (as assumed in the formal definition of source systems in Chapter 2). Formally, let the variables, attributes, etc., of systems ${}^{x}\mathbf{S}$, ${}^{y}\mathbf{S}$ be labeled (identified) by elements of index sets ${}^{x}J$, ${}^{y}J$, respectively. Then, the subsystem relationship between ${}^{x}\mathbf{S}$ and ${}^{y}\mathbf{S}$ is completely described by the subset relationship

$${}^{x}J \subseteq {}^{y}J$$

between their index sets; it is normally assumed that

$${}^{y}J \subseteq N_n.$$

Example 4.1. Let ${}^{1}\mathbf{S}$ be the source system defined in Example 2.3 (a stand of hardwood timber). Then, ${}^{1}J = N_7$. Let ${}^{2}\mathbf{S}$ denote a source system that is defined as a subsystem of ${}^{1}\mathbf{S}$ (${}^{2}\mathbf{S} \prec {}^{1}\mathbf{S}$) by the index set ${}^{2}J = \{1, 2, 3, 7\}$. Then, ${}^{2}\mathbf{S}$ consists of all the entities that are included in ${}^{1}\mathbf{S}$ except v_i, V_i, $\dot{v}_i$, $\dot{V}_i$, a_i, A_i, o_i, e_i for $i = 4, 5, 6$.

► When dealing with directed source systems, the subsystem relationship is also reflected in the respective input/output identifiers. Let ${}^{x}\hat{\mathbf{S}}$, ${}^{y}\hat{\mathbf{S}}$ be directed source systems such that ${}^{x}\hat{\mathbf{S}} \prec {}^{y}\hat{\mathbf{S}}$, and let

$${}^{x}\mathbf{u} = ({}^{x}u(j) \mid j \in {}^{x}J),$$
$${}^{y}\mathbf{u} = ({}^{y}u(j) \mid j \in {}^{y}J)$$

be their input/output identifiers. Then,

$$^{x}u(j) = {}^{y}u(j)$$

for all $j \in {}^{x}J$. To keep track of values j associated with the individual components in the identifiers $^{x}\mathbf{u}$ and $^{y}\mathbf{u}$, it is convenient to assume for any input/output identifier $\mathbf{u}$ that elements $u(j)$ are ordered in increasing order of values j. ◄

The subsystem relationship defined for source systems can be easily extended to data systems. Clearly, given two compatible data systems $^{x}\mathbf{D}$, $^{y}\mathbf{D}$ whose source systems are $^{x}\mathbf{S}$, $^{y}\mathbf{S}$, respectively, $^{x}\mathbf{D}$ is said to be a *data subsystem* of $^{y}\mathbf{D}$, i.e.,

$$^{x}\mathbf{D} \prec {}^{y}\mathbf{D},$$

if and only if

i. $^{x}\mathbf{S} \prec {}^{y}\mathbf{S}$, and
ii. $^{x}\mathbf{D}$ contains only data that are contained in $^{y}\mathbf{D}$ and pertain to variables included in $^{x}\mathbf{S}$.

It is important that the data arrays involved be properly labeled to make the association of their entries with the individual variables unique.

Example 4.2. Let

$$^{1}\mathbf{D} = (^{1}\mathbf{S}, {}^{1}\mathbf{d}) \quad \text{and} \quad ^{2}\mathbf{D} = (^{2}\mathbf{S}, {}^{2}\mathbf{d})$$

denote, respectively, the data system defined in Example 2.6 (a blues tune) and a subsystem of $^{1}\mathbf{D}$ whose source system is defined by the index set $\{1, 2\}$ (i.e., we consider only pitch and rhythm, but not harmony). Then, $^{2}\mathbf{S}$ contains all the entities that are included in $^{1}\mathbf{S}$ except v_3, V_3, $\dot{v}_3$, $\dot{V}_3$, a_3, A_3, o_3, e_3, and the data $^{2}\mathbf{d}$ consist of the matrix in Figure 2.8 without the third row.

► It remains to define the subsystem relationship for the two versions of generative systems—behavior systems and ST-systems. Let

$$^{x}\mathbf{F}_B = (^{x}\mathbf{S}, {}^{x}M, {}^{x}f_B),$$
$$^{y}\mathbf{F}_B = (^{y}\mathbf{S}, {}^{y}M, {}^{y}f_B)$$

be two compatible behavior systems and let ^{x}J, ^{y}J denote the sets of identifiers of variables included in their source systems $^{x}\mathbf{S}$, $^{y}\mathbf{S}$, respectively. Then, $^{x}\mathbf{F}_B$ is said to be a *behavior subsystem* of $^{y}\mathbf{F}_B$, i.e.,

$$^{x}\mathbf{F}_B \prec {}^{y}\mathbf{F}_B,$$

if and only if the following three conditions are satisfied:

i. ${}^{x}J \subseteq {}^{y}J$ so that ${}^{x}\mathbf{S} \prec {}^{y}\mathbf{S}$;
ii. ${}^{x}M \subseteq {}^{y}M$ such that $(v_i, r_j) \in {}^{x}M$ iff $(v_i, r_j) \in {}^{y}M$ and $i \in {}^{x}J$;
iii. ${}^{x}f_B = [{}^{y}f_B \downarrow {}^{x}K]$, where ${}^{x}K$ denotes the set of identifiers of the sampling variables associated with the mask ${}^{x}M$, i.e., ${}^{x}f_B$ is a projection* of ${}^{y}f_B$ with respect to sampling variables of system ${}^{x}\mathbf{F}_B$.

To define

$${}^{x}\mathbf{F}_s = ({}^{x}\mathbf{S}, {}^{x}M, {}^{x}f_s)$$

as a *ST-subsystem* of a compatible ST-system

$${}^{y}\mathbf{F}_s = ({}^{y}\mathbf{S}, {}^{y}M, {}^{y}f_s),$$

conditions (i), (ii) remain the same, while condition (iii) must be replaced by the following, slightly modified condition:

iv. ${}^{x}f_s = [{}^{y}f_s \downarrow {}^{x}K^2]$, where $[{}^{y}f_s \downarrow {}^{x}K^2]({}^{x}\mathbf{c}, {}^{x}\mathbf{c}') = a(\{{}^{y}f_s({}^{y}\mathbf{c}, {}^{y}\mathbf{c}') \mid {}^{y}\mathbf{c} \succ {}^{x}\mathbf{c}, {}^{y}\mathbf{c}' \succ {}^{x}\mathbf{c}'\})$;

a denotes, as in Section 3.6, some aggregation function that is determined by the nature of the function ${}^{y}f_s$ (e.g., sum for probabilistic systems and the min function for possibilistic systems). ◀

Example 4.3. The behavior subsystem/supersystem relationship is illustrated by two behavior systems, ${}^{1}\mathbf{F}_B$ and ${}^{2}\mathbf{F}_B$, defined in Figure 4.2. Both systems are based on the same support set that is totally ordered. Their image systems ${}^{1}\mathbf{I}$, ${}^{2}\mathbf{I}$ contain two-state variables, v_1, v_2, v_3, v_4 and v_1, v_4, respectively. They may be associated with some interpretation (exemplification and observation channels), but it is of no significance for the purpose of this example. Behavior functions of the systems represent probabilistic distributions. We can see that system ${}^{2}\mathbf{F}_B$ satisfies the three conditions of a behavior subsystem with respect to system ${}^{1}\mathbf{F}_B$ and, consequently, ${}^{1}\mathbf{F}_B \succ {}^{2}\mathbf{F}_B$. Since the systems are probabilistic, the aggregation function involved in the projection $[{}^{1}f_B \downarrow \{1, 2, 6\}]$ is the sum function. For example, ${}^{2}f_B(000)$ is obtained by adding the first three probabilities in the table for ${}^{1}f_B$.

The notions of behavior subsystems and ST-subsystems can be easily modified to the other types of generative systems (generative behavior systems, directed behavior systems, etc.). It is solely a matter of convenient identification of the generating, generated, and input variables in both of the systems under consideration.

* The operation of projection is defined and explained in Section 3.6.

BEHAVIOR SYSTEM $^{1}\mathbf{F}_B$

$^{1}\mathbf{I}$: • two–state variables v_1, v_2, v_3, v_4
• totally ordered support set T

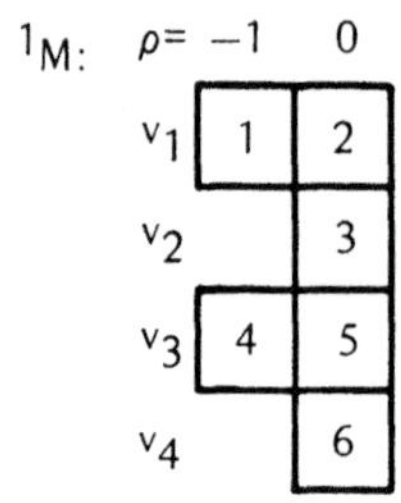

BEHAVIOR SYSTEM $^{2}\mathbf{F}_B$

$^{2}\mathbf{I}$: • two–state variables v_1, v_4
• totally ordered support set T

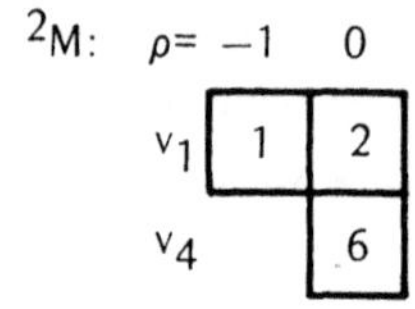

	s_1	s_2	s_3	s_4	s_5	s_6	$^{1}f_B(^{1}\mathbf{c})$
$^{1}\mathbf{c}=$	0	0	0	0	0	0	0.20
	0	0	0	0	1	0	0.05
	0	0	1	1	0	0	0.05
	0	1	0	0	0	0	0.05
	1	1	0	0	1	0	0.10
	1	1	1	0	0	0	0.05
	1	1	1	0	1	0	0.05
	1	1	1	1	0	0	0.10
	1	1	1	1	1	0	0.05
	1	1	1	1	1	1	0.30

Supersystem of $^{2}\mathbf{F}_B$

	s_1	s_2	s_6	$^{2}f_B(^{2}\mathbf{c})$
$^{2}\mathbf{c}=$	0	0	0	0.30
	0	1	0	0.05
	1	1	0	0.35
	1	1	1	0.30

Subsystem of $^{1}\mathbf{F}_B$

$^{2}\mathbf{F}_B \prec {}^{1}\mathbf{F}_B$

Figure 4.2. Example of the behavior subsystem/supersystem relationship (Example 4.3).

4.3. STRUCTURE SOURCE SYSTEMS AND STRUCTURE DATA SYSTEMS

> *Coming back to the general problem of a whole and its parts, we should realize that complication, hence also richness, of arguments concerning this problem stems at least partly from the fact that the same system can be divided into parts in many different ways.*
>
> —SATOSI WATANABE

As previously mentioned, structure systems are basically sets of source, data, or generative systems. Their purpose is to integrate several systems into larger systems. In order to achieve a meaningful integration, the individual systems—or elements of the structure system—must be compatible in the sense that they are of the same type and are defined in terms of the same support set. This is basically the same condition of *compatibility* as the one required for the subsystem relationship (Section 4.2).

In addition to the compatibility condition, it is also desirable to require that no element be a subsystem of another element in the same structure system. The purpose of this requirement is to avoid mixing of refinement levels in individual structure systems so that they can be hierarchically ordered, as hinted in Section 4.1. Furthermore, subsystems of any elements in a structure system are totally redundant in the sense that any information they contain is also contained in and derivable from the elements that are their supersystems. As such, they have no useful role in structure systems. Let this requirement be called an *irredundancy requirement.*

Redundant elements are often used in engineering systems to achieve various error detection or error correction capabilities. As explained in Section 4.4 (Example 4.8), the irredundancy requirement does not by any means exclude structure systems of that sort.

In order to define the various structure systems formally, let us assume that a neutral structure system consists of q elements (neutral systems of some type) that satisfy the compatibility and irredundancy requirements. Let the elements be identified by index x, where $x \in N_q$. In addition, let

$$V = \{v_i \mid i \in N_n\} \tag{4.1}$$

denote the set of all variables included in the elements, and let ${}^xV$ denote the set of variables in a particular element $x\,(x \in N_q)$. Then,

$$V = \bigcup_{x \in N_q} {}^xV. \tag{4.2}$$

For notational convenience, let variables in sets ${}^xV$ be identified by the same index i as the variables in the full set V, defined by (4.1). Then, each element can be uniquely identified by its sets of variables ${}^xV$.

Let each of the various types of structure systems be denoted by the standard symbol representing the type of its elements prefixed by an S. For instance, **SS**, **S$\hat{\mathbf{D}}$**, **SF**$_B$, **S$\hat{\mathbf{F}}$**$_{GS}$ would denote structure systems whose elements are neutral source systems, directed data systems, neutral behavior systems, and directed generative ST-systems, respectively. The prefix S is thus used as an operator which indicates that several systems of the specified type are integrated into a larger system.

The simplest type of structure systems has neutral source systems as elements. It is defined as the set

$$\mathbf{SS} = \{({}^xV, {}^x\mathbf{S}) \mid x \in N_q\}, \tag{4.3}$$

where ${}^x\mathbf{S}$ denotes for each $x \in N_q$ a neutral source system (an element of **SS**); ${}^xV$ denotes the set of variables included in ${}^x\mathbf{S}$ and is used as a convenient identifier of elements in the structure system. The source systems ${}^{x}S$ in (4.3) are required, of course, to satisfy the compatibility and irredundancy requirements, but no additional conditions are imposed.

If two elements in SS, say elements identified by $x, y \in N_q$, share some variables, i.e.,

$$ {}^{x}V \cap {}^{y}V \neq \varnothing, \tag{4.4}$$

we say that the elements are coupled. Let this set of shared variables be called a *coupling* between elements x and y, and let the variables in this set be called *coupling variables.* Couplings are significant traits of structure systems since they represent interactions between their elements. For neutral structure systems, couplings are clearly symmetric, i.e., independent of the order in which the elements are considered. For convenience, let the coupling between neutral elements x and y of a structure system be denoted by the symbol $C_{x,y}$, i.e.,

$$C_{x,y} = {}^{x}V \cap {}^{y}V. \tag{4.5}$$

Example 4.4. Consider a structure source system defined on a potted rosebush from the standpoint of the rosebush grower. The system is defined for the purpose of discovering ways of increasing the total yearly yield of the grower.

All variables involved in this system are based on two supports: a population of individual potted rosebushes that are under investigation, and time. Several populations are actually investigated in parallel, each characterized by some specific properties such as particular kinds of soil, fertilizer and pest control, picking frequency, etc. Observations are required to be made for all individuals in the population every second day for a period of one year; if desirable, the investigation will be extended to several more years.

Six parts are recognized on the whole object of investigation—an individual potted rosebush—as illustrated in Figure 4.3: soil, roots, stems, sap, leaves, blossoms. A source system is defined on each of these parts in terms of specific subsets of the following set of 19 variables (details of the observation and exemplification channels are omitted for the sake of simplicity):

v_1 (soil moisture)—low, medium, high;
v_2 (roots' water absorbing ability)—low, medium, high;
v_3 (roots' mineral absorbing ability)—low, medium, high;
v_4 (stem sap carrying ability)—poor, good;
v_5 (stem blossom bud density)—low, medium, high;
v_6 (stem leaf bud density)—low, medium, high;
v_7 (sap color substances)—low, medium, high;
v_8 (sap odor substances)—low, medium, high;
v_9 (sap growth substances)—low, medium, high;
v_{10} (number of leaves)—sparse, normal, excessive;

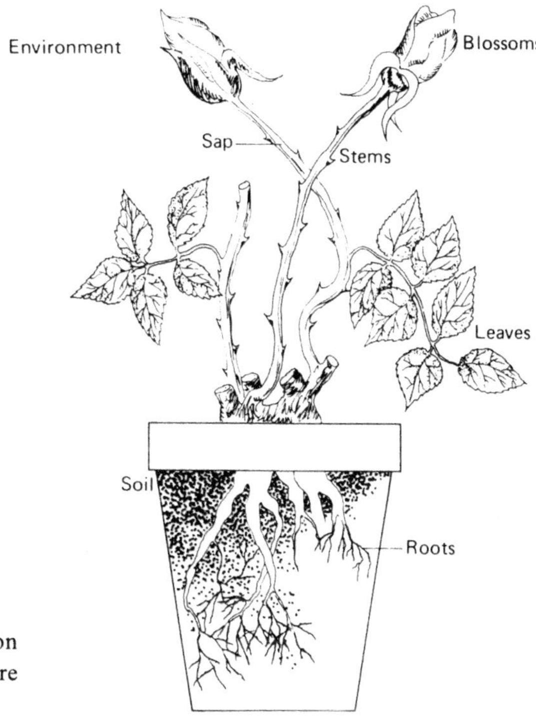

Figure 4.3. Parts of the rosebush on which elements of a structure system are defined (Example 4.4).

v_{11} (leaf color)—poor, good;
v_{12} (leaves sick)—stunted, normal;
v_{13} (blossom color)—pale, normal, intense;
v_{14} (blossom odor)—weak, normal, intense;
v_{15} (blossom size)—small, normal, gigantic;
v_{16} (number of blossoms)—sparse, abundant, profuse;
v_{17} (air temperature in °F)—below 60, 60–69, 70–79, 80–89, 90 or more;
v_{18} (rainfall)—below average, average, above average;
v_{19} (average sunlight in hours per day)—less than 3, 3–6, more than 6.

The six elements of the structure system are defined by the following subsets of this full set of variables;

$x = 1$ (soil)—v_1, v_{17}, v_{18};
$x = 2$ (roots)—v_1, v_2, v_3;
$x = 3$ (stems)—$v_2, v_3, v_4, v_5, v_6, v_{17}, v_{19}$;
$x = 4$ (sap)—$v_2, v_3, v_4, v_7, v_8, v_9, v_{17}, v_{19}$;
$x = 5$ (leaves)—$v_6, v_{10}, v_{11}, v_{12}, v_{17}, v_{18}, v_{19}$;
$x = 6$ (blossoms)—$v_5, v_{13}, v_{14}, v_{15}, v_{16}, v_{17}, v_{18}, v_{19}$.

Couplings between individual elements are easily obtained by taking the appropriate intersections of these sets. For example,

$$C_{1,2} = \{v_1\}, \qquad C_{2,5} = \varnothing, \qquad C_{3,4} = \{v_2, v_3, v_4, v_{17}, v_{19}\}, \quad \text{etc.}$$

Let a structure system $\mathbf{S}\hat{\mathbf{S}}$, whose elements are directed source systems, be defined as the set

$$\mathbf{S}\hat{\mathbf{S}} = \{({}^xX, {}^xY, {}^x\hat{\mathbf{S}}) \mid x \in N_q\}, \tag{4.6}$$

where ${}^xX$, ${}^xY$ denote the sets of input and output variables of element x, respectively. Clearly,

$${}^xX \cup {}^xY = {}^xV. \tag{4.7}$$

Except for distinguishing input and output variables, set (4.6) is quite similar to that defined by (4.3) for neutral structure systems **SS**. However, elements ${}^x\hat{\mathbf{S}}$ of any directed structure system $\mathbf{S}\hat{\mathbf{S}}$ are subject to one additional requirement regarding their input/output identifiers: none of the variables in the set V, as defined by Eq. (4.2), is allowed to be declared as an output variable in more than one of the elements. The reason for this requirement is to guarantee state consistency for all variables at each support instance. Indeed, if a variable were declared as an output variable in more than one element of a structure system, its states would be determined (controlled) at each support instance by all these elements, which would normally lead to inconsistencies (determination of several different states of the variable at the same support instance). Such inconsistencies would be avoided only if all the elements acted on the variable in unison, which is a singular and rather rare case. If it occurs, however, then any one of the elements is sufficient to control the variable and nothing is lost by the demand that only one of them (any one) may be declared as the controlling element. It is appropriate to refer to this requirement, which must be satisfied by all directed structure systems, as the requirement of *control uniqueness*.

The classification of variables of each element of a directed structure system into input and output variables and the requirement of control uniqueness have some important implications for the notion of couplings between elements. Given two elements x, y of a directed structure system, two *directed couplings* must be defined for them. One of the couplings, which is directed from x to y, is denoted by $\hat{C}_{x,y}$ and defined as

$$\hat{C}_{x,y} = {}^xY \cap {}^yX. \tag{4.8}$$

The other one is a coupling from y to x; it is denoted by $\hat{C}_{y,x}$ and defined as

$$\hat{C}_{y,x} = {}^yY \cap {}^xX. \tag{4.9}$$

Since

$$^{x}Y \neq {}^{y}Y \qquad \text{for } x \neq y$$

(due to the requirement of control uniqueness), clearly,

$$\hat{C}_{x,y} \neq \hat{C}_{y,x} \tag{4.10}$$

for different elements x, y.

In addition to the couplings between elements of a directed structure system, there are also couplings between the elements and environment of the system. For convenience, let the environment be viewed as a special element with the unique label $x = 0$. Although the environment is actually not an element of the structure system [as follows from (4.6)], this view enables us to define directed couplings $\hat{C}_{0,x}$ and $\hat{C}_{x,0}$ ($x \in N_q$) between the environment and elements of the structure system in the same manner as the couplings between the elements.

► If a variable is declared as an output variable in some element x of a directed structure system, then this variable is not controlled by the environment (due to the requirement of control uniqueness) and, consequently, it is not included in any $\hat{C}_{0,x}$. If, on the other hand, a variable is not declared as an output variable in any element of the directed structure system, then there is no choice but to view it as a variable that is controlled by the environment. Hence, such a variable must be included in some $\hat{C}_{0,x}$. It follows from these considerations that all variables in any ^{x}X that are not declared in any elements as output variables form the coupling from the environment to element x. Formally,

$$\hat{C}_{0,x} = {}^{x}X \cap \left(V - \bigcup_{y \in N_q} {}^{y}Y \right) \tag{4.11}$$

for each $x \in N_q$.

In order to characterize couplings $\hat{C}_{x,0}$ ($x \in N_q$), let us consider variables in ^{x}Y that are not declared as input variables in any element of the directed structure system under consideration. These variables are not included, by definition, in any of the couplings between elements of the structure system. Hence, there is no choice but to view them as being coupled to the environment, i.e., as being included in $\hat{C}_{x,0}$. The remaining variables may also be included in $\hat{C}_{x,0}$. Whether or not they are actually considered as being coupled to the environment is left to the discretion of the user. Formally

$$^{x}Y \cap \left(V - \bigcup_{y \in N_q} {}^{y}X \right) \subseteq \hat{C}_{x,0} \subseteq {}^{x}Y \tag{4.12}$$

for each $x \in N_q$. ◄

Example 4.5. A directed structure system with five source systems as its elements is defined for the purpose of studying the probation services offered by the State of New York for cases originating from complaints that are processed by criminal courts. The system characterizes the flow of the case workload through the criminal court and probation institutions. It represents a framework for data gathering and processing.

The support in this system is time. Depending on the specific questions to be addressed, observations are made on a monthly, weekly, or even daily basis starting on some fixed date, say January 1, 1970. All variables involved in the system (set V) are defined as follows:

v_1 —the total number of complaints received by the criminal court (during each individual period of observation—month, week, or day);
v_2 —the number of complaints that are carried toward the arraignment;
v_3 —the number of complaints that are dismissed;
v_4 —the number of cases that are held over for sentencing;
v_5 —the number of cases that are acquitted or discharged;
v_6 —the number of cases that are assigned for probation;
v_7 —the number of cases that are not assigned to probation (this includes cases where a fine or restitution is the only punishment, those where imprisonment is assigned, and those where an unconditional or conditional discharge is used);
v_8 —the number of cases which violate the conditions of probation;
v_9 —the number of cases that are discharged from probation;
v_{10}—the number of cases that are discharged from the criminal court institution.

The structure system under consideration consists of five elements. Their sets of input and output variables may be defined by Table 4.1 or, alternatively, by the block diagram in Figure 4.4. Blocks in the diagram, which are given some suggestive names, represent elements of the structure system and its environment. Connections between the blocks represent variables and indicate couplings between the elements (including

TABLE 4.1
Definition of Elements of a Directed Structure System Discussed in Example 4.5 (Equivalent to the Block Diagram in Figure 4.4)

x	${}^xX$	${}^xY$
1	$\{v_1\}$	$\{v_2, v_3\}$
2	$\{v_2\}$	$\{v_4, v_5\}$
3	$\{v_4, v_8\}$	$\{v_6, v_7\}$
4	$\{v_6\}$	$\{v_8, v_9\}$
5	$\{v_3, v_5, v_7, v_9\}$	$\{v_{10}\}$

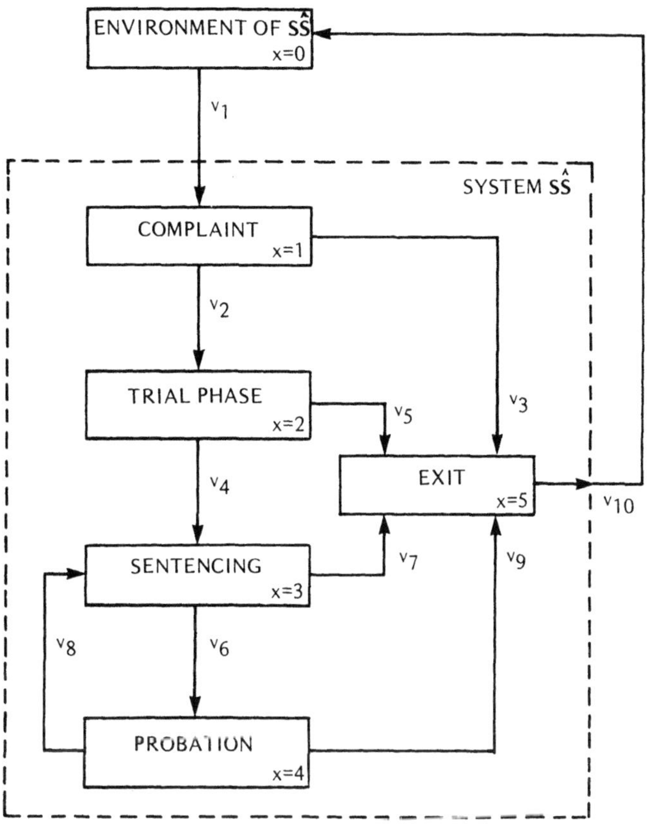

Figure 4.4. Block diagram of the structure system discussed in Example 4.5.

the environment). The full set of couplings $\hat{C}_{x,y}$ $(x, y = 0, 1, \ldots, q)$ is conveniently described by the matrix in Table 4.2. Such a matrix is usually called a *coupling matrix.*

Structure data systems (neutral and directed) are defined in a similar fashion as the structure source systems, namely,

$$\mathbf{SD} = \{[(^xV, {}^x\mathbf{D})|x \in N_q\} \tag{4.13}$$

$$\mathbf{S\hat{D}} = \{(^xX, {}^xY, {}^x\hat{\mathbf{D}})|x \in N_q\}. \tag{4.14}$$

Since each data system contains a source system, the structure data systems must satisfy all the conditions that are required for the corresponding source systems (compatibility, irredundancy, control uniqueness). However, they are normally also required to satisfy a condition of *local data consistency* that is defined as follows: for each coupling variable, data associated with that variable in all elements in which it is included must be exactly the same.

TABLE 4.2
Coupling Matrix for the Structure System Described in Example 4.5

$\hat{C}_{x,y}$	0	1	2	3	4	5
0	∅	$\{v_1\}$	∅	∅	∅	∅
1	∅	∅	$\{v_2\}$	∅	∅	$\{v_3\}$
2	∅	∅	∅	$\{v_4\}$	∅	$\{v_5\}$
3	∅	∅	∅	∅	$\{v_6\}$	$\{v_7\}$
4	∅	∅	∅	$\{v_8\}$	∅	$\{v_9\}$
5	$\{v_{10}\}$	∅	∅	∅	∅	∅

► Formally, if

$$v_i \in C_{x,y} \qquad (\text{or } v_i \in \hat{C}_{x,y}),$$

then

$$^x v_{i,\mathbf{w}} = {}^y v_{i,\mathbf{w}} \qquad \text{for all } \mathbf{w} \in \mathbf{W},$$

where $^x v_{i,\mathbf{w}}$, $^y v_{i,\mathbf{w}}$ ($\mathbf{w} \in \mathbf{W}$) are subsets of data associated with v_i in elements x and y, respectively. ◄

Structure data systems are normally assumed to be locally consistent. However, when the data sets associated with the individual elements are collected independently of each other, possibly by different experimental teams, the resulting family of data sets may not satisfy the condition of local data consistency. A violation of this condition leads to similar inconsistencies at higher epistemological levels and it is necessary to resolve them at some point in the investigation. Procedures for resolving local inconsistencies in structure systems have not been adequately developed as yet. It seems, however, that it is generally easier and more meaningful to resolve local inconsistencies in structure generative systems (introduced in Section 4.4) than in structure data systems. For this reason, the problem of resolving local inconsistencies is discussed in terms of structure generative systems.

4.4. STRUCTURE BEHAVIOR SYSTEMS

> *Every system has an author and the author pursues his own interests within the system.*
>
> —Walter von Lucadou and Klaus Kornwachs

Structure generative systems are defined in the same general form as are the other types of structure systems. They must satisfy the conditions of compatibility and irredundancy that are required for structure source systems. They must also satisfy some additional conditions regarding the masks and behavior or ST-functions of their elements. The fact that the set of sampling variables of a generative system is generally larger than the set of variables in the corresponding source or data systems opens some new possibilities in integrating generative systems into a structure system.

As shown in Section 3.7, every ST-system can be converted to an isomorphic behavior system. It is thus sufficient to discuss structure generative systems under the assumption that their elements are behavior systems, i.e., to discuss them, without any loss of generality, in terms of structure behavior systems.

Suppose that it is desirable to integrate a given set of behavior systems into a structure system. For each behavior system under consideration, identified as element x of the structure system ($x \in N_q$), let ${}^xV$ and ${}^xS$ denote the set of variables in its source system and the set of its sampling variables, respectively, and let

$$V = \bigcup_{x \in N_q} {}^xV = \{v_i \mid i \in N_{|V|}\}, \tag{4.15}$$

$$S = \bigcup_{x \in N_q} {}^xS = \{s_k \mid k \in N_{|S|}\}. \tag{4.16}$$

Clearly,

$$V \subseteq S \quad \text{and} \quad {}^xV \subseteq {}^xS \tag{4.17}$$

for all $x \in N_q$.

The neutral version $\mathbf{SF}_B$ of a *structure behavior system* can now be defined as

$$\mathbf{SF}_B = \{({}^xS, {}^x\mathbf{F}_B) \mid x \in N_q\}. \tag{4.18}$$

To obtain a unique identification of elements x by sets ${}^xS$, it is assumed that sampling variables in all sets ${}^xS$ ($x \in N_q$) are identified by the same index k that is employed as an identifier of the variables in the full set S [Eq. (4.16)]. Couplings $C_{x,y}$ between elements $x, y \in N_q$ of a structure system $\mathbf{SF}_B$ are defined now in terms of the sampling variables by the set intersection

$$C_{x,y} = {}^xS \cap {}^yS. \tag{4.19}$$

In the form (4.18), which does not contain any commitment to a generating order and the resulting partition of sampling variables in sets ${}^xS$ ($x \in N_q$) and S into generating and generated variables, the only additional condition that is normally required for $\mathbf{SF}_B$ is that the equation

$$[{}^xf_B \downarrow {}^xS \cap {}^yS] = [{}^yf_B \downarrow {}^xS \cap {}^yS] \tag{4.20}$$

be satisfied for all pairs of elements $x, y \in N_q$. This condition ensures that projections of behavior functions ${}^xf_B$, ${}^yf_B$ for every pair of elements in $\mathbf{SF}_B$ with respect to variables they share (coupling variables) are equal. This is basically a requirement that variables in different elements that are viewed (defined) as equal actually be equal when abstracted from the context of their respective elements. Let it be referred to as the requirement of *local behavior consistency*.

As previously mentioned, behavior systems ${}^x\mathbf{F}_B$ representing elements of $\mathbf{SF}_B$ are often derived in practical situations from locally inconsistent sets of data and, consequently, do not satisfy the requirement of local behavior consistency. In order to be able to deal with such structure systems in various problem contexts, their inconsistencies must be resolved first. This issue is discussed in Section 4.11. Everywhere else in this chapter, it is assumed that structure behavior systems are locally consistent.

When a generating order is specified for a structure behavior system $\mathbf{SF}_B$, the set S of all sampling variables in $\mathbf{SF}_B$ becomes partitioned into generating and generated variables, denoted $S_{\bar{g}}$ and S_g, respectively. The resulting structure system must satisfy the requirement of control uniqueness. Its meaning in this case is that each variable in S_g must be generated by one and only one of the elements of the structure system. This means, in turn, that sets ${}^xS_g$ $(x \in N_q)$ of generated variables associated with the individual elements of the structure system must form a partition of the set S_g.

Focusing on a particular element x $(x \in N_q)$ of a structure system $\mathbf{SF}_B$, consider now the set

$$({}^xS \cap S_g) - {}^xS_g$$

of variables. It is a set of generated variables (from the global point of view) that are coupled to element x, but are not generated by the element. From the local viewpoint of the individual elements, these are clearly input variables, even though they are generated variables from the global viewpoint of the structure system. This means that the element itself must be considered a directed system while, at the same time, the structure system is viewed as neutral in the sense that the variables in set V are not classified into input and output variables. This does not indicate any inconsistency on the definition of structure behavior systems. It is merely a result of the fact that structure systems encompass two coexisting veiwpoints—the local one (represented by their elements) and the global one (represented by structure systems as wholes). From the local point of view, all elements to which an element is coupled form its environment. It is a sort of *internal environment*, defined solely within the structure system. From the global point of view, no environment (or *external environment*) is recognized. It seems preferable in this case, however, to view the structure system as a directed system in which all variables in set V are declared as output variables.

It remains to discuss the role of the set $S_{\bar{g}}$ of generating variables in the individual elements of a structure system, i.e., the meaning of the variables in the sets

$$S_{\bar{g}} \cap {}^xS$$

for each $x \in N_q$. State of these variables must be made available within the element at each step of the generative process as required by its behavior function. They can be made available either *internally*—i.e., derived in the usual way from their previous states as well as previous states of generated and input variables, or *externally*—i.e., in terms of input variables representing couplings from other elements of the structure system. The variables are thus viewed either as generating variables or as input variables of the element. It is up to the user to decide between the two alternatives. In either case, let these variables be denoted by ${}^xS_{\bar{g}}$. A specification of their actual role is included in the definition of the behavior system representing the element as well as in the definition of directed couplings between the elements.

It follows from the previous discussion that generative behavior systems integrated into a structure system must often be viewed as directed systems. Since this view is always possible, we define

$$\mathbf{S}\hat{\mathbf{F}}_{\mathrm{GB}} = \{({}^xS_{\bar{g}}, {}^xS_e, {}^xS_g, {}^x\hat{\mathbf{F}}_{\mathrm{GB}}) \mid x \in N_q\}, \tag{4.21}$$

with the understanding that no meaningfull definition of $\mathbf{SF}_{\mathrm{GB}}$ is necessary.

This definition includes the special (degenerate) case in which none of the variables in V is declared as an input variable. Symbols ${}^xS_g$ and ${}^xS_e$ denote sets of generated and input variables of elements x, respectively; symbol ${}^xS_{\bar{g}}$ denotes the set of variables introduced and discussed in the previous paragraph. Depending on the alternative chosen, it may contain input or output variables of the element. As in the previous definition of a structure behavior system, it is assumed that variables in each of these three sets and all elements $x \in N_q$ are identified by the index k defined by Eq. (4.16). For each $x \in N_q$, the three sets of variables form a partition of the set ${}^xS$. In addition, for some $y \in N_q \cup \{0\}$, variables in set ${}^xS_g$ participate only in couplings $\hat{C}_{x,y}$, while those in set ${}^xS_e$ participate only in couplings $\hat{C}_{y,x}$; variables in ${}^xS_{\bar{g}}$ may participate in couplings of either direction or in no couplings at all.

Let structure systems of the form (4.18) or (4.21) be called structure behavior systems of the *basic type* and the *generative type*, respectively. Given a structure behavior system of the basic type, it is clear that a family of structure behavior systems of the generative type can be derived from it. Structure systems in this family differ from each other in

i. the partition of S into $S_{\bar{g}}$ and S_g;
ii. the partition of S_g into ${}^xS_g$ $(x \in N_q)$;
iii. the use of variables in sets ${}^xS_{\bar{g}}$ $(x \in N_q)$.

Partitions (i) are determined by the various generative orders (e.g., generative orders oriented to prediction or retrodiction). Partitions (ii) are determined by overlaps of variables in set S_g among the elements of the structure system. More specifically, a variable of set S_g that is included only in one of the elements must obviously be generated by that element. On the other hand, a variable that is shared by several elements can be generated by any one of them, but only one of them. Which one is chosen to generate it is normally decided on the basis of the generative uncertainty (the smaller the generative uncertainties resulting from each choice, the more preferable is the alternative) and, possibly, by some additional preference criteria specified by the user. Alternatives (iii) do not have any influence on the generative uncertainty of the chosen structure system. They are solely variations of the formal representation. The user should be given a chance to express his preference for a particular alternative; if he has no preference, one of the alternatives should be used as a default definition.

Example 4.6. Consider a structure system of the basic type (4.18) that consists of two subsystems based on the same totally ordered support set. Each of the elements contains two binary variables whose constraint is characterized by a probabilistic behavior function. Masks ${}^1M$, ${}^2M$ and behavior functions ${}^1f_B$, ${}^2f_B$ of the elements are specified in Figures 4.5a and 4.5b, respectively. The structure system clearly satisfies the irredundancy requirement. It is also locally consistent as demonstrated in Figure 4.5c. The full mask and block diagram, as well as the two partial masks, are illustrated in

(a)

ELEMENT $x = 1$

$\rho =$	-1	0
v_1	1	2
v_2	3	4

$= {}^1M$

ELEMENT $x = 2$

$\rho =$	-1	0
v_2	3	4
v_3	5	6

$= {}^2M$

(b)

s_1	s_2	s_3	s_4	${}^1f_B(\alpha)$
$\alpha = 0$	0	0	0	0.05
0	0	1	0	0.05
0	0	1	1	0.10
0	1	0	0	0.20
0	1	1	0	0.05
0	1	1	1	0.05
1	0	0	1	0.15
1	0	1	0	0.15
1	1	0	1	0.15
1	1	1	0	0.05

s_3	s_4	s_5	s_6	${}^2f_B(\beta)$
$\beta = 0$	0	0	0	0.05
0	0	1	0	0.10
0	0	1	1	0.10
0	1	0	0	0.05
0	1	0	1	0.20
0	1	1	0	0.05
1	0	0	0	0.05
1	0	0	1	0.10
1	0	1	0	0.10
1	0	1	1	0.05
1	1	1	0	0.05
1	1	1	1	0.10

(c)

s_3	s_4	$[{}^1f_B \downarrow \{s_3, s_4\}](\gamma)$	$[{}^2f_B \downarrow \{s_3, s_4\}](\gamma)$
$\gamma = 0$	0	0.25 (= 0.05 + 0.20)	0.25 (= 0.05 + 0.10 + 0.10)
0	1	0.30 (= 0.15 + 0.15)	0.30 (= 0.05 + 0.20 + 0.05)
1	0	0.30 (= 0.05 + 0.05 + 0.15 + 0.05)	0.30 (= 0.05 + 0.10 + 0.10 + 0.05)
1	1	0.15 (= 0.10 + 0.05)	0.15 (= 0.05 + 0.10)

Figure 4.5. Illustration to Example 4.6.

Figure 4.6a. The remaining block diagrams and masks in Figure 4.6 are characteristic examples of some of the generative types of structure systems that can be derived from this basic type.

To calculate the total number of substantially different structure systems of the generative type that can be derived in this example from the basic type, let us assume that the coupling to the environment contains all generated variables as well as any

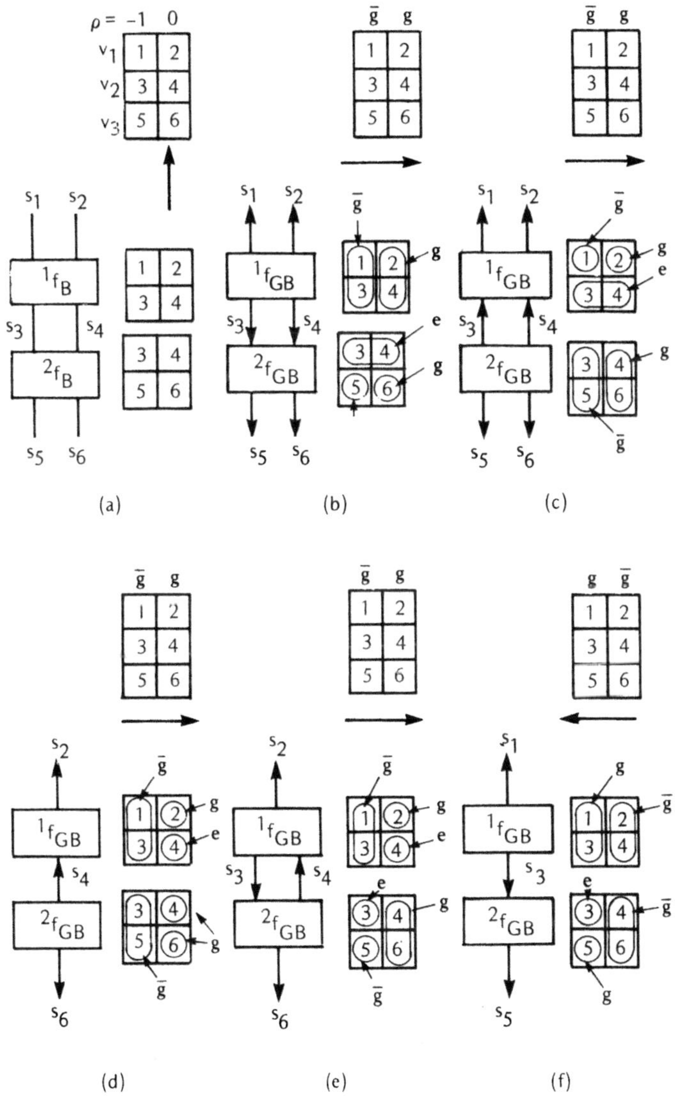

Figure 4.6. Basic type of a structure behavior system and some of the derived generative types (Example 4.6).

other variables that are considered as coupling variables. Then, there are 24 generative structure systems in this example for either of the two generative orders (prediction, retrodiction). They are obtained for the case of prediction by combining the following alternatives for some of the variables involved: two possibilities for variable s_1 and two possibilities for variable s_5 (these variables are either considered as coupling variables or not); three possibilities for variable s_3 (it is either not considered as a coupling variable at all, or else it has one of two possible directions); two possible directions of variable s_4. Counting both of the generative orders, there are thus 48 possible generative structure systems in this example. Further alternatives can be distinguished for each of them by different definitions of couplings to the environment.

Five examples of the 48 alternatives, which illustrate the main issues, are specified in Figures 4.6b–4.6f. For convenience, let us refer to them as systems b, c, . . ., f, respectively. The first four alternatives (systems b–e) are oriented to prediction, while the last one (system f) is oriented to retrodiction.

Structure systems b, c are similar in the sense that in both cases all sampling variables are incorporated in the various couplings. They differ in the role of variables s_3, s_4. In system b, variable s_4 is generated by the first element (in terms of the function ${}^1f_{GB}$, uniquely derivable from ${}^1f_B$ defined in Figure 4.5b) and both variables s_3, s_4 are used as input variables of the second element. In system c, the role of variables s_3, s_4 with respect to the elements is inverted. The two systems can be compared by values of their generative uncertainties associated with variable s_4. They are 0.2427 and 0.6754 for systems b and c, respectively. (Their calculation, which is explained in Section 3.5, is left to the reader as an exercise.) System b is thus preferable because it generates states of variable s_4 with considerably less uncertainty than system c (about 36% of the uncertainty of system c) and, consequently, it is a better predictor than system c.

System d is similar to system c in the sense that they both generate variable s_4 in the same way. They differ in the role of variables s_1, s_3, s_5 as coupling variables and in the formal definition of the first element. In spite of these formal differences, systems c and d generate data in basically the same way. The same holds also for system e. Its only difference from system d is that variable s_3 is used as input variable of the second element rather than its generating variable.

System f is one of the 24 retrodictive structure systems in this example. The main issue regarding these alternatives is to make a decision which of the two elements is preferable to generate variable s_3. The generative uncertainties are 0.8609 (for system f) and 0.9559 (for a system in which s_3 is generated by the second element). Although the generation of s_3 by the first element results in somewhat less uncertainty, the difference is much smaller than for the predictive case.

Example 4.7. To illustrate the flexibility in defining structure systems, assume that four different products—*a*, *b*, *c*, *d*—are manufactured by four different divisions of a manufacturer. Assume further that

i. to produce one unit of product *a* requires two units of product *b* and three units of product *c*;

ii. to produce one unit of product d requires two units of product a, one unit of product c, and four units of product b.

Each division is interested every day in how much of its own product it must manufacture. Let these quantities be represented by variables p_a, p_b, p_c, p_d. They are each determined by their order quantities and inventories and by the amount required by other divisions to manufacture their own products. Let the first two quantities be represented by variables o_a, o_b, o_c, o_d and i_a, i_b, i_c, i_d, respectively. The latter quantities are determined by the values p_a, p_b, p_c, p_d.

Let variables associated with each division (those with the same subscript) be viewed as variables that form an element of a structure system, together with the variables p_a, p_b, p_c, p_d that determine the amount of product of the given division that must be supplied to other divisions. Then, the structure system consists of four elements that are coupled according to the block diagram in Figure 4.7. Input variables of each element, which represent information about the order and inventory quantities, as well as the demands of the other divisions, are determined by the environment (sales division and warehouse) and by the other manufacturing divisions. Each element is a memoryless, deterministic, and directed behavior system whose support is time. Let us define behavior functions of the elements by the following simple equations:

$$
\begin{aligned}
{}^{a}f_B &: \; p_a = o_a - i_a + 2p_d, \\
{}^{b}f_B &: \; p_b = o_b - i_b + 2p_a + 4p_d, \\
{}^{c}f_B &: \; p_c = o_c - i_c + 3p_a + p_d, \\
{}^{d}f_B &: \; p_d = o_d - i_d.
\end{aligned}
$$

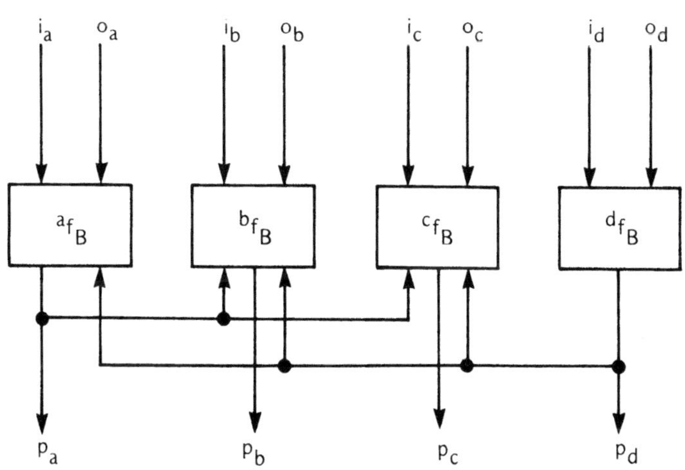

Figure 4.7. Structure system described in Example 4.7.

Example 4.8. The aim of this example is to show that the irredundancy requirement for structure systems is not in conflict with the notion of redundant systems, as used in engineering (particularly computer engineering) for error detection and correction. One of the simplest error correcting schemes consists of three systems that perform the same task in parallel by operating on the same input variables. Such a scheme is illustrated in Figure 4.8 by three serial binary adders operating on variables v_1 and v_2. States of their output variables v_3, v_4, v_5 are equal under normal operation, but one of them may differ from the other two when an accidental error occurs or when one of the units becomes defective. To recognize such situations and allow the whole system to continue its operation, states of the output variables of the three identical units are evaluated by two special systems. One of them is named "majority" in Figure 4.8. It selects the state of the output variables (0 or 1) that is represented by at least two of the variables v_3, v_4, v_5, i.e., the state of its output variable v_6 is equal to the state represented by the majority of its input variables. The aim of the second special subsystem, which is named "error message" in Figure 4.8, is to recognize any disagreement among states of variables v_3, v_4, v_5 and generate an error message in such cases. Its behavior can be described by the statement: $v_7 = 1$ (error message) iff all states of variables v_3, v_4, v_5 are not the same; otherwise, $v_7 = 0$ (normal operation). Error messages that occur rarely provide information about accidental errors; their frequent occurrence is an indicator that one of the three basic units (adders) is defective.

We can see that the irredundancy requirement is not violated by the structure system in Figure 4.8. Although the system contains redundant subsystems (two of the three adders), these are distinguished by their output variables and, under a more

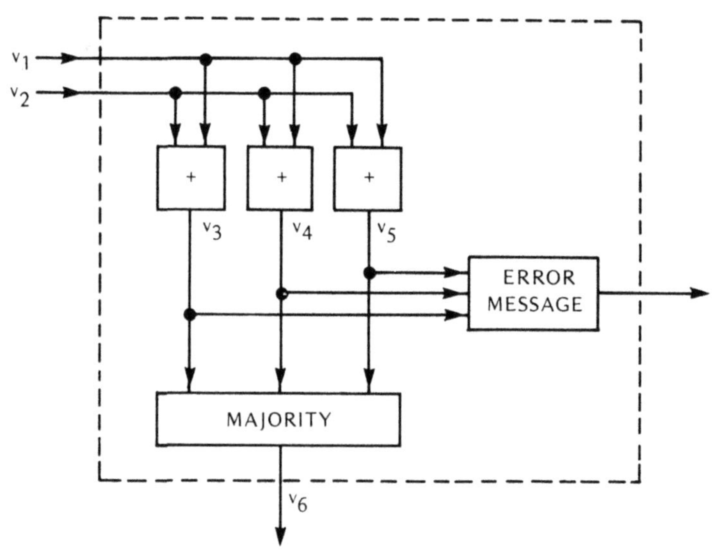

Figure 4.8. Simple error-correcting structure system (Example 4.8).

refined view, by their internal variables. The "engineering" redundancy is thus totally different from the redundancy that is prohibited by the irredundancy requirement. The former meaning of redundancy is that the same operation is executed by several distinct units in parallel; they are distinguished by different output (and internal) variables. The latter meaning of redundancy is that a structure system contains a system that is either indistinguishable from another system or is a subsystem of another system in the structure system.

Example 4.9. The structure system discussed in the previous example and depicted in Figure 4.8 is used in this example to illustrate levels of structure refinement. Let us look in more detail at one of the blocks in Figure 4.8—the serial binary adder. Its behavior is usually described by the equations

$$z_t = x_t + y_t + c_{t-1} \quad \text{(modulo 2)}, \tag{a}$$

$$c_t = (x_t + y_t + c_{t-1} - z_t)/2, \tag{b}$$

where x_t, y_t are states of the input variables at time t that represent digits of two binary numbers (ordered in time by their increasing significance), z_t is a state of the output variable (the sum digit) at time t, and $c_t\,(c_{t-1})$ is a state of an internal variable that represents the carry at time $t\,(t-1)$. According to this description, the binary adder is viewed as a structure system whose block diagram is shown in Figure 4.9a. Its elements, which represent Eqs. (a), (b), can be further refined and viewed also as structure systems. For instance, the element representing Eq. (b) can be represented as a structure system whose elements are standard logic functions of two variables (Figure 4.9b).

We can now see the meaning of the dual proccsses of *structure coarsening* and *structure refinement*. By structure coarsening, the error correcting structure system in Figure 4.8 becomes integrated as an clement in a larger structure system (an arithmetic unit). By further structure coarsening, the latter structure system becomes an element of a structure system still larger (a central processing unit of a computcr), etc., until some ultimate level of integration is reached for a particular purpose. By structure refinement, on the other hand, any element of the error correcting system (e.g., the adders) can be viewed as an appropriate structure system (e.g., the one depicted in Figure 4.9a), any element of the latter structure system can be again viewed as a structure system (Figure 4.9b), etc., until some ultimate level of refinement is reached for a specific purpose.

Two or more levels of structure refinement can also be incorporated in a single system, i.e., a structure system whose elements are structure systems, whose elements are structure systems, whose . . ., etc. This recursion must end, of course, with elements that are not structure systems.

Let structure systems that contain several refinement levels be called *multilevel structure systems* and let such systems be denoted by a generalized operator $\mathbf{S}^k$, where k indicates the number of refinement levels involved. For example, $\mathbf{S}^2\mathbf{F}_B$ denotes a two-level structure behavior system, i.e., a structure system whose elements are structure behavior systems.

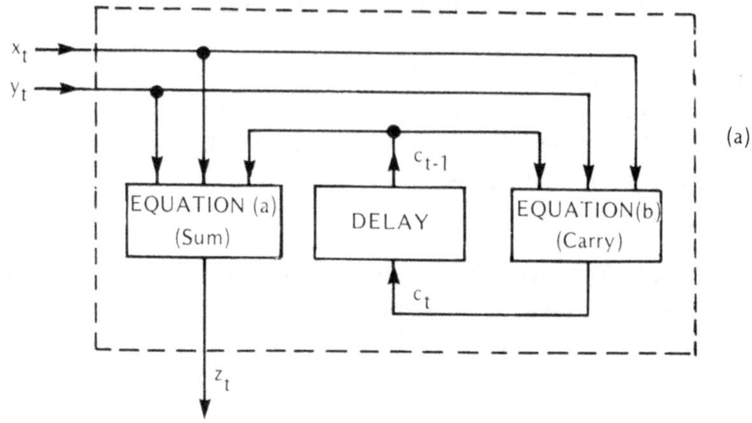

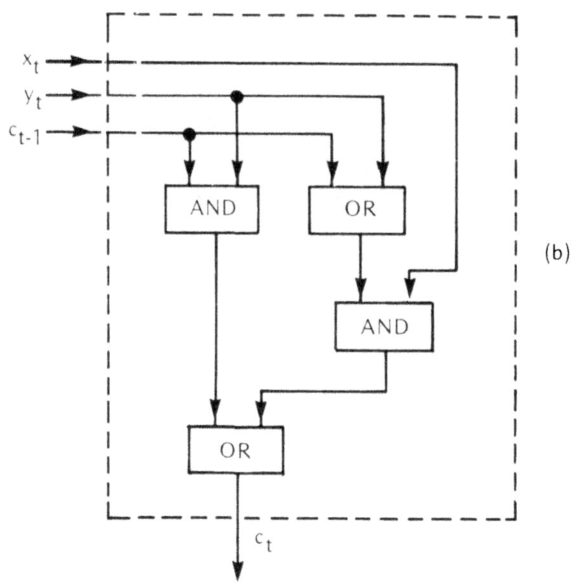

Figure 4.9. Illustration of structure refinement and coarsening (Example 4.9).

4.5. PROBLEMS OF SYSTEMS DESIGN

> *Hierarchy is one of the central structural schemes that the architect of complexity uses.*
>
> —HERBERT A. SIMON

Structure systems are involved in some of the most fundamental systems problems encountered in the contexts of both systems inquiry and systems design. In general, these problems are systems formulations of the various questions associated with the relationship between wholes and parts. The whole–part issues that arise in the context of systems inquiry are substantially different from those connected with systems design. In this section, the role of structure systems is outlined with respect to systems problems involved in systems design. Their role in systems inquiry, which is conceptually more intricate and methodologically less developed, is discussed in more detail in the rest of this chapter.

As discussed in Section 3.10, the first stage in systems design consists in deriving a generative system that represents the task the system under design is supposed to perform. In general, the task is some transformation from states of relevant input variables to states of output (or task) variables. Hence, the derived generative system is always directed. In addition, it is usually a deterministic system. It is often not unique, as illustrated by the following example.

Example 4.10. Consider again the serial binary adder introduced in Example 4.9. Its task is to add two binary numbers that are delivered in time, digit by digit, in increasing order of significance of their digits. The usual behavior system that is employed to represent this task is the one described in Example 4.9. It consists of the mask and behavior specified in Figure 4.10a. In addition to the input and output variables x, y, z, in terms of which the task is fully expressed, the system contains an internal variable c, which is known as the carry. An alternative behavior system for the same task, which does not contain any internal variable, consists of the mask and behavior specified in Figure 4.10b.

We can see that the two behavior systems are totally different, even though both of them perform the same task in terms of the required transformation from the input variables x and y to the output variable z. The differences, which can be recognized quite clearly by comparing their masks, imply necessary differences in structure by which the behaviors are implementable. For example, if we assume that the lagged variables (those defined by $\rho = -1$) are made available in the system directly by delay operators, the two behavior systems imply structure systems in Figures 4.9a and 4.10c, respectively. The two structure systems, which are indistinguishable from the standpoint of the environment (i.e., in their tasks), represent quite different bases for further design. One of them may be chosen by the user according to some preference criteria, or else the design process continues for both of them and a selection is made at some later stage.

After a particular generative system that represents the required task is selected, the aim of the next stage in the design process is to determine a structure system that satisfies the following requirements:

i. it implements the behavior or ST-function of the selected generative system;
ii. all of its elements are generative systems with some specified (admissible) behavior or ST-functions;
iii. it satisfies certain objective criteria specified as desirable;

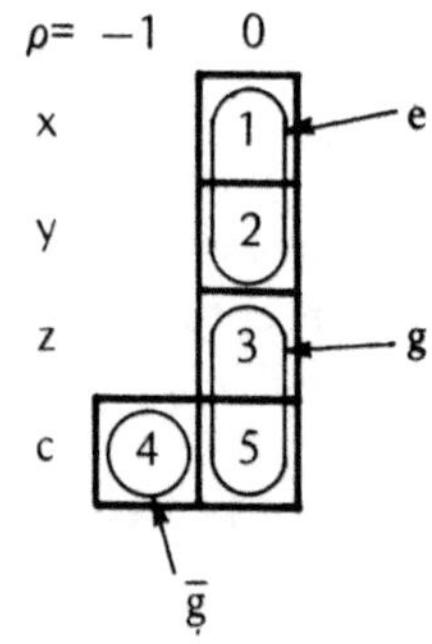

$s_3 = s_1 + s_2 + s_4 \pmod 2$
$s_5 = [s_1 + s_2 + s_4 - (s_1 + s_2 + s_4) \pmod 2]/2$

(a)

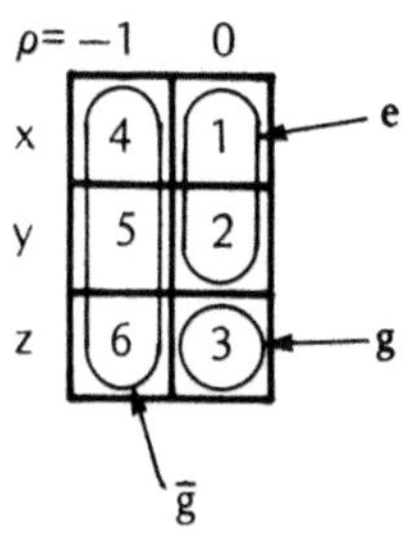

$s_3 = \{s_1 + s_2 + [s_4 + s_5 - s_6 + (s_4 + s_5 - s_6) \pmod 2]/2\} \pmod 2$

(b)

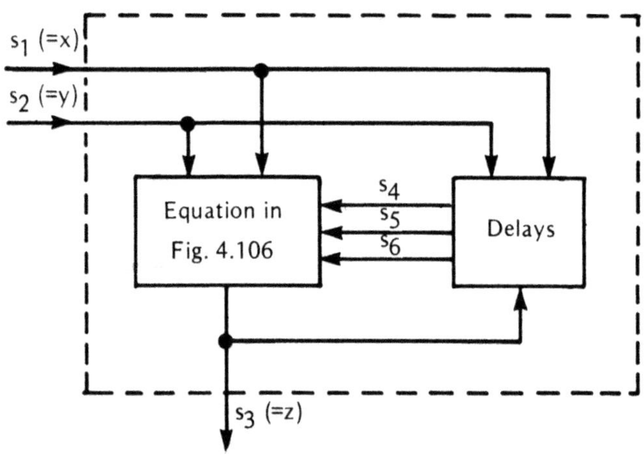

Figure 4.10. Two alternative behavior systems representing the serial binary adder (Example 4.10).

iv. it belongs to a specified class of structure systems (i.e., it satisfies some structure constraints);

Requirement (i) is obvious: since the structure system is supposed to perform the required task and the task is represented by the behavior or ST-function of the chosen generative system, this function must be, in turn, represented by the structure system. Requirement (ii) expresses the available technological resources. It is an inventory of all modules (components, building blocks) that are acceptable for constructing the structure system under design. It is important to make sure that the chosen types of elements are sufficient to implement the given generative system. Requirements (iii) and (iv) are optimization objectives and constraints, respectively. There is a great variety of possible objective criteria and constraints. They are often combinations of several factors connected with cost, complexity, regularity, response time, reliability, testability, maintainability, etc.

The problem of implementing the given behavior or ST-function by the specified types of elements is, in principle, the problem of finding a suitable decomposition (suitable with respect to the objectivity criteria and constraints) of the functions associated with the individual output variables of the given system into functions represented by the given types of elements.

► One approach to this problem, which is applicable only in special cases, is the use of suitable formal rules of an algebra whose operations correspond to the functions represented by the elements. Assuming the system under design is deterministic, this approach involves the following steps:

(a) An algebraic expression is determined for each of the functions associated with the individual output variables. It may, for example, be a convenient canonical form of some sort. These algebraic expressions represent a specific way of composing the functions of the system under design from functions associated with the elements. As such, they satisfy requirements (i) and (ii) of the design problem. If no objective criteria and constraints are required, which is unlikely, the design problem is completed at this point.

(b) The expressions are modified, by employing various rules of the algebra to a form which satisfies the objective criteria and constraints. In general, there are several solutions. It is usually sufficient to determine one of them.

Another approach to the decomposition problem, which is applicable to discrete systems, consists in performing the decomposition independently of any algebraic assumptions, by operating on direct definitions of the functions involved (in their tabular, matrix, or any other convenient form) or to employ appropriate functional equations. To illustrate this approach, let us assume, for the sake of simplicity, that each of the available elements has two input variables and one output variable. Symbols of the input and output variables of the behavior system under design and one of the elements are introduced in Figure 4.11a. It is assumed that each of the output variables is a function of the input variables, namely,

$$\begin{aligned} v_{n+1} &= f_{n+1}(v_1, v_2, \ldots, v_n) \\ v_{n+2} &= f_{n+2}(v_1, v_2, \ldots, v_n) \\ &\cdots \\ v_{n+m} &= f_{n+m}(v_1, v_2, \ldots, v_n) \end{aligned} \tag{4.22}$$

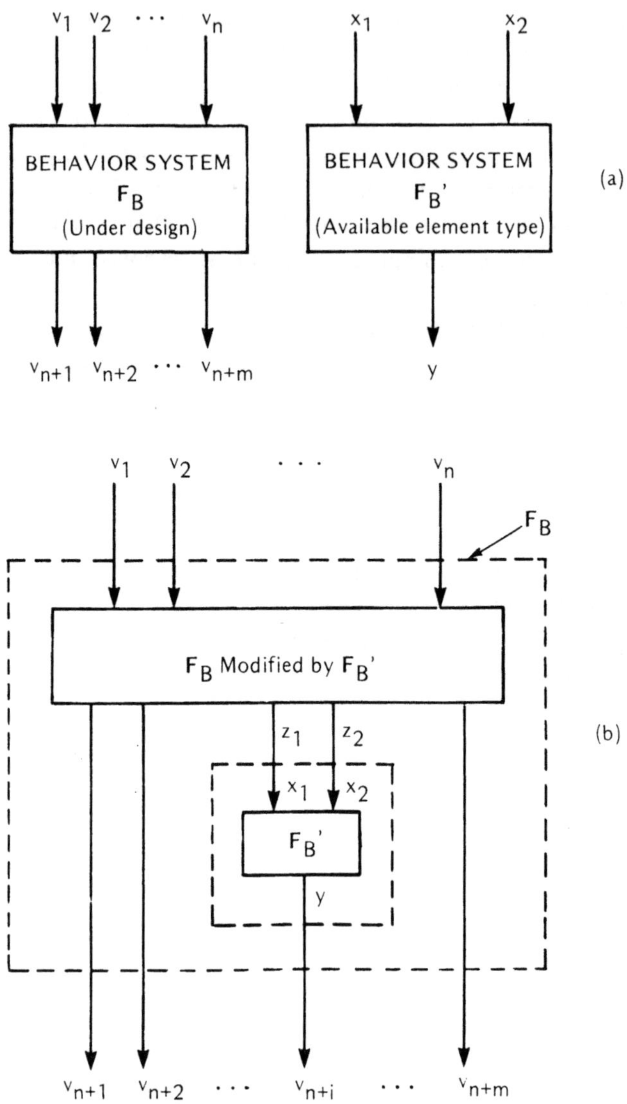

Figure 4.11. Illustration of the decomposition method of systems design.

for the system under design, and

$$y = g(x_1, x_2) \tag{4.23}$$

for the element. Imagine now that the element is incorporated into system F_B in such a way that its output variable is made identical with one of the output variables of F_B, say variable v_{n+i}. This leads to a structure system whose block diagram is shown in Figure 4.11b. It consists of two subsystems, one represented by the element, the other one by a modification of system F_B. The structure system contains two variables, z_1 and z_2, which were not included in the original system F_B. Since variables v_{n+i} and y are viewed in this structure system as identical, the functional equation

$$f_{n+i}(v_1, v_2, \ldots, v_n) = g(x_1, x_2) \tag{4.24}$$

must be satisfied. A solution of this equation consists of two functions

$$\begin{aligned} x_1 &= h_1(v_1, v_2, \ldots, v_n), \\ x_2 &= h_2(v_1, v_2, \ldots, v_n). \end{aligned} \tag{4.25}$$

To qualify as a solution of Eq. (4.24), these functions must, of course, satisfy the equation when substituted for variables x_1, x_2. Since variables x_1, x_2 are in this case viewed as identical with the new variables z_1, z_2, respectively (see Figure 4.11b), we can rewrite Eqs. (4.25) as

$$\begin{aligned} z_1 &= h_1(v_1, v_2, \ldots, v_n) \\ z_2 &= h_2(v_1, v_2, \ldots, v_n). \end{aligned} \tag{4.26}$$

There are usually many solutions of the functional equation (4.24). The main problem is thus to select one of them. First, the set of solutions is reduced by excluding solutions that violate the specified structure constraints. Second, we search for solutions in which functions h_1, h_2 are dependent on the least number of variables $v_1, v_2, \ldots, v_n$. Indeed, the smaller this number, the more powerful is the decomposition and the easier it is to further decompose functions h_1, h_2, if necessary. The decomposition is particularly effective if the functions h_1, h_2 are dependent on disjoint subsets of those input variables on which the function under decomposition depends (function f_{n+i} in our illustration). Third, the remaining solutions are ordered by the objective criteria and those which are not inferior in all criteria (or a subset of them) are accepted as a basis for further decomposition.

The decomposition step illustrated in Figure 4.11b must be repeated for all output variables $v_{n+1}, v_{n+2}, \ldots, v_{n+m}$ and, if necessary, for the new variables $z_1, z_2, \ldots$, which are introduced by the decomposition process. No further decomposition is needed at those decomposition locations at which all new variables become identical with some of the input variables $v_1, v_2, \ldots, v_n$. It is clear that different elements must be tried at each decomposition step and their acceptable decompositions compared.

The repeated decomposition is illustrated in Figure 4.12. For the sake of simplicity, only the type of element introduced in Figure 4.11a is employed. Figure 4.13 illustrates some efficient types of decompositions of a system with a single output variable by elements with two input variables and one output variable. Included are all those

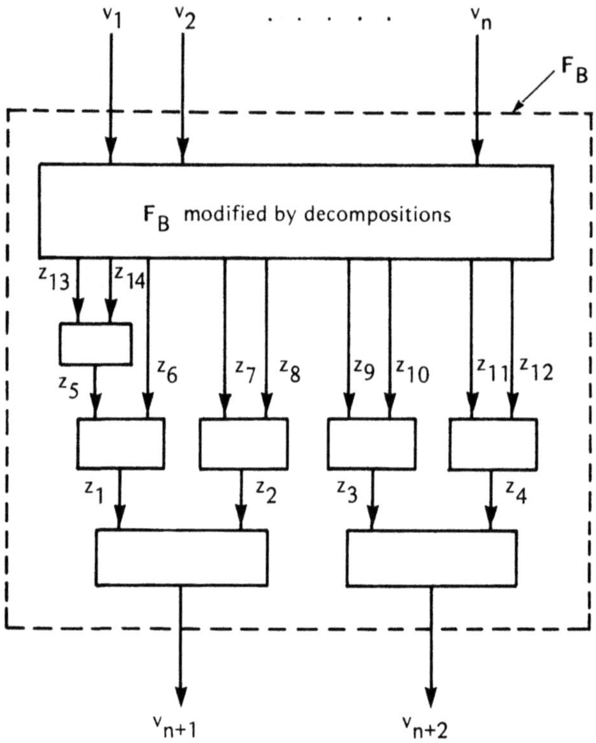

Figure 4.12. Illustration of a possible situation after seven decomposition steps.

decomposition types for $n = 3, 4, 5, 6$ in which the new variables in each decomposition step are dependent on disjoint subsets of the input variables inherited from the preceding decomposition. These are the most desirable decomposition types. Numerals in Figure 4.13 indicate the number of input variables on which each of the output or intermediate variables depends. ◄

It is not hard to see that the computational complexity of the design problem grows extremely rapidly with the number of input and output variables of the system to be designed, as well as with the number of admissible element types. For example, with n input variables, m output variables, and one element type with two input variables and one output variable, the number of decomposition steps of the efficient type illustrated in Figure 4.13 is equal to the product $(n-1)m$, while it is equal to $e^{(n-1)m}$ for e element types ($e \geqslant 2$). Each decomposition is associated, of course, with solving an appropriate functional equation, evaluating and comparing its solutions, and selecting some of them as prospective candidates for the overall design.

There are two principal ways of making a complex design problem manageable.

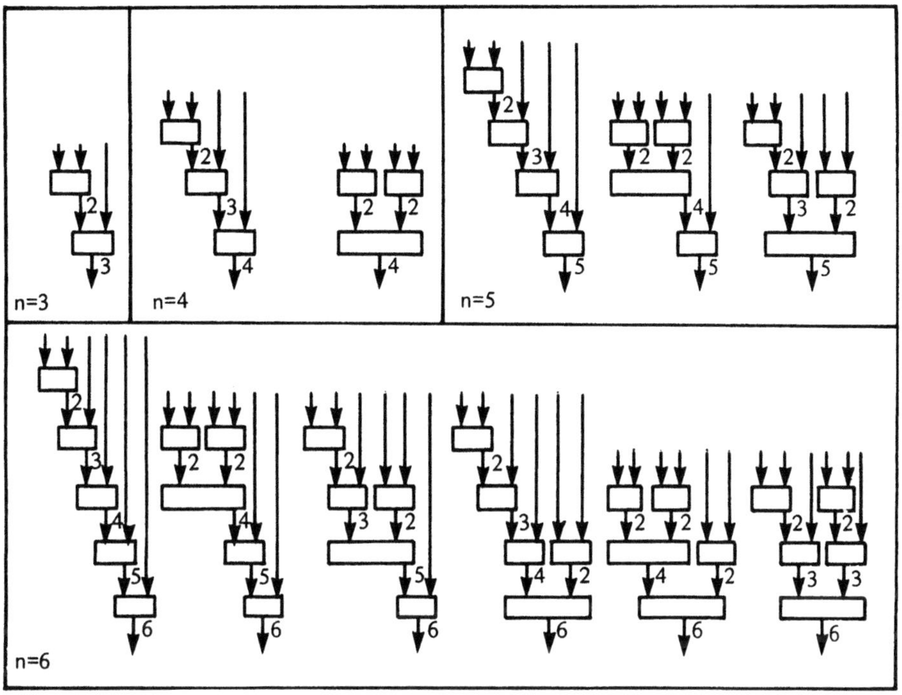

Figure 4.13. Examples of desirable decompositions based on elements with two input variables and one output variable.

One of them is to organize the overall task as a hierarchy of partial tasks in such a manner that the design complexity is modest at each level of the hierarchy. Such an organization usually has some additional advantages connected with reliability, testability, and repairability. The other way of reducing design complexity is to relax the objective criteria. Instead of requiring an optimal design, we accept a "good" one or a "satisfactory" one. This attitude toward systems design, which allows the use of heuristic methods, is called *satisficing*.

Due to the great variety in possible element types, it would be a difficult task to incorporate into the GSPS the various specialized design methods, based on specific algebras or specific types of functional equations. In this respect, the GSPS should serve as an information source. If some specialized method is available for the design problem defined by the user, the GSPS should provide the user with adequate information about it. The facilities of the GSPS for dealing with design problems at the level of structure systems should be developed in terms of the general decomposition approach, as outlined in this section, and in both the optimization sense (for small problems) and the satisficing sense (based on appropriate heuristic methods).

4.6. IDENTIFICATION PROBLEM

> *. . . given the properties of parts and the laws of their interaction, it is not a trivial matter to infer the properties of the whole.*
>
> —HERBERT A. SIMON

Two complementary problems associated with the relationship between overall behavior systems and the various sets of their subsystems arise prominently in systems inquiries. One of them is based on the assumption that a behavior system, which is viewed as an overall system, is given. The aim is to determine which structure systems, each based on a set of subsystems of the given overall systems, are adequate for reconstructing the given overall system with an acceptable level of approximation. In the second problem, a structure behavior system is given and the aim is to make inferences about the unknown overall system.

These problems have been referred to in the literature as the *reconstruction problem* and *identification problem*, respectively. The identification problem is a subject of this section; the reconstruction problem is discussed in the next section.

The identification problem involves two subproblems. The aim of one of them is to determine the set of all overall behavior systems that are represented by the given structure system in the sense that behavior functions of its elements are projections of the behavior function of any of these overall systems. Such a set of overall systems is called the *reconstruction family* of the structure system under consideration. The aim of the second subproblem is to choose one overall system from the reconstruction family that represents, in some specific sense, the best hypothesis of the actual overall system.

RECONSTRUCTION FAMILY

Consider a structure behavior system **SF** of the form (4.18)* whose elements are represented by sets ${}^xS$ of sampling variables and behavior functions ${}^xf\,(x \in N_q)$. We say that a behavior system is comparable with the given structure system **SF** if both systems are defined in terms of the same supports and variables, and both employ the same type of behavior functions (such as probability or possibility distribution functions). Let $\mathscr{C}_{\text{SF}}$ denote the set of behavior functions of all behavior systems that are comparable with **SF** and let $\mathscr{F}_{\text{SF}}$ denote the set of behavior functions of all behavior systems in the reconstruction family of **SF**. Then, $f \in \mathscr{F}_{\text{SF}}$ if and only if $f \in \mathscr{C}_{\text{SF}}$ and

$$[f \downarrow {}^xS] = {}^xf \tag{4.27}$$

* Since no confusion can arise in the rest of this chapter, subscripts B are omitted in symbols associated with behavior systems.

for all $x \in N_q$. For probability or possibility behavior functions, Eq. (4.27) is expressed by a set of equations

$$^{x}f(^{x}\mathbf{c}) = \sum_{\mathbf{c} \succ ^{x}\mathbf{c}} f(\mathbf{c}) \tag{4.28}$$

or

$$^{x}f(^{x}\mathbf{c}) = \max_{\mathbf{c} \succ ^{x}\mathbf{c}} f(\mathbf{c}), \tag{4.29}$$

respectively, where $x \in N_q$. Values $^{x}f(^{x}\mathbf{c})$ are given in these equations; values $f(\mathbf{c})$ are to be determined for all overall states of the variables involved. Let $\mathbf{c} \in \mathbf{C}$ and $^{x}\mathbf{c} \in {}^{x}\mathbf{C}$ for all $x \in N_q$.

To obtain values $f(\mathbf{c})$ that are acceptable as probabilities or possibilities for states $\mathbf{c}$, Eqs. (4.28) or (4.29) must be supplemented with (constrained by) the requirement that $f(\mathbf{c}) \geq 0$ for all $\mathbf{c} \in \mathbf{C}$. Although $f(\mathbf{c})$ must also satisfy some additional requirements, such as $f(\mathbf{c}) \leq 1$ and

$$\sum_{\mathbf{c} \in \mathbf{C}} f(\mathbf{c}) = 1$$

for probabilistic systems, it is easy to see that these additional requirements are in fact implied by the form of the equations and the requirement that $f(\mathbf{c}) \geq 0$ for all $\mathbf{c} \in \mathbf{C}$.

The relationship between a given structure system and the overall behavior systems it implies is thus described by a set of

$$\sum_{x \in N_q} |^{x}\mathbf{C}|$$

equations of either the form (4.28) or form (4.29) in $|\mathbf{C}|$ unknowns, $f(\mathbf{c})$, that are constrained by the inequalities $f(\mathbf{c}) \geq 0$ for all $\mathbf{c} \in \mathbf{C}$. Some of the equations are usually dependent on others and may be excluded from the set. If the given structure system is consistent, then the resulting set of essential equations [e.g., linearly independent equations of the form (4.28)] has at least one solution in the required domain of nonnegative real numbers.

The identification of the overall behavior system from the given structure system is unambiguous if and only if the solution to the constrained set of equations exists (i.e., the structure system is consistent) and is unique. This seems to be a rather rare case. If the solution is not unique, which is considerably more frequent, the identification is ambiguous. This means basically that the actual overall system, while embedding all information regarding the constraint among the variables involved that is available in the structure system, contains some additional information. This fact gives a concrete meaning to the notorious claim of systems science that "the whole is more than the sum of its parts."

TABLE 4.3
Elements of the Structure System in Example 4.11

v_1	v_2	${}^1f({}^1\mathbf{c})$	v_2	v_3	${}^2f({}^2\mathbf{c})$	v_1	v_3	${}^3f({}^3\mathbf{c})$
${}^1\mathbf{c} = 0$	0	0.4	${}^2\mathbf{c} = 0$	0	0.4	${}^3\mathbf{c} = 0$	0	0.4
0	1	0.3	0	1	0.2	0	1	0.3
1	0	0.2	1	0	0.1	1	0	0.1
1	1	0.1	1	1	0.3	1	1	0.2

Example 4.11. Consider a structure system whose elements are memoryless behavior systems, each containing two of three variables v_1, v_2, v_3. Two states, 0 and 1, are recognized for each of these variables. Behavior functions ${}^1f$, ${}^2f$, ${}^3f$ of the elements are probabilistic and are defined in Table 4.3. The structure system is locally consistent, as we can easily verify. For instance, the probability of $v_2 = 0$ is 0.6 (and 0.4 for $v_2 = 1$) regardless whether it is calculated as a projection from ${}^1f$ or ${}^2f$.

Let symbols $p_0, p_1, \ldots, p_7$, defined in Table 4.4, be used for the unknown probabilities of states of the overall behavior systems. Then the reconstruction family of the given structure system is characterized by the inequalities $p_i \geq 0$ $(i \in N_{0,7})$ and 12 equations of the form (4.28):

$$
\begin{array}{llllll}
p_0 + p_1 = 0.4 & (1), & p_0 + p_4 = 0.4 & (5), & p_0 + p_2 = 0.4 & (9),\\
p_2 + p_3 = 0.3 & (2), & p_3 + p_7 = 0.3 & (6), & p_1 + p_3 = 0.3 & (10),\\
p_4 + p_5 = 0.2 & (3), & p_1 + p_5 = 0.2 & (7), & p_5 + p_7 = 0.2 & (11),\\
p_6 + p_7 = 0.1 & (4), & p_2 + p_6 = 0.1 & (8), & p_4 + p_6 = 0.1 & (12).
\end{array}
$$

TABLE 4.4
Symbols Used in Examples 4.11, 4.12, and 4.14

v_1	v_2	v_3	$f(\mathbf{c})$
$\mathbf{c} = 0$	0	0	$f(000) = p_0$
0	0	1	$f(001) = p_1$
0	1	0	$f(010) = p_2$
0	1	1	$f(011) = p_3$
1	0	0	$f(100) = p_4$
1	0	1	$f(101) = p_5$
1	1	0	$f(110) = p_6$
1	1	1	$f(111) = p_7$

By inspecting the equations, we observe that all the unknowns can be expressed in terms of one of them, say p_0. Indeed,

from (1): $p_1 = 0.4 - p_0$;
from (5): $p_4 = 0.4 - p_0$;
from (9): $p_2 = 0.4 - p_0$;
from (2): $p_3 = 0.3 - p_2 = -0.1 + p_0$;
from (3): $p_5 = 0.2 - p_4 = -0.2 + p_0$;
from (6): $p_7 = 0.3 - p_3 = 0.4 - p_0$;
from (8): $p_6 = 0.1 - p_2 = -0.3 + p_0$.

Hence,

$$p_1 = p_2 = p_4 = p_7 = 0.4 - p_0, \quad \text{(a)}$$
$$p_3 = -0.1 + p_0, \quad \text{(b)}$$
$$p_5 = -0.2 + p_0, \quad \text{(c)}$$
$$p_6 = -0.3 + p_0. \quad \text{(d)}$$

Applying now the inequalities, we obtain for each of these equations a restriction upon p_0:

(a) and p_1 (or p_2, p_4, p_7) ≥ 0 implies $p_0 \leq 0.4$;
(b) and $p_3 \geq 0$ implies $p_0 \geq 0.1$;
(c) and $p_5 \geq 0$ implies $p_0 \geq 0.2$;
(d) and $p_6 \geq 0$ implies $p_0 \geq 0.3$.

Since all of these restrictions must be satisfied, we may conclude that the solution range of p_0 is given by the inequalities

$$0.3 \leq p_0 \leq 0.4.$$

Given any value of p_0 within this range, values of all the other unknowns are uniquely determined by the equations (a)–(d). The reconstruction family can thus be defined as indicated in Table 4.5.

Example 4.12. Consider a structure system whose elements are memoryless behavior systems containing two-state variables v_1, v_2 and v_2, v_3, respectively. Behavior functions ${}^1f$, ${}^2f$ of the elements are probabilistic and are defined by the following tables:

	v_1	v_2	${}^1f({}^1\mathbf{c})$
${}^1\mathbf{c} =$	0	1	0.3
	1	0	0.5
	1	1	0.2

	v_2	v_3	${}^2f({}^2\mathbf{c})$
${}^2\mathbf{c} =$	0	0	0.1
	0	1	0.4
	1	1	0.5

TABLE 4.5
Reconstruction Family in Example 4.11

v_1	v_2	v_3	$p_i = f(\mathbf{c})$
$\mathbf{c} = 0$	0	0	$0.3 \leqslant p_0 \leqslant 0.4$ (degree of freedom)
0	0	1	$p_1 = 0.4 - p_0$
0	1	0	$p_2 = 0.4 - p_0$
0	1	1	$p_3 = -0.1 + p_0$
1	0	0	$p_4 = 0.4 - p_0$
1	0	1	$p_5 = -0.2 + p_0$
1	1	0	$p_6 = -0.3 + p_0$
1	1	1	$p_7 = 0.4 - p_0$

To determine the reconstruction family of this structure system, let us use the same symbols for the unknown probabilities as in Example 4.11. Then, the reconstruction family is characterized by the inequalities $p_i \geq 0$ $(i \in N_{0,7})$ and the following eight equations:

$$
\begin{aligned}
p_0 + p_1 &= 0.0 \quad (1) \qquad & p_0 + p_4 &= 0.1 \quad (5) \\
p_2 + p_3 &= 0.3 \quad (2) \qquad & p_1 + p_5 &= 0.4 \quad (6) \\
p_4 + p_5 &= 0.5 \quad (3) \qquad & p_2 + p_6 &= 0.0 \quad (7) \\
p_6 + p_7 &= 0.2 \quad (4) \qquad & p_3 + p_7 &= 0.5 \quad (8)
\end{aligned}
$$

From (1), (7), and the inequalities, we obtain $p_0 = p_1 = p_2 = p_6 = 0$. Then, by trivial considerations, the remaining unknown probabilities are determined: $p_3 = 0.3$, $p_4 = 0.1$, $p_5 = 0.4$, $p_7 = 0.2$. Hence, the identification is in this case unique. That is, the example illustrates one of the rare special cases in which "the whole *is equal* to the sum of its parts."

Example 4.13. To illustrate a more general kind of reconstruction family, let a structure system be given with three elements, each containing two of three variables v_1, v_2, v_3. Let v_1 and v_2 each take states from the set $\{0, 1\}$ and let v_3 take states from the set $\{0, 1, 2\}$. The elements are memoryless and probabilistic behavior systems whose behavior functions ${}^1f$, ${}^2f$, ${}^3f$ are given in Table 4.6. Let symbols defined in Table 4.7 be used for the unknown probabilities of states of the overall system.

Leaving the formulation of the equations that characterize the reconstruction family and the determination of their solution within the constraints $p_i \geq 0$ $(i \in N_{0,11})$ to the reader, let us only mention that the set of 16 original equations in 12 unknowns reduces to ten linearly independent equations with two degrees of freedom. Assuming that the unknowns p_{10} and p_{11} are chosen for the two degrees of freedom, we obtain the

TABLE 4.6
Elements of the Structure System in Example 4.13

v_1	v_2	${}^1f({}^1\mathbf{c})$	v_2	v_3	${}^2f({}^2\mathbf{c})$	v_1	v_3	${}^3f({}^3\mathbf{c})$
${}^1\mathbf{c} = 0$	0	0.25	${}^2\mathbf{c} = 0$	0	0.17	${}^3\mathbf{c} = 0$	0	0.11
0	1	0.18	0	1	0.16	0	1	0.14
1	0	0.20	0	2	0.12	0	2	0.18
1	1	0.37	1	0	0.14	1	0	0.20
			1	1	0.18	1	1	0.20
			1	2	0.23	1	2	0.17

following inequalities:

$$0.06 \leq p_{10} \leq 0.18,$$
$$0.05 \leq p_{11} \leq 0.17,$$
$$0.23 \leq p_{10}+p_{11} \leq 0.34.$$

The range of acceptable values of p_{10} and p_{11}, expressed by these inequalities, is illustrated in Figure 4.14; it is a convex set characterized by four extreme points: (0.06, 0.17), (0.17, 0.17), (0.18, 0.16), (0.18, 0.05). Given any pair of values of p_{10} and p_{11} within this range, values of all the remaining unknowns are uniquely determined as follows:

$$p_0 = 0.34 - p_{10} - p_{11}$$
$$p_1 = -0.04 + p_{10}$$
$$p_2 = -0.05 + p_{11}$$

TABLE 4.7
Symbols Used in Example 4.13

v_1	v_2	v_3	$f(\mathbf{c})$
$\mathbf{c} = 0$	0	0	$f(000) = p_0$
0	0	1	$f(001) = p_1$
0	0	2	$f(002) = p_2$
0	1	0	$f(010) = p_3$
0	1	1	$f(011) = p_4$
0	1	2	$f(012) = p_5$
1	0	0	$f(100) = p_6$
1	0	1	$f(101) = p_7$
1	0	2	$f(102) = p_8$
1	1	0	$f(110) = p_9$
1	1	1	$f(111) = p_{10}$
1	1	2	$f(112) = p_{11}$

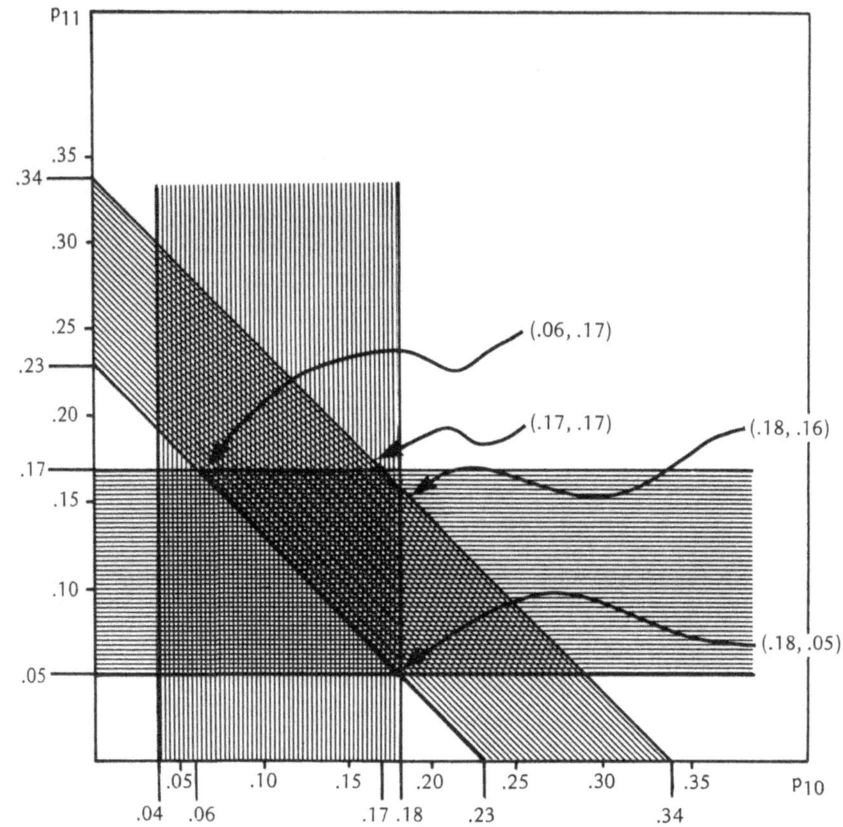

Figure 4.14. A characterization of the reconstruction family in Example 4.13.

$$p_3 = -0.23 + p_{10} + p_{11}$$
$$p_4 = 0.18 - p_{10}$$
$$p_5 = 0.23 - p_{11}$$
$$p_6 = -0.17 + p_{10} + p_{11}$$
$$p_7 = 0.20 - p_{10}$$
$$p_8 = 0.17 - p_{11}$$
$$p_9 = 0.37 - p_{10} - p_{11}$$

► **Example 4.14.** Consider a structure system whose elements are possibilistic behavior systems based on the same sets of variables as the structure system in

TABLE 4.8
Elements of the Structure System in Example 4.14

v_1	v_2	${}^1f({}^1\mathbf{c})$	v_2	v_3	${}^2f({}^2\mathbf{c})$
$\mathbf{c} = 0$	0	0.8	${}^2\mathbf{c} = 0$	0	0.8
0	1	0.5	0	1	0.9
1	0	0.9	1	0	0.8
1	1	0.8	1	1	0.6

Example 4.12. Behavior functions of the elements are defined in Table 4.8. As we can see, they are not normalized.

To characterize the reconstruction family of this structure system, let us use again the symbols introduced in Table 4.4. Then, the reconstruction family is specified by the inequalities $p_i \geq 0$ $(i \in N_{0,7})$ and the following eight equations of the form (4.29):

$$\max(p_2, p_3) = 0.5 \quad (1), \qquad \max(p_0, p_4) = 0.8 \quad (5),$$
$$\max(p_3, p_7) = 0.6 \quad (2), \qquad \max(p_2, p_6) = 0.8 \quad (6),$$
$$\max(p_0, p_1) = 0.8 \quad (3), \qquad \max(p_4, p_5) = 0.9 \quad (7),$$
$$\max(p_6, p_7) = 0.8 \quad (4), \qquad \max(p_1, p_5) = 0.9 \quad (8).$$

Due to (1), neither p_2 nor p_3 can be larger than 0.5; hence, it follows from (6) that $p_6 = 0.8$ and from (2) that $p_7 = 0.6$. Due to (5), p_4 cannot be larger than 0.8; hence, it follows from (7) that $p_5 = 0.9$. Employing these results, the set of equations can now be reduced into two independent subsets of equations,

$$\max(p_2, p_3) = 0.5$$

and

$$\max(p_0, p_1) = 0.8,$$
$$\max(p_0, p_4) = 0.8.$$

The single equation, subject to the constraints $p_2 \geq 0$ and $p_3 \geq 0$, has the solution

$$\text{either} \quad p_2 = 0.5 \quad \text{and} \quad 0 \leq p_3 \leq 0.5,$$
$$\text{or} \quad p_3 = 0.5 \quad \text{and} \quad 0 \leq p_2 \leq 0.5.$$

The pair of equations, subject to constraints $p_0 \geq 0$, $p_1 \geq 0$, and $p_4 \geq 0$, has the solution

$$\text{either} \quad p_0 = 0.8 \quad \text{and both} \quad 0 \leq p_1 \leq 0.8 \quad \text{and} \quad 0 \leq p_4 \leq 0.8,$$
$$\text{or} \quad p_1 = p_4 = 0.8 \quad \text{and} \quad 0 \leq p_0 \leq 0.8.$$

TABLE 4.9
Maximum and Minimal Possibility Distributions of the Reconstruction Family in Example 4.14

v_1	v_2	v_3	f^{SF}	$f_{SF,1}$	$f_{SF,2}$	$f_{SF,3}$	$f_{SF,4}$
c = 0	0	0	0.8	0.8	0.8	0	0
0	0	1	0.8	0	0	0.8	0.8
0	1	0	0.5	0.5	0	0.5	0
0	1	1	0.5	0	0.5	0	0.5
1	0	0	0.8	0	0	0.8	0.8
1	0	1	0.9	0.9	0.9	0.9	0.9
1	1	0	0.8	0.8	0.8	0.8	0.8
1	1	1	0.6	0.6	0.6	0.6	0.6

One maximum and four minimal possibility distributions are readily recognized in this reconstruction family. They are specified in Table 4.9, where symbols f^{SF} and $f_{SF,i}$ $(i \in N_4)$ denote the maximum and minimal distributions, respectively, and SF identifies the structure system under consideration. The reconstruction family consists of these four distributions as well as all distributions that are between the maximum and any of the four minima.

It is known for possibilistic structure systems in general that their reconstruction families have always the form illustrated by Example 4.14. That is, the reconstruction family always contains one *maximum solution* (assuming the given structure system is consistent) and the set of solutions consists of all possibility distributions that are between this maximum and one or several *minimal solutions.* Moreover, the possibility degree of each overall state of the variables involved in any of the minimal solutions is either the same as the possibility degree in the maximum solution or it is equal to zero. This result as well as many additional results regarding the problem of determining the reconstruction family of a given structure behavior system have recently been obtained for both possibilistic and probabilistic systems. Since their full presentation is beyond the scope of this book, a brief survey of the most significant of them is given in Note 4.4.

IDENTIFIABILITY QUOTIENT

It is often desirable to have a suitable measure of the size of the reconstruction family. If adequate, such a measure can be used for measuring the uncertainty associated with the reconstruction of the overall system from the given structure system as well as the degree of identifiability of the actual overall system.

For possibilistic systems, the *size of the reconstruction family* can adequately be expressed by the product

$$\prod_{\mathbf{c} \in \mathbf{A}} [1 + f^{SF}(\mathbf{c})], \tag{4.30}$$

where f^{SF} denotes the maximum element of the reconstruction family $\mathscr{F}_{SF}$ and A stands for the set of all overall states for which the possibility degree in the reconstruction family is ambiguous [i.e., the set of all overall states for which the solution to the constrained set of equations of the form (4.29) is not unique]. Observe that this product is always greater than or equal to 1 and its value is proportional to the size of the set **A** and values f^{SF} (**c**); it is equal to 1 only if **A** is empty (i.e., if the solution is unique).

If the product (4.30) is accepted as a reasonable measure of the size of the reconstruction family, then it is natural to define the *reconstruction uncertainty* u_{SF} associated with structure system **SF** as the logarithm of the product, i.e.,

$$u_{SF} = \log_2 \prod_{\mathbf{c} \in \mathbf{A}} [1 + f^{SF}(\mathbf{c})] = \sum_{\mathbf{c} \in \mathbf{A}} \log_2 [1 + f^{SF}(\mathbf{c})]. \tag{4.31}$$

Clearly,

$$0 \leq u_{SF} \leq |\mathbf{C}|,$$

which means that $|\mathbf{C}|$ is the reconstruction uncertainty of the whole set $\mathscr{C}_{SF}$ of overall systems comparable with SF. The measure

$$\mathbf{I}_{SF} = \frac{|\mathbf{C}| - u_{SF}}{|\mathbf{C}|} = 1 - \frac{u_{SF}}{|\mathbf{C}|}, \tag{4.32}$$

referred to as the *identifiability quotient*, may be then used as a reasonable indication of the ability to identify a unique overall system from a given structure system **SF**. Clearly,

$$0 \leq I_{SF} \leq 1.$$

$I_{SF} = 1$ only if $|\mathscr{F}_{SF}| = 1$; $I_{SF} = 0$ only if $|\mathbf{A}| = |\mathbf{C}|$ and $f^{SF}(\mathbf{c}) = 1$ for all $\mathbf{c} \in \mathbf{C}$.

The identifiability quotient is useful in some systems problems, most notably in comparative studies of competing structure systems. In general, it is much easier to determine the identifiability quotient of a structure system than its reconstruction family: it suffices to determine the maximum solution and the states with unique solutions (i.e., $\mathbf{C} - \mathbf{A}$).

Example 4.15. To determine the identifiability quotient of the structure system specified in Example 4.14, we observe that $|\mathbf{C}| = 8$ and the set **A** consists of the first five states as listed in Table 4.9. Using the values of $f^{SF}(\mathbf{c})$ for these states, we obtain the reconstruction uncertainty

$$u_{SF} = 3 \log_2 1.8 + 2 \log_2 1.5 = 3.714.$$

Hence,

$$I_{SF} = 1 - 3.714/8 = 0.536. \blacktriangleleft$$

UNIQUE SELECTION FROM RECONSTRUCTION FAMILY

Let us discuss now the second subproblem of the identification problem—the problem of selecting, if necessary, one overall system from the reconstruction family as an hypothesis of the actual overall system. The problem is trivial when the reconstruction is unique (i.e., when $I_{SF} = 1$). In other cases (assuming the given structure system **SF** is consistent and, consequently, $\mathscr{F}_{SF} \neq \varnothing$), the choice is totally arbitrary, unless we adopt some criteria of goodness and require that a system be chosen from the reconstruction family that satisfies best these criteria. In the latter case, the selection problem becomes an optimization problem followed by an arbitrary selection from among the best systems. If the optimization criteria employed guarantee a unique optimum, all arbitrariness is eliminated from this problem.

The optimization criteria are always used for some underlying purpose and, as such, must be derived from it. From the epistemological point of view, the most significant purpose is to select an overall system that is *maximally noncommittal* with respect to all matters except the projection requirement (4.27). This purpose can also be stated, more specifically, as follows: given a structure system, select an overall system from its reconstruction family that is based on *all, but no more information* than is contained in the structure system. It is proper to call such an overall system an *unbiased reconstruction*—it is a system that is reconstructed from the structure system without any bias, i.e., by employing all information available while, at the same time, refraining from the use of any additional (unsupported or biased) "information."

The purpose of selecting the unbiased reconstruction is basically the purpose of *inductive inference*. It can be formulated in terms of the following general optimization problem.

Given a structure behavior system **SF**, determine the behavior function f^{SF}, within the set of functions of the reconstruction family $\mathscr{F}_{SF}$, for which the uncertainty measure (Shannon entropy for probabilistic systems, U-uncertainty for possibilistic systems) reaches its maximum subject to the projection constraints (4.27).

For probabilistic systems, this optimization problem is well known in the literature under the name "*the principle of maximum entropy*." It has been justified as a rational principle of inductive inference by several diverse arguments, which are outlined in Note 4.6.

It is well known that the unbiased reconstruction is unique for both probabilistic and possibilistic systems. It represents the weakest possible constraint among the variables involved that conforms to the given structure system. For possibilistic systems, $\mathbf{f}^{SF}$ is the maximum distribution in the reconstruction family $\mathscr{F}_{SF}$ or, in other words, it is a distribution that represents, within the distributions in $\mathscr{F}_{SF}$, the largest fuzzy subset of the set of all overall states of the variables involved.

Although the unbiased reconstruction is epistemologically the most significant, as it is clearly based on a unique and well-justified principle of inductive inference (Note 4.6), other reconstructions may be preferable for alternative purposes. An example of an important purpose, one of a rather practical nature, is to select an overall system for which the largest possible error is minimized. The term "error" is used here in the sense of a distance between the distributions (probabilistic or possibilistic) of the reconstructed overall system and the actual one. A reconstruction of this sort may be well characterized as a *least risk* reconstruction. A specific formulation of the resulting optimization problem depends on the kind of distance employed. Among the many possible kinds of distances, those which measure the loss of information are of special significance. They are discussed later in this chapter, particularly in Sections 4.7 and 4.9.

JOIN PROCEDURES

One of the main results associated with the identification problem is that the unbiased reconstruction can be determined by a computationally simple procedure, without actually solving the optimization problem formulated previously (to maximize the Shannon entropy or U-uncertainty within given constraints). The procedure, referred to as the *join procedure*, is based on the probabilistic or possibilistic version of a *join operation* by which behavior functions characterizing elements of the given structure system are combined in a rather straightforward manner.

▶ Consider two behavior functions

$$^1f: \mathbf{A} \times \mathbf{B} \to [0, 1],$$
$$^2f: \mathbf{B} \times \mathbf{C} \to [0, 1],$$

defined on state sets **A**, **B**, **C**, whose meaning is explained later. Observe that the set **B** is involved in both of the functions. The *join of* ^1f *and* ^2f, denoted by $^1f * {}^2f$, is a function

$$^1f * {}^2f: \mathbf{A} \times \mathbf{B} \times \mathbf{C} \to [0, 1]$$

whose properties depend on the nature of the functions ^1f and ^2f. If they are probability distribution functions, then

$$[^1f * {}^2f]\ (\mathbf{a}, \mathbf{b}, \mathbf{c}) = {}^1f(\mathbf{a}, \mathbf{b}) \cdot {}^2f(\mathbf{c} \mid \mathbf{b}), \tag{4.33}$$

where $^2f(\mathbf{c} \mid \mathbf{b})$ denotes the conditional probability of $\mathbf{c}$ given $\mathbf{b}$; if they are possibility distribution functions, then

$$[^1f * {}^2f]\ (\mathbf{a}, \mathbf{b}, \mathbf{c}) = \min[^1f(\mathbf{a}, \mathbf{b}), {}^2f(\mathbf{b}, \mathbf{c})]. \tag{4.34}$$

Observe that no conditional possibilities are used in (3.34), in analogy with (3.33). This is due to the property

$$\min[^1f(\mathbf{a}, \mathbf{b}), {}^2f(\mathbf{c} \mid \mathbf{b})] = \min[^1f(\mathbf{a}, \mathbf{b}), {}^2f(\mathbf{b}, \mathbf{c})],$$

which can easily be proven.

Assume that the join operation is applied to two elements of a structure system with sampling variables in sets ${}^1S$ and ${}^2S$, and behavior functions ${}^1f$ and ${}^2f$, respectively. Then, the domains of the functions ${}^1f$ and ${}^2f$ must be converted into the forms $\mathbf{A} \times \mathbf{B}$ and $\mathbf{B} \times \mathbf{C}$, respectively, where

- **A** is the set of aggregate states of variables that participate only in the first element, i.e., variables in the set ${}^1S - ({}^1S \cap {}^2S)$;
- **B** is the set of aggregate states of variables that participate in both of the elements, i.e., variables in ${}^1S \cap {}^2S$;
- **C** is the set of aggregate states of variables that participate only in the second element, i.e., variables in ${}^2S - ({}^1S \cap {}^2S)$.

To determine the unbiased reconstruction for a given structure system, the join operation must be applied repeatedly to pairs of its elements. In each of its applications, two elements merge (join) into a larger element of a new structure system. Assume that the join operation is always performed in such order that the result of previously applied join operations enters as the second element, i.e., in terms of the function ${}^2f$ in Eq. (4.33) or (4.34). The procedure terminates when all elements merge into one overall system.

Let the outlined procedure be called a *basic join procedure*. Before formalizing it, two degenerate cases of the join operation must be considered to capture all meaningful situations. In the first case, all of the variables in the first element (associated with ${}^1f$) may be included in the second element (associated with ${}^2f$). This can occur since, in general, the second element is a result of some join operations performed previously. In this case, set **A** is empty and ${}^1f$ assumes a degenerate form

$$ {}^1f\colon \mathbf{B} \to [0, 1]. $$

In the second case, the elements are not coupled. This means that set **B** is empty and the behavior functions assume degenerate forms

$$ {}^1f\colon \mathbf{A} \to [0, 1], $$
$$ {}^2f\colon \mathbf{C} \to [0, 1]. $$

Observe that due to the irredundancy requirement for structure systems (Section 4.3) and the convention that the result of previously performed join operations is always used as ${}^2f$, set **C** cannot be empty.

For probabilistic systems, the two degenerate join operations are defined by

$$ [{}^1f * {}^2f]\,(\mathbf{b}, \mathbf{c}) = {}^1f(\mathbf{b}) \cdot {}^2f(\mathbf{c} \mid \mathbf{b}), \tag{4.35} $$

when $\mathbf{A} = \varnothing$, and

$$ [{}^1f * {}^2f]\,(\mathbf{a}, \mathbf{c}) = {}^1f(\mathbf{a}) \cdot {}^2f(\mathbf{c}), \tag{4.36} $$

when $\mathbf{B} = \varnothing$; for possibilistic systems, they are defined by

$$[{}^1f * {}^2f]\ (\mathbf{b}, \mathbf{c}) = \min\,[{}^1f(\mathbf{b}), {}^2f(\mathbf{b}, \mathbf{c})], \tag{4.37}$$

$$[{}^1f * {}^2f]\ (\mathbf{a}, \mathbf{c}) = \min\,[{}^1f(\mathbf{a}), {}^2f(\mathbf{c})], \tag{4.38}$$

respectively.

Assuming that the symbol ${}^1f * {}^2f$ stands for either the regular join operation or one of its degenerate forms (depending on the context), the basic join procedure can be formalized in terms of the following algorithm.

Basic Join Procedure. Given a locally consistent structure behavior system **SF** (probabilistic or possibilistic) with behavior functions ${}^xf$ ($x \in N_q$), to determine the join of ${}^xf$ for all $x \in N_q$:

1. Let $k = 2$ and $f = {}^1f$;
2. make proper adjustments to the arguments of ${}^kf$ and f and perform the appropriate version of the join operation ${}^kf * f \to f$ (probabilistic or possibilistic, regular or degenerate);
3. if $k < q$, make $k + 1 \to k$ and go to (2);
4. stop.

The following proposition can be proven (Appendix C): if the basic join procedure is applied to a consistent possibilistic structure system **SF**, then it always results in the unbiased reconstruction $\mathbf{f}^{\mathbf{SF}}$ (the maximum possibility distribution in the reconstruction family $\mathscr{F}_{\mathrm{SF}}$). If, however, the procedure is applied to probabilistic systems, it results in the unbiased (maximum entropy) reconstruction only for structure systems of a specific type—the so-called *loopless structure systems*, introduced in Section 4.7. Whether or not the result f of the basic join procedure represents the unbiased reconstruction can be determined directly from the result itself, however, without resorting to the identification of the type of the given structure system. If f satisfies the projections

$$[f \downarrow {}^xS] = {}^xf$$

for all $x \in N_q$, then it is the unbiased reconstruction; otherwise, f does not conform to the given structure system and must be adjusted by the following *iterative join procedure.*

Iterative Join Procedure. Given a locally consistent structure behavior system **SF** with probabilistic behavior functions ${}^jf$ ($j \in N_{0,q-1}$), the result f of the basic join procedure applied to **SF**, and a number $\Delta \in [0, 1]$, to determine the behavior function $f^{\mathbf{SF}}$ of the unbiased (maximum entropy) reconstruction with precision at least equal to Δ:

1. Let $j = 0$, $i = 1$, and $f_0 = f$;
2. make proper adjustments to the arguments of ${}^jf$ and f_{i-1} and perform the join operation ${}^jf * f_{i-1} \to f_i$ of the degenerate form (4.35);

3. if $i \neq 0$ (modulo q), make $i+1 \rightarrow i$, $j+1$ (modulo q) $\rightarrow j$, and go to (2);
4. if $|f_i(\mathbf{c}) - f_{i-q}(\mathbf{c})| > \Delta$ for some $\mathbf{c} \in \mathbf{C}$, make $i+1 \rightarrow i$, $j+1$ (modulo q) $\rightarrow j$, and go to (2);
5. stop.

If $\Sigma_{\mathbf{c}} f_i(\mathbf{c}) = 1$ after the iterative join procedure is executed, then

$$f^{SF}(\mathbf{c}) - \Delta \geq f_i(\mathbf{c}) \leq f^{SF}(\mathbf{c}) + \Delta$$

TABLE 4.10
Sequence of Behavior Functions Obtained by the Basic and Iterative Join Procedures in Example 4.16

	Basic join procedure		Iterative join procedure				
v_1 v_2 v_3	${}^2f * {}^1f$	${}^3f * ({}^2f * {}^1f)$	$i = 1$	$i = 2$	$i = 3$	$i = 4$	$i = 5$
0 0 0	$0.2\overline{6}$	0.312195	0.294647	0.302908	0.309934	0.304982	0.307696
0 0 1	$0.1\overline{3}$	0.111628	0.105353	0.099904	0.096561	0.095018	0.093371
0 1 0	0.075	0.087805	0.095379	0.088024	0.090066	0.092059	0.089834
0 1 1	0.225	0.188372	0.204621	0.210484	0.203439	0.207941	0.209672
1 0 0	0.13	0.084211	0.094444	0.097092	0.089019	0.091490	0.092304
1 0 1	0.06	0.094118	0.105556	0.100096	0.105580	0.108510	0.106629
1 1 0	0.025	0.015789	0.012977	0.011976	0.010981	0.010418	0.010166
1 1 1	0.075	0.105882	0.087023	0.089516	0.094420	0.089582	0.090328

$i = 6$	$i = 7$	$i = 8$	$i = 9$	$i = 10$	$i = 11$	$i = 12$	$i = 13$
0.309608	0.308036	0.308915	0.309511	0.309002	0.309289	0.309481	0.309315
0.092433	0.091964	0.091444	0.091148	0.090998	0.090829	0.090734	0.090685
0.090392	0.091011	0.090314	0.090489	0.090688	0.090462	0.090519	0.090584
0.207567	0.208989	0.209528	0.208852	0.209312	0.209486	0.209266	0.209416
0.090079	0.090826	0.091085	0.090388	0.090626	0.090711	0.090486	0.090563
0.108277	0.109174	0.108556	0.109086	0.109374	0.109171	0.109343	0.109437
0.009921	0.009761	0.009686	0.009612	0.009561	0.009538	0.009514	0.009498
0.091723	0.090239	0.090472	0.090914	0.090439	0.090514	0.090657	0.090502

$i = 14$	$i = 15$	$i = 16$	$i = 17$	$i = 18$	$i = 19$	$i = 20$	$i = 21$
0.309409	0.309472	0.309417	0.309448	0.309469	0.309451	0.309461	0.309468
0.090630	0.090599	0.090583	0.090565	0.090555	0.090549	0.090543	0.090540
0.090510	0.090528	0.090550	0.090525	0.090531	0.090538	0.090530	0.090532
0.209473	0.209401	0.209450	0.209469	0.209445	0.209462	0.209468	0.209460
0.090591	0.090518	0.090543	0.090552	0.090528	0.090536	0.090539	0.090531
0.109370	0.109427	0.109457	0.109435	0.109454	0.109464	0.109457	0.109463
0.009490	0.009482	0.009477	0.009476	0.009472	0.009470	0.009470	0.009469
0.090527	0.090573	0.090523	0.090531	0.090546	0.090530	0.090532	0.090537

TABLE 4.11
Convergence Test in Step (4) of the Iterative Join Procedure in Example 4.16

	$\lvert f_i(\mathbf{c}) - f_{i-q}(\mathbf{c})\rvert$, $i = 0 \pmod q$						
v_1 v_2 v_3	$i = 3$	$i = 6$	$i = 9$	$i = 12$	$i = 15$	$i = 18$	$i = 21$
0 0 0	0.002261	0.000326	0.000097	0.000030	0.000009	0.000003	0.000001
0 0 1	0.015067	0.004128	0.001285	0.000414	0.000135	0.000044	0.000015
0 1 0	0.002261	0.000326	0.000097	0.000030	0.000009	0.000003	0.000001
0 1 1	0.015067	0.004128	0.001285	0.000414	0.000135	0.000044	0.000015
1 0 0	0.004808	0.001060	0.000309	0.000098	0.000032	0.000010	0.000003
1 0 1	0.011462	0.002697	0.000809	0.000257	0.000084	0.000027	0.000009
1 1 0	0.004808	0.001060	0.000309	0.000098	0.000032	0.000010	0.000003
1 1 1	0.011462	0.002697	0.000809	0.000257	0.000084	0.000027	0.000009

for each $\mathbf{c} \in \mathbf{C}$; otherwise, the given structure system **SF** is globally inconsistent (Section 4.11) and no reconstruction of **SF** exists, i.e., $\mathscr{F}_{\mathbf{SF}} = \varnothing$ and, consequently, **SF** is meaningless.

Example 4.16. Consider the structure system discussed in Example 4.11 and defined in Table 4.3. To determine the unbiased reconstruction, we apply the probabilistic version of the basic join procedure first. The intermediate result ${}^2f * {}^1f$ and the final result $f = {}^3f * ({}^2f * {}^1f)$ are shown in the initial part of Table 4.10. We can easily see that the final result f does not conform to the given structure system. For example,

$$[f \downarrow \{v_1, v_2\}]\ (00) = 0.312195 + 0.111628 = 0.423823$$

is not equal to ${}^1f(00) = 0.4$ so that the projection requirement (4.27) is violated Hence, the iterative join procedure must be used. Let $\Delta = 0.00002$. A sequence of behavior functions is generated by the procedure that converges to the unbiased reconstruction. This sequence is shown for $i = 1, 2, \ldots, 21$ in Table 4.10. Numbers associated with the convergence test in step (3) of the procedure are listed in Table 4.11. Hence, if $\Delta > 0.015067$, the procedure would stop for $i = 3$; if $\Delta > 0.004128$, it would stop for $i = 6$, etc. Since $\Delta = 0.00002$, the procedure stops for $i = 21$. ◄

4.7. RECONSTRUCTION PROBLEM

> *The division of the perceived universe into parts and wholes is convenient and may be necessary, but no necessity determines how it shall be done.*
>
> —Gregory Bateson

The *reconstruction problem* can be stated as follows: given a behavior system, viewed as an overall system, determine which sets of its subsystems, each viewed as a reconstruction hypothesis, are adequate for reconstructing the given system with an acceptable degree of approximation, solely from the information contained in the subsystems.

First, we observe that, according to this problem statement, the term "reconstruct" has a specific meaning: to reconstruct by using all, but no more information than is contained in the subsystems involved. This means that the reconstruction is required to be unbiased in the sense discussed in Section 4.6 and, consequently, the appropriate join procedures can be used for this purpose.

In the identification problem, the unbiased reconstruction represents a well justified hypothesis (estimate) of the unknown overall system derived solely from a given structure system. Since the actual overall system is not known, it is not possible to determine the closeness between it and the hypothetical system. In the reconstruction problem, the unbiased reconstruction characterizes the reconstruction capability of the considered reconstruction hypothesis with respect to the given overall system. The closer the unbiased reconstruction is to the actual (given) system, the better the reconstruction hypothesis is.

In general, the *closeness* between two comparable behavior systems may be expressed in terms of an appropriate distance measure defined for behavior functions. There are many possible types of distance measures. For example, the *Minkowski class of distances* is defined by the formula

$$\delta_p(f,f^h) = \left[\sum_{\mathbf{c}\in\mathbf{C}} |f(\mathbf{c}) - f^h(\mathbf{c})|^p \right]^{1/p}, \tag{4.39}$$

where f, f^h denote behavior functions of the given system and the unbiased reconstruction from a reconstruction hypothesis h, respectively, and $p \in \mathbb{N}$ is a parameter whose values characterize the individual types of distances. We obtain, for instance, the Hamming distance for $p = 1$, Euclidean distance for $p = 2$, and upper bound distance for $p = \infty$.

Distances in the Minkowski class are based on point-wise differences

$$|f(\mathbf{c}) - f^h(\mathbf{c})|$$

of probabilities or possibilities, aggregated in the various ways expressed by the general formula (4.39). Although this point-wise characterization of the closeness between f and f^h may be useful for some purposes, it is not well justified theoretically. A better grounded way of viewing the closeness is to express it in terms of the difference between the information that h contains about f or, in other words, in terms of the *loss of information* that takes place when f is replaced with h (a set of projections of f).

Let us call a measure of this loss of information an *information distance* and denote it by $D(f,f^h)$. For probabilistic systems, it is expressed by the well-known formula

$$D(f,f^h) = \frac{1}{\log_2 |\mathbf{C}|} \sum_{\mathbf{c}\in\mathbf{C}} f(\mathbf{c}) \frac{\log_2 f(\mathbf{c})}{\log_2 f^h(\mathbf{c})}, \tag{4.40}$$

where the constant $1/\log_2|\mathbf{C}|$ is used as a normalizing factor to get the property

$$0 \leq D(f, f^h) \leq 1.$$

Since $f^h(\mathbf{c}) = 0$ implies $f(\mathbf{c}) = 0$, the probabilistic information distance is always defined. Observe, however, that it is not a metric distance, due to its fundamental asymmetry; moreover, $D(f^h, f)$ may not even be defined for some instances of f and f^h [when $f^h(\mathbf{c}) > 0$ and $f(\mathbf{c}) = 0$ for some $\mathbf{c} \in \mathbf{C}$].

When the information distance is applied to generative behavior systems, Eq. (4.40) becomes

$$D_G(f, f^h) = \frac{1}{\log_2|G|} \sum_{\bar{g} \in G} f(\bar{\mathbf{g}}) \sum_{g \in G} f(\mathbf{g} \mid \bar{\mathbf{g}}) \frac{\log_2 f(\mathbf{g} \mid \bar{\mathbf{g}})}{\log_2 f^h(\mathbf{g} \mid \bar{\mathbf{g}})} \tag{4.41}$$

The modifications of Eqs. (4.40) and (4.41) for directed behavior systems are obvious.

► For possibilistic systems, the information distance is expressed by the formula

$$D(f, f^h) = \frac{1}{\log_2|\mathbf{C}|} \int_0^1 \log_2 \frac{|c(f^h, l)|}{|c(f, l)|} dl, \tag{4.42}$$

which is the U-uncertainty analog of the probabilistic information distance (4.40). ◄

The use of the information distances for comparing reconstruction hypotheses is described later in this section, after relevant properties of reconstruction hypotheses are sufficiently characterized.

A reconstruction hypothesis of a given overall behavior system is a set of its subsystems. If the overall system consists of n variables, then the number of its subsystems that contain at least one variable is $2^n - 1$ and the total number of sets of these subsystems that contain at least one subsystem is

$$2^{2^n - 1} - 1.$$

This number, whose growth is extremely rapid with n, can be considerably reduced, without any loss of generality, by restricting to irredundant sets of subsystems (Section 4.3).

Another way of reducing the number of reconstruction hypotheses, which is desirable in most cases of systems investigations, is to exclude those sets of subsystems that do not contain all variables of the overall system. This requirement, usually referred to as the *covering requirement*, can be formally stated as

$$\bigcup_k {}^kS = S,$$

where ${}^{k}S$ denote the sets of variables in the subsystems of a reconstruction hypothesis and S denotes the set of variables in the overall system. It is primarily motivated by the necessity to include information about each variable of the overall system in the reconstruction hypothesis to make the reconstruction logically possible. Since the issue of including or excluding sampling variables in the overall system is resolved by the mask analysis (Section 3.6), no generality is lost by the covering requirement.

Let the term "*reconstruction hypothesis*" be used from now on only for sets of subsystems of a given overall system that satisfy both the irredundancy and covering requirements. Reconstruction hypotheses are thus structure behavior systems that are comparable with the overall behavior system. For some purposes, however, it is desirable to deal with all sets of subsystems that satisfy only the irredundancy requirement. Let such sets of subsystems be called *generalized reconstruction hypotheses*. Given an overall behavior system, the set of its reconstruction hypotheses is clearly a subset of its generalized reconstruction hypotheses.

Each reconstruction hypothesis (as well as any generalized reconstruction hypothesis) is fully characterized by two properties: (i) a family of subsets of the variables involved, and (ii) behavior functions associated with the individual subsets of variables. When property (ii) is disregarded, property (i) becomes an invariant of a class of reconstruction hypotheses that differ from each other solely in the behavior functions of their elements. Let this invariant be called a *structure* to distinguish it from the individual reconstruction hypotheses in the class, each of which is a particular structure system. Structures are thus properties of structure systems that are invariant with respect to changes in behavior functions of their elements.

Given a set of variables, say set S, the set of structures that represents all reconstruction hypotheses of any overall system defined in terms of S consists of families of subsets of S that satisfy the irredundancy and covering requirements. For convenience, let us represent all sets of variables with the same cardinality, say n, by a common set of structures, say set $\mathscr{G}_n$, defined in terms of the set N_n of positive integers. Formally, for each $n \in \mathbb{N}$,

$$\mathscr{G}_n = \{G_i \mid G_i \subset \mathscr{P}(N_n), G_i \text{ satisfies the irredundancy and covering requirements}\}.$$

Symbols G_i are used in this formal definition to indicate that elements of $\mathscr{G}_n$ are the most general structures involved in the reconstruction problem (several special types of structures are introduced later for various purposes); subscript i is an identifier of the individual structures in $\mathscr{G}_n$ and, normally, $i \in N_{|\mathscr{G}_n|}$. Set $\mathscr{G}_n$ is trivially interpretable in terms of any set of variables S such that $|S| = n$ by any assignment (one-to-one) of the variables in S to the integers in N_n. For convenience, let structures in sets $\mathscr{G}_n$ be called *G-structures*.

For some purposes, it is convenient to extend the set $\mathscr{G}_n$ to the set $\mathscr{G}_n^+$ of all generalized reconstruction hypotheses. Formally, for each $n \in \mathbb{N}$,

$$\mathscr{G}_n^+ = \{G_i \mid G_i \subset \mathscr{P}(N_n), G_i \text{ satisfies the irredundancy requirement}\}.$$

Although the main focus in the rest of this Chapter is on sets $\mathscr{G}_n$, all results relevant to $\mathscr{G}_n$ can easily be generalized to sets $\mathscr{G}_n^+$.

When the set $\mathscr{G}_n$ for some particular n is given a specific interpretation in the context of an overall behavior system with n variables, the structures in $\mathscr{G}_n$ become unique representations of the reconstruction hypotheses associated with the overall system. This follows immediately from the fact that behavior functions associated with any subsets of the variables become uniquely defined as appropriate projections of the overall behavior function. Reconstruction hypotheses can thus be studied in their abstracted forms of structures. Given a structure in $\mathscr{G}_n$, it becomes a specific reconstruction hypothesis when interpreted in the context of a particular overall behavior system comparable with it (based on n variables).

The main issue in the reconstruction problem is to develop computationally efficient procedures that will allow consideration, evaluation, and comparison of reconstruction hypotheses represented by all structures of a given set of variables. Since the number of structures grows rapidly, as shown later in this section, this is a difficult task. To pursue it successfully, it is essential to utilize appropriate ordering and classification of structures.

First, let us define a natural ordering of structures, referred to as a *refinement ordering*. Given two structures G_i, $G_j \in \mathscr{G}_n$, let G_i be called a *refinement* of G_j (or, alternatively, let G_j be called a *coarsening* of G_i) if and only if for each $x \in G_i$ there is some $y \in G_j$ such that $x \subseteq y$; let $G_i \leq G_j$ denote that G_i is a refinement of G_j.

Consider two structures G_i, $G_j \in \mathscr{G}_n$ such that $G_i \leq G_j$. Then, G_i is called an *immediate refinement* of G_j (or, alternatively, G_j is called an *immediate coarsening* of G_i) if and only if there is no $G_k \in \mathscr{G}_n$ such that $G_i \leq G_k$ and $G_k \leq G_j$. Given a particular structure $G_i \in \mathscr{G}_n$, we define its *structure neighborhood* as the set of all its immediate refinements and immediate coarsenings in $\mathscr{G}_n$.

It is easy to see that the refinement relation is a partial ordering. Furthermore, the pair $(\mathscr{G}_n, \leq)$ is a lattice. This can be demonstrated by the following facts: (i) there exists a universal upper bound—set $\{N_n\}$; (ii) there exists a universal lower bound—set $\{\{x\} \mid x \in N_n\}$; (iii) for each pair G_i, $G_j \in \mathscr{G}_n$, the greatest common refinement is the irredundant counterpart of the set $\{x \cap y \mid x \in G_i, y \in G_j\}$; (iv) for each pair G_i, $G_j \in G_n$, the least common coarsening is the irredundant counterpart of the set $G_i \cup G_j$. Let us refer to these lattices as the *refinement lattices* of G-structures (one for each $n \in \mathbb{N}$). Observe that the refinement ordering can readily be applied to sets $\mathscr{G}_n^+$ and forms lattices $(\mathscr{G}_n^+, \leq)$.

It is obvious that a refinement lattice or any desirable part of it can be generated by a repeated application of a procedure through which all immediate refinements are generated for any given structure in the lattice. One possible procedure of this sort is defined as follows.

Refinement Procedure for G-Structures (or *RG-procedure*). Given a G-structure $G_i = \{{}^kS \mid k \in N_q\} \in \mathscr{G}_n$, to determine all its immediate refinements:

1. Let $k = 0$;
2. if $k < q$, make $k + 1 \rightarrow k$; else, go to (5);

3. if $|{}^kS| \geq 2$, make $(G_i - \{{}^kS\}) \cup X \rightarrow R$, where $X = \{x | x \subset {}^kS, |x| = |{}^kS| - 1\}$; else, go to (2);
4. make $R \rightarrow Q$, where Q is the irredundant counterpart of R, record Q as an immediate refinement of G_i, and go to (2);
5. stop.

Observe that the condition $|{}^kS| \geq 2$ in step (3) guarantees that the generated structures satisfy the covering requirement; replacing it with condition $|{}^kS| \geq 1$ would allow us to deal with sets $\mathscr{G}_n^+$ of generalized reconstruction hypotheses. Step (4) guarantees that they satisfy the irredundancy requirement. The fact that the smallest possible change in G_1 is made in step (3)—only one element of G_i is excluded and replaced by immediately smaller elements (subsets)—is a guarantee that the generated structures are immediate refinements of G_i.

Example 4.17. Given $G_i = \{{}^1S = \{1, 2, 3\}, {}^2S = \{2, 3, 4\}, {}^3S = \{1, 4\}\}$, we can immediately see that $|{}^kS| \geq 2$ for all $k \in N_3$ and, consequently, the RG-procedure can operate on each of the three elements of G_i; hence, there are three immediate refinements of G_i. Set ${}^1S$ is replaced with the sets $\{1, 2\}, \{1, 3\}, \{2, 3\}$, but the third one is a subset of ${}^2S$ and is excluded in step (4); this results in the immediate refinement

$$\{\{1, 2\}, \{1, 3\}, \{2, 3, 4\}, \{1, 4\}\}.$$

By a similar replacement of ${}^2S$, the second immediate refinement

$$\{\{1, 2, 3\}, \{2, 4\}, \{3, 4\}, \{1, 4\}\}$$

is obtained. Finally, set ${}^3S$ is replaced with the sets $\{1\}$ and $\{4\}$, both of which are redundant and excluded in step (4); hence, the third immediate refinement is

$$\{\{1, 2, 3\}, \{2, 3, 4\}\}.$$

To reduce the computational complexity involved in the generation of reconstruction hypotheses, it is useful to partition the refinement lattices into convenient equivalence classes at several computational levels. The equivalence classes can be then represented by appropriate canonical structures and refinement procedures developed to deal only with these canonical representations at the various computational levels. To illustrate this issue, let only two levels of computation be described, referred to as local and global.

The *local level of computation* is represented by the RG-procedure just described. To develop a *global level of computation*, let us define functions

$$r_n: \mathscr{G}_n \rightarrow \mathscr{R}_n,$$

$n \in \mathbb{N}$, where $\mathcal{R}_n$ is the set of all symmetric and reflexive binary relations defined on the set N_n (also called compatibility relations, tolerance relations, or undirected graphs with loops), and $r_n(G_i)$ is the binary relation in which integers a and b ($a, b \in N_n$) are related if and only if they both belong to at least one of the subsets of N_n included in G_i. Formally,

$$r_n(G_i) = \{(a, b) \mid (\exists x \in G_i)(a \in x \text{ and } b \in x)\}.$$

Let us refer to elements of $\mathcal{R}_n$ as *graphs*. We should keep in mind, however, that all these graphs are undirected (symmetric) and with loops (reflexive). Some examples illustrating functions r_4 and r_5 are shown in Figure 4.15. In depicting the graphs, we omit the obvious loops on the nodes.

Functions r_n are clearly onto and for $n \geq 3$ are also many-to-one. As such, they impose the following equivalence relation on the respective sets $\mathcal{G}_n$ of G-structures:

$$G_i \overset{r}{\equiv} G_j \quad \text{iff} \quad G_i, G_j \in \mathcal{G}_n \quad \text{and} \quad r_n(G_i) = r_n(G_j)$$

for some particular $n \in \mathbb{N}$. If $G_i = G_j$, we say that structures G_i and G_j are *r-equivalent*. Let us use the standard symbol $\mathcal{G}_n / r_n$ to denote the set of equivalence classes imposed upon $\mathcal{G}_n$ by r_n.

For each $n \in N_n$, the set $\mathcal{R}_n$ together with the subset relation (or, alternatively, the operations of set union and intersection) form a Boolean lattice. The obvious one-to-one correspondence between $\mathcal{G}_n / r_n$ and $\mathcal{R}_n$ imposes then an isomorphic Boolean lattice on the set $\mathcal{G}_n / r_n$. This isomorphism enables us to generate equivalence classes in $\mathcal{G}_n / r_n$

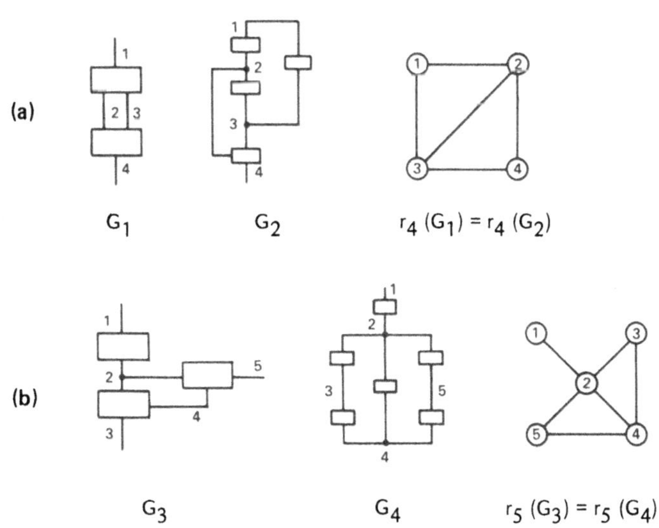

Figure 4.15. Examples illustrating functions r_n.

by appropriately operating on graphs in $\mathscr{R}_n$. It is desirable, however, that each equivalence class in $\mathscr{G}_n/r_n$ be uniquely represented by some canonical structure. For that purpose, we introduce for each $n \in \mathbb{N}$ the following subsets of $\mathscr{G}_n$:

- $\mathscr{C}_n$, consisting of those G-structures G_i in the set $\mathscr{G}_n$ which contain only and *all* maximal compatibility classes associated with the graph $r_n(G_i)$ or, in another terminology, are based on cliques of the graph. Such structures are also known as complete covers in $r_n(G_i)$. Let us denote structures in sets $\mathscr{C}_n$ by C_j and refer to them as *C-structures*. Structures G_1 and G_3 in Figure 4.15 are examples of C-structures.
- $\mathscr{P}_n$, consisting of those G-structures G_i in $\mathscr{G}_n$ whose elements consist of the pairs of integers that are connected in the graph $r_n(G_i)$ and the single integers that are isolated in the graph. Let structures in sets $\mathscr{P}_n$ be called *P-structures* and denoted by P_k. Structure G_4 in Figure 4.15 is an example of a P-structure.

It follows immediately from these definitions and the fact that the set of all maximal compatibility classes in every undirected graph is unique that each equivalence class in $\mathscr{G}_n/r_n$ contains exactly one C-structure from $\mathscr{C}_n$ and one P-structure from $\mathscr{P}_n$. C-structures and P-structures can thus be viewed as two *canonical representations* of the r-equivalence classes of G-structures. Each r-equivalence class, represented by a graph, is polarized by the two canonical structures: the C-structure and P-structure are the least refined and most refined structures in the equivalence class, respectively. The one-to-one correspondence between $\mathscr{R}_n$ and $\mathscr{C}_n$ (or $\mathscr{P}_n$) imposes a Boolean lattice on $\mathscr{C}_n$ (or $\mathscr{P}_n$) that is isomorphic with the natural Boolean lattice defined on $\mathscr{R}_n$.

As an example, Figure 4.16 illustrates the lattice $(\mathscr{G}_3, \leq)$ as well as the Boolean lattices (mutually isomorphic) defined on $\mathscr{R}_3, \mathscr{C}_3, \mathscr{P}_3$, and $\mathscr{G}_3/r_3$. For describing larger lattices, it is useful to define an equivalence relation $\overset{i}{\equiv}$ on the sets $\mathscr{G}_n$ in the following way:

$$G_i \overset{i}{\equiv} G_j \quad \text{iff} \quad G_i, G_j \in \mathscr{G}_n \quad \text{and} \quad G_i, G_j \text{ are isomorphic}$$

(i.e., one of them can be obtained from the other solely by a permutation of the integers in N_n). Let $\overset{i}{\equiv}$ be referred to as *i-equivalence* and let $\mathscr{G}_n/i$ denote the set of equivalence classes of isomorphic structures (or permutation equivalence classes) defined on $\mathscr{G}_n$.

An example illustrating the meaning of the i-equivalence is shown in Figure 4.17, where the boldface symbols $\mathbf{G}_k$ $(k \in N_5)$ denote i-equivalence classes in $\mathscr{G}_3/i$. Figure 4.17a describes the lattice $(\mathscr{G}_3/i, \leq)$. It is the same lattice as the one shown in Figure 4.16, but simplified in the sense that isomorphic structures are not distinguished. This is done by deleting labels of entries in the block diagrams and including only one block diagram for each permutation equivalence class. Structures in each of these classes can easily be determined by permuting integers 1, 2, 3 along entries of the blocks. The simplified lattice shown in Figure 4.17a is a homomorphic image of the full lattice in

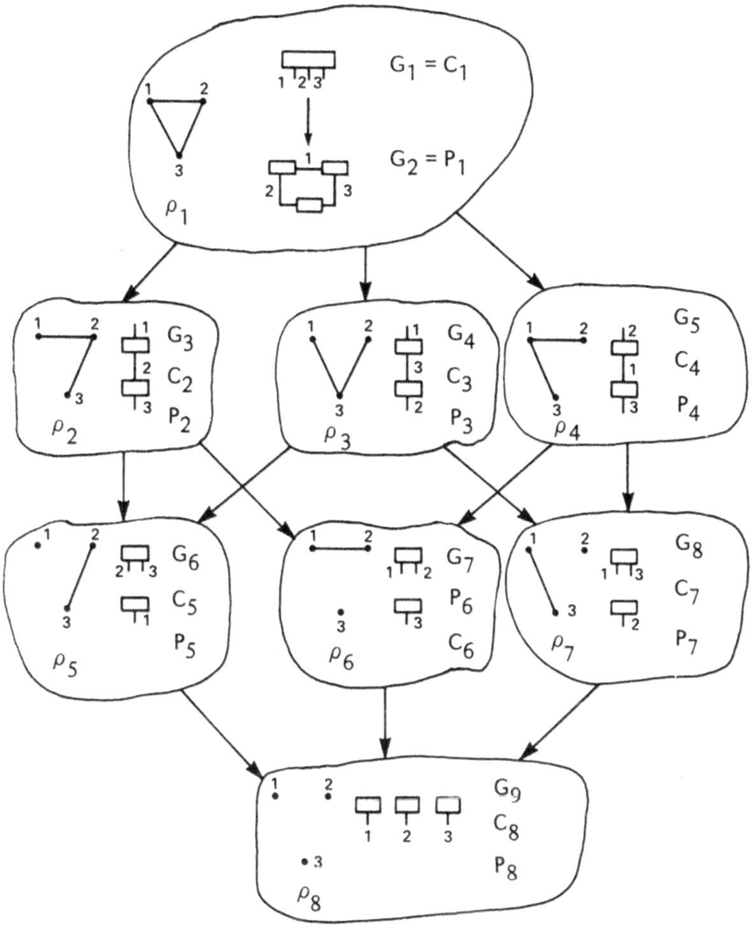

Figure 4.16. Lattice $(\mathscr{G}_3, \leq)$ and Boolean lattices defined on $\mathscr{R}_3$, $\mathscr{C}_3$, $\mathscr{P}_3$, and $\mathscr{G}_3/r_3$.

Figure 4.16. The homomorphic mapping, which is the basis for this simplification, is specified in Figure 4.17b.

The more complicated lattice $(\mathscr{G}_4/i, \leq)$ is specified in Figure 4.18. Permutation equivalence classes of G-structures as well as C-structures and P-structures are again denoted by boldface symbols and grouped together in the r-equivalence classes. Each of the latter classes is associated with a graph ρ_k and the two canonical structures C_k and P_k $(k \in N_{11})$. To show more clearly the overall properties of this lattice, it is summarized in Figure 4.19, where the individual permutation equivalence classes of G-structures are denoted only by their identifiers and the r-equivalence classes are more emphasized.

The lattice $(\mathscr{C}_4/i, \leq)$, which is a sublattice of $(\mathscr{G}_4/i, \leq)$, is described in Figure 4.20. The number placed next to each block diagram indicates the number of different

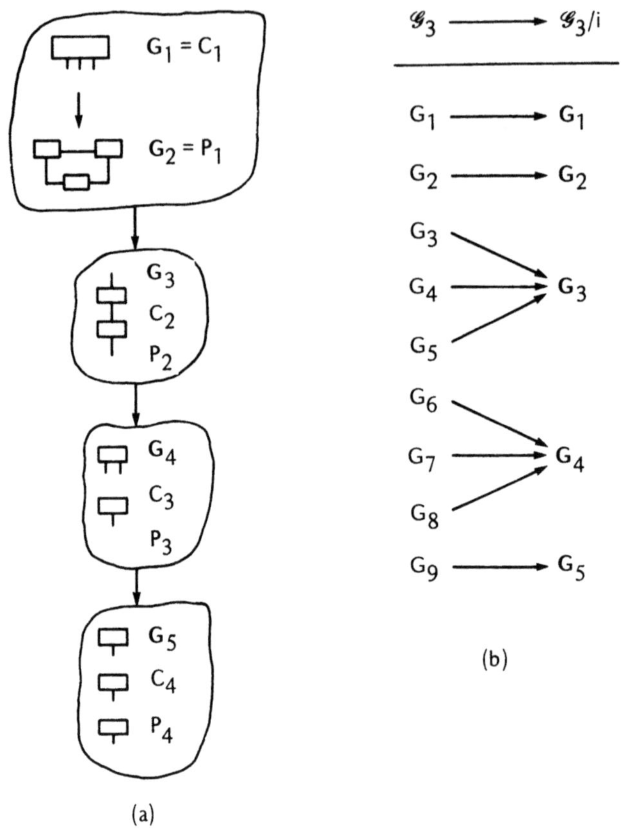

Figure 4.17. Lattice $(\mathcal{G}_3/i, \leq)$ and the homomorphic mapping $\mathcal{G}_3 \rightarrow \mathcal{G}_3/i$.

C-structures in the permutation equivalence class expressed by that block diagram; the number placed next to each arrow indicates the number of immediate refinements of each C-structure of one permutation class in the other class. As explained earlier, this lattice is isomorphic with the lattices defined on $\mathcal{R}_4/i$, $\mathcal{P}_4/i$, and $(\mathcal{G}_4/i)/r$.

While the full lattices $(\mathcal{G}_n, \leq)$ represent the basis for the local level of computation in the reconstruction problem, the lattices $(\mathcal{C}_n, \leq)$ or their isomorphic counterparts are the basis for the global level of computation. To operate at the global computational level, a procedure is required by which all immediate refinements in the lattices $(\mathcal{C}_n, \leq)$ are generated for any given C-structure $C_k \in \mathcal{C}_n$ $(n \in \mathbb{N})$. One such procedure, which utilizes the graph representation of C-structures, is described as follows.

Refinement Procedure for C-structures (or *RC-Procedure*). Given a C-structure $C_k \in \mathcal{C}_n$ and the corresponding graph $r_n(C_k)$, to determine all immediate refinements of C_k in the set $\mathcal{C}_n$:

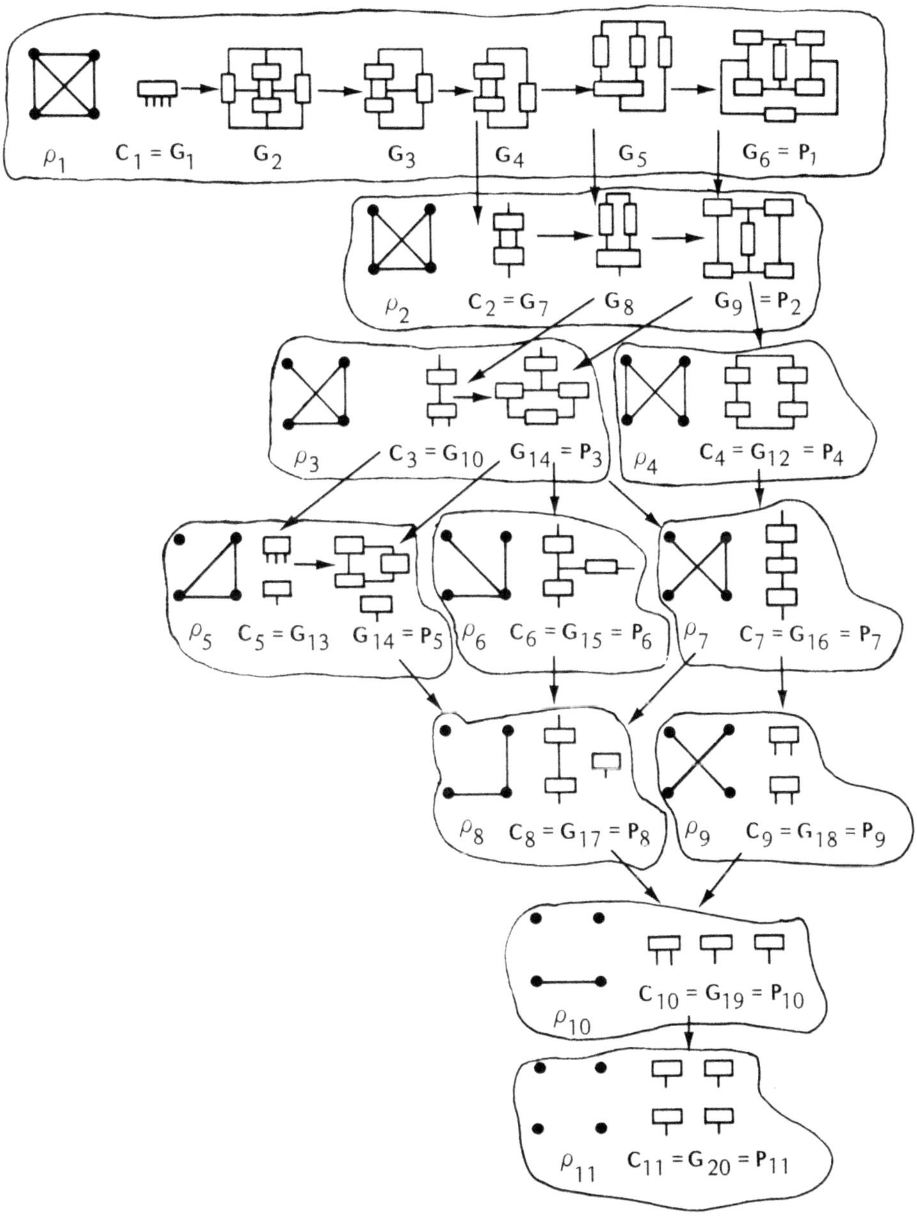

Figure 4.18. Lattice ($\mathscr{G}_4/i, \leq$) with the indication of r-equivalence classes and canonical C-structures and P-structures.

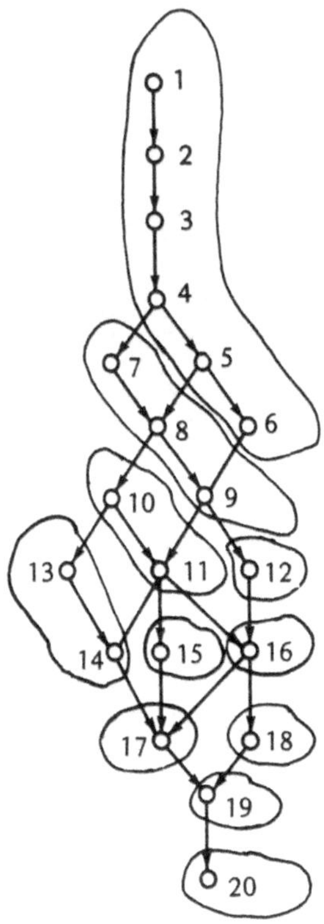

Figure 4.19. Summary of lattice ($\mathscr{G}_4/i, \leq$) described fully in Figure 4.18.

1. Exclude one edge from the graph $r_n(C_k)$, say edge (a, b);
2. split each element x of C_k that contains both a and b into two elements, $x_a = x - \{b\}$ and $x_b = x - \{a\}$, and replace x in C_k with x_a and x_b;
3. exclude all x_a's and x_b's generated in step (2) that are redundant and record the result as an immediate refinement of C_k in the lattice $(\mathscr{C}_n, \leq)$;
4. repeat steps (1)–(3) for all edges of the graph $r_n(C_k)$ and, then, stop.

The procedure is justified by the following facts: (i) there is a one-to-one correspondence between sets $\mathscr{R}_n$ and $\mathscr{C}_n$ and, hence, each change in a graph is reflected by a change in the corresponding C-structure; (ii) the smaller the number of edges in a graph, the more refined the corresponding C-structure is; (iii) since no loop on a vertex may be excluded from a graph without violating the covering requirement for the corresponding C-structure, the smallest possible reduction of the graph is to exclude one of its edges. The number of edges in this graph indicates thus the number of immediate refinements of the corresponding C-structure.

Example 4.18. Consider the graph ρ_1 and the corresponding C-structure C_1 specified in Figure 4.21a. The graph has six edges and, hence, there are six immediate

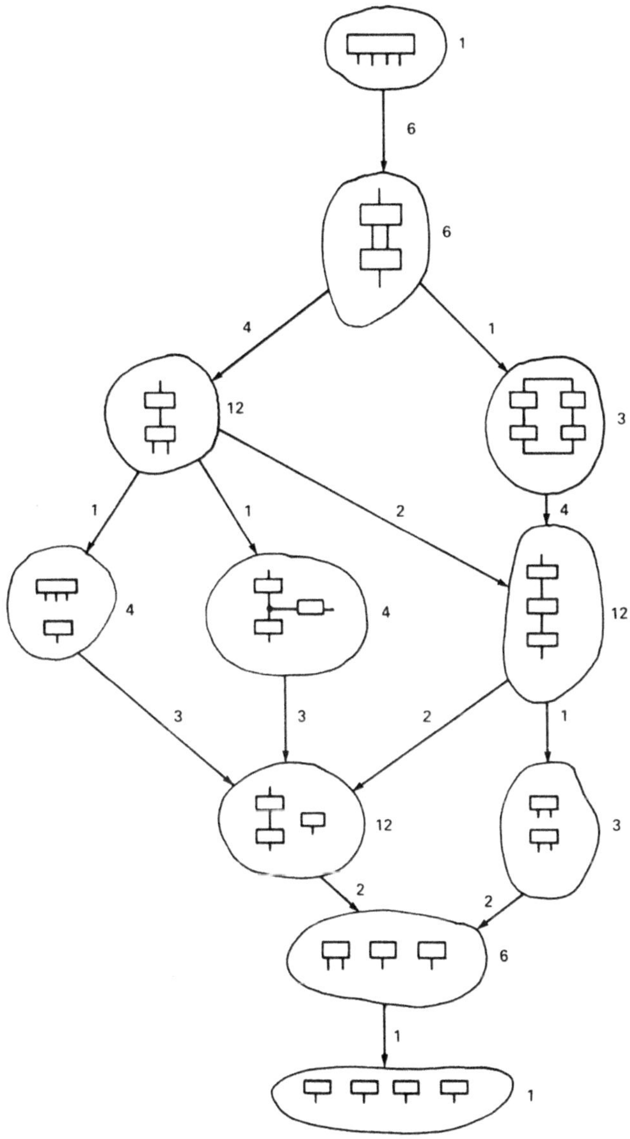

Figure 4.20. Lattice ($\mathscr{C}_4/i, \leq$).

refinements of the C-structure. They are shown in Figure 4.21b. The refinement C_7, for example, is derived by the RC-procedure as follows: (1) edge (4, 5) is excluded from ρ_1 so that graph ρ_7 is obtained; (2) element $\{2, 4, 5\}$ of C_1 (the only element of C_1 that contains both 4 and 5) is split into elements $\{2, 5\}$ and $\{2, 4\}$; (3) since the element $\{2, 4\}$ is the only redundant element $\{2, 4\} \subset \{2, 3, 4\}$), it is excluded and the result $C_7 = \{\{1, 2\}, \{2, 5\}, \{2, 3, 4\}\}$ is recorded as an immediate refinement of C_1.

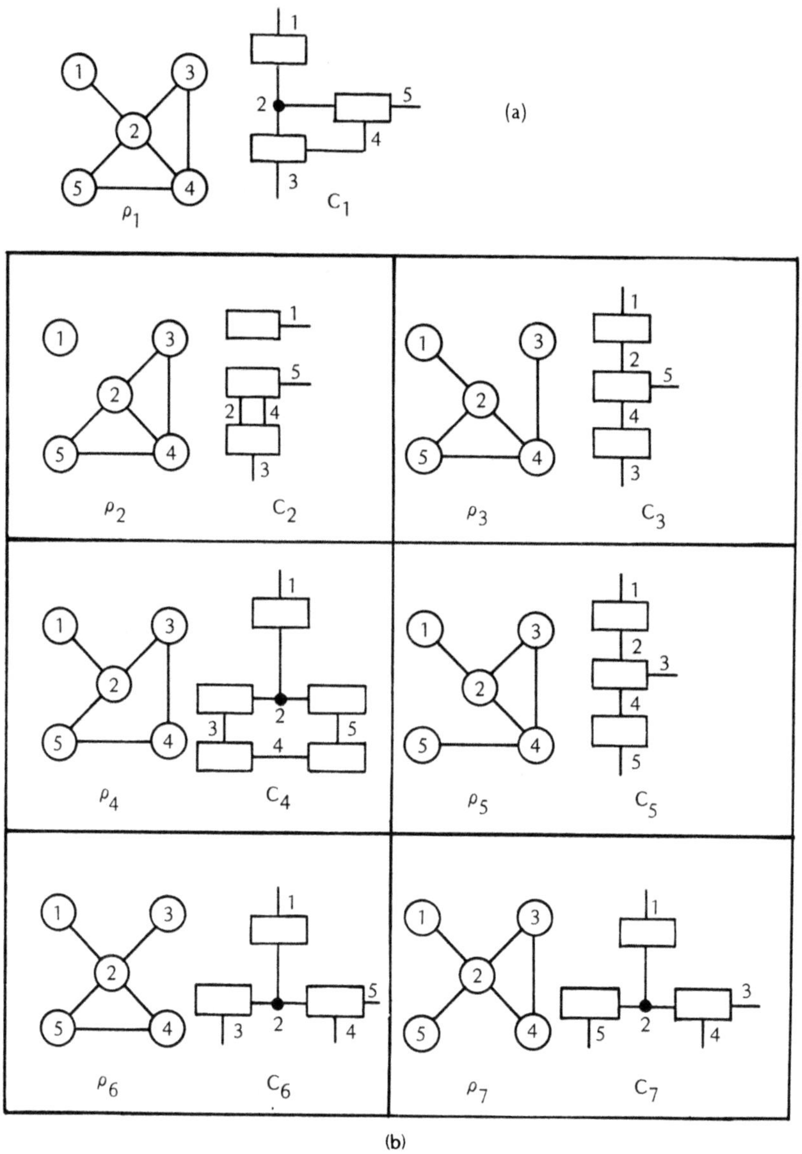

Figure 4.21. Illustration of the RC-procedure: (a) given graph and the corresponding C-structure; (b) immediate refinements of C_1.

Since elements of P-structures represent directly edges of the corresponding graphs, a refinement procedure for P-structures (or ***RP-procedure***) is rather trivial. It consists of excluding individual edges from the given graph [as in step (1) of the RC-structure] and interpreting the results as P-structures.

Procedures by which all immediate coarsenings are determined within a set of G-structures, C-structures, or P-structures are also useful, especially for the purpose of determining the full structure neighborhood of a given structure, but we leave their formulations to the reader as an exercise (see also Note 4.10). Examples of structure neighborhoods are given for the three types of structures in Figures 4.22–4.24. The given structures are denoted in these examples by symbols G, C, and P, respectively. Their immediate refinements are distinguished by subscripts, their immediate coarsenings by superscripts. We can see in Figure 4.22 that the structure neighborhood of a given G-structure may include G-structures that are not in the same r-equivalent class as the given structure (structure G_3 in the figure). In order to be restricted only to immediate refinements in the same r-equivalence class, the RG-procedure can be trivially modified by changing the requirement $|{}^kS| \geq 2$ in step (3) into $|{}^kS| > 2$. Then, indeed, no element of the given structure that contains only two integers would be allowed to change and, consequently, the graph of the given G-structure would remain intact.

The notion of immediate refinements (or coarsenings) of structures of the various types can be employed for partitioning the corresponding set of structures into blocks of structures that are equivalent in their refinement level, i.e., are reached from the universal upper bound $\{N_n\}$ of the respective refinement lattice through sequences of

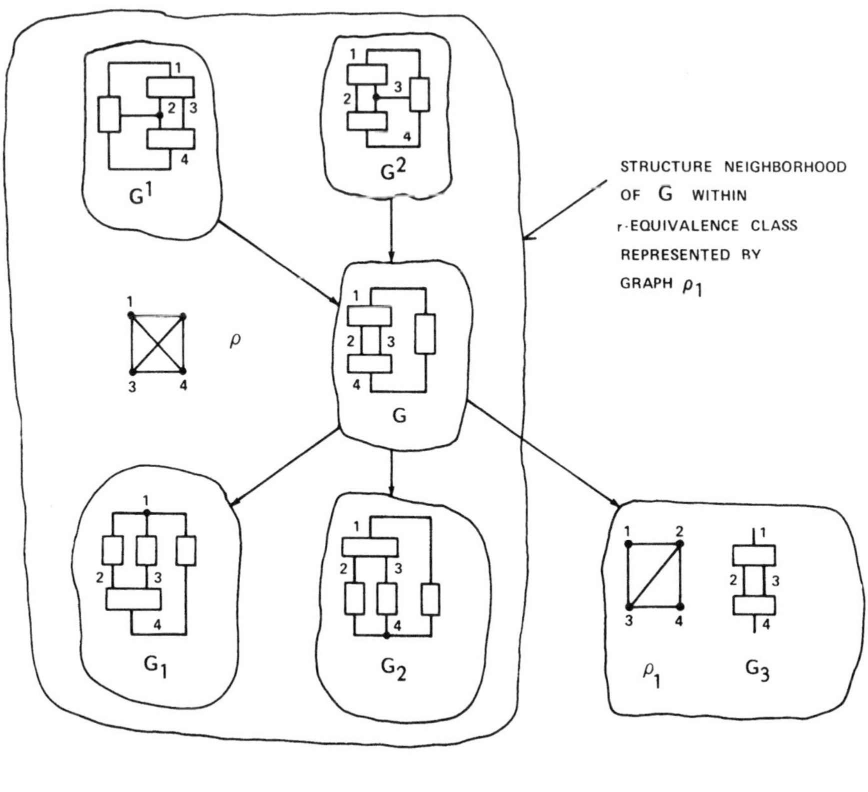

Figure 4.22. Structure neighborhood of G-structure G.

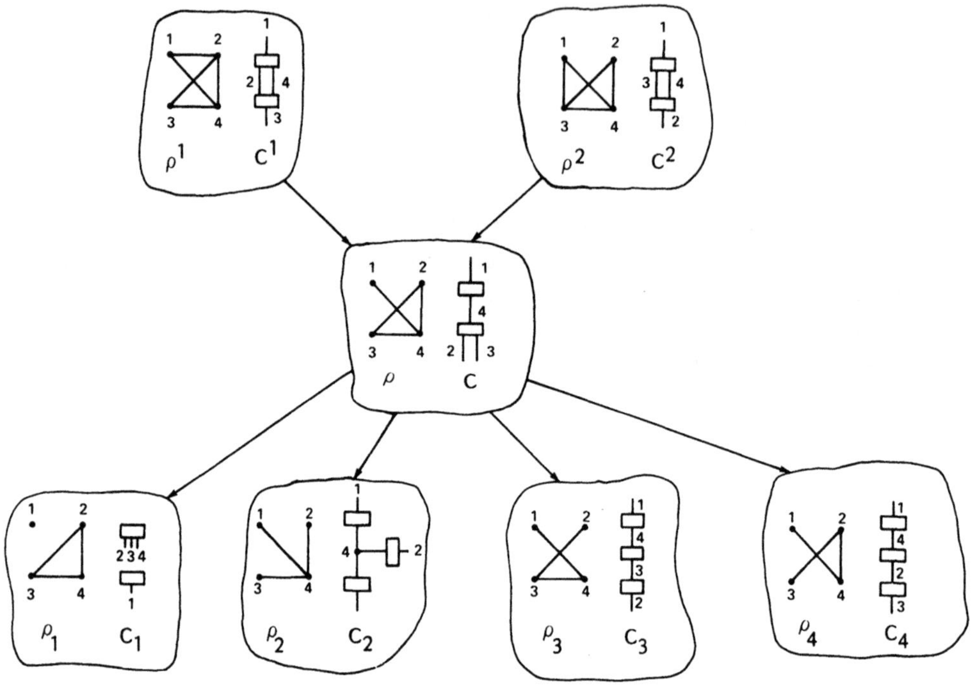

Figure 4.23. Structure neighborhood of C-structure C.

immediate refinements of the same length. Let us call this equivalence a *refinement level equivalence* or *l-equivalence* and denote it by $\overset{l}{\equiv}$. For instance, structures G_1, G_2, G_3 in Figure 4.22 are l-equivalent G-structures in the set $\mathscr{G}_4$; structures specified in Figure 4.21b are l-equivalent C-structures in the set $\mathscr{C}_5$; structures P_1, P_2, P_3, P_4 in Figure 4.24 are l-equivalent P-structures in the set $\mathscr{P}_4$.

To get some feeling about the growth of the numbers of structures of the three types with n, as well as the numbers of their i-equivalence and l-equivalence classes, the known results for $n \leq 7$ are summarized in Table 4.12. It is clear that

$$|\mathscr{C}_n| = |\mathscr{P}_n| = |\mathscr{R}_n|$$

and, obviously,

$$|\mathscr{R}_n| = 2^{n(n-1)/2},$$

where the exponent $n(n-1)/2$ represents the total number of possible edges in graphs defined on N_n. Clearly, $|\mathscr{C}_n/l| = n(n-1)/2 + 1$. The counting of $|\mathscr{R}_n/i|$ for a given n is more complicated, but this combinational problem has been solved in graph theory

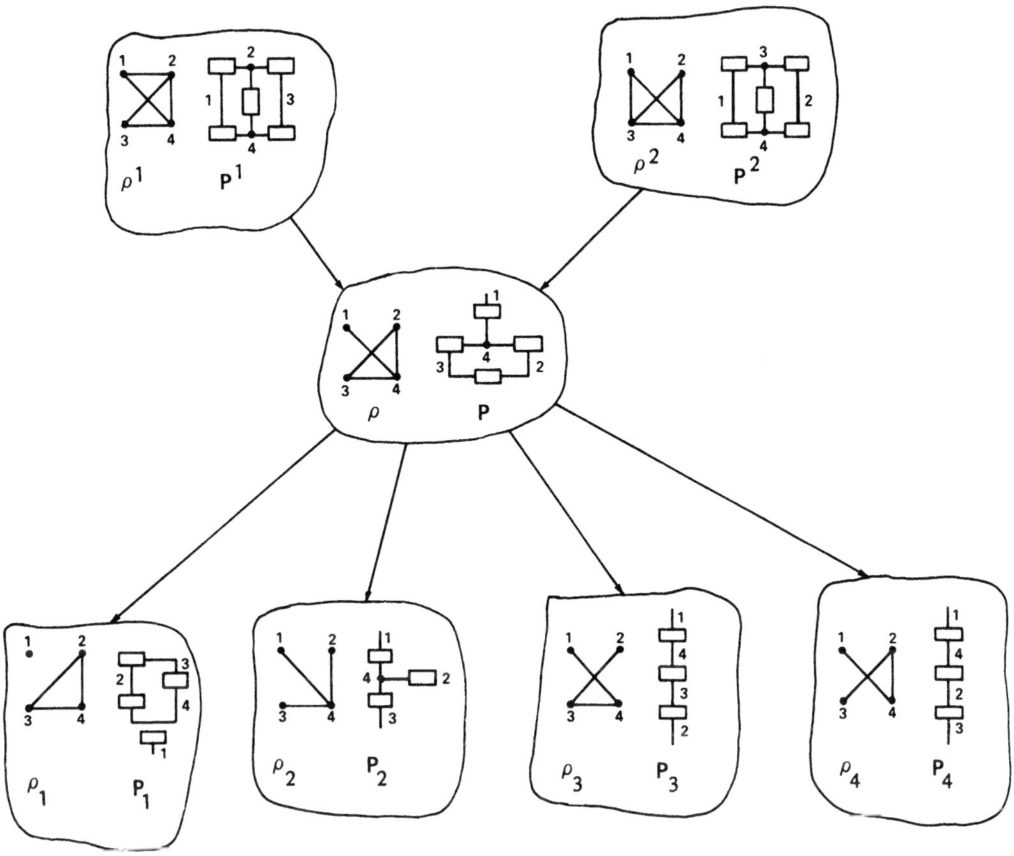

Figure 4.24. Structure neighborhood of P-structure P.

TABLE 4.12
Numbers of G-Structures in $\mathcal{G}_n^+$ and $\mathcal{G}_n$, C-Structures, and their Isomorphic Equivalence Classes and Refinement Level Classes for $n \leq 7$

n	1	2	3	4	5	6	7
$\lvert\mathcal{G}_n^+\rvert$	1	4	18	166	7,579	7,828,352	2,414,682,040,996
$\lvert\mathcal{G}_n\rvert$	1	2	9	114	6,894	7,785,062	2,414,627,396,434
$\lvert\mathcal{C}_n\rvert$	1	2	8	64	1,024	32,768	2,097,152
$\lvert\mathcal{G}_n^+/i\rvert$	1	3	8	28	208		
$\lvert\mathcal{G}_n/i\rvert$	1	2	5	20	180		
$\lvert\mathcal{C}_n/i\rvert$	1	2	4	11	34	156	1,044
$\lvert\mathcal{G}_n^+/l\rvert$	1	3	7	15	31	63	127
$\lvert\mathcal{G}_n/l\rvert$	1	2	5	12	27	58	121
$\lvert\mathcal{C}_n/l\rvert$	1	2	4	7	11	16	22

(Note 4.11). The numbers given in Table 4.11 for $|\mathcal{G}_n^+|$, $|\mathcal{G}_n^+/i|$, $|\mathcal{G}_n^+/l|$, $|\mathcal{G}_n|$, $|\mathcal{G}_n/i|$, and $|\mathcal{G}_n/l|$ are known only for $n \leq 7$.

Complete catalogs of lattices $(\mathcal{G}_n/i, \leq)$ as well as lattices $(\mathcal{C}_n/i, \leq)$ for $n = 3, 4, 5$, which were determined exhaustively by the RG-procedure and an isomorphism test on a computer, are given in Appendix D.

► Structures of one additional type, mentioned previously, should be introduced. These are structures for which the iterative join procedure is not needed to determine the unbiased reconstruction when dealing with probabilistic systems. They are usually called *loopless structures.*

Loopless structures of a special kind, which may be called *strict loopless structures* or L-structures, are of particular interest in the reconstruction problem. We say that a G-structure $G_i \in \mathcal{G}_n$ is also a L-structure if and only if none of the pairs $(a, b) \in N_n^2$ is *both* included in some elements of G_i *and* connected through several coupled elements. Let us denote the set of all L-structures for each n by $\mathcal{L}_n$.

To define the set $\mathcal{L}_n$ formally, assume that a G-structure $G_i \in \mathcal{G}_n$ is given. For each pair $(a, b) \in N_n^2$, let

$$X_{a,b} = \{x \mid x \in G_i, \{a, b\} \subset x\}.$$

Then, G_i is a L-structure ($G_i \in L_n$) if and only if no pair $(a, b) \in N_n^2$ is an element of the transitive closure of $r_n(G_i - X_{a,b})$.

For example, all structures in the set $\mathcal{G}_3$ (Figure 4.16) are obviously L-structures except the structure $G_2 = \{\{1, 2\}, \{2, 3\}, \{3, 1\}\}$. In the set $\mathcal{C}_4$, there are only three structures that are not L-structures. They are in the same i-equivalence class, which may be represented, e.g., by the C-structure $C = \{\{1, 2\}, \{2, 3\}, \{3, 4\}, \{4, 1\}\}$. Indeed, the transitive closure of $r_4(C - X_{1,2})$ is N_4 and, hence, the pair (1, 2) is an element of it.

The significance of L-structures is that the iterative join procedure is not needed for them, independently of the order in which elements of the L-structure under consideration enter the join operations. Loopless structures that are not L-structures do not possess this convenient property. The iterative join procedure can be avoided for them only for some specific orders in which their elements are employed in the basic join procedure. This requires further tests, computationally quite involved, to determine at least one of the proper orders; the methodological significance of these structures is thus considerably reduced. ◄

Using the various concepts introduced in this section, the reconstruction problem can now be discussed in more specific terms. Given an overall system and a set of its reconstruction hypotheses specified by the user (based, e.g., on the set $\mathcal{G}_n$, $\mathcal{C}_n$, $\mathcal{P}_n$, or $\mathcal{L}_n$ of structures), the reconstruction problem amounts to selecting a subset of reconstruction hypotheses from the given set according to some requirements. Normally, it is required that (i) distances associated with the selected reconstruction hypotheses be as small as possible, and (ii) the hypotheses themselves be as refined as possible. Either of these requirements imposes an ordering on the set of reconstruction hypotheses involved. The ordering based on requirement (ii) is fixed—it is the partial ordering of structure refinement with the lattice properties described previously. The ordering based on requirement (i), which may be appropriately called a *distance ordering*, is not fixed. It depends on the given overall system as well as the chosen kind of distance, and can be determined only by calculating the unbiased reconstructions and distances of the individual reconstruction hypotheses.

If the distance expressed by formula (4.40) for probabilistic systems [or formula (4.42) for possibilistic systems] is used, which measures the amount of information lost when the overall system is replaced by a reconstruction hypothesis, then there exists a specific *distance preordering*:* the information distance is monotonically nondecreasing with increasing refinement of reconstruction hypotheses. In addition, both versions of the information distance are *additive* along any path in the refinement lattice involved. This means that

$$D(f^x, f^z) = D(f^x, f^y) + D(f^y, f^z) \tag{4.43}$$

for any three reconstruction hypothesis x, y, z of the same overall systems such that $x \geq y \geq z$. The properties of preordering and additivity are very useful in dealing with the reconstruction problem and give the information distance a special significance. In our further discussion of the reconstruction problem, we always assume the use of the relevant version of the information distance (i.e., probabilistic or possibilistic and basic or generative).

When combined, the distance ordering and refinement ordering form a joint preference ordering associated with the reconstruction problem. The solution set in the reconstruction problem is then characterized in terms of this combined ordering as follows: it consists of such a subset of reconstruction hypotheses from the given set that contains no hypothesis inferior to any other hypothesis from the viewpoint of the combined ordering. The term "inferior" is used here in the usual sense: h_1 is inferior to h_2 if and only if either h_1 is less refined and its distance is not smaller than that of h_2, or h_1 has a larger distance than h_2 and, at the same time, it is not more refined than h_2. Let elements of the solution set be called *admissible reconstruction hypotheses.*

We can now observe a striking similarity of the reconstruction problem with two problems discussed previously, the problem of determining admissible behavior systems (Sections 3.4 and 3.6) and the problem of determining admissible simplifications of a given behavior system (Section 3.9). When the uncertainty and complexity orderings in these problems are compared with the distance and refinement orderings in the reconstruction problem, respectively, the similarities among these three problems become obvious. We leave it to the reader to utilize these similarities and define the combined ordering and solution set for the reconstruction problem formally.

We see that problems associated with climbing up the epistemological hierarchy as

* The term "preordering" is not used here in the technical sense of a reflexive and transitive relation; it is, in fact, a partial ordering.

well as problems of systems simplification form a major problem category, which has the following general characterization:

GIVEN:

—a set X of considered (acceptable) systems;
—a set of preference ordering relations $\overset{a}{\leq}, \overset{b}{\leq}, \overset{c}{\leq}, \ldots$, on X.

SOLUTION SET:

$$X_s = \{x \in X \mid (\forall y \in X)(y \overset{*}{\leq} x \Rightarrow x \overset{*}{\leq} y)\},$$

where $\overset{*}{\leq}$ is a join preference ordering on X defined for all $x, y \in X$ as

$$x \overset{*}{\leq} y \quad \text{iff} \quad x \overset{a}{\leq} y \quad \text{and} \quad x \overset{b}{\leq} y \quad \text{and} \quad x \overset{c}{\leq} y \quad \text{and} \ldots.$$

Three sets of procedures are required in the process of solving the reconstruction problem:

i. procedures through which all desirable reconstruction hypotheses can be generated;
ii. procedures through which the generated reconstruction hypotheses can be evaluated and compared with respect to the objectives of the reconstruction problem;
iii. procedures through which it can be decided, at relevant points in the solution process, which of the generated reconstruction hypotheses should be accepted as members of the solution set, which of them should be used as a basis for generating further reconstruction hypotheses, and whether the solution process should continue or terminate.

The way in which these three sets of procedures are integrated in the overall solution process is shown schematically in Figure 4.25. The kernel of the process is the generation of all desirable reconstruction hypotheses. This can be done conveniently by generating relevant structures, which are then interpreted as reconstruction hypotheses in terms of the given overall system. The interpretation is made by assigning the variables of the overall system to the integers involved in the structures and calculating then the required projections of the overall behavior function. The generation of structures may be restricted in various ways to reduce computing time and the corresponding cost or, possibly, for some other reasons. For example, it may be restricted to a subset of all relevant G-structures, such as relevant C-structures or

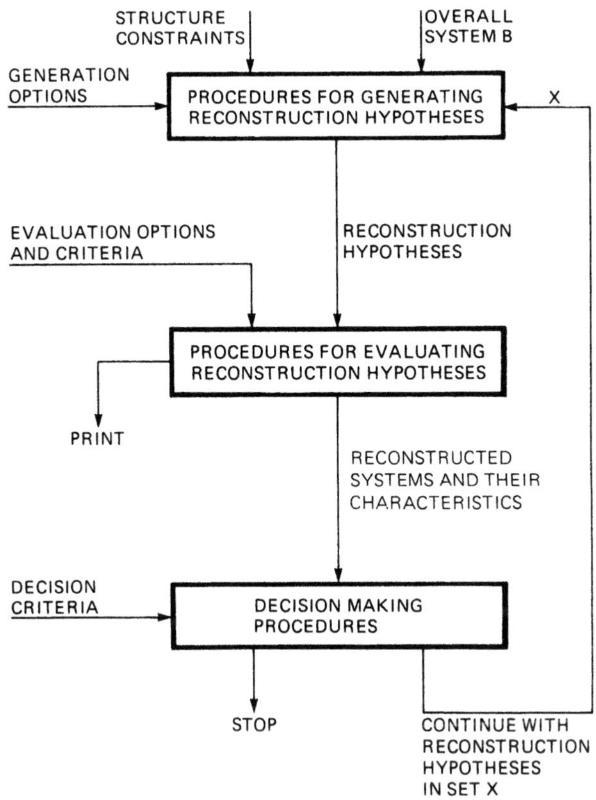

Figure 4.25. General scheme of the solution process associated with the reconstruction problem.

L-structures, or it may be restricted by considering only levels of refinement (l-equivalence classes) for which the loss of information does not exceed some acceptable value specified by the user. The generation of reconstruction hypotheses may be thus restricted either by constraining the set of structures considered or by committing to a special generation option.

A multitude of generation options should be made available in the GSPS to allow flexibility in dealing with the reconstruction problem (Note 4.10), but this aspect is beyond architectural considerations. The essence of the various procedural options is to generate appropriate refinements (or coarsenings) of given reconstruction hypotheses, as exemplified by the RG-procedure and RC-procedure (and their coarsening counterparts).

As mentioned previously, the generation of structures can be also organized at several computational levels. For example, the RC-procedure may be used at the *global level* to deal only with r-equivalence classes of G-structures. The RG-procedure modified by replacing the condition $|^kS| \geq 2$ in step (3) by $|^kS| > 2$ is then used at the *local level* to generate immediate refinements in some significant r-equivalence classes determined at the global level.

It is often necessary to generate refinements or coarsenings of several structures defined at the same level of refinement. In such cases, a procedure is needed by which the generation of duplicate structures is prevented.

The input to the procedures for evaluating reconstruction hypotheses represented by the second block in Figure 4.25, consists of the generated reconstruction hypotheses and various evaluation options and criteria specified by the user. The latter include definitions of the desirable distance (as well as other required characteristics such as the identifiability quotient, reconstruction uncertainty, or confidence degree of some sort) and a principle upon which the reconstruction should be based (unbiased reconstruction, min-max reconstruction, etc.). The information distance and unbiased reconstruction, which are theoretically well justified, should normally be used as the default options. The received reconstruction hypotheses are evaluated in the required way and compared. When results relevant to user's interest are obtained, particularly those pertaining to the solution set, they are printed.

The decision-making procedures, illustrated by the third block in Figure 4.25, use information regarding the evaluated reconstruction hypotheses and make various decisions, according to decision criteria specified by the user. The most fundamental decisions are whether to continue or terminate the solution process and, if the process continues, which of the considered reconstruction hypotheses should be used in the next stage (set X in Figure 4.25).

Let us illustrate now the various issues of the reconstruction problem, as outlined in this section, by several examples.

Example 4.19. Consider a possibilistic behavior system that was derived from data obtained by monitoring four variables defined on a computer complex. The objective is to find conditions under which the utilization of CPU (central processing unit) is high. The monitored variables represent the utilization of CPU and three communication channels, say channels C1, C2, and C3. The monitoring was performed in the period of one hour of a typical workload and in each interval of 1 sec the utilization of each of the observed units was recorded. Hence, 3,600 observations were made. If it was observed during a particular interval of 1 sec that the utilization of a unit was smaller than a certain threshold defined by the investigator (and based on previous experimental studies), it was considered low (L); if it was higher than that threshold, it was considered high (H).

The investigator decided to use the possibilistic and memoryless methodological options for deriving a behavior system from the data. Only six states of the 16 states defined for the variables were actually observed. Since they occurred in the data with approximately the same frequencies, the investigator decided to discriminate by the behavior function only between observed and unobserved states. He thus declared the observed states as the only possible states and assigned to each of them the possibility degree of 1; the remaining states (which were not observed) were then assigned the

TABLE 4.13
Possibilistic Behavior Functions of the Overall System and Some Unbiased Reconstructions Discussed in Example 4.19

CPU	$C1$	$C2$	$C3$	f	$f^1=f^4=f^6$	f^2	f^3	f^5	$f^7=f^8=f^9$	$f10$
L	L	L	L	1	1	1	1	1	1	1
L	L	L	H	1	1	1	1	0	1	1
L	L	H	L	0	0	1	0	0	0	0
L	L	H	H	0	0	1	1	1	0	1
L	H	L	L	0	0	0	1	0	0	0
L	H	L	H	0	0	0	0	1	0	0
L	H	H	L	0	0	0	0	0	1	0
L	H	H	H	1	1	1	1	1	1	1
H	L	L	L	0	0	1	1	1	0	1
H	L	L	H	0	0	1	0	1	0	0
H	L	H	L	1	1	1	1	1	1	1
H	L	H	H	1	1	1	1	1	1	1
H	H	L	L	1	1	1	1	1	1	1
H	H	L	H	0	0	0	0	0	1	0
H	H	H	L	0	0	0	0	1	0	0
H	H	H	H	0	0	0	1	0	0	0

possibility degree of 0. The result is the behavior function f specified in Table 4.13. It provides the investigator with an important insight: the utilization of CPU can be kept high by some changes in the computer organization under which the utilization of the three channels is mixed in one of the following three ways:

$C1$	$C2$	$C3$
L	H	L
H	L	L
L	H	H

To further enhance this insight, the investigator decided to explore the reconstruction properties of the overall behavior function f. He was interested only in reconstruction hypotheses with no loss of information.

The variables are obviously strongly constrained in this case (only 6 out of 16 states defined for the variables are possible). However, we can easily determine that projections of f associated with any pair of the four variables (there are six such pairs) are totally unconstrained, i.e., the possibility degree is 1 for all four states defined for the respective pair of variables. Hence, the variables are pair-wise independent and the overall behavior function cannot be reconstructed solely from its two-dimensional projections.

To determine whether f can be reconstructed from any projections at all, it is useful to consider reconstruction hypotheses based on C-structures first. Using the

RC- procedure, we obtain six reconstruction hypotheses at the first level of refinement. Their block diagrams and the associated graphs are specified in Figure 4.26a, each labeled by an integer at the left top corner of the respective box, and their unbiased reconstructions f^h ($h \in N_6$) are given in Table 4.13. Hypotheses 1, 4, and 6 reconstruct f exactly and are thus prospective candidates for the solution set. Each of the remaining hypotheses produces four incorrect states of the overall system. Their information distances, calculated by formula (4.42), are thus

$$(\log_2 10 - \log_2 6)/\log_2 16 = (3.32 - 2.58)/4 = 0.185.$$

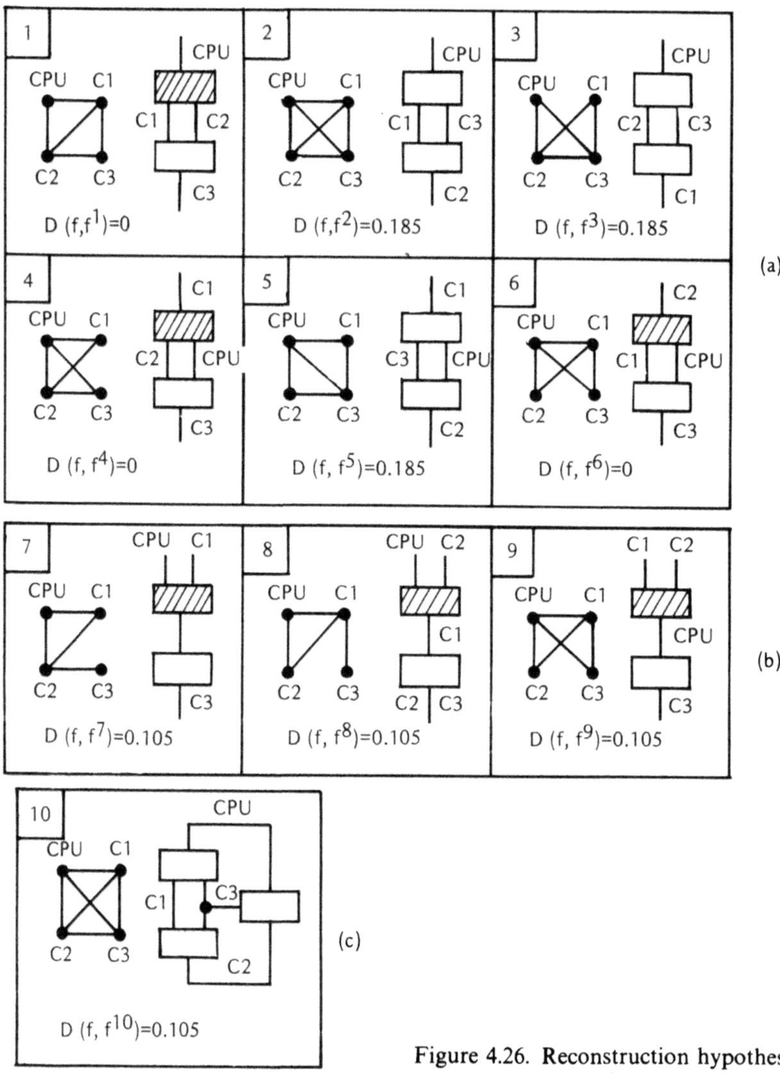

Figure 4.26. Reconstruction hypotheses evaluated in Example 4.19.

The unbiased reconstructions are determined, of course, by the possibilistic version of the join procedure. It is illustrated for reconstruction hypothesis 1 in Figure 4.27. Connections in the diagram indicate states of the individual three-dimensional projections whose possibility degrees are 1. The result of the join procedure, in which the join operation is performed only once, consists in this case of all quadrupples of states L and H that lie on paths in the diagram connecting its left and right nodes.

When inspecting the three successful reconstruction hypotheses in Figure 4.26a, we see that all of them contain the subsystem based on variables CPU, $C1$, and $C2$, as indicated by the shaded block. Any potentially successful hypotheses at the next refinement level must thus contain this subsystem. There are only three hypotheses of this kind, which are specified in Figure 4.26b. As indicated in Table 4.13, their unbiased reconstructions are equal. This is due to the fact that the two-dimensional projections do not contain any information. The reconstructions are not perfect: eight instead of six states are reconstructed and the distances are equal to 0.105. They are thus not acceptable according to the requirements of this problem.

Coarsenings of the three successful reconstruction hypotheses need not be considered since they are clearly inferior: they are less refined (by definition) and their distances cannot be smaller than those of the successful hypotheses (i.e., they cannot be smaller than 0). However, coarsenings of the hypotheses 2, 3, and 5 in Figure 4.26a (the unsuccessful ones) must be considered. Taking advantage of Figure 4.18, which describes the relevant lattice of G-structures, we can see that immediate refinements of the structures under consideration are the G-structures in the isomorphic class $\mathbf{G}_4$. The two subsystems with three variables are chosen from within any of the subsystems shown in Figure 4.26a except the subsystem represented by the variables CPU, $C1$, and $C2$ (and shaded in the figure), which is associated with the successful hypotheses. We know, however, that these pairs of subsystems are not successful, and also that no subsystem based on two variables adds in this case any information. Hence, all the

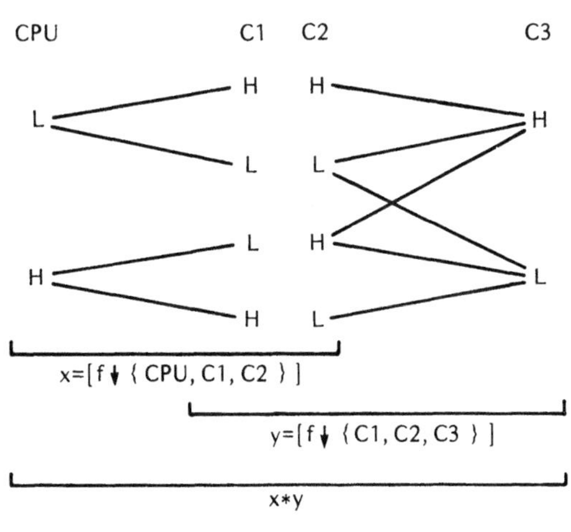

Figure 4.27. Illustration of the join procedure for the reconstruction hypothesis 1 in Figure 4.26.

reconstruction hypotheses represented by the isomorphic class $\mathbf{G}_4$ can be rejected without even calculating their unbiased reconstructions and distances.

It remains to consider reconstruction hypotheses based on the G-structures in the isomorphic class $\mathbf{G}_3$. Since the subsystems represented by the variables CPU, $C1$, and $C2$ cannot again be included, there is only one hypothesis to be considered; it is specified in Figure 4.26c and its reconstructed behavior function f^{10} is given in Table 4.13. We see that this hypothesis is not perfect; its distance is 0.105 and, hence, it must be rejected.

Since the only reconstruction hypothesis based on $\mathbf{G}_2$ is a coarsening of the successful hypotheses, it need not be considered. This leads to the conclusion that the solution set consists of the reconstruction hypotheses 1, 4, and 6 specified in Figure 4.26a.

This result should improve the insight of the investigator by focusing his attention on the critical subsystem based on the variables CPU, $C1$, and $C2$, which represents the successful hypotheses. According to this subsystem, the utilization of CPU can be kept high by any arrangement in the computer complex by which the utilization of the channels $C1$ and $C2$ is prevented from being high or low for both of them simultaneously.

Example 4.20. This example is based on data collected in a study of premarital contraceptive usage (Note 4.13). States of the following binary variables were determined for a population of 414 undergraduate female university students:

v_1—attitude on extramarital coitus (0—always wrong, 1—not always wrong);
v_2—use of the university contraceptive clinic (0—yes, 1—no);
v_3—virginity (0—virgin, 1—nonvirgin).

Frequencies $N(c)$ of the individual states and the corresponding probabilistic behavior function f are given in Table 4.14a.

When reconstruction hypotheses based only on C-structures are considered, we can use the RC-procedure to obtain hypotheses at the first level of refinement. Their block diagrams, graphs, and information distances (obtained as a result of their evaluation) are given in Figure 4.28. Using the information distance preordering and the distances at the first level of refinement, we can determine lower bounds of distances for all reconstruction hypotheses at the second level of refinement, as indicated in the figure. For instance, $D^6 \geq 0.0637$ since hypothesis 6 is a refinement of hypothesis 3 and $D^3 = 0.0637$. These lower bounds of distances directly imply that hypothesis 2 is a member of the solution set.

We evaluate now hypothesis 4, which has the least lower bound among the competing hypotheses at the second level of refinement, and obtain the actual distance $D^4 = 0.0127$. Since it is smaller than any of the lower bounds of the other hypotheses and $D^7 \geq 0.0637$, hypothesis 4 is a member of the solution set. Observe that we arrived at this conclusion without actually evaluating either the competing hypotheses or the successor. If we are interested in hypothesis 7, we can determine that $D^7 = 0.0802$ and,

TABLE 4.14
Behavior Functions in (a) Example 4.20, and (b) Example 4.21

(a)					(b)				
v_1	v_2	v_3	$N(\mathbf{c})$	$f(\mathbf{c})$	s_1	s_2	s_3	s_4	$f(\mathbf{c})$
$\mathbf{c} = 0$	0	0	23	0.056	$\mathbf{c} = 0$	0	0	0	1/3
0	0	1	127	0.307	0	0	1	0	2/3
0	1	0	23	0.056	0	0	2	0	1/3
0	1	1	18	0.043	0	0	3	0	1/3
1	0	0	29	0.070	0	1	0	0	1
1	0	1	112	0.270	0	0	3	1	1/3
1	1	0	67	0.162	0	1	0	1	2/3
1	1	1	15	0.036	0	1	1	1	1/3
					0	1	2	1	1
					0	1	3	1	1/3
					0	0	2	2	1/3
					1	0	1	2	1
					1	0	2	2	1/3
					1	0	3	2	2/3
					1	1	3	2	1/3
					0	1	0	3	1/3
					1	1	1	3	2/3
					1	1	2	3	2/3
					1	1	3	3	1/3

obviously, we must include it in the solution set since it is more refined than any other hypothesis in the refinement lattice.

Members of the solution set are shown in Figure 4.28 by the shaded boxes. We can see that they are totally ordered in this case by the combined ordering relation. We can see that variables v_1 (attitude) and v_2 (use of clinic) are more determined by variable v_3 (virginity) than by each other. The relationship is particularly strong between v_2 and v_3.

Example 4.21. The aim of this example is to illustrate some issues that arise in the reconstruction problem where the given behavior system is memory dependent. The system represents three variables defined for an individual person (v_1—job performance, v_2—overall health condition, v_3—stress) whose support is time (totally ordered). Observations were made each day for some period of time. The constraint among the variables is expressed in terms of the possibilistic behavior function in Table 4.14b, which is defined on the state set of the following sampling variables:

$$s_{1,t} = v_{1,t}, \qquad s_{2,t} = v_{2,t},$$
$$s_{3,t} = v_{3,t}, \qquad s_{4,t} = v_{3,t-1}.$$

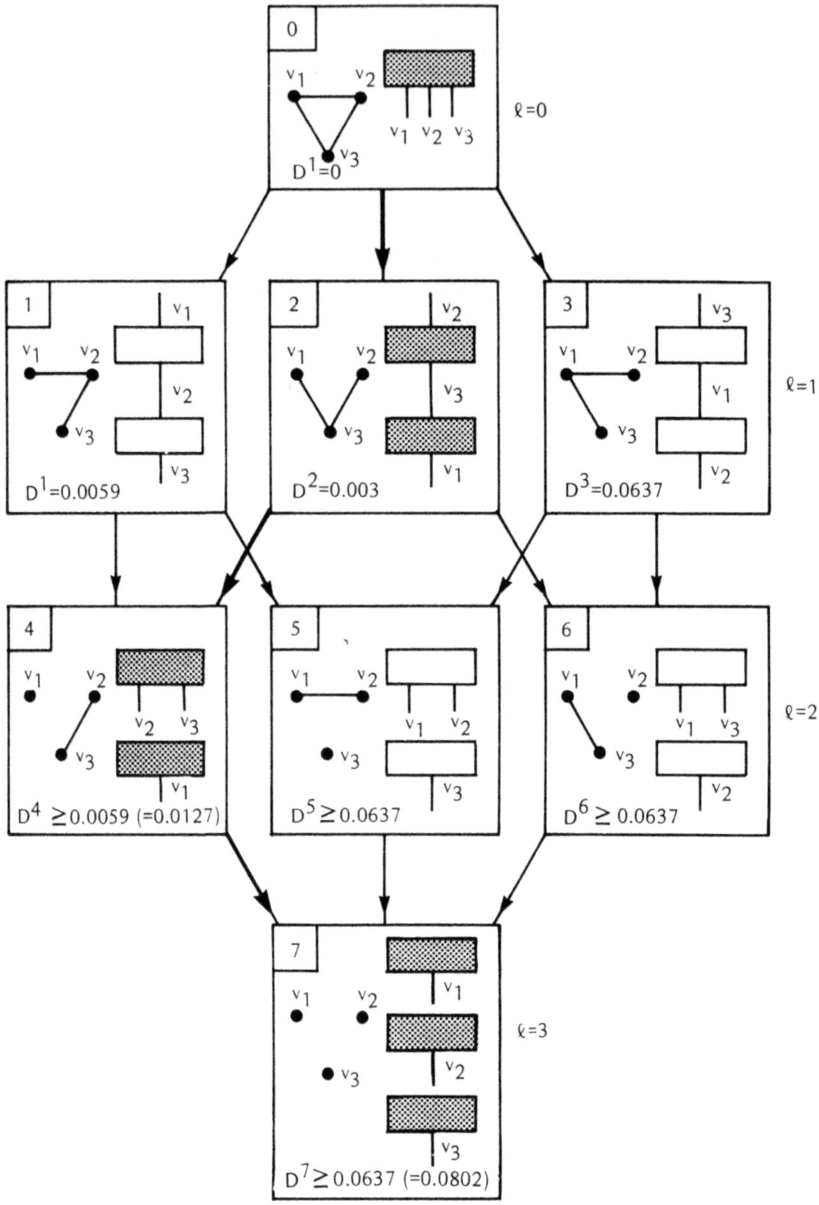

Figure 4.28. Illustration of the reconstruction problem discussed in Example 4.20.

The behavior function was derived from data by the mask evaluation method described in Section 3.6. Without going into further details regarding previous stages of this investigation, let us focus on the reconstruction problem of the given behavior function. Assume the standard formulation of the problem based on the concepts of unbiased reconstruction and information distance. Assume further that reconstruction hypotheses based only on C-structures are requested and that the maximum acceptable distance is 0.1.

First, we generate and evaluate reconstruction hypotheses based on C-structures at the first level of refinement. They are specified in Figure 4.29 (hypotheses 1–6), where the sampling variables s_k are represented by their identifiers $k\,(k \in N_4)$ and subsets of variables are separated by slashes. The evaluation of these hypotheses consists of determining their unbiased reconstructions (by the possibilistic version of the join procedure) and calculating their distances [by formula (4.42)].

Since hypothesis 4 has the smallest distance ($D^4 = 0$), we generate and evaluate all its immediate C-refinements. There are five of them, labeled 7–11. The smallest distance in this group is $D^{10} = 0.021$. It follows from the monotonicity of the information distance that the distance of each hypothesis at the first level of refinement is also a lower bound of distances of all its refinements. Hence, hypothesis 5, whose distance is smaller than 0.021 ($D^5 = 0.0179$), is the only one at the first level of refinement that has the potential of being a source of refinements with distances smaller than or equal to 0.021. However, we can easily find that each immediate refinement of hypothesis 5 is also a refinement of one of the other hypotheses at the first level. This implies that each immediate refinement of hypothesis 5 is either among hypotheses 7–11 or among those with lower bounds of distances greater than 0.021. Hence, hypothesis 10 is the best one at the second level. Its immediate refinements are hypotheses 12–15, among which hypothesis 13 has the smallest distance.

To make sure that hypothesis 13 is the best one at the third level, we have to evaluate all remaining hypotheses at this level except those which are refinements of

1 123/234 D^1=.0774	2 123/134 D^2=.058	3 123/124 D^3=.0952	4 124/234 D^4=0	5 124/234 D^5=.0179	6 124/134 D^6=.0449	ℓ=1
7 234/13 D^7=.086	8 134/23 D^8=.0681	9 14/24/13/23 D^9=.122	10 234/14 D^{10}=.021	11 134/24 D^{11}=.0477	ℓ=2	
12 234/1 D^{12}=.1626	13 14/24/34 D^{13}=.065	14 14/24/23 D^{14}=.1354	15 14/34/23 D^{15}=.0887	16 12/14/34 D^{16}=.1188	ℓ=3	
17 12/13/4 D^{17}=.2381	18 12/23/4 D^{18}=.1743	19 12/24/3 D^{19}=.2056	20 13/34/2 D^{20}=.2381	ℓ=4		

Figure 4.29. Reconstruction hypotheses evaluated in Example 4.21.

hypotheses 1, 3, 7, and 8, whose lower bounds exceed D^{13}. Since hypothesis 7 is a refinement of hypothesis 1, we can neglect it. Graphs of hypotheses 1, 3, and 8 are shown in Figure 4.30a. Graphs of hypotheses that are not their refinements at the third level must contain edges (1, 4), (3, 4), and either edge (1, 2) or edge (2, 4). There are only two graphs that satisfy these conditions. They are shown in Figure 4.30b. The first one is actually the graph of hypothesis 13; the second one represents hypothesis 12/14/34, which is the only potential competitor of hypothesis 13.

After evaluating this potential competitor, which is labeled in Figure 4.29 as hypothesis 16, we see that $D^{16} = 0.1188 > D^{13}$. Hence, hypothesis 13 is the best one at the third level of refinement.

Since the smallest distance at the third level ($D^{13} = 0.065$) is smaller than the largest acceptable distance (0.1), we have to explore level 4. There are 15 hypotheses at this level (represented by all pairs of edges in the graphs with four nodes), but only four of them are not refinements of hypotheses 9, 12, 14, and 16, whose distances exceed the critical value 0.1. They are: 12/13/4, 12/23/4, 12/24/3, and 13/34/2. Their labels and distances are given in Figure 4.29. Since all of the distances are greater than 0.1, none of these hypotheses is a member of the solution set and no further refinements are necessary. The solution set is totally ordered and consists of hypotheses 4, 10, and 13 (and, possibly, hypothesis 0—the overall system 1234).

Observe that by utilizing the preordering of the information distance, we were able to solve this problem (with complete certainty) by evaluating only 20 out of 63 possible reconstruction hypotheses, i.e., less than one third. For systems with larger number of variables, the utilization of the information distance preordering tends to be even more significant. In general, the more discriminated (by their distances) are the reconstruction hypotheses evaluated at the individual refinement levels, the more effective is the preordering.

It is often useful to inspect the increments in the minimal distance associated with adjacent refinement levels. For that purpose, we determine the distance of the most

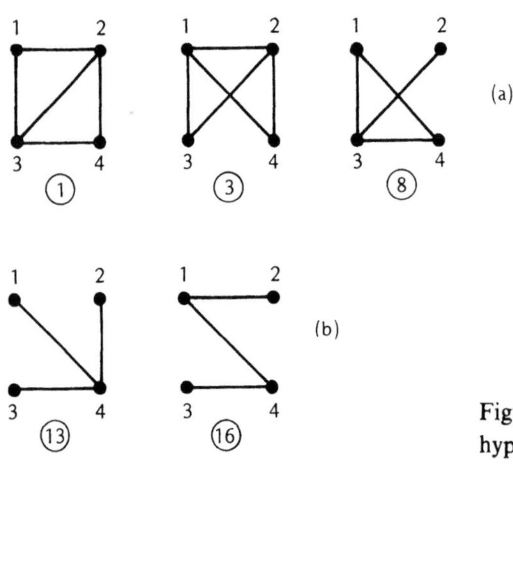

Figure 4.30. Graphs of several reconstruction hypotheses discussed in Example 4.21.

refined hypothesis and calculate the average distance increment by dividing this largest distance by the total number of refinement levels. In this example, the distance of the most refined hypothesis 1/2/3/4 is 0.4591 and the average distance increment is thus 0.4591/6 = 0.0765. Extrapolating from the known distance values, we can then draw a plot characterizing the dependence of the minimal distance, say D_l, on the refinement level l. For the example discussed, such a plot is shown in Figure 4.31. The plot is exact for $l = 0, 1, 2, 3, 6$, approximate for $l = 4$ (we know that $0.1188 \leq D_4 \leq 0.1748$), and estimated for $l = 5$.

It remains to resolve the issue of control uniqueness for each member of the solution set (Section 4.4, Example 4.6). As shown in Figure 4.32a, variables 1, 2, 3 are obviously generated variables, while variable 4 is the only generating variable. Each of the generated variables must be controlled (determined) by exactly one subsystem in each reconstruction hypothesis. In the case of hypothesis 134/234, variables 1 and 2 are clearly controlled by the subsystems 134 and 234, respectively, but variable 3 can be controlled by either of them. Which of the subsystems is chosen to control variable 3 may be decided by their generative uncertainties. The one that is able to generate the variable with smaller uncertainty is normally preferred. In our example, we calculate the conditional U-uncertainties $U(3|1, 4) = 0.834$ and $U(3|2, 4) = 0.679$ associated with subsystems 134 and 234, respectively. Since $U(3|2, 4) < U(3|1, 4)$, the second subsystem is chosen to control variable 3.

We must also decide how to represent the generating variable 4 in the subsystems. There are three options: the variable can be stored in either of the subsystems or in both of them. If it is stored only in one of them, then it must be used as input variable in the other one. It should be emphasized, however, that differences between these options are more stylistic than functional and, consequently, the choice is, in fact, rather arbitrary. In our example, let variable 4 be stored in subsystem 234 and viewed as an input variable in subsystem 134.

The result of the two decisions regarding the roles of variables 3 and 4 in the reconstruction hypothesis 134/234 is expressed by the block diagram in Figure 4.32b.

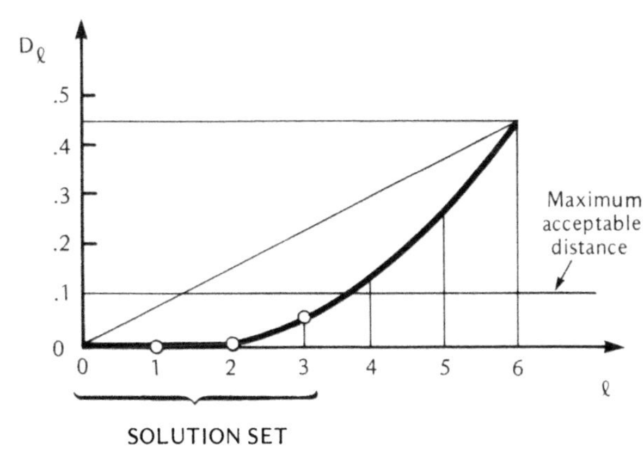

Figure 4.31. Dependence of the minimal distance D_l on the refinement level l (Example 4.21).

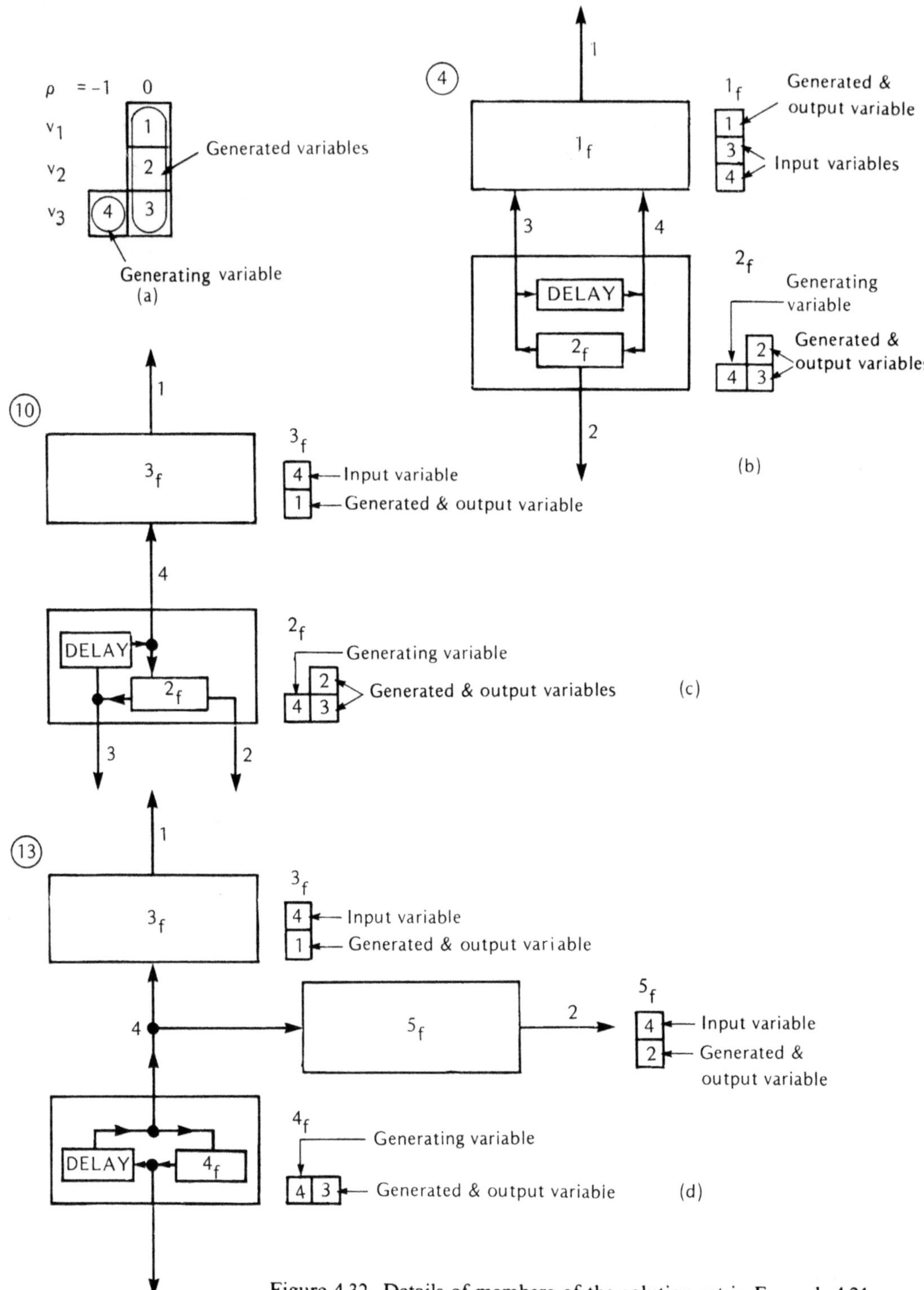

Figure 4.32. Details of members of the solution set in Example 4.21.

Also shown in the block diagram are the masks associated with the individual subsystems, in which the generated, generating, and input variables are indicated.

The final block diagrams of the remaining members of the solution set—hypotheses 14/234 and 14/24/34—are shown in Figures 4.3c and 4.3d, respectively. In either case, roles of all sampling variables, as specified in the block diagrams, are obviously unique.

It is important to realize that only some reconstruction hypotheses are meaningful when dealing with memory-dependent systems. Indeed, a hypothesis is clearly not meaningful when a generating variable is not included in at least one subsystem of the hypothesis that contains the associated generated or input variable, or another generating variable (defined in terms of the same basic variable) from which it can be determined by storing. Such a generating variable would be left undetermined (in limbo) since it could neither be generated (due to its generating role) nor derived by storing another variable that itself is determined in some specific way. For instance, when the overall system is characterized by the mask in Figure 4.32a, all hypotheses that do not include both variables 3 and 4 in at least one subsystem are meaningless. This implies that exactly one half of reconstruction hypotheses based on C-structures are meaningless in this case; these are hypotheses whose graphs do not contain the edge (3, 4), e.g., 123/124, 14/24/13/23, 123/4, 13/24, etc. Although the solution set in Example 4.21 does not contain any meaningless hypotheses, the solution process could have been simplified when only meaningful hypotheses were evaluated (eight of the 20 evaluated hypotheses were evaluated unnecessarily).

Assuming that the support set involved is totally ordered, the notion of a meaningful reconstruction hypothesis for memory-dependent overall behavior systems can formally be defined as follows. A reconstruction hypothesis h is *meaningful* if and only if each generating variable s_k, which is defined by the equation

$$s_{k,t} = v_{i,t+a},$$

is included in at least one subsystem of h that contains a variable s_j defined by the equation

$$s_{j,t} = v_{i,t+b},$$

where $b > a$ when the variables are generated in the increasing order of t (in the predictive manner), and $b < a$ when they are generated the other way around (in the retrodictive manner). This notion of meaningful reconstruction hypotheses can be easily generalized to memory-dependent systems based on two or more totally ordered support sets (such as two-dimensional or three-dimensional Cartesian spaces), but the formalization becomes considerably more complicated for such systems, primarily due to the large increase in the number of possible generative orders.

4.8. RECONSTRUCTABILITY ANALYSIS

> *The squirming facts exceed the squamous mind,*
> *If one may say so. And yet relation appears,*
> *A small relation expanding like the shade*
> *Of a cloud on sand, a shape on the side of a hill.*
>
> —WALLACE STEVENS

Reconstructability Analysis is an example of a package of methodological tools within the GSPS that deals with a significant class of problem types characterized by a common theme: the relationship between overall systems and their various subsystems. This class of problem types involves two epistemological systems types: generative systems and structure generative systems, both represented usually by their behavior forms. It is naturally divided into two subclasses, which differ from each other in the epistemological type of the initial (given) system. Problems in which the initial system is a generative structure system are referred to as *identification problems;* those in which the initial system is a generative system are called *reconstruction problems.*

General types of identification and reconstruction problems, in which no properties are recognized in state sets of the variables involved, are formulated and discussed in Sections 4.6 and 4.7, respectively. Basic issues (or subproblems) associated with these problems that are independent of specific methodological distinctions are the subject of *general reconstructability analysis.* They are depicted in Figure 4.33 and the following is their list:

- a determination of the *reconstruction family* for a given structure behavior system;
- a determination of the *identifiability quotient* (or reconstruction uncertainty) for a given structure behavior system;
- a determination of the *unbiased reconstruction* for a given structure behavior system;
- a determination of the *least risk reconstruction* or, perhaps, some other kind of reconstruction for a given structure behavior system;
- a resolution of *local inconsistencies* in a given structure behavior system (Section 4.11);
- a generation of desirable *reconstruction hypotheses* for a given behavior system;
- a calculation of desirable *projections* of a given behavior system;
- a calculation of the *distance* between the given behavior system and the one reconstructed from a reconstruction hypothesis;
- an ordering of relevant reconstruction hypotheses and determining the *admissible reconstruction hypotheses* (the solution set in the reconstruction problem);
- a determination of the control of the variables involved.

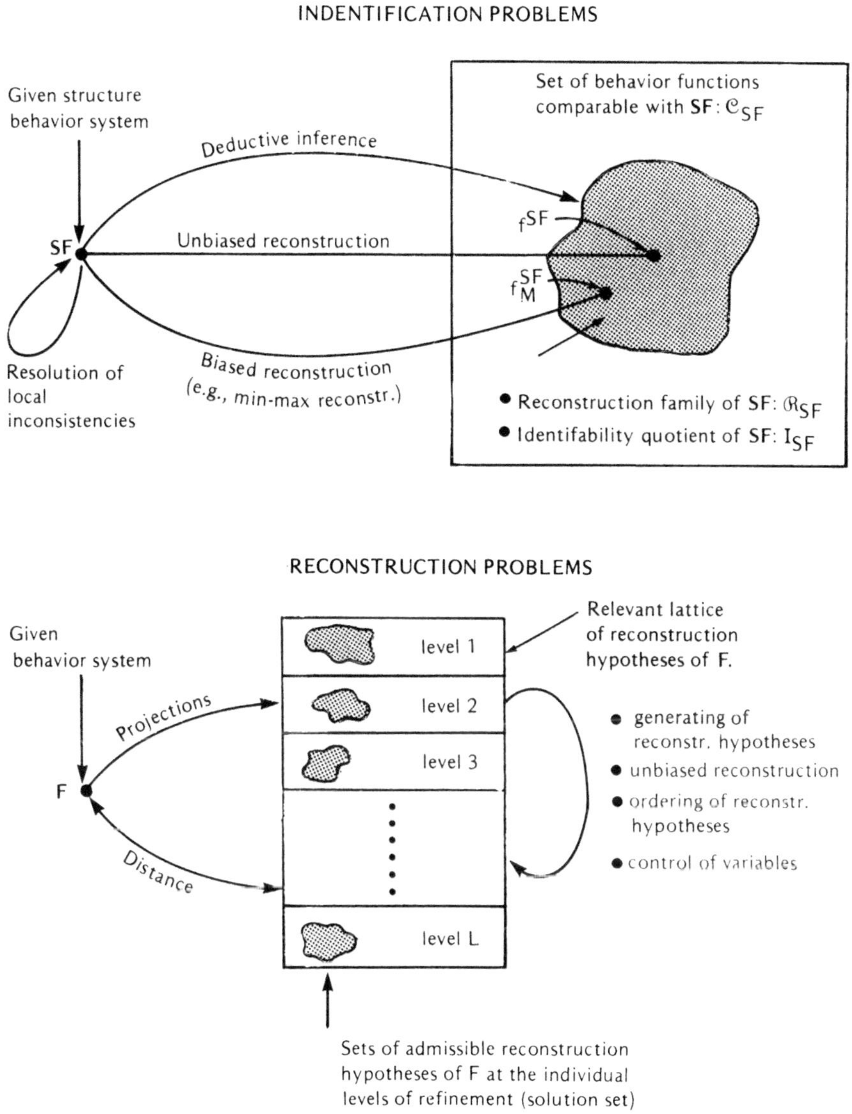

Figure 4.33. Summary of basic issues that are the subject of reconstructability analysis.

From the standpoint of the epistemological hierarchy of systems, we can easily see that reconstructability analysis deals with sequences of problem types that belong to the four categories of problems that are characterized in Figure 4.34 by the labeled arrows. The following is a specific listing of the subproblems associated with reconstructability analysis that are subsumed under each of these categories:

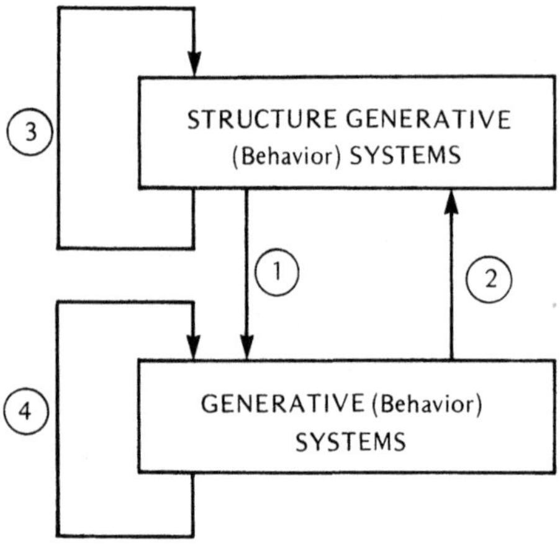

Figure 4.34. Categories of problem types involved in reconstructability analysis.

1—reconstruction family, identifiability quotient, unbiased or least risk reconstruction;
2—projections;
3—resolution of local inconsistencies, generation and ordering of reconstruction hypotheses, control of variables;
4—distance.

Alternative types of identification or reconstruction problems emerge for different methodological distinctions. For instance, when dealing with continuous variables, projections from an overall behavior system not only depend on the chosen subsets of variables, but also can be influenced by transformations in the coordinates. The three-dimensional solid object shown in Figure 4.35a, for example, can be totally reconstructed from three of its two-dimensional (planar) projections (views), say left side, front, and bottom (Figure 4.35b), based on the Cartesian coordinate system specified in the figure. Although this reconstruction property is preserved under displacements of the origin of the coordinate system, it is obviously not preserved under its rotations. In fact, there is a continuum of projections of the same object, which corresponds to the continuum of rotations of the coordinate system or, alternatively, to the continuum of rotations of the object in the same coordinate system. In addition, auxiliary projections can be made into any plane defined within the coordinate system employed. On top of all this variety in projections, we should also realize that the projections illustrated in Figure 4.35 are only one special kind of projections—so-called *orthographic projections*, obtained by erecting perpendiculars from every point of the object to the respective projection planes. Another kind of projections, usually called *shadows*, are obtained by connecting every point of the object with a fixed point (called projection point or light source) and taking intersections of these straight lines with the chosen projection plane. Clearly, there is again a continuum of projections, which corresponds to the continuum of locations of the projection points. Furthermore, intersections of the object with

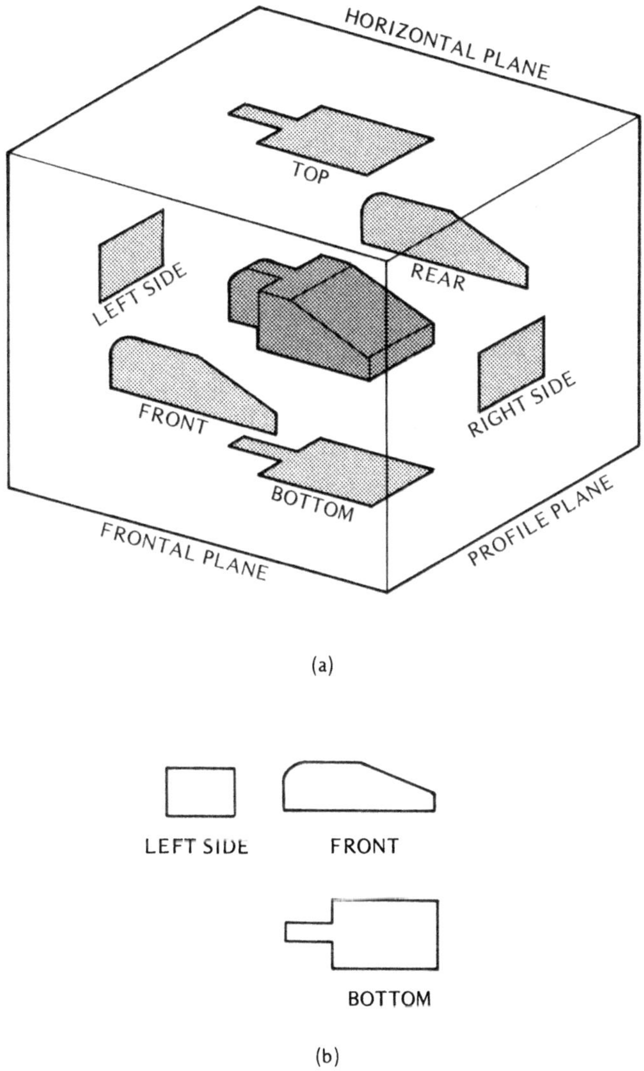

Figure 4.35. Solid object reconstructable from three of its two-dimensional orthographic projections.

various planes (so-called *slides* or *slices*) can also be used as its two-dimensional representations.

In spite of the tremendous variety of possible projections for systems of continuous variables, the basic issues regarding the relationship between wholes (say three-dimensional solid objects) and parts (their various two-dimensional or one-dimensional projections)—those of reconstruction family, unbiased reconstruction, local consistency, distance, etc.—remain the same, even though they may take appropriate special forms. For example, the join procedure, by which the unbiased reconstruction is determined from two-dimensional orthographic projections, can be performed by erecting an unrestricted cylinder for each of the available projections and, then, determining the common volume of all these cylinders (i.e., set intersection of their points). There are, of course, some additional problems when dealing with continuous variables, associated basically with the choice of appropriate projections. Although these problems are of great importance to areas such as optical and mechanical engineering, cartography, and tomography, they are special problems, not applicable to all systems. As such, they are beyond the scope of this book, even though the GSPS should be equipped to deal with them. Some of these problems have been studied in descriptive geometry and, more recently, in the area of image processing.

A methodological package, such as the one for reconstructability analysis, should be available in an *interactive mode* as well as in an *automatic* (*batch*) *mode*. When operating under the interactive mode, the user may employ the relevant procedures in any order and piecemeal. His decisions are based on the intermediate results from steps taken previously as well as his background knowledge. In dealing with a reconstruction problem, for instance, he may start with an initial structure system as a feasible reconstruction hypothesis, evaluate it, and, if desirable, compare it with its immediate refinements, with all reconstruction hypotheses in its structure neighborhood, with other feasible hypotheses that are not in its neighborhood, or he may proceed in any other way. In the identification problem, he may compare several competing structure systems based on the same variables by calculating their identifiability quotients. Depending on the results, he may then decide to determine reconstruction families or some specific kinds of reconstructions only for some of them. The fact that the interactive mode allows the user to focus on certain specific questions and take full advantage of his background knowledge makes it attractive for dealing with large systems, where the full processing is practically impossible due to unacceptable computational demands.

When operating under the automatic mode, a number of alternatives of sequencing the various procedures should be made available to the user to allow him to deal with meaningful variations of the problem involved. In the reconstruction problem, for example, one sequence of procedures, which are repeated at each level of refinement, may consist of the RC-procedure, join procedure, calculation of distance, and decision procedure regarding the continuation; in another procedure, the RC-procedure may alternate with the RG-procedure (restricted to refinements in the same r-equivalence class), each followed by the other three procedures (the join procedure, etc.); still other sequences may be based on coarsening rather than refining procedures, etc. One of the sequences must be adopted as a default option, perhaps the simple sequence based on the RC-procedure.

In addition to its main role in analyzing natural systems, reconstructability analysis can be utilized in some problems associated with man-made systems as well. For instance, the concept of structure neighborhood is directly applicable to the problem of identifying defects in connections between elements of a structure system in situations where direct observations of the connections is not possible. When properly used, it may also be of great help in systems design. For example, it can be utilized for determining the whole set of structure refinements that totally preserve a given behavior system to be designed. Such refinements serve then as a base for performing the design in a natural piecemeal fashion, making it thus more manageable. The term "natural" is used here to indicate that the refinements contain only the input and output variables included in the given behavior system, i.e., no additional (or artificial) variables are introduced at this stage. In each refinement, at least some of the given variables participate in several subsystems and assume thus several different roles. In the highest refinements, this multiple role utilization of each variable reaches its limit. Further variables must be then introduced, where necessary, by the usual decomposition methods or other appropriate way of systems design.

In spite of the indicated utilization of reconstructability analysis in the area of man-made systems, it should be emphasized that its main role is in the investigation of natural systems. This results from the fact that the whole–part relationship is far more intricate in natural systems than in man-made systems. For instance, every man-made structure system is also a definition of the associated overall system. This is a direct consequence of the fact that coupling variables of any man-made system have no meaning other than being either variables that represent a unique behavior system (given in the design problem) directly, or being artificially introduced as coupling variables for the sole purpose of representing this unique system indirectly. Hence, the reconstruction family of every man-made structure system is unique and consists solely of the join of the behavior functions of its elements. That is, the overall system of every man-made structure system is always represented by the unbiased reconstruction of the structure system.

This one-to-one correspondence between man-made structure systems and the overall systems associated with them is undoubtedly one reason, perhaps the most important one, why the relationship between comparable behavior and structure systems in systems inquiries (i.e., inquiries of natural systems) is often not properly understood, particularly by people with engineering education or experience. Indeed, many large systems have been described in the literature that are supposed to characterize various natural phenomena and are constructed by interconnecting smaller systems (subsystems). Using the resulting structure system in a particular case, inferences are then made regarding various properties of the overall system in a manner analogous to man-made systems, i.e., by joining or composing behavior functions of relevant elements. Such inferences are obviously based on the assumption that the structure system represents the overall system in the same way as in man-made systems. This unjustified and usually incorrect assumption, which is never stated explicitly in such studies, is apparently taken for granted, as a result of the invalid and misleading analogy with man-made systems.

It is interesting to compare a different, quite illuminating way in which Robert Rosen arrives at basically the same conclusion [RO6]:

> By *analysis* we mean here the resolution of a system into a family of subsystems somehow "simpler" than the original system from which they were extracted, and

attempting to infer the properties of the original system from the properties of the subsystems. The extraction of a subsystem corresponds formally to a process of *abstraction* in which a number of degrees of freedom of the original system (i.e., potential interactive capabilities) are excluded, and only a limited number are retained. This process of abstraction can be physically implemented (as when a molecular biologist extracts a fraction of molecular species from a cell, thereby creating an abstract cell) or they can be purely formal (as when an ecologist represents a population of real organisms in terms of predation relations). The basic requirements of such abstractions are the following:

(1) The subsystems so obtained must be "simpler" than the original system from which they are abstracted;

(2) The subsystems must be obtained by "natural" means (i.e., utilizing familiar and justifiable procedures); and

(3) The properties of the subsystems so obtained must permit the determination of the properties of the original system.

The property (1) is obviously crucial; nothing is gained if we extract systems as intractable as the original system. This has long been recognized implicitly in scientific modes of analysis. Of equal importance is the property (3); any property of isolated subsystems not bearing on the properties of the original system is an *artifact*. The property (2), however, is a purely subjective matter, and refers only to the manner in which we find it convenient to interact with the original system. It thus stands on a different footing from (1) to (3).

Nevertheless, in many empirical modes of system analysis, the greatest weight is placed upon condition (2). It seems to be intuitively hoped that, by relying on procedures which satisfy (2), the conditions (1) and (3) will automatically be satisfied. At the very least, it is hoped that (1) + (2) will imply (3). However, from what we have already said, this is plainly absurd, in general. Indeed, what we learn from the above is that the crucial properties (1) and (3), which must be satisfied by any useful means of analysis of systems, must be allowed to determine what we are to regard as "natural." Indeed, "naturality" must not be allowed to be posited in advance, but only in terms of its bearing on the problems under discussion in a particular context.

A simple example may make this clear. In physics, the three-body problem is complex in a well-defined sense; the dynamical equations governing a system of three gravitating masses in an arbitrary configuration cannot be integrated directly. We could hope to approach this kind of problem by analysis into a family of "simpler" subsystems, which will allow us to solve the problem. Intuitively, the subsystems available to us are two-body systems and one-body systems. These are indeed "simpler" than the original system, and are abstracted from that system in "natural" ways. However, it is clear that we cannot solve a three-body problem in this fashion, for the act of decomposing the original system into isolated simpler subsystems destroys irreversibly the dynamics in which we are originally interested (here again we see the inability of physics to deal with arbitrary interaction). Thus, from the standpoint of solving the three-body problem, our apparently "natural" decompositions are useless; if analysis is to be successful in this kind of problem at all, the appropriate subsystems (i.e., those which satisfy (3)) must necessarily be of a kind which would appear most "unnatural" in terms of what we find it convenient to do physically to a system of particles.

4.9. SIMULATION EXPERIMENTS

> *Experimentation in the computer is not merely possible but may give information that is otherwise unobtainable.*
>
> —W. Ross Ashby

As an example of metamethodological considerations within the GSPS, simulation experiments are described in this section by which some fundamental characteristics of reconstructability analysis (associated with the reconstruction problem) have been determined. The purpose of these characteristics is threefold: (i) to get a deeper understanding of reconstructability analysis; (ii) to help GSPS users to properly utilize reconstructability analysis in their overall systems investigations; and (iii) to evaluate new principles such as the new principle of inductive inference discussed in Section 4.10.

In a typical experiment, a reconstruction hypothesis was selected for a given number of variables and cardinalities of their state sets. The process of generating data by this hypothesis was then simulated on a computer. In most of the experiments, sequences of 2,000 data points were generated. Reconstructability analysis was performed, according to rules described later, on ten different segments of each of these data sequences, containing 10, 20, 40, 80, 160, 320, 640, 1000, 1500, and 2000 data points. Results obtained for each data segment were then compared with the given reconstruction hypothesis.

For a given number of variables and their state sets, sufficient number of different data sequences were generated and analyzed. Average results of these experiments were then employed in determining the various characteristics. For the sake of simplicity, the experiments were restricted to C-structures. They were performed for sets $\mathscr{C}_3$, $\mathscr{C}_4$, and $\mathscr{C}_5$; in each of them, all refinement levels were properly represented. Comparable experiments were repeated for state sets of equal cardinality (2, 3, 4, or 5) for all variables involved, as well as some specific mixtures of different cardinalities. Since the distinctions between variables characterized by the concepts of mask and environment, which are very important in overall systems investigations, are of no significance for the reconstructability analysis proper, the experiments were performed only for neutral and memoryless systems.

Each data sequence was generated with the help of a random generator according to a specific probabilistic structure system (representing a C-structure). It was then analyzed in probabilistic as well as possibilistic fashion. In fact, one of the purposes of the experiments was to compare results of these two analyses and identify their complementary ranges of applicability.

It was observed at the initial stage of the experimentation that the possibilistic analysis shows a tendency to naturally cluster reconstruction hypotheses at each level of refinement into good and bad ones, i.e., into hypotheses with small distances and large distances, respectively. It was also observed that the correct hypothesis (the one by which the analyzed data were generated) often does not have the smallest distance, but it almost always belongs to the good cluster. Due to these observations, the two analyses (probabilistic and possibilistic) were performed according to slightly different rules.

In probabilistic analysis, each generated data sequence was analyzed twice, for two different search procedures in the relevant refinement lattice. According to the first procedure, only structures with the minimum distance were refined at each level of refinement. According to the second procedure, all structures whose distances did not exceed the minimum distance by more than 100% were refined. Separate characteristics were determined for either of these two procedures.

▶ In possibilistic analysis, structures at each level of refinement were clustered into good and bad ones, and only the good structures were further refined. Each data sequence was analyzed twice, for two different clustering procedures. To describe the clustering procedures, let

$$R = \{(C_i, d_i) | i \in N_r\}$$

denote the set of all C-structures C_i that were evaluated at some refinement level of a particular experiment and their distances d_i. Assume that $d_i \le d_{i+1}$ for all $i \in N_{r-1}$ and let

$$G = \{C_1, C_2, \ldots, C_c\}$$

and

$$B = \{C_{c+1}, C_{c+2}, \ldots, C_r\}$$

denote the clusters of good and bad structures, respectively, where

$$1 \le c \le r$$

(i.e., G is always nonempty while B may be empty in special instances).

In the first clustering procedure, c is determined by the smallest value of i for which the difference $d_i - d_{i-1}$ exceeds the average difference in R for all $i \in N_{r-1}$. That is,

$$d_i - d_{i-1} \le \frac{d_r - d_1}{r}$$

for $i \in N_c (d_0 = 0)$ and

$$d_{c+1} - d_c > \frac{d_r - d_1}{r}.$$

Let us call this procedure a *clustering by average difference* or *AD-clustering*.

In the second clustering procedure, c is determined by that value of $k \in N_r$ for which the expression

$$\frac{1}{a_2 - a_1}\left(\sum_{i=1}^{c} |d_i - a_1| + \sum_{i=c}^{r} |d_i - a_2|\right)$$

reaches its minimum, where

$$a_1 = \frac{1}{c} \sum_{i=1}^{c} d_i$$

and

$$a_2 = \frac{1}{r-c} \sum_{i=c+1}^{r} d_i.$$

This procedure is based on the natural clustering requirement that distances between clusters should be large while distances within clusters should be small; let us call it a *clustering by inside and outside distance* or *IOD-clustering*. ◄

Using the diagram in Figure 4.36 as a guide, the complete procedure involved in one experiment can be now summarized. It begins with a selection of a structure behavior system ${}^T\mathbf{SF}$ (viewed as the true system in the experiment), which is based on a C-structure. This structure system, which represents on overall behavior system ${}^T\mathbf{F}$ (obtained from ${}^T\mathbf{SF}$ by the join procedure), is simulated on a computer and used for generating data. Once the data are generated, an overall memoryless behavior system ${}^D\mathbf{F}$ (probabilistic or possibilistic) is derived from the corresponding data system **D**. Reconstructability analysis is then performed for system ${}^D\mathbf{F}$ according to one of the search procedures mentioned (based, e.g., on one of the two kinds of clustering for possibilistic systems). The result is a sequence of sets of C-structures (and their distances) that are evaluated at the individual refinement levels of the relevant refinement lattice, say sets

$$E_l = \{(C_{i_l}, d_{i_l}) \mid i_l \in I_l\}$$

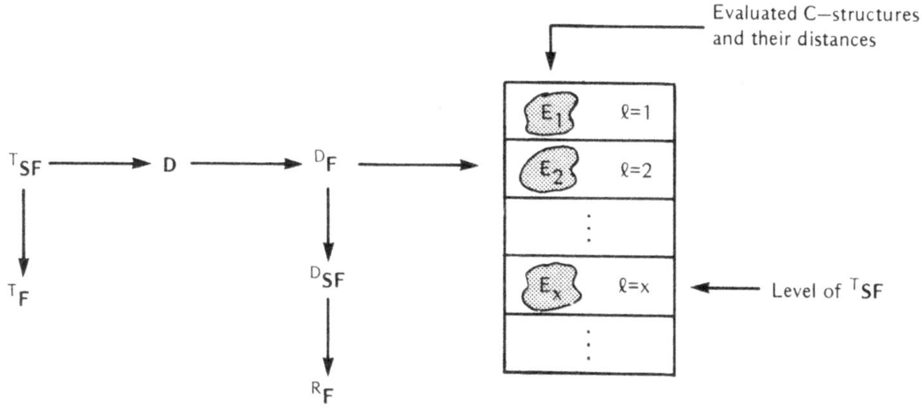

Figure 4.36. Summary of a simulation experiment.

for $l = 1, 2, \ldots, n(n-1)/2$, where n is the number of variables involved. Set E_l associated with the same refinement level as ${}^{T}\mathbf{SF}$ is of particular interest. System ${}^{D}\mathbf{F}$ is also used for deriving a structure system ${}^{D}\mathbf{SF}$ based on the same C-structure as the given structure system ${}^{T}\mathbf{SF}$, and this system (${}^{D}\mathbf{SF}$) represents an overall behavior system ${}^{R}\mathbf{F}$ (a reconstructed overall system).

Set $E_l (l \in N_{n(n-1)/2})$ and the overall behavior systems ${}^{T}\mathbf{F}$, ${}^{D}\mathbf{F}$, and ${}^{R}\mathbf{F}$, obtained for all experiments of the same kind (certain number of variables, certain state sets, probabilistic or possibilistic option, etc.), are the resources from which the various characteristics of reconstructability analysis can be determined. Let me now illustrate these characteristics by a few examples for systems with three variables; a more complete set of characteristics based on the described experiments for $n = 3, 4, 5$ and several cardinalities of state sets is presented in Ref. [HA1].

For probabilistic systems (with three variables), some basic characteristics are expressed by the plots in Figure 4.37. They are based on the search procedure in which only the structures with minimum distance are refined.

Plots (a) characterize the effect of the number of observations (data size $|d|$) on the performance of reconstructability analysis for two different state sets—two states and five states per variable. The performance is expressed by the percentage of those experiments in which the correct structure was reached by the search procedure and emerged as a structure with minimum distance at the respective level of refinement. We can see that the performance of 100% is reached rather quickly in both of these cases. Although a convergence to 100% performance with increasing number of observations is a general trend in all investigated cases, the rate of convergence somewhat decreases with increasing number of variables. This is primarily caused by the high selectivity of the search procedure involved. We can also observe that variables with five states (upper plot) perform better than those with two states (lower plot). This, again, is a general trend: increase in the cardinalities of the state sets involved results in improved performance. For any particular number of variables, the performance characteristic representing systems with binary variables can be thus viewed as the worst case.

The remaining plots in Figure 4.37 are based only on binary variables. Plots (b) characterize how much the correct structure is discriminated by the information distance from other structures considered at the same level of refinement. The lowest plot represents $D({}^{D}f, {}^{R}f)$, i.e., the distance of the correct structure; the middle plot represents the smallest distance of structures that compete with the correct structure at the same level of refinement; the highest plot represents the average distance of all structures that compete with the correct structure (according to the search procedure) at the same level of refinement. Although shapes of these plots are affected by the number of variables and cardinalities of the state sets involved, as well as the kind of distance measure employed, the distances always decrease with the increasing number of observations and the distance of the correct structure converges to zero.

Plots (c) compare the information distances between the true system ${}^{T}\mathbf{F}$ and systems ${}^{D}\mathbf{F}$ and ${}^{R}\mathbf{F}$, respectively. Since the respective pairs of probability

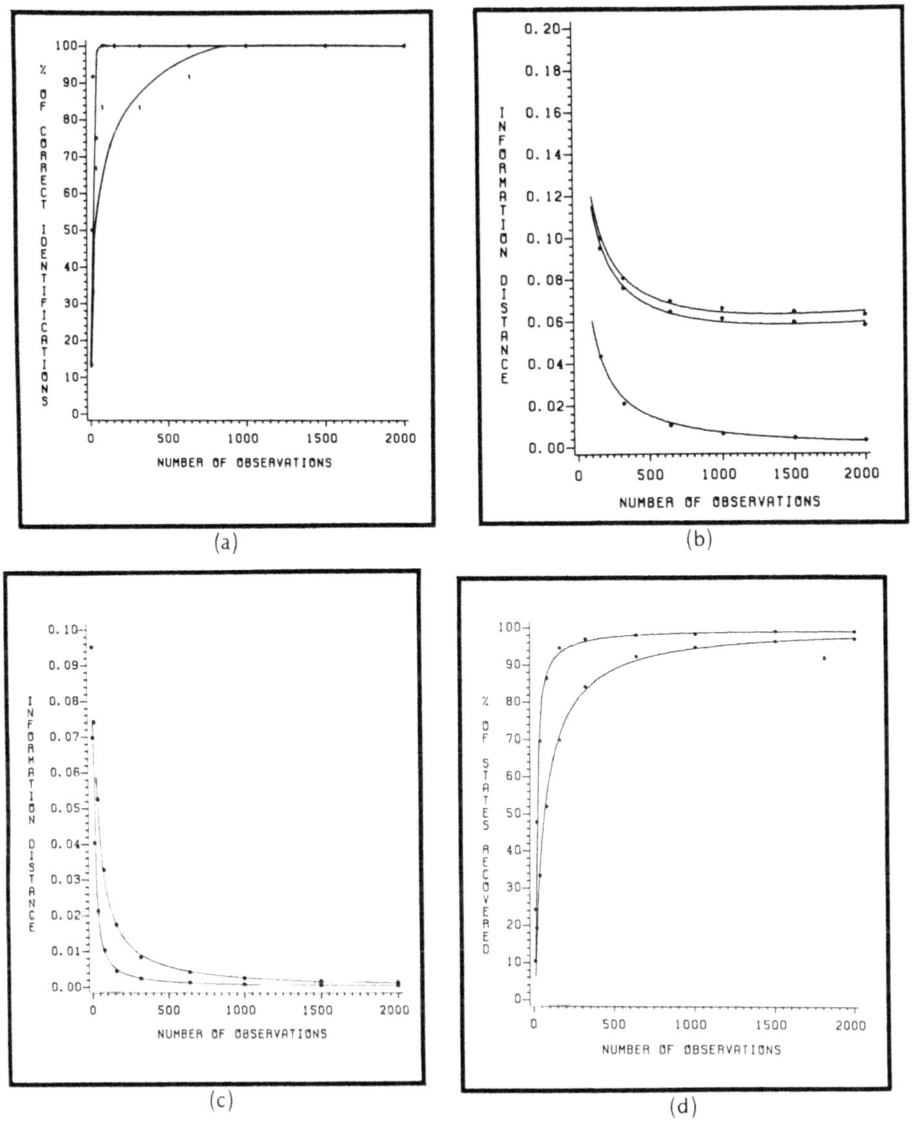

Figure 4.37. Some characteristics of reconstructability analysis for probabilistic systems.

distributions—$^T\mathbf{f}$, $^D\mathbf{f}$ and $^T\mathbf{f}$, $^R\mathbf{f}$—are arbitrary, a general information distance measure is needed. Such a distance measure, say G, is defined by the formula

$$G(^1\mathbf{f}, {}^2\mathbf{f}) = D\left({}^1\mathbf{f}, \frac{{}^1\mathbf{f} + {}^2\mathbf{f}}{2} \right) + D\left({}^2\mathbf{f}, \frac{{}^1\mathbf{f} + {}^2\mathbf{f}}{2} \right), \tag{4.44}$$

where ${}^1\mathbf{f}$ and ${}^2\mathbf{f}$ are arbitrary probability distributions defined on the same finite set of states, D is the special information distance defined by Eq. (4.40), and $({}^1\mathbf{f} + {}^2\mathbf{f})/2$ denotes the probability distribution obtained by taking the average of each pair of corresponding probabilities in ${}^1\mathbf{f}$ and ${}^2\mathbf{f}$. The lower and upper plots in (c) represent $D({}^Tf, {}^Rf)$ and $D({}^Tf, {}^Df)$, respectively. The reconstructed system ${}^R\mathbf{F}$ is thus closer to the true system ${}^T\mathbf{F}$ than the system ${}^D\mathbf{F}$ that is based solely on the available data. The significance of this rather surprising result is discussed in Section 4.10.

Plots (d) characterize the relationship among state sets with nonzero probabilities in the three behavior systems involved in the simulation experiments—systems ${}^T\mathbf{F}$, ${}^D\mathbf{F}$, and ${}^R\mathbf{F}$; let us denote these state sets by ${}^TX$, ${}^DX$, and ${}^RX$, respectively. The lower plot represents the percentage of those states in ${}^T\mathbf{F}$ that are recognized in ${}^D\mathbf{F}$ (due to scarcity of data), i.e., $({}^DX/{}^TX) \times 100$; the upper plot represents the percentage of those states in ${}^TF$ that are recognized in ${}^RF$, i.e.,

$$\frac{{}^RX}{{}^TX} \times 100.$$

These plots clearly indicate that

$${}^DX \subseteq {}^RX \subseteq {}^TX. \tag{4.45}$$

This property has a similar significance as the one expressed by plots (c) and, hence, its discussion is left for Section 4.10.

► Possibilistic counterparts of the described characteristics are given in Figure 4.38. They are based on the IOD-clustering. Since reconstructability analysis of possibilistic systems is based on dealing with clusters of structures rather than individual structures, the correspondence between the probabilistic characteristics and their possibilistic counterparts is not direct.

Plot (a) in Figure 4.38 summarizes the performance of possibilistic reconstructability analysis for different state sets (between two and five states per variable). A summary is used in this case because the differences for different state sets are small and no obvious trend emerges from them. The performance is expressed by the percentage of those experiments in which the correct structure is included in the cluster of good structures. The remaining plots in Figure 4.38 are based only on binary variables.

Plots (b) characterize the upper and lower information distances for the two clusters of structures. As such, they are quite different from their probabilistic counterparts. Plots (c) and (d), on the other hand, are quite similar to their probabilistic counterparts. A possibilistic version of general information distance, on which plots (c) are based, is defined by the formula

$$G({}^1\mathbf{f}, {}^2\mathbf{f}) = D({}^1\mathbf{f}, {}^1\mathbf{f} \vee {}^2\mathbf{f}) + D({}^2\mathbf{f}, {}^1\mathbf{f} \vee {}^2\mathbf{f}), \tag{4.46}$$

where ${}^1\mathbf{f}$ and ${}^2\mathbf{f}$ are arbitrary possibility distributions defined on the same finite set of

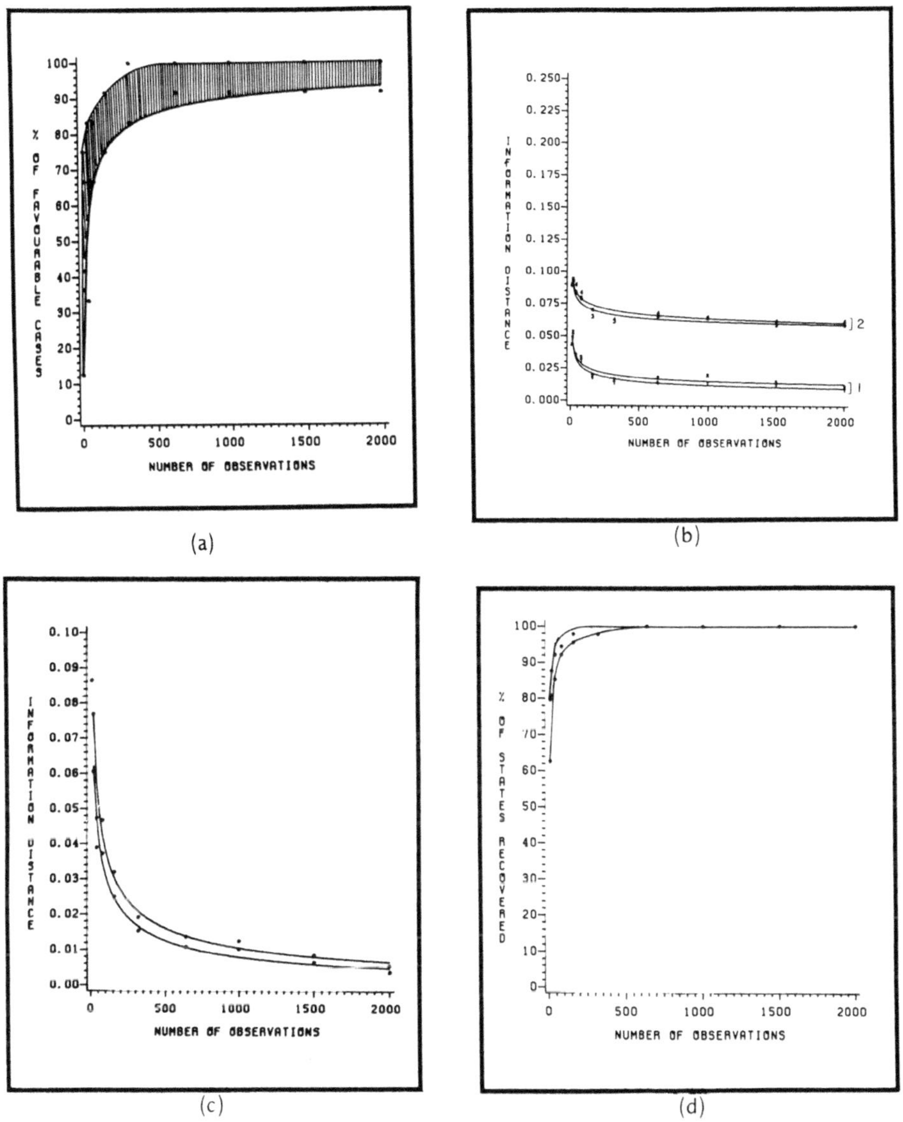

Figure 4.38. Some characteristics of reconstructability analysis for possibilistic systems.

states, D is the special information distance defined by Eq. (4.42), and ${}^1\mathbf{f} \vee {}^2\mathbf{f}$ denotes the possibility distribution obtained by taking the maximum of each pair of corresponding possibilities in ${}^1\mathbf{f}$ and ${}^2\mathbf{f}$ (see Note 4.8). ◀

All the simulation experiments for which the described characteristics were determined are based on the assumption that the data are generated by specific structure systems. Their objective is to determine how scarcity of data affects reconstructability analysis. Although such idealized experiments are valuable and represent a natural first stage in the evaluation of reconstructability analysis, it is highly desirable to extend them to more general and realistic situations. Let me describe, as an example, generalized experiments that are currently under preparation.

Similarly to the idealized simulation experiments, the generalized experiments will be performed in groups, each one characterized by a particular number of variables and specific cardinalities of their state sets. A particular distribution will be chosen for each experiment in which the number of occurrences will be specified for each overall state of the variables involved. Some of the distributions will be selected from various data archives as well as literature, others will be generated by a random process. The two classes of experiments will be analyzed separately to determine whether distributions based on real-world data possess some special reconstruction properties when compared with the randomly generated distributions.

Each distribution selected will be used in two ways. First, its reconstruction properties will be analyzed using both probabilistic and possibilistic methods. Second, it will be used for generating data, typically with 2,000 observations. Experimental distributions derived from various segments of the data will be then analyzed in the same manner as the original (true) distribution and by using the same method (probabilistic or possibilistic). Finally, relevant experimental results obtained for each segment of data will be compared with the corresponding theoretical properties obtained for the original distribution. The aim of this comparison is to determine how well the theoretical (true) properties are preserved in their various experimental counterparts. For each property viewed as significant, the final outcome of each group of experiments will be the dependence of average degree to which the property is preserved (and the variation of this degree) on the size of the analyzed data segment and the method employed.

Simulation experimentation, such as that described in this section for reconstructability analysis, is a fundamental tool of systems science for metamethodological studies. The GSPS should not only provide the user with methods for dealing with the various problem types, but it should also provide him with metamethodological characterizations of the methods. The set of characteristics of reconstructability analysis described in this section is a simple example of such a metamethodological characterization.

4.10. INDUCTIVE REASONING

> *If hypotheses do not spring from the brain of Zeus, where do they come from? The partial answer is that they come from some hypothesis-generating process.*
>
> —HERBERT A. SIMON

Although inductive reasoning emerged explicitly only in connection with the identification and reconstruction problems discussed earlier in this chapter, it is in fact involved, in one way or another, in virtually every problem associated with the discovery approach to systems inquiry. It is thus desirable to overview the main issues associated with it. This is one of two aims of this section; the other aim is to introduce a novel principle of inductive inference based on some characteristics of reconstructability analysis.

Rather than to view inductive reasoning in its traditional, narrow sense as the inference from particular cases to a general conclusion, I intend to use the name "*inductive reasoning*" as broadly as "to cover all cases of nondemonstrative argument, in which the truth of the premises, while not entailing the truth of the conclusion, purports to be a good reason for belief in it" (*Encyclopedia of Philosophy*, Macmillan, 1967).

The notion of inductive reasoning has been a subject of great controversy in philosophical circles for centuries, especially after the publication of David Hume's classical analysis of this notion in 1739*. Although quite a number of arguments have been invented to overcome Hume's scepticism about the possibility of justifying inductive inference, each of these turns out in the final analysis to contain some flaws. In some cases, it is a hidden circularity (i.e., a justification of induction by induction) which destroys the argument; in other cases, it is the dependence of the argument on some metaphysical assumptions (such as the uniformity of nature) which makes it self-defeating.

The main difficulty of all the arguments which try to resolve the Humean challenge is that they take it seriously as a meaningful problem. An alternative approach is to reject the notion of justifying inductive reasoning by deductive standards, which is implicitly included in Hume's analysis, and reformulate inductive reasoning as a *process of truth estimation in the face of imperfect information*. One of the most promising strategies for justifying inductive reasoning within the latter approach is based on the ideas of *methodological pragmatism* as recently proposed by Nicholas Rescher [RE7]. Due to its focus on methods as the key in the justification of inductive reasoning, methodological pragmatism is clearly relevant to the theme of this book; let me briefly summarize its main aspects.

Rescher views inductive reasoning as a method for "truth-estimation through systematization with experience that effects the optimally plausible blending of conjecture with information-in-hand" [RE10]. He recognizes that our only access to information about nature is through our interaction with it, and if one is not prepared to rely on such interaction, then there is no choice but to abandon the whole project of inquiry into nature. Although the aims of this knowledge-producing inquiry are both cognitive and practical, the ultimate testing standard must be based on the success of the produced knowledge in the effective guidance of human actions. This is due to the fact that, in contrast with purely theoreticocognitive settings, practical situations require urgent decisions in the attempt to achieve desirable goals (e.g., to avoid death, injury, disease, pain, frustration, etc.). The necessity of inductive reasoning for practical and

*David Hume, *A Treatise of Human Nature*, William Collins, Glasgow, 1962.

action-oriented decision making, essential for the survival and well-being of the decision maker, is discussed in great depth in Rescher's book *The Primacy of Practice* [RE4].

Since inductive reasoning is viewed as a method for question resolution in the face of incomplete information, its justification is a matter of *justifying a method with respect to its pragmatic success*; hence the name "methodological pragmatism."

Rescher's justification of a method for inductive reasoning proceeds in two stages, referred to as the initial and ultimate justification, respectively. The *initial justification*, which is noninductive, requires a demonstration that the method offers a relatively optimal prospect or potential of success when compared with compatible alternatives. The *ultimate justification* is a matter of the actual effectiveness of the method, i.e., it is required that its success exceeds that of any other available competitors.

An important aspect of Rescher's initial justification of a method of inductive reasoning is the consideration of a degree in which results produced by the method can be integrated into a system based on previous experience. Everything else being the same, various *parameters of systematicity* (completeness, cohesiveness, consonance, simplicity, etc.) are used as arbiters in making the initial justification of the method. This aspect follows from Rescher's coherence theory of truth [RE5] and is discussed at great length in his book "Cognitive Systematization" [RE9].

Rescher also argues that inductive reasoning "lies at the very root of man's communicative use of language. It is the natural language that closes the evidential gap between the claims at issue and the actual evidence by a fact-transcending imputational process of inductive nature" [RE10].

The relationship between inductive reasoning and natural language, quite compatible with the basic ideas of methodological pragmatism, is recognized in writings of a few of contemporary philosophers. One of the most explicit statements in this regard is by Max Black. He says [BL2]:

> I find it natural to think of induction as an institution and, indeed, as a rule-governed one. That is to say, as a system of human activities, involving appropriate terminology and also involving distinctive rules for the derivation of judgements. The inductive institution commits its participants to labeling certain situations in prescribed ways, to drawing inferences in prescribed fashions, and, notably, to adopting certain cognitive attitudes preparatory to taking appropriate actions . . . . Roughly speaking: inductive rules tell us what to say, how to think, and within limits, how to act . . . . There is an a priori aspect of the rules and the practices that are demanded of those properly using those rules; given our present language and the system of concepts that it embodies, we are logically unable to imagine wholesale deviation from them. But this does not mean that we have to be dogmatic: the constitutive rules of the inductive institution allow for considerable play in the differential judgements we make concerning inductive conclusions, the reliability of rules, and so on. Now it is the purpose of appeal to past experience to supply just such a basis of rational grounds for reinforcing or, within modest limits, for modifying the inductive institution and its components. Appeal to past experience can, however, be only gradualist and revisionist (to use political language): for revolutions in our modes of thought we must look elsewhere.

Inductive reasoning can be well formulated in terms of information theory. One of the most sophisticated contributions in this respect was made by Ronald Christensen [CH5–9]. He defines an *induced proposition* as a proposition which represents all, but no more than, the available information. Although induced propositions do not necessarily follow from the available information by deduction, instances in which deduction is applicable are also included as special (extreme) cases of induced propositions. Two general principles of inductive reasoning follow then from the notion of induced propositions:

i. our beliefs should represent *no more information than is available* to us;
ii. our beliefs should represent *all of the information that is available* to us.

It is fascinating that these very fundamental principles were clearly recognized by a Chinese philosopher Lao Tsu as early as the sixth century B.C., and are beautifully expressed in his book *Tao Te Ching*† by the following two simple statements:

Knowing ignorance is strength.
Ignoring knowledge is sickness.

To develop a particular methodology for inductive reasoning based on these general principles, one has to commit to a particular meaning of the term "*information*." Different meanings are, of course, applicable under different contexts. In the context of systems inquiry, where information is viewed as a measure of the degree of constraint among variables of concern, the meaning is determined by the way in which the constraint is expressed. Within the framework of probability theory, the two general principles of inductive reasoning become principles of maximum and minimum entropy, respectively. For alternative frameworks, they are expressed by appropriate counterparts of these principles. Within possibility theory, for instance, they became principles of maximum and minimum U-uncertainty.

The *principle of maximum entropy* is employed for estimating unknown probabilities (which cannot be derived deductively) on the basis of the available information. According to this principle, the estimated probability distribution should be such that its entropy reaches maximum within the constraints of the situation, i.e., constraints that represent the available information. This principle thus guarantees that no more information is used in estimating the probabilities than available.

The *principle of minimum entropy* is employed in the formulation of resolution forms and related problems. According to this principle, the entropy of the estimated probability distribution, conditioned by a particular classification of the given events (e.g., states of the variable involved), is minimum subject to the constraints of the situation. This principle thus guarantees that all available information is used, as much as possible within the given constraints (e.g., required number of states), in the estimation of the unknown probabilities. It is basically a general principle for pattern recognition, as well characterized by Watanabe [WA7]:

† Vintage Books (Random House), Chapter 71, New York, 1972.

> Pattern recognition is an intellectual adaptation, in the presence of a number of events, aiming at revealing a "form" in them. The nearest mathematical translation of this theme would be that pattern recognition consists of formulating, reformulating, modifying our frame of reference in such a way as to minimize, within the inevitable constraints, the entropy suitably defined according to this frame of reference.

A proper use of both of the complementary methodological principles—the maximum and minimum entropy principles—forms the Christensen methodology for inductive reasoning; it is referred to as the *entropy minimax methodology*. In justifying this methodology, Christensen analyzes the grammatical and morphemic structure of the natural language and its evolution. He argues that

> Just as our physical measurements are relative to our physical frame of reference, our inductive judgements are relative to our conceptual frame of experience. A major portion of this experience is submerged in the structure of our language and the meanings of the words, phrases and sentences we use. The search for a solid "principle of induction" on which to anchor our generalizations is as futile as the search for an ultimate frame of reference for Newtonian mechanics . . . [CH5, p. 599].
>
> Each generalization made by a person depends not only upon the particular data from which it is immediately drawn, but also depends less immediately upon the entire history of experience behind the evolution of the language in which the generalization is obtained . . . . The ultimate use of all propositions entertained by human beings is in the aid of making a decision of some kind or other when faced with alternative courses of action. Now what the decision will be is controlled by the beliefs and the value judgements of the decision-maker, where his beliefs include his assessment of the nature of the world external to him. The values and the beliefs are influenced both by the external world and by the individual himself . . . .
>
> Suppose that we assume that inductive reasoning is used generally by members of a society speaking a common language. Then adopting the first condition in the definition of an induced proposition, namely, that it represents no more information than is available, we have seen that the nondecreasing entropy law of thermodynamics is a consequence. From this law we conclude that future experience will tend to be at least as simply expressible in the currently prevailing language as is past experience.
>
> Adopting the second aspect of the definition of an induced proposition, namely that it represents all the information that is available, together with the assumption that inductive reasoning is used generally in the society, we arrive at the *principle of the evolution of language*: [*A language will tend to evolve in a direction which will lead to a simpler description of the experiences of the members of the society using the language.*] This implies that past experience can be simply described in the currently prevailing language. But this means that inductive reasoning will yield reliable representation of physical reality. Thus we have demonstrated a contingent validity of inductive reasoning, contingent upon the general use of inductive reasoning in society. In other words, the validity of inductive reasoning depends upon whether or

> not it is conducted in a "living language." But, and here is the essential point, the contingency is within the control of those people who are making the decisions whether or not to believe the induced propositions. By basing their decisions upon the results of processes of inductive reasoning, they are validating the very thing upon which the validity of inductive reasoning is contingent. In this sense, decision-making upon the basis of inductive reasoning, conducted in the living language is a self-justifying process [CH9, pp. 168, 345–347].

When considering all the arguments contributing to the justification of the entropy minimax methodology as a methodology for inductive reasoning—the four diverse arguments described in Note 4.6 that support the principle of maximum entropy and Christensen's justification of both principles in terms of the principle of the evolution of language—the methodology can be considered as well justified. It is likely that counterparts of this methodology for other classes of fuzzy measures (e.g., *U*-uncertainty minimax methodology) will be developed and properly justified.

After this brief overview of some fundamental issues of inductive reasoning, let me now describe a novel principle of inductive inference. Since this principle is embedded in the reconstruction problem, let me name it the *reconstruction principle of inductive inference.*

Assume that the constraint of the given overall behavior system was determined from some empirical data by appropriate inductive reasoning, say the entropy minimax methodology. Since all data are limited, often severely limited, this constraint is only an estimate of the way the variables involved are actually constrained; it is an unbiased estimate that is based on all information which the data base contains about the actual constraint.

Suppose now that the actual constraint is such that it can be reconstructed from some particular set of its projections. The estimated constraint may not show this property, due to the limited data. However, it is likely that the reconstruction hypothesis based on the subsystems will be more successful in reconstructing the estimated overall constraint than its competitors at the same level in the refinement lattice. This superiority will then be exhibited by all coarsenings of this successful reconstruction hypothesis at each lower level in the lattice. Now we come to a crucial argument. If, indeed, the correct reconstruction hypothesis and/or its coarsenings are identified as superior at the various refinement levels, then any of them has the potential of reconstructing some of those overall states which the variables are actually able to assume, but which are not included in the available data and, consequently, are not included in the constraint of the given overall system. This is well documented by extensive experimental results exemplified by the plots in Figures 4.37d and 4.38d. Moreover, since each subsystem is associated with a smaller state set than the state set of the overall system, its constraint is generally better characterized by the data than the constraint of the overall system (e.g., the ratio between the number of observations and number of potential states is greater). This means that the superior reconstruction hypotheses have the ability to improve our original estimate of the overall constraint.

This again is well documented by experimental results exemplified by the plots in Figures 4.37c and 4.38c. These results clearly indicate that the inequality

$$G({}^T\mathbf{f}, {}^R\mathbf{f}) < G({}^T\mathbf{f}, {}^D\mathbf{f})$$

is always satisfied for finite data, regardless of the data size. This means that the overall system reconstructed from the correct reconstruction hypothesis is always information-wise closer to the true overall system than the one derived solely from the given data. A more direct evidence that the reconstructed constraint ${}^R\mathbf{f}$ is a better estimate of the true constraint ${}^T\mathbf{f}$ than the one based only upon data (${}^D\mathbf{f}$) is expressed by the inequality

$$\delta_1({}^T\mathbf{f}, {}^R\mathbf{f}) < \delta_1({}^T\mathbf{f}, {}^D\mathbf{f}),$$

where δ_1 denotes the Hamming (or city-block) distance. This inequality is also documented by results obtained by the simulation experiments. An example of actual experimental results for three variables and five states per variable is given in Tables 4.15a and b for probabilistic and possibilistic systems, respectively. For comparison, the table also contains results based on the information distance.

Whether or not we actually take the reconstruction from any of the superior reconstruction hypotheses at some level of refinement as an improved estimate of the overall constraint depends on our belief that the reconstruction hypothesis in question does indeed reflect some underlying reconstruction property of the variables involved. How can the investigator be helped to rationally form his belief in this respect? I offer this answer: he can be helped by being provided with useful reconstruction characteristics prepared by extensive experimentation simulated on the computer, as described in Section 4.9. These characteristics enable him to evaluate his individual situations and develop the relevant beliefs. The characteristics can even be combined with appropriate

TABLE 4.15

An Example of Experimental Support of the Reconstruction Principle of Inductive Inference

$\|d\|$	10	20	40	80	160	320	640	1,000	1,500	2,000
					(a) Probabilistic system					
$\delta_1({}^T\mathbf{f}, {}^D\mathbf{f})$	0.0132	0.0111	0.0086	0.0063	0.0044	0.0031	0.0023	0.0018	0.0015	0.0012
$\delta_1({}^T\mathbf{f}, {}^R\mathbf{f})$	0.0106	0.0073	0.0052	0.0036	0.0025	0.0018	0.0014	0.0011	0.0009	0.0008
$G({}^T\mathbf{f}, {}^D\mathbf{f})$	0.0952	0.0741	0.0526	0.0328	0.0174	0.0084	0.0042	0.0025	0.0017	0.0012
$G({}^T\mathbf{f}, {}^R\mathbf{f})$	0.0697	0.0404	0.0213	0.0104	0.0045	0.0024	0.0013	0.0008	0.0005	0.0004
					(b) Possibilistic system					
$\delta_1({}^T\mathbf{f}, {}^D\mathbf{f})$	0.2779	0.2479	0.2188	0.1859	0.1508	0.1100	0.0883	0.0746	0.0627	0.0565
$\delta_1({}^T\mathbf{f}, {}^R\mathbf{f})$	0.2645	0.1999	0.1470	0.1216	0.1021	0.0764	0.0648	0.0544	0.0457	0.0403
$G({}^T\mathbf{f}, {}^D\mathbf{f})$	0.0999	0.0894	0.1002	0.0872	0.0776	0.0599	0.0526	0.0450	0.0379	0.0369
$G({}^T\mathbf{f}, {}^R\mathbf{f})$	0.0972	0.0742	0.0691	0.0647	0.0568	0.0450	0.0383	0.0334	0.0260	0.0261

guidelines of how to form the beliefs or, eventually, some justifiable belief functions can be developed and one of them declared in the GSPS as a default option. It is obvious that any belief function should be expressed in terms of the reconstruction characteristics of type (a) and (b) in Figures 4.37 or 4.38 (performance and discrimination characteristics).

Our ability to deal with the reconstruction problem offers thus an unorthodox approach to inductive reasoning. It proceeds in two stages. In the first stage, an overall constraint is derived from the available data by using the usual principles of inductive reasoning (e.g., the entropy minimax). The second stage consists of three steps:

i. superior reconstruction hypotheses are determined for the overall system at the various refinement levels;
ii. beliefs of various degrees that these superior hypotheses reflect the actual reconstruction properties of the variables involved are formed, on the basis of relevant experimental characteristics, guidelines, or a specific belief function;
iii. the given overall constraint is supplemented with (or replaced by) the constraints reconstructed by the superior reconstruction hypotheses, each associated with the respective degree of belief.

While using only the information included in the available data, this two-stage method allows us to include in the estimated overall constraint certain features (e.g., overall states) which are not directly derivable from the data. Hence, it allows us, for instance, to predict or retrodict, with a specific degree of belief (credibility), certain states of the investigated variables which are not included in the data available at the time of making the prediction or retrodiction.

4.11. INCONSISTENT STRUCTURE SYSTEMS

> *. . . the distinction between reason and unreason can be decoupled from that between consistency and inconsistency. . . . one can maintain as rigid a line as ever between rationality and irrationality even in the face of inconsistency. . . . Inconsistency can be tolerated in the objects of thought and assertion, while, ultimately, discussion about them can and should be consistent at the meta-level of our cognitive commitments.*
>
> —Nicholas Rescher and Robert Brandom

Systems consistency is perhaps the most fundamental criterion for classifying structure systems. For structure behavior systems, it is closely associated with the identification problem (Section 4.6): when a structure behavior system is consistent, its reconstruction family is nonempty; when it is inconsistent, its reconstruction family is empty.

There are two kinds of inconsistencies in structure behavior systems. They are usually referred to as local and global inconsistencies. A structure system is *locally inconsistent* if it does not satisfy the requirements of local consistency, expressed by Eq. (4.20); it is *globally inconsistent* if it is locally consistent and, yet, its reconstruction family is empty.

Example 4.22. Consider a structure behavior system whose elements are characterized by the probabilistic behavior functions specified in Table 4.16a. When the projections of these functions with respect to the coupling variable v_2 are calculated, we obtain

	v_2	$[{}^1f\downarrow\{v_2\}](\alpha)$
$\alpha =$	0	0.6
	1	0.4

	v_2	$[{}^2f\downarrow\{v_2\}](\alpha)$
$\alpha =$	0	0.55
	1	0.45

We can see that the system is locally inconsistent since

$$[{}^1f\downarrow\{v_2\}](\alpha) \neq [{}^2f\downarrow\{v_2\}]\ (\alpha).$$

Example 4.23. The structure behavior system whose probabilistic behavior functions are specified in Table 4.16b is clearly locally consistent. From simple inspection of the equations that define the reconstruction family, it is evident, however, that the reconstruction family is empty. For example, probabilities of states 001 and 011 (written in the order v_1, v_2, v_3) are required to be 0 by ${}^1f$ and ${}^2f$ while, at the same time, their sum is required to be 0.3 by ${}^3f$. The system is thus globally inconsistent.

Local inconsistencies in a structure system may be (and usually are) caused by the

TABLE 4.16
Examples of Structure Behavior Systems that are Inconsistent

(a) Locally inconsistent structure system

	v_1	v_2	${}^1f({}^1\mathbf{c})$
${}^1\mathbf{c} =$	0	0	0.5
	0	1	0.2
	1	0	0.1
	1	1	0.2

	v_2	v_3	${}^1f({}^2\mathbf{c})$
${}^2\mathbf{c} =$	0	0	0.4
	0	1	0.25
	1	0	0.15
	1	1	0.2

(b) Globally inconsistent structure system

	v_1	v_2	${}^1f({}^1\mathbf{c})$
${}^1\mathbf{c} =$	0	1	0.7
	1	0	0.3

	v_2	v_3	${}^1f({}^2\mathbf{c})$
${}^2\mathbf{c} =$	1	1	0.3
	1	0	0.7

	v_1	v_3	${}^3f({}^3\mathbf{c})$
${}^3\mathbf{c} =$	0	0	0.4
	0	1	0.3
	1	0	0.3

fact that behavior functions associated with its elements are only estimates, each derived from limited experimental data. Such inconsistencies (in contrast with a global inconsistency) do not imply that the variables investigated are themselves inconsistent. They simply reflect the fact that we have only incomplete information regarding each of the subsets of variables involved in the structure system. It is this incompleteness (i.e., our ignorance) that creates the local inconsistencies and, consequently, it is meaningful and desirable to accept locally inconsistent structure systems and deal with them.

A global inconsistency, on the other hand, is more serious. Its meaning is that the structure system is ill conceived; it is a mathematical artifact that has no meaning in the real world. Excellent examples of globally inconsistent wholes whose elements are locally consistent can be found in the world of graphic arts. I have in mind, for example, some drawings by M. C. Escher (such as his lithograph Belvedere) and, particularly the many drawings by the Swedish artist Oscar Reutersvärd, which are referred to as impossible figures or perspective japonaise.

Locally inconsistent structure systems are obviously resistant to the use of regular logical procedures. Two attitudes toward them can be recognized. According to one of them, such structure systems should be rejected on the basis of the fact that they do not represent any overall system. According to the other attitude, the local inconsistencies should be resolved by adjusting the given behavior functions in such a way that the new behavior functions are locally consistent and as close to the original ones as possible in some specific sense, usually in terms of the information distance. The resulting structure system is then used instead of the original system.

Given a behavior structure system

$$\mathbf{SF} = \{({}^xS, {}^x\mathbf{F}) \mid x \in N_q\}$$

with probabilistic behavior functions ${}^xf\,(x \in N_q)$ that are locally inconsistent, the following is a possible formulation of the problem of resolving these inconsistencies:

Determine behavior functions ${}^xf_c$ of the same form as given functions ${}^xf\,(x \in N_q)$ such that the function

$$\sum_{x \in N_q} D({}^xf, {}^xf_c) \tag{4.47}$$

reaches its minimum subject to the constraints

$$[{}^xf_c \downarrow {}^xS \cap {}^yS] = [{}^yf_c \downarrow {}^xS \cap {}^yS] \tag{4.48}$$

for all $x, y \in N_q$, and

$${}^xf\,({}^x\mathbf{c}) \neq 0 \Rightarrow {}^xf_c({}^x\mathbf{c}) \neq 0 \tag{4.49}$$

for all states ${}^x\mathbf{c}\,(x \in N_q)$. Let me call this problem a *problem of optimal resolution of local inconsistencies.*

TABLE 4.17
Solution to the Problem of Optimal Resolution of Local Inconsistencies for Behavior Functions Specified in Table 4.16a

v_1	v_2	${}^1f_c({}^1\mathbf{c})$	v_2	v_3	${}^2f_c({}^2\mathbf{c})$
${}^1\mathbf{c} = 0$	0	0.5208	${}^2\mathbf{c} = 0$	0	0.3846
0	1	0.1875	0	1	0.2404
1	0	0.1042	1	0	0.1732
1	1	0.1875	1	1	0.2018

Equations (4.48) express the requirement that the resulting structure system be locally consistent. Statements (4.49) require that any state that is possible under the original formulation must not be rejected in the modified, locally consistent formulation; this requirement makes it possible to use the simple information distance D given by Eq. (4.40) or Eq. (4.42) in the objective function (4.47).

Example 4.24. Consider the locally inconsistent structure behavior system with two elements whose behavior functions are specified in Table 4.16a. When the problem of optimal resolution of local inconsistencies is solved for these behavior functions, behavior functions ${}^1f_c$ and ${}^2f_c$ specified in Table 4.17 are obtained. We can easily verify that these functions are locally consistent and that

$$D({}^1f, {}^1f_c) + D({}^2f, {}^2f_c) = 0.00245.$$

Interest in the problem of resolving local inconsistencies in structure systems has been shown only recently. At this time, the problem is methodologically undeveloped and a subject of active research.

NOTES

4.1. The relationship between wholes and parts enjoys rich coverage in the literature. A few representative references will help the reader, if interested, to get more information about the main issues involved [BAI, GO1 LA1, LE1, TR1].

4.2. Holistic ideas can be found in the thinking of some ancient Greek philosophers, particularly Aristotle, and can even be traced back into the Chinese *Book of Changes* (*I Ching*). However, the methodological doctrine of holism is usually attributed to Jan C. Smuts [SM1]. Reductionism is predominantly associated with science since about the sixteenth century, but its roots can also be found in thinking of some ancient Greek philosophers. The holistic view and a criticism of reductionism is extensively covered in the book *Beyond Reductionism*, edited by A. Koestler and J. R. Smythies [KO2], as well as in Koestler's own book *The Ghost in the Machine*

[KO1]. Some claims of holism, particularly extreme holism, are critically analyzed by D. C. Phillips [PH1].

4.3. The general decomposition method for systems design, which is outlined in Section 4.5, is best exemplified in the literature by the area of switching circuits, i.e., systems with binary variables [CE1, 2, KL5]. A generalization to arbitrary discrete systems was developed by Givone [GI1]. An unorthodox decomposition method for systems with binary variables was proposed by Brown [BR8]. For continuous systems, functional equations are well covered in an extensive monograph by Aczel [AC3]. A comprehensive methodology that encompasses the whole process of systems design was developed and successfully applied by Wymore [WY2]. This methodology seems to be a good candidate for being integrated into the GSPS to deal with the problems of systems design.

4.4. The problems of determining reconstruction families from structure behavior systems have been studied for both probabilistic and possibilistic systems. For probabilistic systems, two methods based on matrix algebra were proposed, one developed jointly by Roger Cavallo and me [CA6], one by Bush Jones [JO1]. For possibilistic systems, a powerful method was developed in terms of fuzzy relation equations jointly by Masahiko Higashi, Michael Pittarelli, and me [HI4].

4.5. Reconstruction uncertainty, characterized by Eq. (4.31), and the associated identifiability quotient [Eq. (4.32)] are given extensive theoretical justification in the paper mentioned in Note 4.4 [HI4]. For probabilistic systems, the measure of reconstruction uncertainty should reflect several characteristics of the reconstruction family, including the number of overall states for which the probabilities are not unique, the number of degrees of freedom, their ranges, and the extent to which they are mutually interdependent. There are various ways in which these characteristics can be incorporated into a single measure, but none of them seems to show a clear superiority over the others in its intuitive appeal. Unless some measure with strong and universal intuitive appeal emerges, it is best to encourage the GSPS users to define their own measures and adopt one of the possible measures as a provisional default option.

4.6. The principle of maximum entropy has been justified by at least three diverse arguments:

(1) The maximum entropy probability distribution is the only *unbiased distribution*, i.e., the distribution that takes into account all available information, but no additional (unsupported) information (bias). This follows directly from the facts that (i) all available information (but nothing else) is required to form the constraints of the optimization problem, and (ii) the chosen probability distribution is required to be the one that represents the maximum uncertainty (entropy) within the constrained set of probability distributions. Indeed, any reduction of uncertainty is an equal gain of information. Hence a reduction of uncertainty from its maximum value, which would occur when any distribution other than the one with maximum entropy were chosen, would mean that some information was implicitly added.

This argument of justifying the maximum entropy principle is covered in the literature quite extensively. Its best and most thorough presentation is perhaps given in a paper by E. T. Jaynes [JA2], which also contains an excellent historical survey of related developments in probability theory, and in a book by R. Christensen [CH5]. Both of these publications also contain extensive bibliographies, which cover the literature relevant to the principle of maximum entropy almost completely.

(2) It was shown by E. T. Jaynes [JA1], strictly on combinatorial grounds, that the maximum entropy probability distribution is the *most likely* distribution. Given a reconstruction

hypothesis, each element of the reconstruction family of that hypothesis could have been generated by some number of actual data sets. The largest number of possible data sets that are mutually comparable and compatible with the given reconstruction hypothesis are those with the maximum entropy overall probability distribution.

(3) It was shown by J. E. Shore and R. W. Johnson that the principle of maximum entropy is deductively *derivable from* the following *consistency axioms* for inductive reasoning [SH5]:

- *uniqueness*: the result should be unique;
- *invariance*: the choice of coordinate system (permutation of variables) should not matter;
- *system independence*: it should not matter whether one accounts for independent information about independent systems separately in terms of marginal probabilities or together in terms of joint probabilities;
- *subset independence*: it should not matter whether one treats an independent subset of system states in terms of separate conditional probabilities or in terms of full system probabilities.

The rationale for choosing these axioms is expressed by Shore and Johnson as follows: any acceptable method of inference must be such that different ways of using it to take the same information into account lead to consistent results. Using the axioms, they derive the following proposition: given some information in terms of constraints regarding the probabilities to be estimated, there is only one probability distribution satisfying the constraints which can be chosen by a method that satisfies the consistency axioms; this unique distribution can be attained by maximizing entropy (or any other function that has exactly the same maxima as the entropy function) subject to the given constraints.

In addition to these classical arguments, which are extensively covered in the literature, a novel argument justifying the maximum entropy principle, first suggested in 1981 [CA6], is based on properties of man-made structure systems. As argued at the end of Section 4.8, every man-made structure system is associated with a unique overall system—the one represented by its unbiased reconstruction. That is to say, if a probabilistic structure system is given and we know that it is a man-made system, the maximum entropy reconstruction is the only one possible. Or, in other words, it is not possible to design a real-world probabilistic structure system whose actual reconstruction is different from the maximum entropy reconstruction. It is understood, of course, that systems are designed in the usual way to function in any possible environment. However, if a system were designed to function only in a particular environment, the argument would still hold. In this case, the join procedure would involve not only elements of the designed structure system but its environment (known in this case) as well.

For possibilistic systems, the counterpart of the maximum entropy principle is a *principle of maximum U-uncertainty*. Although its derivation from relevant consistency axioms has not been demonstrated as yet, it is justified by possibilistic counterparts of the other arguments. In particular, it is known that the possibilistic join procedure leads to the maximum uncertainty, i.e., unbiased reconstruction [CA9].

4.7. Two of my previous papers, coauthored with Roger Cavallo, contain the basic results regarding the relationship between the results of the join procedures and either the maximum entropy reconstruction [CA6] or the maximum U-uncertainty reconstruction [CA9]. In fact, it was proved first by P. M. Lewis II [LE2] that the probabilistic version of the basic join procedure leads to the maximum entropy reconstruction when applied to probabilistic structure systems that are consistent and represented by the L-structures. The convergence of the iterative join procedure to the maximum entropy reconstruction was proved by D. T. Brown [BR7]; the proof is also well covered in a book by Bishop *et al.* [BI1]. The proof that no iterative join procedure is needed for possibilistic systems is included in one of my papers mentioned in this note previously [CA9].

4.8. The concepts of general symmetric distances that characterize information closeness between arbitrary possibility or probability distributions were developed by M. Higashi and me [HI3]. In their general form, they are essential for the simulation experiments described in Section 4.9. In their special forms (4.40) and (4.42), they measure information loss (or gain) involved between pairs of comparable probability or possibility distributions, respectively.

4.9. In mathematical terminology, G-structures are irredundant hypergraphs. A *hypergraph* is defined as a family of subsets of a given set (say set N_n in our case) that satisfies the covering requirement and does not contain the empty set [BE5].

4.10. Refinement and coarsening procedures for various classes of structures, as well as procedures for preventing the generation of duplicate structures, are described in more detail than in this chapter in two of my papers [CA5, KL16].

4.11. A procedure for calculating $|\mathscr{R}_n/i| = |\mathscr{C}_n/i|$ was developed by Polya [PO2]. All isomorphic classes of undirected graphs for $n \leq 6$ are listed in a book by Harary [HA8]; Sloane [SL1] gives the numbers $|\mathscr{R}_n/i|$ for $n \leq 15$.

4.12. The numbers $|\mathscr{G}_n^+|$ (Table 4.12) are given in Sloane's *Handbook* [SL1] for $n \leq 7$ (sequence No. 1439), together with relevant references. These are the only values of $|\mathscr{G}_n^+|$ known at this time; no formula for calculating $|\mathscr{G}_n^+|$ has been found. It is known, however, that $|\mathscr{G}_n^+|$ is equal to the number of monotone Boolean functions of n variables. The numbers $|\mathscr{G}_n|$ can be calculated by the formula

$$|\mathscr{G}_n| = \sum_{k=0}^{n-1} (-1)^k \binom{n}{n-k} |\mathscr{G}_{n-k}^+|,$$

which is based on the combinatorial principle of inclusion and exclusion. It is also known that $|\mathscr{G}_n^+/1| = 2^n - 1$ and $|\mathscr{G}_n/1| = 2^n - n$.

4.13. Behavior function used in Example 4.20 (Table 4.14a) is based on data collected for a study on premarital contraceptive usage; they are analyzed in a book by Fienberg [FI1], where the original source is given (p. 121).

4.14. The general reconstruction problem was first recognized by W. Ross Ashby [AS4]. He suggested the concept of *cylindrance* for multidimensional relations (i.e., crisp possibilistic systems in our terminology) and developed an algorithm through which it is possible to determine whether a given relation of dimension n can be reconstructed from its projections of a particular dimension $k < n$. The reconstruction, if possible, is accomplished by the set intersection of extensions (cylinders) of all k-dimensional projections of the given relation. It can easily be shown that the Ashby procedure is a special case of the join procedure [CA6]. Ashby (jointly with R. F. Madden) was also first in recognizing the general identification problem discussed in Section 4.6 [MA2].

One of the Ashby's main contribution is that he recognized the great significance of the reconstruction and identification problems for systems inquiries, problems that had been largely neglected before him. His work on these problems, which is rather restrictive from the methodological point of view, was a primary stimulus for my own work in this direction. My first ideas about a more comprehensive approach to the reconstruction problem developed in the mid-1970s, during my stay at the Netherlands Institute for Advanced Study in Wassenaar [KL7]. My work on the reconstruction problem and, later, also the identification problem continued in various directions and is still on-going. Results, often produced in cooperation with some of my colleagues and assistants, are published in a series of papers [CA5–7, CA9, KL10, KL12, KL14–16]. The reconstruction problem has also been investigated by Gerrit Broekstra [BR3–5], Roger Conant [CO5–7], Klaus Krippendorff [KR2], and others [DU4]. Bush Jones contributed to the identification problem [JO1].

The best source of information regarding the status of reconstructability analysis in the early 1980s is a *Special Issue on Reconstructability Analysis of the International Journal of General Systems* (Vol. 7, No. 1, 1981, pp. 1–107), which also contains an extensive bibliography.

4.15. Initial and rather limited simulation experiments regarding reconstructability analysis were performed in 1976 and 1977 by Hugo Uyttenhove [KL14, 15]. More comprehensive experiments, which are described in Section 4.9, have been performed since fall 1981 by Abdul Hai [HA1]. They are still in progress, particularly those of the general form outlined at the end of Section 4.9.

4.16. The reconstruction principle of inductive inference (Section 4.10) was presented first at my Presidential Address at the 27th Annual Meeting of the Society for General Systems Research in Washington, D.C., January 5–9, 1982 [KL13].

4.17. The problem of resolving local inconsistencies in systems (Section 4.11) has been studied by N. Rescher [RE3, RE6, RE11]; it is fair to say that methods for dealing with this problem have not been sufficiently developed as yet.

4.18. The question of the effect of different degrees of data quantization on reconstructability analysis of probabilistic systems was investigated by R. E. Valdes-Perez and R. C. Conant [VA1]. The following quote summarizes the conclusions of this investigation:

> Assuming a probabilistic characterization of the system under study, we find that although there is an incremental benefit from quantizing data more finely, since by so doing a higher proportion of original information is retained, this benefit decreases rapidly and becomes nearly insignificant for Q (number of states per variable) greater than about 4 or 5. On the other hand, quantizing more finely carries a very rapidly accelerating cost in the computer space and time needed to carry out reconstructability analysis. These factors seem to indicate that ternary and higher quantization is significantly better than binary, while Q greater than about 3 or 4 bears a much higher computer cost and brings only marginal benefits.

These conclusions, which are based on a mathematical analysis, are virtually the same as those based on the simulation experiments described in Section 4.9.

EXERCISES

4.1. Consider an overall system of n variables ($n \geq 3$), each of which has k states. Determine for some small values of k, say $k = 2, 3, 4$, the value of n for which it becomes less economical (in terms of the required memory cells) to store the overall system than
(a) to store all its two-variable subsystems;
(b) to store all its three-variable subsystems.

4.2. Let the notation introduced in Table 4.4 be used for probabilities of overall states of behavior systems with three binary variables. Given any probability distribution $\{p_i | i \in N_{0,7}\}$, show that
(a) it is fully reconstructable from subsystems $\{v_1, v_2\}$ and $\{v_2, v_3\}$ if and only $p_0 p_5 = p_1 p_4$ and $p_2 p_7 = p_3 p_6$;
(b) it is fully reconstructable from subsystems $\{v_1, v_3\}$ and $\{v_2, v_3\}$ if and only if $p_0 p_6 = p_2 p_4$ and $p_1 p_7 = p_3 p_5$.
(c) it is fully reconstructable from subsystems $\{v_1, v_2\}$ and $\{v_1, v_3\}$ if and only if $p_0 p_3 = p_1 p_2$ and $p_4 p_7 = p_5 p_6$.

4.3. Determine reconstruction families and values of the identifiability quotients for
(a) all members of the solution set in Example 4.19;
(b) reconstruction hypotheses 2, 7, and 10 in Example 4.19;
(c) all reconstruction hypotheses of the possibilistic overall system whose behavior function is specified in Table 4.18a.

4.4. Develop some reasonable identifiability quotient for probabilistic behavior systems.

4.5. Determine reconstruction families and values of the identifiability quotient of your choice (Exercise 4.4) for
(a) all members of the solution set in Example 4.20;
(b) hypothesis 3 in Example 4.20;
(c) hypothesis 18 in Example 4.21;
(d) a structure systems with the same elements as those specified in Table 4.3 except that ${}^k f({}^k\mathbf{c}) = 0.25$ for all $k \in N_3$.

4.6. Prove that min $[{}^1 f(\mathbf{a}, \mathbf{b}), {}^2 f(\mathbf{c} | \mathbf{b}) = \min [{}^1 f(\mathbf{a}, \mathbf{b}), {}^2 f(\mathbf{b}, \mathbf{c})]$, where ${}^1 f$ and ${}^2 f$ are possibility distribution functions defined on the Cartesian products $\mathbf{A} \times \mathbf{B}$ and $\mathbf{B} \times \mathbf{C}$, respectively.

4.7. Let f and f^h denote the behavior functions of an overall system and its unbiased reconstruction from a reconstruction hypothesis h, respectively.
(a) Prove that $f^h(\mathbf{c}) = 0$ implies $f(\mathbf{c}) = 0$ for all $\mathbf{c} \in \mathbf{C}$ when the systems are probabilistic.
(b) Construct a counter example to demonstrate that it is not true for probabilistic systems that $f(\mathbf{c}) = 0$ implies $f^h = 0$.
(c) Prove that $f^h(\mathbf{c}) \geq f(\mathbf{c})$ for all $\mathbf{c} \in \mathbf{C}$ when the systems are possibilistic.

4.8. Develop procedures for determining all immediate coarsenings of
(a) a given G-structure;
(b) a given C-structure in the appropriate set $\mathscr{C}_n$;
(c) a given P-structure in the appropriate set $\mathscr{P}_n$.

4.9. Utilizing the similarity with the combined ordering defined by Eqs. (3.72) and (3.106), define formally the combined ordering associated with the reconstruction problem. Compare properties of the individual orderings, preorderings, and the combined orderings in the three problem types.

4.10. Utilizing the similarity with the solution sets defined by Eqs. (3.73) and (3.107), define formally the solution set (the set of admissible reconstruction hypotheses) for the reconstruction problem.

4.11. Assume that the RC-procedure is applied to several C-structures at the same level of refinement. Develop a procedure which, when appropriately combined with the RC-procedure, would prevent a generation of duplicate C-structures. Show also how these two procedures must be combined.

4.12. Calculate the unbiased reconstructions and distances for
(a) reconstruction hypotheses 1, 2, 3, 5, and 6 in Example 4.20;
(b) the reconstruction hypothesis with all two-variable subsystems in Example 4.20 (assume $\Delta = 0.0005$);
(c) all members of the solution set and reconstruction hypothesis 9 in Example 4.21.

4.13. Utilizing the result of Exercise 4.12(b), draw a plot of the dependence of D_l on l for the overall system in Example 4.20.

4.14. Consider a modification of Example 4.19 in which only C-structures are of interest, but it is not required that the reconstruction be perfect.
(a) Determine the solution set for this modification.
(b) Draw a plot of the dependence of D_l on l for this modification.

4.15. Determine all G-structures (not only isomorphic equivalence classes) in several r-equivalence classes in G_4, say those represented by the fully connected graph and a graph without one edge.

4.16. Utilizing results of Exercise 4.8, determine the structure neighborhood of
(a) C-structures 123/234/456, 12/23/34/14, 123/234/345/456/156, and 123/234/345/456/156/135/246;
(b) P-structures 14/15/45/24/13/35/25/12/ and 12/13/15/24/25/45;
(c) G-structures 13/14/125/245/345, 12/24/25/135/145, and 1234/35/45/346;
(d) the same G-structures as in (c), but under the assumption that the structure neighborhood is restricted to the same r-equivalence class.

4.17. Determine the number of structures isomorphic to each of the structures specified in Exercise 4.16. Check also some of the numbers $\# g$ of isomorphic structures given in the tables of refinement lattices in Appendix D.

4.18. Using Eq. (D.1) in Appendix D, determine
(a) the number of immediate predecessors of each C-structure in the individual i-equivalence classes depicted in Figure 4.20 in the other i-equivalence classes (label the connection in the diagram by the number of predecessors of the respective type);
(b) the number of immediate predecessors of each structure in the i-equivalence classes $g = 79$–82 of $\mathcal{G}_5$ (Appendix D) in the i-equivalence classes $g = 73$–78.

4.19. Supplement tables of lattices $(\mathcal{G}_3/i, \leq)$ and $(\mathcal{G}_4/i, \leq)$ in Appendix D by the relevant information regarding predecessors.

4.20. Determine all admissible reconstruction hypotheses for the possibilistic behavior systems whose behavior functions are specified in Table 4.18a, b. Draw a plot of D_l versus l for each of the systems.

4.21. Determine reconstruction families and values of the identifiability quotient for all members of the solution set in Exercise 4.20 and compare them with corresponding results of one reconstruction hypothesis that is not admissible.

4.22. W. Ross Ashby proposed a procedure for determining the unbiased reconstruction of multidimensional relations (or crisp possibilistic systems in our terminology) from their projections. His procedure is based on the concept of a *cylindric extension* of a projection with respect to variables of the overall relation that are not involved in it. Let $[{}^kf \uparrow S - {}^kS]$ denote the extension of a projection represented by a possibilistic behavior function (crisp) ${}^kf$. Then, the extension is a function from the set $\mathbf{C}$ of all overall states into $\{0, 1\}$ such that

$$[{}^kf \uparrow S - {}^kS](\mathbf{c}) = \begin{cases} 1, & \text{if } {}^kf({}^k\mathbf{c}) = 1 \quad \text{for some } {}^k\mathbf{c} \prec \mathbf{c}, \\ 0, & \text{otherwise.} \end{cases}$$

Given a reconstruction hypothesis h with projections ${}^kf\,(k \in N_q)$, the *Ashby procedure* can be expressed either by the formula

$$f^h(\mathbf{c}) = \min_k \, [{}^kf \uparrow S - {}^kS](\mathbf{c})$$

or by the formula

$$f^h(\mathbf{c}) = \max_k \, [{}^k\bar{f} \uparrow S - {}^kS](\mathbf{c}),$$

where ${}^k\bar{f}({}^k\mathbf{c}) = 1 - {}^kf({}^k\mathbf{c})$. Using either of the formulas, determine the unbiased reconstructions for all reconstruction hypotheses of each system specified in Table 4.18 and compare the results with the reconstructions obtained by the join procedure.

4.23. Determine

(a) all admissible reconstruction hypotheses for the probabilistic behavior system specified in Table 4.18c and draw the plot of D_l versus l;

TABLE 4.18
Illustrations to Exercises

(a)

	v_1	v_2	v_3	$f(\mathbf{c})$
$\mathbf{c} =$	0	0	0	0.8
	0	0	2	1
	0	1	1	0.6
	0	1	2	1
	1	0	0	0.7
	1	1	0	0.8
	1	1	1	0.6
	1	1	2	1

(b)

	v_1	v_2	v_3	$f(\mathbf{c})$
$\mathbf{c} =$	0	0	0	1
	0	0	1	1
	0	0	1	1
	1	0	1	1

(c)

	v_1	v_2	v_3	$f(\mathbf{c})$
$\mathbf{c} =$	0	0	0	0.1
	0	0	1	0.4
	1	0	0	0.05
	1	0	1	0.2
	1	1	0	0.25

(d)

	v_1	v_2	v_3	v_4	$f(\mathbf{c})$
$\mathbf{c} =$	0	0	0	0	0.1
	0	1	1	0	0.2
	0	1	1	1	0.1
	1	0	0	0	0.1
	1	0	1	0	0.1
	1	0	1	1	0.1
	1	1	1	0	0.2
	1	1	1	1	0.1

(c) all admissible reconstruction hypotheses that are based only on C-structures and do not exceed the distance of 0.1 for the probabilistic behavior system specified in Table 4.18d.

4.24. Using any of the probabilistic or possibilistic methodological distinctions, determine all admissible reconstruction hypotheses for the memoryless systems with three binary variables defined on various populations whose frequency distributions are specified in the first part of Table 4.19a. The systems are based on studies published in the literature. The following are brief characterizations of the studies and relevant references, where the systems and experimental circumstances are fully described. States of variables are given in the order 0, 1; variables are listed in the order v_1, v_2, v_3.

(a) Epidemologic investigation of a food poisoning outbreak [BI1, p. 90]. Support: 304 persons attending a picnic; variables: illness (presence or absence), crabmeat (eaten or not eaten), potato salad (eaten or not eaten).

(b) Response (favorable or unfavorable) to three different drugs observed on 46 persons [BI1, p. 308].

TABLE 4.19
Frequency Distributions of Systems in Exercises 4.24 and 4.25

					Exercise 4.24				
v_1	v_2	v_2	(a)	(b)	(c)	(d)	(e)	(f)	(g)
0	0	0	120	6	84	1	11	86	2
0	0	1	4	2	8	4	2209	35	16
0	1	0	22	2	22	2	48	32	1
0	1	0	0	6	2	6	239	11	6
1	0	0	80	16	25	12	0	73	48
1	0	1	31	4	12	1	111	70	8
1	1	0	24	4	7	3	72	61	36
1	1	1	23	6	14	1	2074	41	6

					Exercise 4.25				
v_1	v_2	v_3	v_4	(a)	(b)	(c)	(d)	(e)	(f)
0	0	0	0	12	187	350	554	387	20
0	0	0	1	16	256	150	281	36	2
0	0	1	0	27	15	60	87	876	9
0	0	1	1	32	42	112	49	250	2
0	1	0	0	8	42	26	338	383	6
0	1	0	1	22	34	23	531	270	1
0	1	1	0	22	40	19	56	381	4
0	1	1	1	30	62	80	110	1712	1
1	0	0	0	47	177	1,878	97	955	38
1	0	0	1	14	194	1,022	75	162	7
1	0	1	0	46	14	148	182	874	24
1	0	1	1	9	27	404	140	510	6
1	1	0	0	14	30	111	85	104	25
1	1	0	1	23	52	161	184	176	6
1	1	1	0	25	63	22	171	91	23
1	1	1	1	15	121	265	458	869	42

(c) Factors effecting tromboembolism [BI1, p. 112]. Support: 116 women; variables: tromboembolism (absence or presence), smoking (no or yes), use of oral contraceptives (no or yes).

(d) Response of lymphoma patients to combination chemotherapy by sex and cell type [BI1, p. 148]. Support: 30 patients; variables: type of disease (modular, diffuse), sex (male, female), response (no, yes).

(e) A study of death penalties in Florida in 1973–1979 [KR2, p. 75]. Support: 4,764 murder cases; variables: killer (black, white), victim (black, white), penalty (death, other).

(f) Structural habitat categories of Anolis lizard of Bimini [FI1, p. 27]. Support: 409 lizards; variables: species (sagrei adult male, distichus adult and subadult), perch height (≤ 4.75 ft, > 4.75 ft), perch diameter (≤ 4 in., > 4 in.).

(g) A study of diabetic patients [FI1, p. 53]. Support: 123 patients; variables: age at onset (< 45, ≥ 45), family history of diabetes (no, yes), dependence on insulin injections (no, yes).

4.25. Repeat Exercise 4.24 for systems with four binary variables whose frequency distributions are specified in the second part of Table 4.19. Variables are listed in the order v_1, v_2, v_3, v_4.

(a) A study of patients receiving psychiatric care [FI1, p. 90]. Support: 362 patients, variables: validity (psychasthetic, energetic), solidity (hysteric, rigid), stability (extroverted, introverted), acute depression (no, yes).

(b) A study of the nature of sexual symbolism [FI1, p. 89]. Support: 1,356 male subjects; variables: actual anatomical measuring (male, female), responses based on exposure rates 1/1000 or 1/5 second (male, female), purpose of experiments (not explained, explained).

(c) An investigation of car accidents [FI1, p. 90]. Support: 4,831 car accidents; variables: car weight (small, standard), driver ejected (no, yes), severity (not severe, severe), accident type (collision, rollover).

(d) A study of the attitudes toward the "leading crowd" among schoolboys [GO3, p. 424]. Support: 3,398 schoolboys; variables: membership (out, in) and attitud toward (unfavorable, favorable) the "leading crowd" during first interview and second interview.

(e) Preference of World War II recruits for location of training camp [BI1, p. 138]. Support: 8,036 soldiers; variables: race (black, white), region of origins (north, south), location of present camp (north, south), preferred location (north, south).

(f) Particularistic versus universalistic values in role conflict [GO3, p. 404]. Support: 216 persons; variables represent four different situations of role conflict; their states represent tendencies toward particularistic or universalistic values.

5

METASYSTEMS

No single frame either explains
Nor foretells the whole continuity—
The picture of the caterpillar
Does not foretell the butterfly,
Nor does one picture of a butterfly
Show that a butterfly flies.

—R. BUCKMINSTER FULLER

5.1. CHANGE VERSUS INVARIANCE

> *The scientist does not so much seek descriptions of the changeless as changeless descriptions of the changing.*
>
> —G. Spencer Brown

One of the most fundamental human capabilities, perhaps the most fundamental, is the capability of recognizing *differences*. Its most primitive manifestation is the *making of distinctions* by human beings, as well depicted by Goguen and Varela [GO1]:

> A distinction splits the world into two parts, "that" and "this," or "environment" and "system," or "us" and "them," etc. One of the most fundamental of all human activities is the making of distinctions. Certainly, it is the most fundamental act of system theory, the very act of defining the system presently of interest, of distinguishing it from its environment.
>
> Distinctions coexist with purposes. A particularly basic case is a system defining its own boundaries and attempting to maintain them; this seems to correspond to what we think of as self-consciousness. It can be seen in individuals (ego or identity maintenance) and in social units (clubs, subcultures, nations). In such cases, not only is there a distinction, but an *indication*, that is, a marking of one of the two distinguished states as being primary ("this," "I," "us," etc.); indeed, it is the very purpose of the distinction to create this indication.
>
> A less basic kind of distinction is one made by a distinction for some purpose of his own. This is what we generally see explicitly in science, for example, when a discipline "defines its field of interests," or a scientist defines a system which he will study.
>
> In either case, the establishment of system boundaries is inescapably associated with what we will call a *cognitive point of view*; . . . in particular, it is associated with some notion of value, or interest. It is also linked up with the cognitive capacities (sensory capabilities, knowledge background) of the distinctor. Conversely, the distinctions made reveal the cognitive capabilities of the distinctor. It is in this way that biological and social structures exhibit their coherence, and make us aware that they in fact have cognitive capacities or that they are "conscious" in some degree.

The recognition of differences is thus closely connected with the making of distinctions, and is of two types. We can recognize either that two things are different or that the same thing has changed with time (or, using the GSPS terminology, with respect to the relevant backdrop or the associated support). The two meanings are intimately interrelated and complement each other. Those aspects of things that are viewed as *permanent* (invariant, constant) are captured by the first meaning; those viewed as *temporary* (varying, changing) are in the domain of the second one.

The importance of the notion of *change*, which is one of the derivatives of the notion of difference, has been expressed in literature in many imaginative ways. For example, Heraclitus, the ancient Greek philosopher, expressed it by his famous statement

> Nothing is permanent except change;

John Wilmot, a seventeenth century British poet, conveyed a similar message in one of his poems:

> Since 'tis Nature's law to change,
> Constancy alone is strange;

and Edmund Burke, a British statesman, expressed the same sentiments in terms of politics in his *Speech on American Taxation* in 1774:

> A state without the means of some change is without the means of its conservation.

Regardless of whether or not we subscribe to the view expressed by Heraclitus, the fact remains that we have a fundamental need to view certain things in our environment as invariant, at least on pragmatic grounds. If this were not possible, we would not be able to communicate, as there would be no identifiable entities, and, indeed, we would not even be able to act in any meaningful way, as there would be nothing in our environment we could take for granted.

There are several reasons why we can recognize invariance in our environment. One obvious reason is that certain things in our environment change at a considerably slower rate than the rate of our perception, cognition, and acting. Such changes are thus negligible for our practical purposes, or we are not even aware of them. Another reason is that some changes occur at a level of resolution that is beyond the human scale. We do not notice such changes and, unless they are manifested in some fashion within our scale, they are irrelevant to us.

We can also recognize invariance of a different sort, one associated with the process of change rather than the thing which changes. In the GSPS framework, this kind of invariance is exemplified by the notion of the generative system. Its variables are subject to changes, but the manner in which they change (as expressed by its behavior or ST-function) is support invariant, i.e., it is permanent (changeless) within the support set considered.

The search for invariances is the very essence of science, as is well described in this delightful quote from a book by G. Spencer Brown [BR9]:

Science is concerned with the discovery of constants: it is the study of the changeless. If I drop a bomb from my top storey window, it will fall to the ground with an ever-increasing speed. This change of speed is anathema to the scientist. He may not rest content until he has found a way of picturing it changelessly. In this case he has not far to seek. The speed of the bomb may change, but the rate at which it changes (called the acceleration) does not. The function 32 *feet per second per second* is a constant which describes not only the behaviour of my bomb, but also that of other bombs dropped in the vicinity.

We talk of the function 32 *ft per sec*^{2} as if it were *absolutely* constant, but a little reflection shows that it is not so. The mass of the earth is slowly increasing as it picks up meteorites and interstellar dust. We may thus expect g, the acceleration due to gravity, to increase as time goes on. If we formalize this increase in terms of a further "constant," we have no reason to suppose that this further "constant" may not itself be changing. Our attempt at a perfect description of the acceleration due to gravity has ended in a regress. It may seem that the regress could be broken by the following means. We suppose that statements involving the concept g are statements dependent on given masses, distances, and other factors known to be "relevant." Given all the relevant factors, we are in a position to formulate a constant which does not change. But the problem is now seen to be purely linguistic; any change in the constant made necessary by observation and experiment can be blamed upon our faulty assessment of the *relevant* conditions under which the constant should be observed. In other words there is always a "real" constant to which our observations tend: it just happens that when we think we have found it we discover afterwards that what we have found is only an approximation of it.

This latter way of talking is analogous to the philosophy of the thing-in-itself, or "the reality beneath the appearance." It could be called "the constant beyond the approximation." Such an assumption is indeed part of the scientific attitude and its convenience for some purposes remains undoubted. We shall discuss its usefulness later, but for the moment we must emphasize that the laws of nature are merely the descriptions we have made of structures which have been found to change only very slowly. We have in fact no evidence for the existence of any structure which does not change at all . . . .

Exactly what we notice can plausibly be ascribed to how, and especially how fast, we ourselves can change. We notice, for example, things which change as slowly as or more slowly than we do, but not in general things which change much more quickly. Thus the faster we can change, the more we can notice.

If we take a cinematograph of a plant at, say, one frame a minute, and then show this moving picture speeded up to 30 frames a second, the plant appears to behave like an animal. When something is placed near it, it clearly perceives it and reacts to it. It is obviously a sentient being. Why, then, does it not ordinarily appear conscious? The answer is, perhaps, because it thinks too slowly. To beings which reacted eighteen hundred times as quickly as we reacted, we might appear as mere unconscious vegetables. Indeed, the beings who moved so quickly would be justified in calling us unconscious, since we should not normally be conscious of their behaviour. Such glimpses of it as might appear from time to time would mean nothing. A tree can no more perceive me walking past it than I can see a bullet flying past me. I might perceive certain events in the wake of the bullet, such as a broken

> arm; and similarly, if my passage were destructive enough, the tree might eventually perceive certain events in my wake, such as a broken branch. But what is fast for a tree is slow and boring for me, whereas what is normal speed for me is something out of this world for a tree. . . . Anyone who could move infinitely fast would be in a position to know everything, because to him nothing would move. He would have an infinite time in which to learn it. And if he were also allowed to move bits of the universe himself, he would not only be omniscient, but also omnipotent, since he would have as long as he liked to beetle about altering things.
>
> We have seen that science seeks to reduce change to an unchanging formula. Wherever there *has been* change, such formulae can always be found; but they do not always apply to the future. When the change itself changes, we need a new formula.
>
> We are now ready to differentiate between the task of the historian and that of the scientist. The scientist, we have seen, is concerned with recording in a changeless way phenomena which are still changing; whereas the historian as such is concerned only with recording changes that have already stopped. The historian is not concerned to find a formula which will work from henceforward for all time. If he ever found such a formula, no more records would be necessary and he would lose his job. It is not history which repeats itself, but science. The scientist begins by looking at the welter of change and fixing in formulae whatever parts of it he can. History is what is left over after the scientist has taken his pick.
>
> History is therefore more fundamental than science. It is our first appreciation of things. But its study is not urgent. That which does not change, such as the past, is not dangerous. As such, it cannot harm us. But we must beware of what changes. And in order that we may adapt ourselves to it, our sense must be quick to sense it.

It is clear that the search for invariances, which is so basic to science, should be one of the main capabilities of the GSPS. Some aspects of support invariance are associated with generative systems and structure generative systems. These are discussed in Chapters 3 and 4, respectively. Such aspects are only special cases of a more general concept of support invariance, as discussed in the next section.

5.2. PRIMARY AND SECONDARY SYSTEMS TRAITS

> *The law of identity does not permit you to have your cake and eat it, too.*
>
> —Ayn Rand

To exist means to have an identity, and that, in turn, means to have an identifier. A system is given its identity when some of its traits are defined. Let these traits, which form an identifier of the system, be called its *primary traits*. It is then natural to call any other traits of the system, which are not involved in the identifier, its *secondary traits*.

The set of all primary traits of a system thus forms its definition. It is a general property of the epistemological hierarchy of systems that the set of primary traits associated with a particular level is a subset of the set of primary traits associated with any higher level. This is illustrated for neutral systems up to the level of structure behavior systems in Figure 5.1. The figure can be easily modified to structure data systems or structure source systems by excluding the traits that are not included in their definitions (masks, behavior functions, or data functions); a modification to directed systems is trivial.

A necessary condition for dealing with a system in some problem situation is that it keeps its identity. This means that its primary traits are required to remain unchanged. No such restriction is, of course, required for secondary traits. For example, a given data system may be supplemented by a behavior function derived from its data. This function is obviously a trait of the data system. Since, however, there are many different behavior functions that can be derived from the same data system for different masks and different ways of expressing the constraint among variables, the behavior function cannot be employed to identify the data system. It is its secondary trait and, as such, it is allowed to change. We may switch from one behavior function to another and that does not change the identity of the data system. The example can also be inverted. A given behavior system can generate different data sets for different initial conditions. Each of these data sets is a secondary trait of the behavior system. Regardless of which of them is under consideration, the identity of the behavior system remains unchanged.

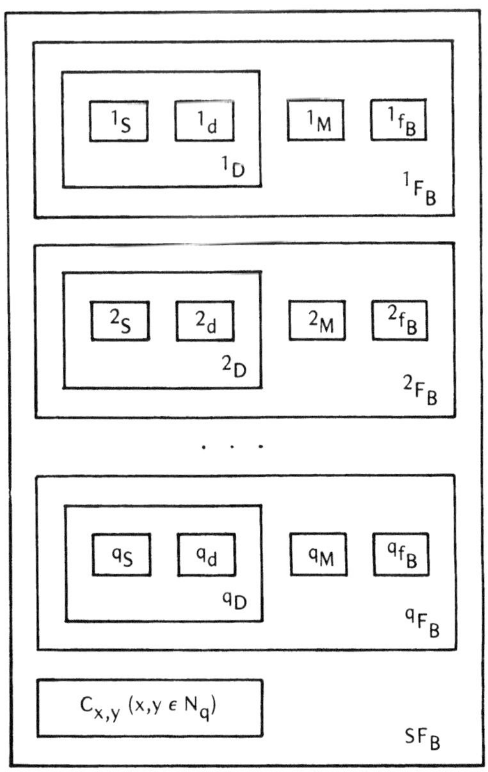

Figure 5.1. Subset relationship of primary traits in the epistemological hierarchy of systems.

When dealing with a system, it is a common practice to redefine it, at an appropriate stage of the problem-solving process, in the sense that some of its secondary traits are accepted as primary traits. In empirical investigation, for example, the system is initially defined as a source system and this definition is maintained during the data-gathering stage. When the investigator becomes confident that the data obtained are sufficient to properly characterize the variables of the source system, he may accept the data array as a primary trait. This means that he redefines the source system, making it a data system. Such a change represents an *inductive step* because it results from an inductive act based on assumptions such as, for example: any state pattern that it is possible for the variables to form is among the patterns included in the data or it can be derived from them. The decision to redefine the system from a source system to a data system reflects in this case the investigator's belief that the information contained in the data is sufficient for the purpose of the investigation. Factors that contribute to forming this belief include not only the data, but also the purpose of the investigation, the envisioned manner in which the data will be processed, a comparison with similar investigations performed previously, as well as subjective qualities of the investigator (experience in the area of investigation, intuition, and the like).

After the source system is redefined as a data system, the investigation becomes theoretical. The first objective is to find a generative system that will adequately represent the data system. Each of the behavior or ST-functions that are derived from the data for different masks is a secondary trait of the data system. Once the investigator develops sufficient confidence in any one of them, he may accept it as a primary trait and thereby redefine the data system as a generative system. This again involves an inductive step because the support-invariant nature of the accepted behavior or ST-function extends the claims made by the generative system beyond the range of the given support set as well as to different initial conditions. Similarly, the generative system can be redefined as an appropriate structure generative system, in another inductive step, after its reconstruction properties have been adequately analyzed.

Systems are also redefined in the process of systems design. Assume that the aim is to implement a given behavior system by a structure system that consists of specific types of elements and satisfies some given requirements (objectives, constraints). Traits of prospective structure systems that are determined during the design process are viewed as secondary traits. They do not change the identity of the given behavior system; its source system, mask, and behavior function are kept as the only primary traits during the entire course of the design process. The problem often results in several structure systems. Once one of them is accepted, the behavior system may be redefined as a structure behavior system; the latter then serves as a basis for the implementation of the design.

Primary traits, as a means of identifying a system, must be completely known and support-invariant. No such requirements are imposed upon the secondary traits. They may be completely unknown or known only partially, and need not be support invariant. If it happens that a primary trait of a system changes, then, by definition, the system loses its identity and a new system emerges. On the other hand, the identity of the

that a procedure be given that represents one particular function of the form (5.2), even though it may be impossible or impractical to determine which function it actually is. One possibility is, of course, to define the replacement function explicitly. In such a case, the replacement procedure is identical with (defined by) the replacement function. These are illustrated later in this section by a few examples.

Equation (5.1), which defines a *metasystem of neutral behavior systems*, can be easily modified to other types of metasystems by replacing symbols $\mathbf{MF}_B$ and $\mathscr{F}_B$ with symbols representing the other systems. As a practical convention, let a set of systems of some type always be denoted by the capital script version of the letter symbol that denotes systems of that type. Then, for example,

$$\mathbf{MSF}_S = (\mathbf{W}, \mathscr{S}\mathscr{F}_S, r),$$
$$\mathbf{M\hat{D}} = (\mathbf{W}, \hat{\mathscr{D}}, r)$$

are definitions of a *metasystem of structure ST-systems* (neutral) and a *metasystem of directed data systems*, respectively. It is trivial to obtain definitions of all the remaining types of metasystems.

In general, metasystems can also be defined on a set of systems of different types. Let such a general type be denoted by **MX**. Then,

$$\mathbf{MX} = (\mathbf{W}, \mathscr{X}, r) \tag{5.3}$$

where $\mathscr{X}$ is an arbitrary set of systems whose support sets are subsets of $\mathbf{W}$; r is again a replacement procedure, which is required to implement a specific replacement function

$$r: \mathbf{W} \rightarrow \mathscr{X}. \tag{5.4}$$

From the standpoint of this general formulation, we can view metasystems whose elements are systems of the same type as special cases of (5.3). These special cases are then characterized by

$$\mathscr{X} \in \{\mathscr{S}, \mathscr{D}, \mathscr{F}_B, \mathscr{F}_S, \mathscr{S}\mathscr{S}, \mathscr{S}\mathscr{D}, \mathscr{S}\mathscr{F}_B, \mathscr{S}\mathscr{F}_S\},$$
$$\hat{\mathscr{X}} \in \{\hat{\mathscr{S}}, \hat{\mathscr{D}}, \hat{\mathscr{F}}_B, \hat{\mathscr{F}}_S, \mathscr{S}\hat{\mathscr{S}}, \mathscr{S}\hat{\mathscr{D}}, \mathscr{S}\hat{\mathscr{F}}_B, \mathscr{S}\hat{\mathscr{F}}_S\}$$

for neutral systems, and in a similar way (using symbols with carets) for directed systems. Let such metasystems be called *homogeneous metasystems*.

Replacement procedures, each of which is obviously a primary trait of a metasystem, can be defined in many different ways. They may even include random decisions. The only requirement is that each replacement procedure implement a particular replacement function of the general form (5.4). Some typical ways of defining replacement functions are illustrated by the following examples.

Example 5.1. This example describes the functioning of a set of traffic lights at an intersection for a 24-h period as a homogeneous metasystem that consists of three elements defined as data systems. Each of the three elements contains the same variables and state sets. Variables describing the lights for traffic bound north–south, south–north, east–west, and west–east are denoted by NS, SN, EW, and WE, respectively, and those describing the lights for left turns for traffic bound north–east, south–west, east–south, and west–north are denoted by NE, SW, ES, and WN, respectively. The support is time; 1 sec is the smallest recognizable interval of time in terms of which all other relevant time intervals are defined.

Data matrices $\mathbf{d}_1, \mathbf{d}_2, \mathbf{d}_3$ of the three elements $\mathbf{D}_1, \mathbf{D}_2, \mathbf{D}_3$ are defined in Figure 5.2a; their time sets are specified directly by the relevant time intervals. The data matrices are periodical and are defined by their first periods. As indicated in the figure, the individual systems $\mathbf{D}_1$, $\mathbf{D}_2$, $\mathbf{D}_3$ represent the traffic control at night, at periods of normal traffic during day, and during rush hours, respectively. The systems, viewed as elements of a metasystem, replace each other at specific times during each period of 24 h. A convenient manner of defining the replacement function is in this case the labeled diagram in Figure 5.2b. Its nodes represent the three elements of the metasystem, each arrow from $\mathbf{D}_i$ to $\mathbf{D}_j$ $(i, j = 1, 2, 3)$ indicates that $\mathbf{D}_i$ is replaced by $\mathbf{D}_j$, and the label attached to the arrow specifies the time at which the replacement is made. The metasystem is thus the triple

$$\mathbf{MD} = (T, \mathcal{D} = \{\mathbf{D}_1, \mathbf{D}_2, \mathbf{D}_3\}, r),$$

where $\mathcal{D}$ and r are fully specified in Figure 5.2, and T consists of 5,760 defined time intervals for each period of 24 h (420 periods of d_1, 390 periods of d_2, and 240 periods of d_3).

Example 5.2. Consider a patient whose kidneys do not function properly at times. His condition is monitored in terms of several variables. When necessary, the functioning of his kidneys is replaced by the so-called hemodialysis machine. The same monitoring continues even when the machine is used, but some additional variables must be observed during such periods. Two source systems, say $\mathbf{S}_1$ and $\mathbf{S}_2$, can thus be recognized for the purpose of monitoring the patient. One of them is associated with the periods during which the natural kidneys function adequately, while the other one represents periods during which the hemodialysis machine is employed. System $\mathbf{S}_1$ contains the following four variables:

v_1—water in urine (measures to an accuracy of 0.1 liters in the range of 0–1 liter);
v_2—glucose in urine (measured to an accuracy of 20 g in the range of 0–200 g);
v_3—urea in urine (measured to an accuracy of 5 g in the range of 0–50 g);
v_4—blood urea nitrogen (only two states are defined by the observation channel, say states 1 and 0, depending on whether or not the actual value reaches at least 150 mg per 100 ml of blood).

ELEMENT D_1: night traffic control.

$t_i \in$	[0, 20)	[20, 30)	[30, 50)	[50, 60)
NE = SW	g	y	r	r
NS = SN	g	y	r	r
ES = WN	r	r	g	y
EW = WE	r	r	g	y

ELEMENT D_2: normal traffic control.

$t_i \in$	[0, 15)	[15, 25)	[25, 55)	[55, 65)	[65, 80)	[80, 90)	[90, 110)	[110, 120)
NE = SW	g	y	r	r	r	r	r	r
NS = SN	r	r	g	y	r	r	r	r
ES = WN	r	r	r	r	g	y	r	r
EW = WE	r	r	r	r	r	r	g	y

ELEMENT D_3: traffic control during rush hours.

$t_i \in$	[0, 30)	[30, 40)	[40, 50)	[50, 60)
NE = SW	r	r	r	r
NS = SN	g	y	r	r
ES = WN	r	r	r	r
EW = WE	r	r	g	y

(a)

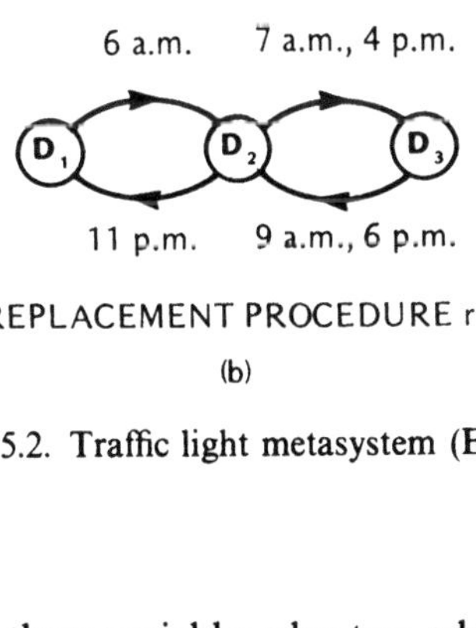

REPLACEMENT PROCEDURE r

(b)

Figure 5.2. Traffic light metasystem (Example 5.1).

System S_2 contains all of these variables plus two additional variables:

v_5—temperature of blood (measured to an accuracy of 0.2°F in the range of 97–100°F);

v_6—blood pressure (measured to an accuracy of 2 mm of mercury column in the range of 110–130 mm).

These variables are essential for the hemodialysis machine, which must maintain each of them in a narrow range. All of the introduced variables are observed in time. The actual time set (frequency of observation) depends on the seriousness of patient's condition as well as other factors and there is no need to define it for the purpose of this example.

The two source systems can be viewed as a metasystem under the following replacement procedure r: if $v_4 = 1$, replace $\mathbf{S}_1$ by $\mathbf{S}_2$; if $v_4 = 0$, replace $\mathbf{S}_2$ by $\mathbf{S}_1$. The metasystem is thus the triple

$$\mathbf{MS} = (T, \mathcal{S} = \{\mathbf{S}_1, \mathbf{S}_2\}, r),$$

where T is the union of the time sets defined for the individual systems $\mathbf{S}_1$, $\mathbf{S}_2$.

Example 5.3. Consider a structure system whose elements are arranged in an $n \times n$ array. Assume that each of the elements, which are often called *cells* of the array, is coupled only to cells that are adjacent to it in the array. For example, a 5×5 array is shown in Figure 5.3. Individual cells in the array can be identified conveniently by two integers $i, j \in N_{0,n-1}$ that label rows and columns of the array, respectively. As indicated in Figure 5.3, they can be also identified by a single integer

$$c = ni + j.$$

Let c be called a *cell identifier*.

Assume that the *internal environment* of each cell c (except the boundary cells) consists of its four adjacent cells, as shown in Figure 5.4. The cell has four input

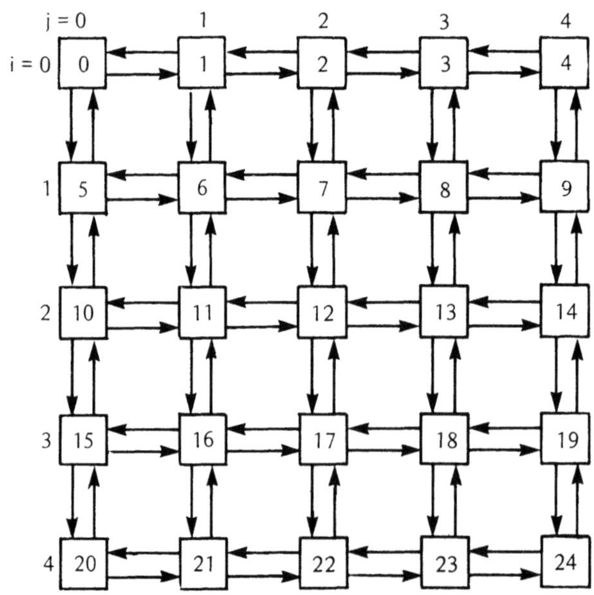

Figure 5.3. 5×5 cellular array (Example 5.3).

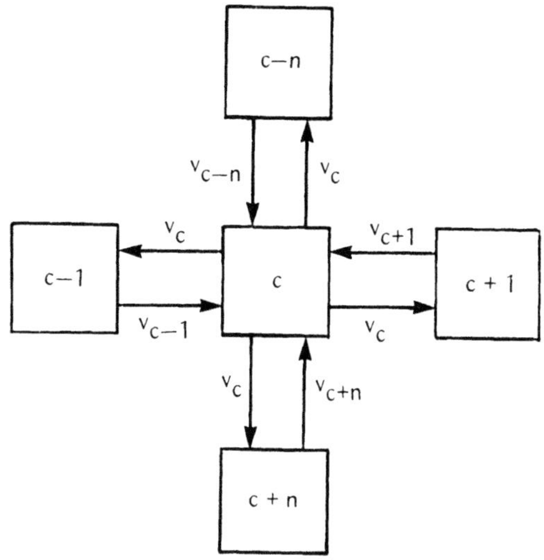

Figure 5.4. Internal environment of a cell in a cellular array (Example 5.3).

variables v_{c-n}, v_{c-1}, v_{c+1}, v_{c+n}, one from each of the adjacent cells. It has one output variable, which is coupled to all the adjacent cells. The internal environment of each of the boundary cells (cells in rows $i = 0, n-1$ and columns $j = 0, n-1$) is degenerated in an obvious way.

Let all cells in a cellular array, say the one in Figure 5.3, be deterministic directed ST-systems defined by the same totally ordered time set T and a ST-function

$$v'_c = f_c(v_{c-n}, v_{c-1}, v_c, v_{c+1}, v_{c+n}), \tag{5.5}$$

where v'_c represents the next state of v_c and $c \in N_{0,24}$; it must, of course, be specified how (5.5) is interpreted for the boundary cells, where some of the input variables are not present. Assume further that only two states, 0 and 1, are distinguished for each of the variables. When $v_c = 1$ ($v_c = 0$), let cell c be called *active* (*inactive*, respectively).

Given a cellular array, a set of structure systems can be defined on it, each characterized by a subset of its cells. For example, there are 2^{25} (more than 3.3×10^7) structure systems for the cellular array in Figure 5.3. It is sometimes desirable to integrate structure systems in this set, say set $\mathcal{S}\hat{\mathcal{F}}_S$, into a metasystem:

$$\mathbf{MS\hat{F}}_S = (T, \mathcal{S}\hat{\mathcal{F}}_S, r) \tag{5.6}$$

by a suitable replacement procedure. As a simple example, let the replacement procedure r in (5.6) be defined as follows: cell $c\,(c \in N_{0,24})$ is included in the structure system if and only if it is active or at least one cell in its internal environment is active, i.e., if and only if

$$v_{c-n} + v_{c-1} + v_c + v_{c+1} + v_{c+n} \geq 1.$$

To show a specific metasystem of the kind characterized by (5.6), let us use the proposed replacement procedure and let

$$v_c' = [(v_{c-5} + v_{c-1} + v_{c+1}) \pmod 2) + v_c + v_{c+5}] \pmod 2)$$

be the ST-function of cells in the 5 × 5 cellular array. Variables that are not available in the internal environment of a cell are simply excluded from the formula. This metasystem generates sequences of structure systems, one for each initial structure system. Short segments of three such sequences are shown in Figure 5.5, where the black and gray squares identify cells that are included in the individual structure systems, and distinguish active cells (black squares) from inactive cells (gray squares); white squares identify cells that are not included in the various structure systems.

Variations of the metasystem introduced in this example are possible by using different ST-functions for the cells, or different replacement procedures. More radical variations can be produced by using different arrays, generally k-dimensional, where $k \geqslant 1$. The members of this class of metasystems are usually referred to in the literature as *tessellation automata.*

Example 5.4. Consider an image system that consists of a single variable v, state set V, and a single support t, which is viewed as an index identifying the location in

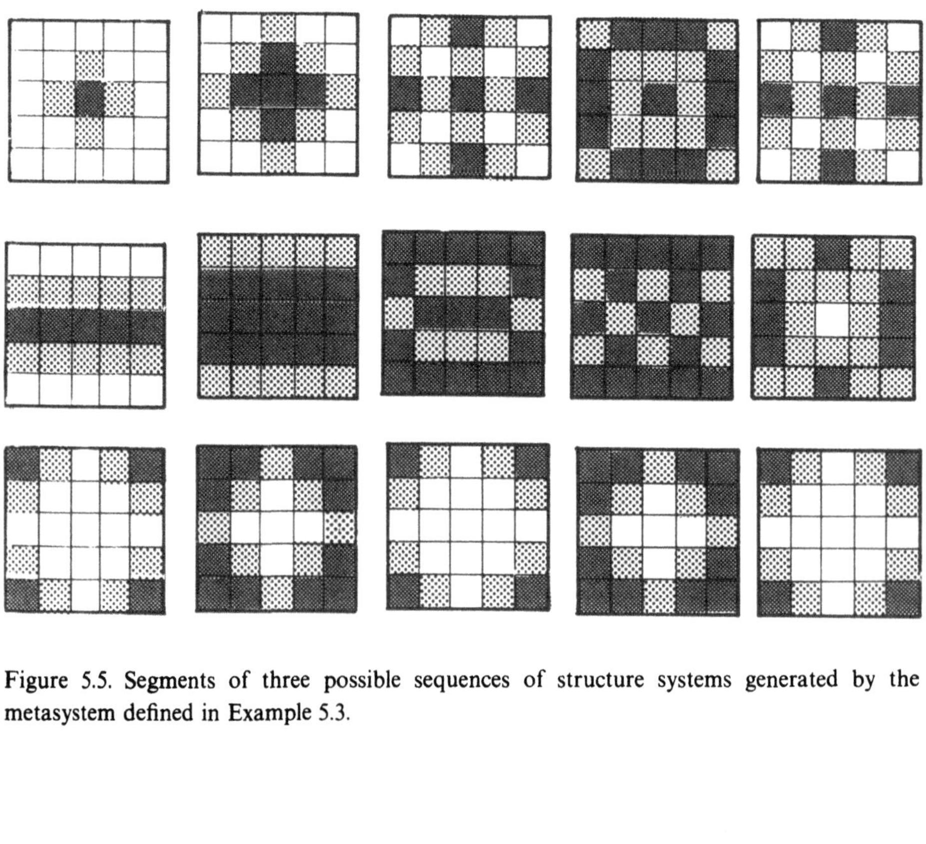

Figure 5.5. Segments of three possible sequences of structure systems generated by the metasystem defined in Example 5.3.

strings of states from V. The support set T is totally ordered and represented by the set of nonnegative integers. A class of metasystems of the type

$$\mathbf{MD} = (T, \mathscr{D}, r)$$

can be defined on this image system in such a way that

- $\mathscr{D}$ is the set of all data systems that can be formed by variable v for all possible subsets N_n of T ($n = 1, 2, \ldots$), and
- r is a replacement procedure defined in the following general form: given a data system $\mathbf{D} \in \mathscr{D}$ with a data array $\mathbf{d}$, scan $\mathbf{d}$ in increasing order of its support (normally from left to right) and replace $\mathbf{d}$ with $\mathbf{d}'$ by substituting for each state α in $\mathbf{d}$ a string $p(\alpha)$ of states, as specified by a function

$$p: V \to V \cup V^2 \cup \cdots \cup V^k \tag{5.7}$$

for some finite k; $\mathbf{d}'$ defines a new system $\mathbf{D}' \in \mathscr{D}$.

These metasystems are usually referred to in the literature as *developmental OL-systems* (or Lindenmayer systems without interactions). Pairs α, $p(\alpha)$ are usually called *production rules* and are often denoted by $\alpha \to p(\alpha)$.

As a specific example of a deterministic OL-system (a metasystem in our terminology), let $V = N_{0,9}$ and let the production rule function

$$p: V \to V \cup V^2$$

be defined by the table

α	0	1	2	3	4	5	6	7	8	9
$p(\alpha)$	12	93	49	61	25	87	78	34	9	9

Then, for example, if the initial data system has the data array [0], the metasystem generates a sequence of data systems with the following data arrays:

[0]
[1 2]
[9 3 4 9]
[9 6 1 2 5 9]
[9 7 8 9 3 4 9 8 7 9]
[9 3 4 9 9 6 1 2 5 9 9 3 4 9]
[9 6 1 2 5 9 9 7 8 9 3 4 9 8 7 9 9 6 1 2 5 9]
[9 7 8 9 3 4 9 8 7 9 9 3 4 9 9 6 1 2 5 9 9 3 4 9 9 7 8 9 3 4 9 8 7 9]

For nondeterministic OL-systems, production rules are not defined by a production function of the form (5.7), but by any set of pairs (α, β) taken from the Cartesian product

$$V \times (V \cup V^2 \cup \cdots \cup V^k).$$

The selection of individual production rules may be based on conditional probabilities of β given α.

5.4. METASYSTEMS VERSUS STRUCTURE SYSTEMS

> *Limited by space, a frog in a well cannot understand what is an ocean.*
> *Limited by time, an insect in summer cannot understand what is ice.*
>
> —CHUANG-TZU

Structure systems and metasystems represent two schemes for integrating other systems (source, data, or generative systems) into larger units. They are different schemes, independent of each another, and neither is superior to the other. They can also be combined, i.e., applied to each other.

Structure systems integrate other systems with respect to the sets of their variables and under the assumption that they are all based on the same support set. Elements of structure systems are thus systems with different sets of variables, but with the same support sets.

Metasystems, on the other hand, integrate other systems with respect to their support sets, regardless of whether or not they are all based on the same set of variables. Elements of metasystems are thus systems with different local support invariances defined in terms of an overall support set; they may be defined for a single overall set of variables.

As illustrated in Section 5.3, metasystems can be employed for integrating structure systems, which, in turn, are employed to integrate other systems. For instance, a class of metasystems of structure ST-systems (tessellation automata) is discussed in Example 5.3. Such systems are denoted by the symbol $\mathbf{MS\hat{F}}_S$, in which both of the integrating operators, **M** and **S**, are used.

It is also possible to use structure systems for integrating metasystems defined for different sets of variables. To represent such systems symbolically, the operator **S** must precede the operator **M**. For example, symbol $\mathbf{SMF}_B$ denotes a structure metasystem whose elements (i.e., elements of the metasystems integrated in the structure system) are neutral behavior systems, while symbol $\mathbf{MSF}_B$ denotes a metasystem whose elements are structure behavior systems. The operators are thus noncommutative. Systems **MSX** and **SMX** (for any **X**) are not only different, but also noncomparable in terms of the ordering represented by the epistemological hierarchy of systems.

Example 5.5. Consider a directed structure system that consists of two metasystems whose elements are directed behavior systems. The metasystems (elements of the structure system), denoted by symbols ${}^1\mathbf{M\hat{F}}_{GB}$ and ${}^2\mathbf{M\hat{F}}_{GB}$, are coupled according to the block diagram in Figure 5.6a. Each of them consists of two directed behavior

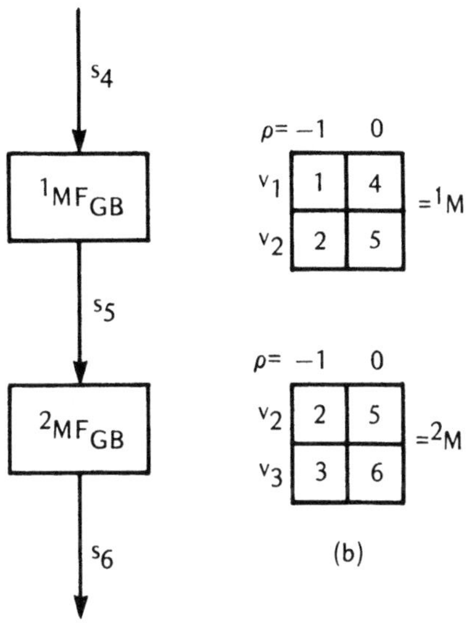

$\alpha =$

s_1	s_2	s_4	s_5 (${}^{11}f_{GB}(\alpha)$)	s_5 (${}^{12}f_{GB}(\alpha)$)
0	0	0	0	0
0	0	1	1	0
0	1	0	1	0
0	1	1	0	1
1	0	0	1	0
1	0	1	0	1
1	1	0	0	1
1	1	1	1	1

$\beta =$

s_2	s_3	s_5	s_6 (${}^{21}f_{GB}(\beta)$)	s_6 (${}^{22}f_{GB}(\beta)$)
0	0	0	1	1
0	0	1	0	1
0	1	0	0	1
0	1	1	1	1
1	0	0	0	0
1	0	1	1	1
1	1	0	1	1
1	1	1	0	0

(c)

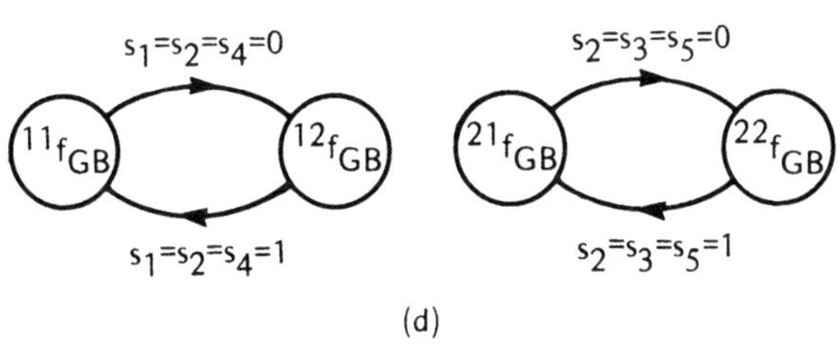

(d)

Figure 5.6. Illustration of a structure metasystem (Example 5.5).

systems that are based on variables with two states, 0 and 1, and the same totally ordered support set. Behavior systems of each metasystem are defined in terms of the same mask, but differ in their behavior functions.

The masks ${}^1\hat{M}$ and ${}^2\hat{M}$ are defined in Figure 5.6b. As required for structure systems, sampling variables that occur in both of them have the same identifiers. Each mask is used in both behavior systems of its metasystem.

The behavior functions (generative) are deterministic, i.e., the generated variable in each of the behavior systems is a function of the generating and input sampling variables. They are defined in Figure 5.6c. Symbols ${}^{11}\hat{f}_{GB}$, ${}^{12}\hat{f}_{GB}$ denote behavior functions involved in the first metasystem, and symbols ${}^{21}\hat{f}_{GB}$, ${}^{22}\hat{f}_{GB}$ denote those in the second metasystem.

Both metasystems use a replacement procedure of the same type: when all the input and generating variables are in state 0, replace the first behavior system with the second one; when all the input and generating variables are in state 1, replace the second behavior system with the first one. When this description is applied to the appropriate variables, procedures ${}^1r$ and ${}^2r$ are obtained, which are shown in terms of their replacement diagrams in Figure 5.6d.

A sample of data generated by this structure metasystem for the initial condition $s_1 = s_2 = s_3 = 0$ and a particular input sequence (sequence of states of variable v_1) is shown in Figure 5.7. It is indicated which of the behavior systems of the two metasystems is used at each support instance t.

Identifier of behavior function in ${}^1MF_{GB}$		1	1	2	2	2	2	2	1	1	1	1	2	2	2	2	1	
t=	1	2	3	4	5	6	7	8	9	10	11	12	13	14	15	16	17	
v_1	0	1	0	0	1	1	0	1	1	0	1	0	0	0	1	1	1	...
v_2	0	1	0	0	0	1	1	1	1	0	1	0	0	0	0	1	1	...
v_3	0	0	0	1	1	1	0	1	0	0	0	0	1	1	1	1	0	...
Identifier of behavior function in ${}^2MF_{GB}$		1	1	2	2	2	1	1	1	1	1	1	2	2	2	2	1	
	initial condition																	

Figure 5.7. A sample of data generated by the structure metasystem defined in Figure 5.6 (Example 5.5).

5.5. MULTILEVEL METASYSTEMS

> *Insofar as identities are preserved, they are orderly laws of nature. Insofar as identities decay, these laws are subject to modification. But the modification itself may be lawful. The change in the individual may exhibit a law of change—and yet such laws of change are themselves liable to change.*
>
> —ALFRED NORTH WHITEHEAD

We may now define multilevel metasystems in a manner similar to that in which multilevel structure systems are defined (Section 4.4). These are metasystems whose elements are metasystems, whose elements . . ., etc. This recursion must terminate with elements that are not metasystems.

Following the analogy with structure systems, let multilevel metasystems be denoted by a generalized operator $\mathbf{M}^k$, whre k indicates the number of levels involved. For example, symbol $\mathbf{M}^3\mathbf{D}$ denotes a three-level metasystem whose final elements are data systems; it is a metasystem whose elements are metasystems, whose elements are again metasystems, whose elements are neutral data systems. Two-level metasystems may for convenience be called *meta-metasystems.*

Formally, a k-level metasystem is defined by the triple

$$\mathbf{M}^k\mathbf{X} = (\mathbf{W}^k, \mathcal{M}^{k-1}\mathcal{X}, r^k), \tag{5.8}$$

where $\mathbf{W}^k$ and r^k denote its support set and replacement procedure, respectively, and $\mathcal{M}^{k-1}\mathcal{X}$ denotes a set of its elements (metasystems of level $k-1$) whose final elements are systems in set $\mathcal{X}$, which are not metasystems.

In a multilevel metasystem, say $\mathbf{M}^k\mathbf{X}$, only the highest-level replacement procedure r^k is a primary trait through which the identity of the system is recognized. Replacement procedures of all the lower-level metasystems are secondary traits since they represent only local support invariance from the standpoint of the metasystem $\mathbf{M}^k\mathbf{X}$.

Example 5.6. Consider a two-level metasystem (or a meta-metasystem)

$$\mathbf{M}^2\mathbf{S}\hat{\mathbf{F}}_s = (T, \mathcal{M}\mathcal{S}\hat{\mathcal{F}}_S = \{{}^1\mathbf{MS}\hat{\mathbf{F}}_S, {}^2\mathbf{MS}\hat{\mathbf{F}}_S\}, r^2),$$

whose support set T and first element ${}^1\mathbf{MS}\hat{\mathbf{F}}_S$ are the same as those of the metasystem defined in Example 5.3 (a tessellation automaton). The second element (metasystem) has the form

$${}^2\mathbf{MS}\hat{\mathbf{F}}_S = (T, \mathcal{S}\hat{\mathcal{F}}_S, {}^2r),$$

where T, $\mathcal{S}\hat{\mathcal{F}}_S$ are the same components as those in metasystem ${}^1\mathbf{MS}\hat{\mathbf{F}}_S$, and ${}^2r$ is the following replacement procedure: make one randomly selected active cell in the structure system inactive and exclude from the structure system those cells in its internal environment, including the cell itself, for which it is the case that neither they nor any of their adjacent cells are active.

The two metasystems are integrated in the meta-metasystem by the following second-level procedure (or metaprocedure) r^2: if the structure system recognized at support instant $t-1$ is different from the one recognized at support instant $t-2$, then use metasystem ${}^1\mathbf{MS\hat{F}}_S$; otherwise, use metasystem ${}^2\mathbf{MS\hat{F}}_S$.

Example 5.7. Consider a developmental OL-system (Example 5.4), which is defined as a meta-metasystem. It consists of two metasystems

$$ {}^1\mathbf{MD} = (T, \mathscr{D}, {}^1r) $$
$$ {}^2\mathbf{MD} = (T, \mathscr{D}, {}^2r), $$

which differ from each other only in their replacement procedures ${}^1r$, ${}^2r$. Their components T and $\mathscr{D}$ are the same as defined in Example 5.4 except that $V = \{0, 1, 2, 3\}$. Replacement procedures ${}^1r$, ${}^2r$ are defined by the following production rule functions ${}^1p$, ${}^2p$, respectively:

α	0	1	2	3
${}^1p(\alpha)$	01	21	30	32

α	0	1	2	3
${}^2p(\alpha)$	01	20	30	32

The meta-metasystem $\mathbf{M}^2\mathbf{D}$ is then defined as

$$ \mathbf{M}^2\mathbf{D} = (T, \{{}^1\mathbf{MD}, {}^2\mathbf{MD}\}, r^2), $$

where the metaprocedure r^2 is defined as follows: scan the last data array obtained; if at least one half of the entries are 0, use metasystem ${}^1\mathbf{MD}$ (function ${}^1p$); otherwise, use metasystem ${}^2\mathbf{MD}$ (function ${}^2p$).

For example, given the initial data array [0], metasystem ${}^1\mathbf{MD}$ generates the following sequence of data arrays:

[0]
[0 1]
[0 1 2 1]
[0 1 2 1 3 0 2 1]
[0 1 2 1 3 0 2 1 3 2 0 1 3 0 2 1]
[0 1 2 1 3 0 2 1 3 2 0 1 3 0 2 1 3 2 3 0 0 1 2 1 3 2 0 1 3 0 2 1]
. . .

Metasystem ^{2}**MD** generates another sequence of data arrays:

[0]
[0 1]
[0 1 2 0]
[0 1 2 0 3 0 0 1]
[0 1 2 0 3 0 0 1 3 2 0 1 0 1 2 0]
[0 1 2 0 3 0 0 1 3 2 0 1 0 1 2 0 3 2 3 0 0 1 2 0 0 1 2 0 3 0 0 1]
. . .

Meta-metasystem **M**^{2}**D** generates still another sequence of data arrays (it is indicated in each case which of the production rule functions is used):

[0], ^1p
[0 1], ^1p
[0 1 2 1], ^2p
[0 1 2 0 3 0 2 0], ^1p
[0 1 2 1 3 0 0 1 3 2 0 1 3 0 0 1], ^2p
[0 1 2 0 3 0 2 0 3 2 0 1 0 1 2 0 3 2 3 0 0 1 2 0 3 2 0 1 0 1 2 0], ^2p
. . .

Multilevel metasystems can be combined with multilevel structure systems in any sequence. The only requirement is that elements of the metasystem or structure system at the lowest level be one of the three basic system types—source, data, or generative systems. Systems such as $\mathbf{MSMSF}_B$, $\mathbf{SM}^2\mathbf{S}^2\mathbf{D}$, or $\mathbf{M}^2\mathbf{S}^2\mathbf{MSS}$ are thus perfectly acceptable in the GSPS language.

5.6. IDENTIFICATION OF CHANGE

> *We understand change only by observing what remains invariant, and permanence only by what is transformed.*
>
> —GERALD M. WEINBERG

The great variety of ways in which metasystems with different types of elements and different numbers of levels can be defined is illustrated by some representative examples in Sections 5.3–5.5. However, the problem of determining, if desired, an appropriate metasystem characterization of the investigated variables from their data is one of the most difficult and methodologically least developed systems problems.

The problem stems from a fundamental question associated with systems inquiries: should the constraint among the investigated variables be viewed as support invariant or rather as varying according to some support-invariant rules of change (a replacement procedure)? One of the difficulties is that there is no absolute answer to this question. Which of the views should be taken depends not only on the nature of the variables themselves, but also on the objectives of the inquiry, the way in which the constraint is expressed (mask, constraint measure), whether the data are complete or not, and other factors, some of which may be connected with the investigative contexts.

When restricted to a particular way of expressing the constraint (e.g., probability measures, and the largest acceptable mask), and under the usual objective criteria of generative uncertainty and complexity, the problem becomes one of identifying a change in the behavior or ST-function. In other words, it becomes a problem of identifying significant local constraints among the variables within the support set involved.

If a behavior function representing the whole support set (a global function) does not differ much from behavior functions corresponding to various subsets of the support set (local functions), then there is no need for resorting to a metasystem formulation. If, however, substantial differences between the functions are detected, the metasystem formulation should be seriously considered. This common-sense conclusion involves, unfortunately, some conceptual as well as practical difficulties.

First, the term "substantial difference between behavior functions" must be defined in some operational manner. That is, an appropriate distance function must be selected for behavior functions of the kind involved to give a specific meaning to the term "difference." In addition, some threshold value of the distance must be specified to give a meaning to the term "substantial." Although these specifications should be left to the discretion of the user, the GSPS should offer (upon request) some options and use one of them as a default option.

Second, the difference (distance) between the global and local behavior functions is significant only if the local function is determined for a sufficiently large subset of the support set. Once again, a decision must be made to specify the size of the smallest subsets of the support set that are considered significant (acceptable) for determining a meaningful local behavior function. The significance depends on the number of states distinguished by the variables, the measure in terms of which their constraint is characterized, the mask involved, and possibly other factors.

In addition to the conceptual difficulties mentioned , the problem of identifying significant local constraints involves considerable practical difficulties. These are primarily connected with the fact that the number of subsets of the support set, which must be inspected in the process of searching for significant local constraints, grows exponentially with the size of the support set. As a consequence, the amount of computation involved grows explosively with the size of the support set so that the problem becomes intractable for support sets of even modest size.

As an illustration of this problem area, a simple procedure for the identification of local constraints is described in the rest of this section. Let us call it a *metasystem identification procedure*. The procedure is based on the assumption that the support set T is totally ordered and that the variables are characterized by a data system. It either does not produce any metasystem (if no significant local constraints are found) or it produces a metasystem that consists of a sequence of behavior systems in the support set. The systems replace each other at specific instances of the support set, which are determined by the procedure.

Given a data system with totally ordered support set $T = N_n$, a specific mask (usually the largest acceptable mask), and a particular way of representing constraint among the variables (with its measure of generative uncertainty), the essence of the metasystem identification can be described as follows:

i. Let m, Δ be a given integer and a given rational number, respectively, and let $t = 1$, $k = 1$.
ii. Determine the behavior function for the subset of data that corresponds to segment $[t, t+m]$ of the support set and calculate its generative uncertainty U_1.
iii. Increment k by 1; if $t+km \notin T$, go to (vi).
iv. Determine the behavior function for the subset of data that corresponds to segment $[t, t+km]$ of the support set and calculate its generative uncertainty U_k.
v. If $|U_k - U_{k-1}|/\max(U_k, U_{k-1}) < \Delta$, go to (iii); otherwise, record $t+(k-1)m$ as an approximate point of replacement of elements of a metasystem, $t = (k-1)m$, make $k = 1$, and go to (ii).
vi. Stop.

The procedure is based on the following observation: if there are no significant local constraints in the data, the generative uncertainties of local behavior functions based on segments $[1, t]$ of the support set converge quickly with increasing t (after some initial erratic changes) to values within a small interval Δ; if, on the other hand, the data contain a significant local constraint, say within the segment $[t_1, t_2]$ of the support set, the generative uncertainties are likely to exhibit changes substantially larger than Δ around the parameter instances t_1, t_2 while it converges to another small interval inside the segment $[t_1, t_2]$. A large change in the generative uncertainty, after it converged to some small interval, is thus a basis for viewing the system as a metasystem, each element of which is associated with a subset of the support set.

The sensitivity and computational complexity of the procedure depend considerably on the chosen values of m and Δ. To help the user in this respect, the GSPS should be equipped with relevant characteristics obtained by appropriate simulation experiments, similar to those described in Chapter 4 for reconstructability analysis. It is also possible to repeat the procedure for several values of m and Δ and average the obtained results. In any event, the procedure indicates only an approximate location of a feasible replacement point. To locate it more precisely and evaluate its significance, one must search for its natural meaning in the context of the overall investigation.

Example 5.8. This example describes an application of the metasystem identification procedure in the area of performance evaluation of aircraft pilots during their training on flight simulators. The source system consists of four variables describing relevant characteristics of a jet aircraft and time as a support. To avoid technical details, let each variable be characterized by its label, descriptive name of the corresponding attribute, and a set of states:

s—speed, state set N_5;
a—altitude, state set N_7;
r—TACAN (typical tactical air navigation approach) radial, state set N_4;
d—TACAN DME (distance-measuring equipment) direction, state set N_4.

Observation channels and a data matrix for these variables are given in Table 5.1. The data matrix, which is presented in increasing order of time, describes a typical (ideal, correct) tactical navigation landing approach by a jet aircraft at some specific approach

TABLE 5.1
Specifications of the System Discussed in Example 5.8

Observation channels

		Speed	Altitude	Radial	Direction
Attribute:		Speed	Altitude	Radial	Direction
Variable:		s	a	r	d
Units:		Knots	Feet above sea level	Degrees relative to magnetic north	Nautical miles
Meaning of states	1	[150, 170)	[1,700–1,840)	[0–93)	[0–2.6)
	2	[170, 220)	[1,840–2,400)	[93–113)	[2.6–14)
	3	[220, 250)	[2,400–2,700)	[113–122)	[14–21)
	4	[250–270)	[2,700–4,900)	[122–360)	> 21
	5	≥ 270	[4,900–6,900)	—	—
	6	—	[6,900–22,800)	—	—
	7	—	> 22,800	—	—

Data matrix

s	444455555555555555555555555555555555555555555554444444333
a	777666666666666666666666666666666666666666666666666655555
r	211111111111111111111111111111111111111111111111111111222
d	444444444444444444444444444444444444433333333333333333333

s	333333333333332222222222222111111111111111111111111111111
a	555555555555444444444444444444333333333333333333333333333
r	222333333333333333333333333333333333333333333333333333333
d	333333333333333322222222222222222222222222222222222222222

s	1111111111111112222233333333333333333333333333333
a	3222222111111112222223334444444444444444444444444
r	3333333333333333333344444444444444444444444444444
d	2222222222111111111111111111222222222222222222222

site. The full time period of the landing approach was divided into 163 equal time intervals and for each of them the appropriate states of the variables were defined.

The data matrix in Table 5.1 is a reference with respect to which performances of various student pilots are compared. Based on this comparison, students are evaluated and graded. It is desirable, from the standpoint of both students and instructors, to determine natural subtasks in the overall task so that weaknesses in students' skills become identified more specifically by being localized.

The metasystem identification procedure was applied to the data matrix in Table 5.1 for a probabilistic behavior function, a two-column mask, and $\Delta = 0.1$. It was repeated for several values of m. The replacement points obtained for different values of m were then averaged. This resulted in three elements of a metasystem, which are represented by the following segments of the time set: 1–70, 71–122, 123–163. These elements can be, in fact, given a natural interpretation in terms of the actual landing task. They correspond to the penetration descent, the flight along the arc and interception of the final approach course, and the final approach, respectively.

Mask evaluation (Section 3.6) and reconstruction procedure (Section 4.7) were then performed on each of the elements of the metasystem. The results obtained are summarized in Figure 5.8. The final system is thus of the type MSF_B. Clearly, this system would permit a much more focused evaluation of students than would be possible given only a single overall behavior system.

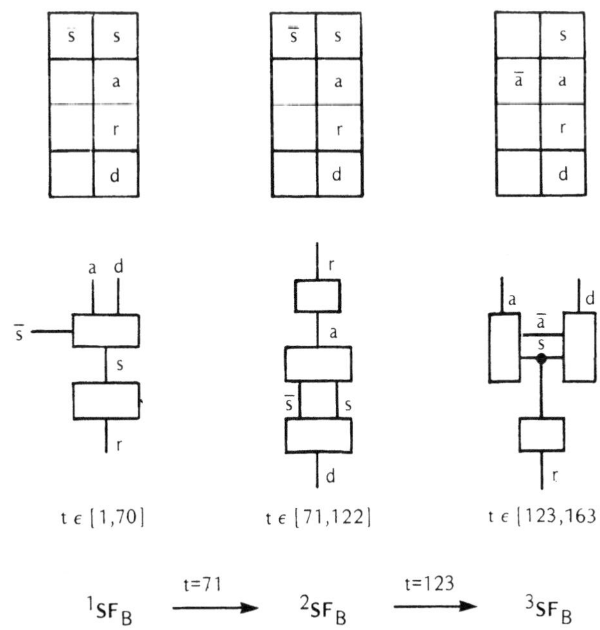

Figure 5.8. Metasystem identified for a typical landing approach by a jet aircraft (Example 5.8).

NOTES

5.1. The concept of tessellation automata (Examples 5.3, 5.6) was introduced by Yamada and Amoroso in 1969 [YA3]. Their paper initiated research into the mathematical properties of tesselation automata [AM2, YA2, YA4, 5] as well as their applications [HA4, OS1].

5.2. The concept of developmental systems (Examples 5.4, 5.7) was introduced by Lindenmayer [LI2], primarily as a convenient mathematical formalism for describing various phenomena of biological growth. It has become a very active area of mathematical as well as applied research [LI3–6, HE2, RO9].

5.3. The metasystem identification procedure described in Section 5.6 was proposed by Uyttenhove [UY1], who also performed some simulation experiments to identify desirable values of m and Δ. Example 5.8 is described in more detail in a paper by Comstock and Uyttenhove [CO1].

EXERCISES

5.1. Make a list of all traits included in systems types **S**, **D**, $\mathbf{F}_B$, $\mathbf{F}_S$, **SS**, **SD**, $\mathbf{SF}_B$, **MS**, **MD**, $\mathbf{MF}_B$, $\mathbf{SMF}_B$, $\mathbf{MSF}_B$, and determine for each of the types which of the traits are primary and which are secondary.

5.2. Define at least one real-world metasystem in the area of your interest.

5.3. Extend Example 5.3 in the following ways:

(a) generate sequences of structure systems for an initial system different from those in Figure 5.5;

(b) repeat the example for a different ST-function of cells;

(c) repeat the example for a different replacement procedure.

5.4. Consider a probabilistic metasystem that represents a developmental OL-system (Example 5.4). Its production rules $\alpha \to \beta$ are associated with conditional probabilities $p(\beta \mid \alpha)$ as follows:

α	0	0	1	1	2	2
β	0	00	1	02	2	11
$p(\beta \mid \alpha)$	0.25	0.75	0.75	0.25	0.75	0.25

Generate sequences of data systems for various initial systems by determining each choice, for example, by throwing two coins.

5.6. Determine the total number of all epistemological types of systems under the assumption that the maximum number of combined levels of structure systems and metasystems is n. (Count only generative systems, rather than behavior and ST-systems; count only source systems rather than object and image systems.)

6

COMPLEXITY

> *. . . the Scientist, like the Pilgrim, must wend a straight and narrow path between the Pitfalls of Oversimplification and the Morass of Oversimplification.*
>
> —RICHARD BELLMAN

6.1. COMPLEXITY IN SYSTEMS PROBLEM SOLVING

> . . . *the fruits of science are simple fruits, or more precisely, fruits of simplification.*
> —GERALD M. WEINBERG

The notion of complexity appears in this book on numerous occasions and in different contexts. It also appears in different forms, depending primarily on the type of system or problem to which it is applied. This clearly indicates that complexity plays an important role in systems problem solving. In fact, it is emerging as a fundamental concept of systems science, one that is perhaps as fundamental as the concept of energy is to the natural sciences.

There are many facets to complexity. Starting with a common dictionary, we find that complexity is "the quality or state of being complex," i.e., "having many varied interrelated parts, patterns, or elements and consequently hard to understand fully "or" marked by an involvement of many parts, aspects, details, notions, and necessitating earnest study or examination to understand or cope with" (*Webster's Third New International Dictionary*). This general characterization of complexity does not contain any qualification regarding the kind of things to which it is applicable. As such, it can be applied to virtually any kinds of things, material or abstract, natural or man-made, products of art or science; it can be applied to systems, problems, methods, theories, laws, games, languages, machines, organisms, organizations, or any other things we may name. Regardless of what it is that is actually considered as being complex or simple, its degree of complexity is, in general, associated with the number of recognized parts as well as the extent of their interrelationship. In addition, complexity has a somewhat subjective connotation since it is related to the ability to understand or cope with the thing under consideration. Thus a thing that is complex for one person may be simple for someone else.

From the standpoint of the GSPS, the common-sense meaning of complexity is expressed by the interaction between the investigator and the object of investigation, through which a source system emerges. Complexity in this sense is thus not an intrinsic property of the investigated object; it is rather a result of the way in which the investigator interacts with it. That is to say, we do not attempt to deal with complexities of objects, only complexities of systems defined on objects. This point is well characterized by Ross Ashby in one of his last writings [AS12]:

The word "complex," as it may be applied to systems, has many possible meanings, and I must first make my use of it clear. There is no obvious or pre-eminent meaning, for although all would agree that the brain is complex and a bicycle simple, one has also to remember that to a butcher the brain of a sheep is simple while a bicycle, if studied exhaustively (as the only clue to a crime) may present a very great quantity of significant detail.

Without further justification, I shall follow, in this paper, an interpretation of "complexity" that I have used and found suitable for about ten years. I shall measure the degree of *"complexity" by the quantity of information required to describe the vital system.* To the neurophysiologist the brain, as a feltwork of fibres and a soup of enzymes, is certainly complex; and equally the transmission of a detailed description of it would require much time. To a butcher the brain is simple, for he has to distinguish it from only about thirty other "meats," so not more than $\log_2 30$, i.e. about 5 bits, are involved. This method admittedly makes a system's complexity purely *relative to a given observer*; it rejects the attempt to measure an absolute, or intrinsic, complexity; but this acceptance of complexity as something in the eye of the beholder is, in my opinion, the only workable way of measuring complexity.

Others have expressed their views on this important issue differently, though in the same spirit. The following two imaginative quotes should reinforce the point I want to make here:

We can only hope for explicit models of the world and not for reality itself or even a small part of it. (Patrick Suppes [SU3])

One of the functions of the experimental method is to substitute simple artificial systems for the complex systems that Nature presents to us. (Herbert A. Simon [SI4])

At the source system level, the notion of systems complexity is somewhat primitive. It is expressed solely in terms of the cardinalities of the sets involved—the set of variables, set of supports, state sets, and support sets—since no relationship among the sets is available. At higher epistemological levels, the notion of systems complexity becomes more meaningful. It is, of course, different for different system types.

As discussed in Chapters 2–5, the same source system can be described at the various higher epistemological levels in many different ways. Complexity is usually one of the criteria involved in defining an overall preference ordering on such sets of alternative systems.

In some contexts, complexity is a desirable property, i.e., we search, within given constraints, for systems with a high degree of complexity. Cryptography and the design of random number generators are two typical examples of such contexts. In some situations, a certain degree of complexity is a necessary condition for obtaining some specific systems properties, usually referred to as *emergent properties*. Self-reproduction, learning, and evolution are examples of such properties.

In other contexts, which seem to predominate in systems problem solving, we search for simple systems or attempt to simplify existing systems. The importance of systems simplicity and simplification methods is well depicted by Herbert Simon [SI4]:

> The human species has survived and thrived in the world, simple or complex as it may be, not so much through the speed and power of its computational capacities, as by exploiting the fact that the systems of interest to it represent highly special cases that can often be analyzed by relatively simple means, provided their underlying structure is detected. This argues for a strategy of searching for that structure, of pattern induction—a skill that is rather highly developed in the animal kingdom—followed by special analysis and heuristic problem-solving search, rather than brute-force analysis of very general classes of highly interconnected complex systems.

Similar sentiments are expressed by Edward Teller, a physicist[TE1]:

> Simplicity, for me, is best characterized in a story from the art traditionally the favorite of mathematicians and scientists: music. When Mozart was fourteen years old, he listened to a secret mass in Rome, Allegri's *Miserere*. The composition had been guarded as a mystery; the singers were not allowed to transcribe it on pain of excommunication. Mozart heard it only once. He was then able to reproduce the entire score.
>
> Let no one think that this was exclusively a feat of prodigious memory. The mass was a piece of art and, as such, had threads of simplicity. The structure is the essence of art. The child who was to become one of the world's greatest composers may not have been able to remember the details of this complicated work, but he could identify the threads, remember them and reinvent the details having listened once with consummate attention. These threads are not easily discovered in music or in science. Indeed, they usually can be discerned only with effort and training. Yet the underlying simplicity exists and once found makes new and more powerful relationships possible.

Gerald Weinberg goes even further when he suggests defining systems science as a science of simplification [WE3]. After describing the remarkable simplification which Newton successfully developed for mechanics, Weinberg summarized the importance of simplifications as follows:

> Newton was a genius, but not because of superior computational power of his brain. Newton's genius was on the contrary his ability to simplify, idealize, and streamline the world so that it became, in some measure, tractable to the brains of perfectly ordinary men. By studying the methods of simplification which have succeeded and failed in the past, the general systems theorist hopes to make the progress of human knowledge a little less dependent of genius.

Complexity and its inverse—simplicity—are associated with many profound philosophical, mathematical, computational, and psychological issues. Some of them, which are of a general nature and are relevant to systems problem solving, are touched on in the subsequent sections of this chapter.

6.2. THREE RANGES OF COMPLEXITY

> *Sometimes it is simplicity which is hidden under what is apparently complex; sometimes, on the contrary, it is simplicity which is apparent, and which conceals extremely complex realities. . . . No doubt, if our means of investigation became more and more penetrating, we should discover the simple beneath the complex, and then the complex from the simple, and then again the simple beneath the complex, and so on, without ever being able to predict what the last term will be. We must stop somewhere, and for science to be possible we must stop where we have found simplicity. That is the only ground on which we can erect the edifice of our generalizations.*
>
> —HENRI POINCARE

Before discussing how complexity can be measured, what are its various forms, and how it is expressed within the GSPS framework, it is desirable to continue for a while with its common sense connotation and take a global view over the entire range of its possible meanings. A historical view seems most appropriate for identifying the main methods of our dealing with complexity. How has complexity been dealt with in science, engineering, and other areas of human endeavor?

When looking at the history of modern science(since the seventeenth century), we can clearly see that prior to the twentieth century science had been predominantly occupied with very simple systems, usually systems with two variables. The recorded history about the main discoveries in science from the seventeenth through the nineteenth century consists basically of variations on the same theme: a discovery of hidden simplicity in a situation that appears complex. Situations of this sort are characterized by an extreme discrimination between a small number of significant factors ând a large number of negligible factors. This allows the scientist to introduce strong, but experimentally acceptable, simplifying assumptions and thus consider the investigated attributes as being "isolated" from all the negligible factors.

Situations in which a few significant factors can be isolated from a large number of presumably negligible factors are plentiful in physics, but they are in short supply in other areas of science. That is why physics was so successful, leaving the other sciences far behind. It was primarily Newton who paved the way to these successes by showing the feasibility of drastic simplifications in physics. His law of universal gravitation, which is still considered one of the greatest achievements of the human mind, is a result of extreme simplifying assumptions. And yet, it is adequate to calculate, fairly precisely, the orbits of the planets.

The actual calculations based on the Newton law require, of course, appropriate mathematical apparatus. Such an apparatus—the calculus and differential equations—is again due to Newton (and, perhaps, also to Leibniz). He developed it as a convenient tool for solving problems associated with simple physical systems such as those obeying his law. It might be said that he tailored it for dealing with the physical simplicitv he himself uncovered.

Until about 1900, science had been predominantly under the influence of Newton's achievements. His drastic simplification had been tried in other contexts. It succeeded in the study of some physical phenomena, such as electricity, magnetism or fluid mechanics, but it failed in virtually all other areas, most notably biology and medicine. Problems with which science was concerned or rather equipped to deal with, were largely problems involving deterministic systems with only two or three variables. They were treated analytically, usually in terms of differential equations. Problems with these characteristics—very small number of variables, high degree of determinism, suitable for analytic treatment—are usually referred to as problems of *organized simplicity*.

In the late nineteenth century, certain physicists became interested in the investigation of systems representing the motions of gas molecules in a closed space. Such a system would typically consist of, say, 10^{23} molecules. The molecules have tremendous velocities and their paths, affected by incessant impacts, assume the most capricious shapes. No one would deny that this is an extremely complex system! No one would deny either that Newton's law of gravitation is of no use in analyzing these systems, in spite of all the tremendous simplifications behind it.

The problem of analyzing the motions of gas molecules in a closed space, a system that is extremely complex and disorganized, is certainly hopeless when viewed from the standpoint of ideas and tools developed for dealing with situations of organized simplicity. If anything can be used at all, a radically new approach to the problem must be found. That happened, when some creative thinkers, most notably Ludwig Boltzmann and Josiah Willard Gibbs, developed powerful statistical methods to deal with problems that involve very large numbers of variables acting in a highly random manner. Such problems are now usually called problems of *disorganized complexity*.

Statistical methods do not produce results regarding the individual variables (e.g., motions of the individual molecules). Their purpose is to calculate a small number of average properties. To explain this point, let me quote from a famous paper by Warren Weaver [WE1]:

> The classical dynamics of the nineteenth century was well suited for analyzing and predicting the motion of a single ivory ball as it moves about on a billiard table. In fact, the relationship between positions of the ball and the times at which it reaches these positions forms a typical nineteenth-century problem of simplicity. One can, but with a surprising increase in difficulty, analyze the motion of two or even of three balls on the billiard table. There has been, in fact, considerable study of the mechanics of the standard game of billiards. But, as soon as one tries to analyze the motion of ten or fifteen balls on the table at once, as in pool, the problem becomes unmanageable, not because there is any theoretical difficulty, but just because the actual labor of dealing in specific detail with so many variables turns out to be impracticable.
>
> Imagine, however, a large billiard table with millions of balls rolling over its surface, colliding with one another and with the side rails. The great surprise is that

> the problem now becomes easier, for the methods of statistical mechanics are applicable. To be sure the detailed history of one special ball cannot be traced, but certain important questions can be answered with useful precision, such as: On the average how many balls per second hit a given stretch of rail? On the average how far does a ball move before it is hit by some other ball? On the average how many impacts per second does a ball experience?
>
> Earlier it was stated that the new statistical methods were applicable to problems of disorganized complexity. How does the word "disorganized" apply to the large billiard table with the many balls? It applies because the methods of statistical mechanics are valid only when the balls are distributed, in their positions and motions, in a helter-skelter, that is to say a disorganized, way. For example, the statistical methods would not apply if someone were to arrange the balls in a row parallel to one side rail of the table, and then start them all moving in precisely parallel paths perpendicular to the row in which they stand. Then the balls would never collide with each other nor with two of the rails, and one would not have a situation of disorganized complexity.

Since their inception at the beginning of this century, statistical methods have been successful in dealing with many problems of disorganized complexity in science as well as other areas. Statistical mechanics, thermodynamics, and statistical (or quantitative) genetics are well-known examples of their success in science. In engineering, they have played a major role in the design of large-scale telephone networks and time-sharing computer systems, in dealing with problems of engineering reliability, and other issues. In business, they have clearly been essential for dealing with problems of marketing, insurance and the like.

While analytical methods, developed for organized simplicity, become impractical for even a modest number of variables, say five variables, the relevance and precision of statistical methods increase with an increase in the number of variables involved. These two types of methods are thus highly complementary. They cover the two extremes of the complexity spectrum. Although complementing each another, the two kinds of methods cover in fact only a tiny fraction of the whole spectrum of complexity. This means, in turn, that the whole complexity spectrum except its extreme ends is methodologically underdeveloped in the sense that neither analytical nor statistical methods are adequate to cope with it. Problems that are associated with this methodologically underdeveloped middle region in the complexity spectrum are usually called problems of *organized complexity* for reasons which are well described by Warren Weaver [WE1]:

> This new method of dealing with disorganized complexity, so powerful an advance over the earlier two-variable methods, leaves a great field untouched. One is tempted to oversimplify, and say that scientific methodology went from one extreme to the other—from two variables to an astronomical number—and left untouched a great middle region. The importance of this middle region, moreover, does not depend primarily on the fact that the number of variables involved is moderate—large

compared to two, but small compared to the number of atoms in a pinch of salt. The problems in this middle region, in fact, will often involve a considerable number of variables. The really important characteristic of the problems of this middle region, which science has as yet little explored or conquered, lies in the fact that these problems, as contrasted with the disorganized situations with which statistics can cope, show the essential feature of *organization*. In fact, one can refer to this group of problems as those of *organized complexity* . . . . These new problems, and the future of the world depends on many of them, requires science to make a third great advance, an advance that must be even greater than the nineteenth-century conquest of problems of simplicity or the twentieth-century victory over problems of disorganized complexity. Science must, over the next 50 years, learn to deal with these problems of organized complexity.

Instances of problems with the characteristics of organized complexity are abundant, particularly in the life, behavioral, social, and environmental sciences, as well as in applied fields such as modern technology or medicine. Some of the problem areas that involve organized complexity are especially profound, such as cancer research, the study of aging, or the rich area of difficult and diverse problems associated with modern technology. This last area is well characterized by George B. Dantzig in his 1979 Distinguished Lecture at the International Institute for Applied Systems Analysis in Laxenburg, Austria, an institute that has played an important role in this new thrust of science into the domain of organized complexity:

It is not easy to paint a picture of just how complex modern technology is. One way to start is to list the activities of a small town. By using the classified section of the telephone directory, I can list a few activities of the town of Richmond, California. Here are those that begin with the letters *Br*: Bridge Builders, Bridge Tables, Broadcasting Stations, Brochures, Brokers, Bronze, Brushes, Brooches, Brakes, Brandies, Brazing, Bricks, Brick Stain, Bric-a-Brac. I counted over 6,000 activities in all.

Another way to see the diversity of the material side of life is to look at a catalog of electronic supply items that are for sale. There are thousands upon thousands of different kinds of resistors, condensers, vacuum tubes, transistors, cables, sockets, knobs, switches, dials, circuit boards, cabinets. Look up the number of different items listed in a chemical supply catalog or a Sears, Roebuck catalog, and again the number of different items runs into many thousands. A modern university can have a hundred different departments. The United States Government has nearly 2,000 different kinds of offices in San Francisco alone, each presumably carrying out a different function for the public good. So far we have spoken only of diversity, but complexity has other dimensions.

The Leontief input–output model of the national economy of the United States classifies industries into about 400 major types and requires data for each of these industries about how much it shipped (or received) from every other industry. The resulting 400 × 400 table contains 160,000 numbers. Each region of the country has such an input–output table, and there are many regions. Each number in an input–output table expresses a dependency of one industry upon another; the transactions between regions and industries represent further dependencies; there are a great number of cross combinations. Countries depend on each other in the same way.

There are also *time* dependencies: facilities are built and maintained for future use; material is stockpiled for future use; people are trained for future jobs. There are *locational* dependencies as well: men, material, and facilities are moved to new locations, not only on the surface of the globe but below and above.

While we may easily understand the ins and outs of each small part of this vast web of activities, the problem is how to track all the interactions at once. We know that the powerful forces of population growth, shortages of raw materials, food, energy, growing affluence, and so on, are rapidly reshaping this complexity. There is a fear that the structure that interconnects these activities may not hold up very well under these stresses. We see the possibility of all kinds of system failure if we let the changes go on uncontrolled.

By definition, systems with the characteristics of organized complexity are rich in factors that cannot easily be justified as negligible. And, by the same token, they are not sufficiently complex and random to yield meaningful statistical averages. That means that they are susceptible to neither of the two simplification strategies invented by science. And, yet, simplification is unavoidable in most instances. Even if a problem regarding a highly complex system can be successfully handled without any simplification by a computer, the solution must be eventually reduced to a level of complexity that is acceptable to the mind of, say, a decision maker who is in a position to utilize it. Since neither the Newtonian nor statistical simplification strategies are applicable, new avenues to the simplification of systems are needed. In general, a good simplification should minimize the loss of relevant information with respect to the achieved reduction of complexity. Some ideas along these lines are discussed in Chapters 3 and 4.

One way of dealing with very complex systems that possess the characteristics of organized complexity, perhaps the most significant one, is to allow imprecision in describing properly aggregated data. Here, the imprecision is not of a statistical nature, but rather of a more general modality, even though the possibility of imprecise statistical descriptions is included as well. The mathematical apparatus for this new modality, which is recognized under the name "theory of fuzzy sets," has been under development since the mid-1960s. To describe the essence and significance of this new theory, I can do no better than to quote Lotfi A. Zadeh, its founder [ZA4]:

Given the deeply entrenched tradition of scientific thinking which equates the understanding of a phenomenon with the ability to analyze it in quantitative terms, one is certain to strike a dissonant note by questioning the growing tendency to analyze the behavior of humanistic systems as if they were mechanistic systems governed by difference, differential, or integral equations.

Essentially, our contention is that the conventional quantitative techniques of system analysis are intrinsically unsuited for dealing with humanistic systems or, for that matter, any system whose complexity is comparable to that of humanistic systems. The basis for this contention rests on what might be called the *principle of incompatibility*. Stated informally, the essence of this principle is that as the complexity of a system increases, our ability to make precise and yet significant statements about its behavior diminishes until a threshold is reached beyond which precision and significance (or relevance) become almost mutually exclusive characteristics.[1] It is in this sense that precise quantitative analyses of the behavior of humanistic systems are not likely to have much relevance to the real-world societal, political, economic, and other types of problems which involve humans either as individuals or in groups.

An alternative approach . . . is based on the premise that the key elements in human thinking are not numbers, but labels of fuzzy sets, that is, classes of objects in which the transition from membership to non-membership is gradual rather than abrupt. Indeed, the pervasiveness of fuzziness in human thought processes suggests that much of the logic behind human reasoning is not the traditional two-valued or even multivalued logic, but a logic with fuzzy truths, fuzzy connectives, and fuzzy rules of inference. In our view, it is this fuzzy, and as yet not well-understood, logic that plays a basic role in what may well be one of the most important facets of human thinking, namely, the ability to *summarize* information—to extract from the collections of masses of data impinging upon the human brain those and only those subcollections which are relevant to the performance of the task at hand.

By its nature, a summary is an approximation to what it summarizes. For many purposes, a very approximate characterization of a collection of data is sufficient because most of the basic tasks performed by humans do not require a high degree of precision in their execution. The human brain takes advantage of this tolerance for imprecision by encoding the "task-relevant" (or "decision-relevant") information into labels of fuzzy sets which bear an approximate relation to the primary data. In this way, the stream of information reaching the brain via the visual, auditory, tactile, and other senses is eventually reduced to the trickle that is needed to perform a specific task with a minimal degree of precision. Thus, the ability to manipulate fuzzy sets and the consequent summarizing capability constitute one of the most important assets of the human mind as well as a fundamental characteristic that distinguishes human intelligence from the type of machine intelligence that is embodied in present-day digital computers.

Viewed in this perspective, the traditional techniques of system analysis are not well suited for dealing with humanistic systems because they fail to come to grips with the reality of the fuzziness of human thinking and behavior. Thus to deal with systems radically, we need approaches which do not make a fetish of precision, rigor, and mathematical formalism, and which employ instead a methodological framework which is tolerant of imprecision and partial truths.

[1] A corollary principle may be stated succinctly as, "The closer one looks at a real-world problem, the fuzzier becomes its solution."

Methodological tools for dealing with systems problems in the categories of organized simplicity or disorganized complexity have reached a fairly satisfactory stage of development, and are readily available in terms of various statistical, numerical analysis, and symbol manipulation packages of computer software. On the other hand, systems problems in the category of organized complexity are still methodologically underdeveloped. It is the main purpose of the GSPS to improve this situation.

6.3. MEASURES OF SYSTEMS COMPLEXITY

> *Our views regarding the concept of "complexity" have tended to be as richly varied as complexity itself.*
>
> —ROBERT ROSEN

In the context of systems problem solving, complexity has two roles. In its first role, it represents a property of systems; in its second role, it is a property of systems problems. Let these two kinds of complexity be referred to as *systems complexity* and *problem complexity*, respectively.

This section is devoted solely to systems complexity. Some aspects of problem complexity, which is often called *computational complexity* in the literature, are discussed in Sections 6.4 and 6.5.

In the GSPS, systems have many different faces, each represented by one of the epistemological system types introduced in Chapters 2–5. As a consequence, the complexities associated with these types of systems have many faces as well. That is, different system types in the epistemological hierarchy give the notion of complexity different meanings, each of which requires a special treatment.

Notwithstanding the differences in complexities of the various system types, *two general principles of systems complexity* can be recognized; they are applicable to any of the system types and can thus be utilized as guidelines for a comprehensive study of systems complexity.

According to the first general principle, the complexity of a system (of any type) should be proportional to the amount of information required to describe the system. Here, the term "information" is used solely in a syntactic sense; no semantic or pragmatic aspects of information are employed. One way of expressing this descriptive complexity, perhaps the simplest one, is to measure it by the number of entities involved in the system (variables, states, components) and the variety of interdependence among the entities. Indeed, everything else being the same, our ability to understand or cope with a system tends to decrease when the number of entities involved or the variety of their interconnections increase. There are, of course, many different ways in which descriptive complexity can be expressed. Each of them, however, must satisfy some general requirements formulated as follows.

Let X denote the set of all systems of a particular epistemological type, let $\mathcal{P}(X)$ denote the power set of X, and let C_X denote a *measure of descriptive complexity* within the set X. Then, C_X is a function

$$C_X: \mathscr{P}(X) \rightarrow \mathbb{R}$$

that satisfies the following requirements (axioms):

(C1) $C_X(\emptyset) = 0$;
(C2) if $A \subset B$, then $C_X(A) \leq C_X(B)$;
(C3) if A is a homomorphic image of B, then $C_X(A) \leq C_X(B)$;
(C4) if A and B are isomorphic, then $C_X(A) = C_X(B)$;
(C5) if (i) $A \cap B = \emptyset$, (ii) A and B do not interact with each other, and (iii) neither A nor B is a homomorphic image of the other, then $C_X(A \cup B) = C_X(A) + C_X(B)$.

Requirements (C1) and (C2) guarantee that complexity of any system be characterized by a non-negative number. Requirements (C2) and (C3) deal with fundamental properties of monotonicity: complexity should not increase when the given set of systems is reduced or when less detail is distinguished in the systems. Requirement (C4) is obvious: when some (any) entities in the given systems are relabelled, everything else remaining the same, the complexity should not change. Requirement (C5) describes a desirable property of additivity: if two sets of systems are taken together that in all relevant respects have nothing in common (no common systems, no interaction, no morphic relationship), then the total complexity should be equal to the sum of the two individual complexities.

According to the second general principle, systems complexity should be proportional to the amount of information needed to resolve any uncertainty associated with the system involved. Here, again, syntactic information is used, but information that is based on a relevant *measure of uncertainty* (Sec. 3.5).

Systems complexity is primarily studied for the purpose of developing sound methods by which systems that are incomprehensible or unmanageable can be simplified to an acceptable level of complexity. When we simplify a system, we want to reduce both the complexity based on descriptive information and the complexity based on the uncertainty information. Unfortunately, these two complexities conflict with each other. In general, when we reduce one, the other increases or, at best, remains unchanged. Based on these considerations, a general problem of simplification can be formulated as follows:

Given a system of some particular epistemological type, let $\mathscr{X}$ denote the set of all its meaningful simplifications. Let $\overset{d}{\leq}$ and $\overset{u}{\leq}$ denote the two complexity orderings on $\mathscr{X}$, based on the descriptive information and uncertainty information, respectively. Both of these orderings are, in general, only weak orderings (i.e., reflexive and transitive relations on $\mathscr{X}$). Let $\overset{\alpha}{\leq}, \overset{\beta}{\leq}, \ldots$ denote other (optional) orderings on $\mathscr{X}$ (weak, partial or total), which express special preferences specified by the user of the given system. In terms of all the orderings involved, we define a *joint preference ordering* $\overset{*}{\leq}$ by the following formula

$$(\forall x, y \in \mathscr{X})(x \overset{*}{\leq} y \Leftrightarrow x \overset{d}{\leq} y \text{ and } x \overset{u}{\leq} y \text{ and } x \overset{\alpha}{\leq} y \text{ and } x \overset{\beta}{\leq} y \text{ and } \ldots)$$

The *solution set* $\mathscr{X}_s$ of the simplification problem consists of those systems in $\mathscr{X}$ that are either equivalent or incomparable in terms of the joint preference ordering $\overset{*}{\leq}$. Formally,

$$\mathscr{X}_s = \{x \in \mathscr{X} \mid (\forall y \in X)(y \overset{*}{\leq} x \Rightarrow x \overset{*}{\leq} y)\}. \tag{6.1}$$

A reader familiar with previous chapters of this book should now be able to associate this general formulation of the simplification problem with these special cases of the problem:

—the problem of determining admissible behavior systems (Sec. 3.6);

—the problem of simplifying generative systems by resolution coarsening (Sec. 3.9);

—the reconstruction problem (Sec. 4.7).

All these problems conform to the general simplification problem, even though they differ from each other in the set $\mathscr{X}$ and mathematical properties of the preference orderings $\overset{d}{\leq}$ and $\overset{n}{\leq}$ defined on $\mathscr{X}$.

The basic philosophy of the GSPS is to allow the user to define his own preference orderings for systems under consideration. If the user indicates that complexity is one of the preference orderings, but he does not define his own measure of complexity, the GSPS should offer him a list of possible options. If the user does not make a choice, the GSPS should use some measure of complexity as a designated default option.

The complexity measures offered to the user as possible options should be measures that are well established and, of course, must be applicable to the problem for which they are considered. An overview of the literature dealing with various types of complexity measures is included in the Notes to this chapter.

6.4. BREMERMANN'S LIMIT

> *No data processing system, whether artificial or living, can process more than 2×10^{47} bits per second per gram of its mass.*
>
> —HANS J. BREMERMANN

This conjecture is the central theme of a paper by Hans Bremermann, which was published in 1962 [BR1]. He derives it by the following considerations, in which the phrase "processing of N bits" means the transmission of that many bits over one or several channels within the computing system.

► It is obvious that information which is to be acted upon by a machine must be physically encoded in some manner. Assume that it is encoded in terms of energy levels within the interval $[0, E]$ of energy of some sort; E is viewed as the total energy available for this purpose. Assume further that energy levels can be measured with an accuracy of only ΔE. Then, the most refined encoding is defined in terms of markers by which the whole interval is divided into $N = E/\Delta E$ equal subintervals, each associated with the energy amount ΔE. If at each instant no more than one of the levels (represented by the markers) is occupied, then

$$\log_2 (N+1)$$

is the maximum number of bits that are representable by energy E; $N+1$ is used here to account for the case in which none of the levels is occupied. If, instead of one marker with energy levels in $[0, E]$, K markers $(2 \leq K \leq N)$ are used simultaneously, then

$$K \log_2 (1 + N/K)$$

bits become representable. The optimal utilization of the available amount of energy E is obtained when N markers with levels in the interval $[0, \Delta E]$ are used. In this optimal case, N bits of information can be represented.

In order to represent more information by the same amount of energy, it is desirable to reduce ΔE. This is possible only to a certain extent since the resulting levels must be distinguished by some measurement process which, regardless of its nature, always has some limited precision. The extreme case is expressed by the Heisenberg principle of uncertainty: energy can be measured to the accuracy of ΔE if the inequality

$$\Delta E \Delta t \geq h$$

is satisfied, where Δt denotes the time duration of the measurement, $h = 6.625 \times 10^{-27}$ ergs/sec is Planck's constant, and ΔE is defined as the mean deviation from the expected value of the energy involved. This means that

$$N \leq \frac{E \Delta t}{h} \tag{6.2}$$

Now, the available energy E can be expressed in terms of the equivalent amount of mass m by Einstein's formula

$$E = mc^2,$$

where $c = 3 \times 10^{10}$ cm/sec is the velocity of light in a vacuum. If we take the upper (most optimistic) bound of N in (6.2), we get

$$N = \frac{mc^2 \Delta t}{h}. \tag{6.3}$$

Substituting the numerical values for c and h, we obtain

$$N = 1.36 m \Delta t \times 10^{47}. \tag{6.4}$$

For a mass of 1 gram ($m = 1$) and time of 1 sec ($\Delta t = 1$), we obtain the value

$$N = 1.36 \times 10^{47},$$

which implies the conjecture. ◀

Using the limit of information processing obtained for one gram of mass and one second of processing time, Bremermann then calculates the total number of bits processed by a hypothetical computer the size of the Earth within a time period equal to the estimated age of the Earth. Since the mass and age of the Earth are estimated to be less than 6×10^{27} g and 10^{10} y, respectively, and each year contains approximately 3.14×10^7 sec, this imaginary computer would not be able to process more than 2.56×10^{92} bits or, when rounding up to the nearest power of ten, 10^{93} bits. The last number—10^{93}—is usually referred to as *Bremermann's limit* and problems that require processing more than 10^{93} bits of information are called *transcomputational problems*.

Bremermann's limit seems at first sight rather discouraging for system problem solving, even though it is quite conservative (more reasonable assumptions would lead to a number smaller than 10^{93}). Indeed, many problems dealing with systems of even modest size exceed it in their information processing demands. Consider, for example, a system of n variables, each of which can take k different states. The set of all overall states of the variables consists clearly of k^n states. In each particular system, however, the actual overall states are restricted to a subset of this set. There are 2^{k^n} such subsets. Suppose we need to select, identify, distinguish, or classify one system from the set of all possible systems of this sort. Then, under the assumption that the most efficient method of searching is used, in which each bit of information (the answer to a dichotomous question) allows us to cut the remaining choices in half,

$$\log_2 2^{k^n} = k^n$$

bits of information have to be processed. The problem becomes transcomputational when

$$k^n > 10^{93}.$$

That happens, e.g., for the following values of k and n:

k	2	3	4	5	6	7	8	9	10
n	308	194	154	133	119	110	102	97	93

The problem of transcomputationality arises in various contexts. One of them is pattern recognition. Consider, for example, a $q \times q$ spatial array of the chessboard type, each square of which can have one of k colors. There are clearly k^n color patterns, where $n = q^2$. Suppose we want to determine the best classification (according to certain criteria) of these patterns. This requires a search through all possible classifications of the patterns. In the case of only two classes, the problem becomes isomorphic to the previous one. For two colors, for example, the problem becomes transcomputational when the array is 18×18; for a 10×10 array, the problem becomes transcomputational when nine colors are used. This pattern recognition problem is directly relevant to physiological studies of the retina, but its complexity is tremendous. The retina contains about a million light-sensitive cells. Even if we consider (for simplicity) that each of the cells have only two states (active and inactive), the attempt to study the retina as a whole would require the processing of

$$2^{1,000,000} \doteq 10^{300,000}$$

bits of information. This is far beyond Bremermann's limit.

Another context in which the same problem occurs is the area of testing large-scale integrated digital circuits. These are tiny electronic chips with considerable complexity and a large number of inputs and outputs. For properly defined electric signals (each, usually, with two ideal states), the individual outputs should represent some specific logic functions of the logic variables associated with the inputs. To test a particular integrated circuit means to analyze it as a "black box": to determine the actual logic functions it implements, solely by manipulating the input variables and observing the output variables. For each output variable, the testing problem is thus basically the same as the problem previously discussed for $k = 2$ (unless a multiple valued logic is used). It follows that testing of circuits, for example, with 308 inputs and one output is a transcomputational problem. However, it is well known that the practical complexity limits of this testing problem are considerably lower. Some currently manufactured large-scale integrated circuits cannot be in fact completely tested. The focus is thus on developing testing methods that can be practically implemented and guarantee only that the testing be *almost complete*, that, say, well over 90% of all possibilities be tested.

A more detailed characterization of the complexity of this problem, from the practical domain to Bremermann's limit, is expressed by Figure 6.1. The figure shows the dependence of the time (in years) required to select (identify, classify, distinguish, etc.) one logic function of n variables under the consideration of different information processing rates in the range from 10 through 10^{100} bits per second. Two significant

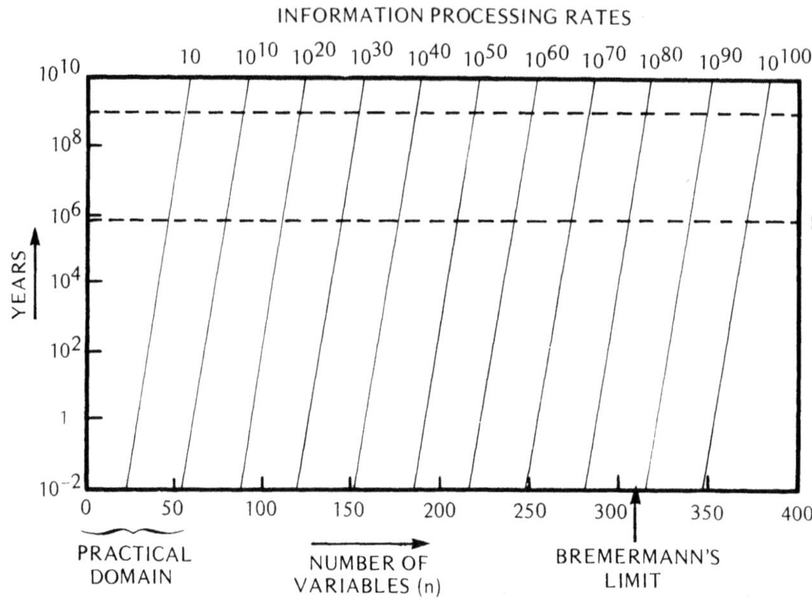

Figure 6.1. Time required to select or identify one logic function of n variables for information processing rates of 10, 10^{10}, . . . , 10^{100} bits per second.

values of time are also shown in the figure: L indicates the approximate age of the oldest fossil records of life on the Earth; M shows the approximate time since men first appeared on the Earth.

The testing example is in no way exceptional. Genuine systems problems are notorious for their huge demands on information processing capabilities. This point is illustrated by specific examples on various occasions elsewhere in this book. It is also well depicted by Bremermann in the conclusion of his paper (BR1]:

> The experiences of various groups who work on problem solving, theorem proving and pattern recognition all seem to point in the same direction: These problems are tough. There does not seem to be a royal road or a simple method which at one stroke will solve all our problems. My discussion of ultimate limitations on the speed and amount of data processing may be summarized like this: Problems involving vast numbers of possibilities will not be solved by sheer data processing quantity. We must look for quality, for refinements, for tricks, for every ingenuity that we can think of. Computers faster than those of today will be a great help. We will need them. However, when we are concerned with problems in principle, present-day computers are about as fast as they ever will be.

If a problem is transcomputational, it is obvious that it can be dealt with only in some modified form. It is desirable to modify it no more than is necessary to make it manageable. The most natural way of modifying a problem is to soften its requirements. For instance, a requirement of getting the best solution may be replaced with a requirement of getting a good solution, instead of requiring a precise solution we may accept an approximate solution, and so on. Such softening of requirements permits the use of heuristic methods, in which vast numbers of unpromising possibilities are ignored, or approximate (fuzzy) methods, in which substantial aggregation takes place.

The GSPS should be able to assist the user in estimating the complexity of his problem and, if the problem is found unmanageable within the available computational resources, should suggest some feasible modifications in it. Bremermann's limit allows one to make only the most rudimentary categorization of systems problems by their complexities. It does not say anything about the actual, practical computational limits. Nevertheless, it is a useful benchmark for a preliminary evaluation of each problem situation, as emphasized by Ashby [AS12]:

> One of its most obvious consequences, yet one almost universally neglected today, is that, before the study of a complex system is undertaken, at least a rough estimate of its informational demands should be made. Should the estimate be 2000 bits we have little to worry about, but should it prove to be 10^{300} bits we would know that our whole strategic approach to the system needs re-formulating.

This simple benchmark—10^{93}—must be supplemented, of course, by sharper bounds on problem complexity, derived for specific computer systems.

As emphasized in this book on a number of occasions, one of the main goals of the GSPS is to achieve a powerful man–machine (user–computer) symbiosis, in which the two symbionts would complement each other in their capabilities to perform certain problem–solving tasks efficiently. Although such a symbiosis, when properly implemented, seems to be the best way in dealing with complex problems, Bremermann's limit still cannot be overcome thereby. In fact, the limit is an indicator of fundamental limits to our knowledge, as Ashby explains [AS7]:

> The most obvious fact is that we, and our brains, are themselves made of matter, and are thus absolutely subject to the limit. Not only are we subject as individuals, but the whole cooperative organization of World Science is also made of matter, and is therefore subject to it. Thus both the total information that I can use personally, and the information that World Science can use, are limited, on any ordinary scale, to about 10^{80}* bits. Whatever our science will become in the future, all will lie below this ceiling.
>
> We cannot claim any special advantage because of our pre-eminent position in the world of organisms. We have been shaped, and selected to be what we are, by the process of natural selection. As a selection, this process can be measured by an

* Ashby derives the value of 10^{80} from the Bremermann limit for one second and one gram by considering "centuries of time and tons of computers" (e.g., about ten thousand centuries and 10^{15} tons of mass). It is not important for the argument whether we take 10^{80} or 10^{93}.

information-measure; it is therefore subject to its limits. In any type of selection, under any planetary conditions, a planetary surface made of matter cannot produce adaptation faster than the rate of the limit. However good we may think we are, 10^{80} measures something that we do not exceed. The science of the future will be built by brains that cannot have had more than 10^{80} bits used in their preparations, and they themselves will advance only by something short of 10^{80}. This is our informational universe: what lies beyond is unknowable.

6.5. COMPUTATIONAL COMPLEXITY

> *There is no doubt that the results about what can and cannot be effectively computed or formalized in mathematics have had a profound influence on mathematics, and, even more broadly, they have influenced our view of our scientific methods.*
>
> —JURIS HARTMANIS

Bremermann's limit (discussed in the previous section) works well, as a simple benchmark, for problems whose information processing demands exceed it, but it does not say much about the remaining problems. Even if a problem is not rejected by Bremermann's limit, it may still be practically intractable. A more refined understanding of the notion of problem complexity is thus needed.

Computational properties of problems are studied under the general theory of algorithms. This general theory includes three large subject areas: the theory of computability, design of algorithms, and the theory of computational complexity. It is beyond the scope of this book to cover these areas in any depth. It is desirable, however, to provide the reader with a brief overview, focusing primarily on computational complexity, of those results and issues that are of particular significance to systems problem solving. No proofs of the summarized results are presented here. However, a guide to the literature on computational complexity, where the proofs and other details can be found, is given in the Notes to this chapter.

An *algorithm* is understood intuitively as a set of instructions, expressed in some language, for executing a sequence of operations for solving a problem of some specific type. Algorithms are required to be *finite*, i.e., each algorithm must terminate after a finite number of steps (operations) have been executed.

The intuitive notion of an algorithm was formalized in several ways, including formalizations based on the concepts of Turing machines, Markov algorithms, and recursive functions, which were all proven to be equivalent. One of the concepts—that of a *Turing machine*—is envisioned as a simple device that consists of a finite-state *control unit* and a *tape*. The control unit has a *memory*, which makes it capable of being in any one of a finite set of states, say set $Z = \{z_1, z_2, \ldots, z_n\}$. The tape is potentially infinite in both directions, and is marked off along its length into spaces of equal size.

Each of these spaces, referred to as *cells*, has written on it a symbol from a finite set of symbols, say set $X = \{x_0, x_1, \ldots, x_m\}$. One of the symbols, say symbol x_0, is always interpreted as a *blank space* (empty cell). Communication between the control unit and tape is provided by a *read–write head*, which is capable of reading symbols from the tape and writing over the symbols that are written on it. Only one cell of the tape is accessible to the head at any time.

The control unit of a Turing machine operates in discrete steps. In each step it replaces the current state with a new one, and performs a single operation of one of the following three types:

i. it replaces the current symbol on the tape with a new one;
ii. it moves the tape by one cell to the left or right;
iii. it stops the computation (the so-called halt operation).

The new state as well as the operation performed are uniquely determined by the current state and the symbol read on the tape.

Let z_c, z_n denote the current and next state of a Turing machine, respectively, let x_r denote the symbol that is read on the tape, and let y_p denote the operation performed. Then, given an initial string of symbols on the tape (any cell for which a symbol is not given is assumed to be blank) and a particular initial state, a computation on the Turing machine is defined by an ordered set of quadruples

$$(z_c, x_r, z_n, y_p).$$

If no two quadruples in the set are allowed to begin with the same pair z_c, x_r, the Turing machine is said to be *deterministic*; otherwise, it is said to be *nondeterministic*.

A hypothesis that has become known as *Church's thesis* (or the Church–Turing thesis), and which has been generally accepted, states that any function regarded naturally as computable can be computed on a deterministic Turing machine. According to this hypothesis, a Turing machine is taken to be a precise formal equivalent of the intuitive notion of an algorithm. The hypothesis cannot be proven mathematically, but it is well justified by informal arguments and empirical evidence. It can be overthrown only by proposing an alternative formalization of computation, generally acceptable on intuitive grounds and capable of describing computation that are beyond the capabilities of Turing machines. The existence of such a formalization is considered highly unlikely.

In general, a problem is considered unsolvable if no algorithm exists by means of which a solution can be obtained. The notion of deterministic Turing machines together with Church's thesis have made possible the study of the existence of algorithms for various problems in a formal manner. To prove that a problem is unsolvable, it is sufficient to prove that it cannot be solved by a Turing machine. Such proofs of unsolvability have been obtained for a number of problems.

Results regarding problem unsolvability should be incorporated in the GSPS only insofar as they concern problems that can be identified within its framework. If an identifiable problem that is known to be unsolvable is requested by a user, the GSPS must reject the problem, but, at the same time, it should provide the user with adequate information regarding its unsolvability, including relevant references to the literature.

Unsolvable problems form one of three primary classes of problems. The second class consists of problems that have not been proven unsolvable, but for which no algorithms are known for solving them. These are thus problems whose solvability status has not been resolved as yet. If any identifiable problem of this kind is requested by a user of the GSPS, he should be informed about its uncertain status and nothing more (unfortunately) can be done for him.

All remaining problems are solvable. That is, they are solvable in principle. In practice, however, many of them cannot be solved due to their excessive demands on computing resources such as computing time and memory size. Since the required computing time is usually the single factor that determines whether or not a problem is practically solvable, computational complexity has been predominantly studied in terms of this single resource.

The practical solvability of a problem depends on

i. the algorithm employed for solving the problem;
ii. the size of the particular systems involved in the problem;
iii. the computational power of the computing resources available.

Given a particular algorithm for solving a problem, it is convenient to express its time requirements in terms of a single variable that represents the size of the systems involved in the problem. This variable, which is often called the *size of a problem instance*, is supposed to express the amount of input data needed to describe the particular systems.

Given a particular systems problem instance, let n denote its size. Then, the time requirements of a specific algorithm for solving the problem are expressed by a function

$$f: \mathbb{R} \rightarrow \mathbb{R} \tag{6.5}$$

such that $f(n)$ is the largest amount of time needed by the algorithm to solve a problem instance of size n. Function f is usually called a *time complexity function.*

It has been recognized that it is useful to distinguish two classes of algorithms by the rate of growth of their time complexity functions. One class consists of algorithms whose time complexity functions can be expressed in terms of a polynomial. They are called *polynomial time algorithms.* Since the degree of each polynomial is considerably more significant, especially for large values of n, than its coefficients and lower-order terms, it is useful to classify polynomial time complexity functions by their order. A function f is said to be of complexity $O(n^k)$, where k is a positive integer, if and only if there is a constant $c > 0$ such that

$$f(n) \leq c\,n^k$$

for all $n \geq n_0$, where n_0 is a positive integer that usually represents the smallest size of the problem instances involved. For example, function

$$f(n) = 25n^2 + 18n + 31$$

is of complexity $O(n^2)$ since

$$f(n) \leq 74n^2$$

when $n_0 = 1$, or

$$f(n) \leq 42n^2$$

when $n_0 = 2$, etc.

The second class of algorithms consists of those whose time complexity functions are not bounded by complexity $O(n^k)$ for some k. They are usually referred to as *exponential time algorithms.*

The distinction between the polynomial and exponential time algorithms is significant, especially when considering large problem instances. This is illustrated in Table 6.1 by showing differences in growth rates for several time complexity functions.

TABLE 6.1

Illustration of Growth Rates of Several Polynomial and Exponential Time Complexity Functions

Time complexity function	Problem instance size: n						
	1	10	20	30	40	50	100
n	0.000001 sec	0.00001 sec	0.00002 sec	0.00003 sec	0.00004 sec	0.00005 sec	0.0001 sec
n^2	0.000001 sec	0.0001 sec	0.0004 sec	0.0009 sec	0.0016 sec	0.0025 sec	0.01 sec
n^5	0.000001 sec	0.1 sec	3.2 sec	24.3 sec	1.7 min	5.2 min	2.8 hr
n^{10}	0.000001 sec	2.8 h	118.5 days	18.7 yr	3.3 centuries	31.0 centuries	3.2×10^4 centuries
2^n	0.000002 sec	0.001 sec	1.0 sec	17.9 min	12.7 days	35.7 yr	4×10^{14} centuries
3^n	0.000003 sec	0.059 sec	58 min	6.5 yr	3,855 centuries	2×10^8 centuries	1.6×10^{32} centuries
10^n	0.00001 sec	2.8 hr	3.2×10^4 centuries	3.2×10^{14} centuries	3.2×10^{24} centuries	3.2×10^{34} centuries	3.2×10^{84} centuries
2^{2^n}	0.000004 sec	5.7×10^{292} centuries	$10^{3 \cdot 10^5}$ centuries	$10^{3 \cdot 10^8}$ centuries	$10^{3 \cdot 10^{11}}$ centuries	$10^{3 \cdot 10^{14}}$ centuries	$\sim 10^{3 \cdot 10^{29}}$ centuries
n^n	0.000001 sec	2.8 h	3.3×10^{10} centuries	6.5×10^{28} centuries	3.8×10^{48} centuries	$\sim 2.8 \times 10^{69}$ centuries	$\sim 3.2 \times 10^{184}$ centuries
$n!$	0.000001 sec	3.6 sec	771.5 centuries	8.4×10^{16} centuries	2.6×10^{32} centuries	$\sim 9.6 \times 10^{48}$ centuries	$\sim 2.9 \times 10^{142}$ centuries

The computing times in this table are based on the assumption that the computing is performed at a rate of one million operations per second. When comparing, for instance, n^2 with n^{10}, we can see that the degree of a polynomial time complexity function has a considerable effect on practical limitations of the corresponding algorithms. However, polynomial time algorithms are substantially more responsive than exponential time algorithms to increases in computing power (except for small values of n). This can be seen by comparing plots of some polynomial and exponential time complexity functions in Figure 6.2 and, even more explicitly, by examining the actual increases in the ranges of applicability due to increases in computing speed, as illustrated by the formulas in Table 6.2.

Due to the essential differences between polynomial and exponential time complexity functions, polynomial time algorithms are considered *efficient*, while exponential time algorithms are considered *inefficient*. As a consequence, problems for which it can be proven that they are not solvable by polynomial time algorithms are viewed as *intractable*, while problems for which polynomial time algorithms are known are viewed as *tractable*. The latter problems are usually called *P-problems* (i.e., solvable in polynomial time); the set of all such problems is called the *problem class P*.

It is known that differences among standard schemes used in practice for encoding problems as well as differences in the computer types used do not affect the classification of problems into tractable and intractable. Standard encoding schemes and computer types are known to differ from each other at most polynomially. Alternative encoding schemes or computer types may thus influence the practical range of solvability of a problem, but they do not affect its tractability status.

It turns out that for most of the problems encountered in practice, neither is a

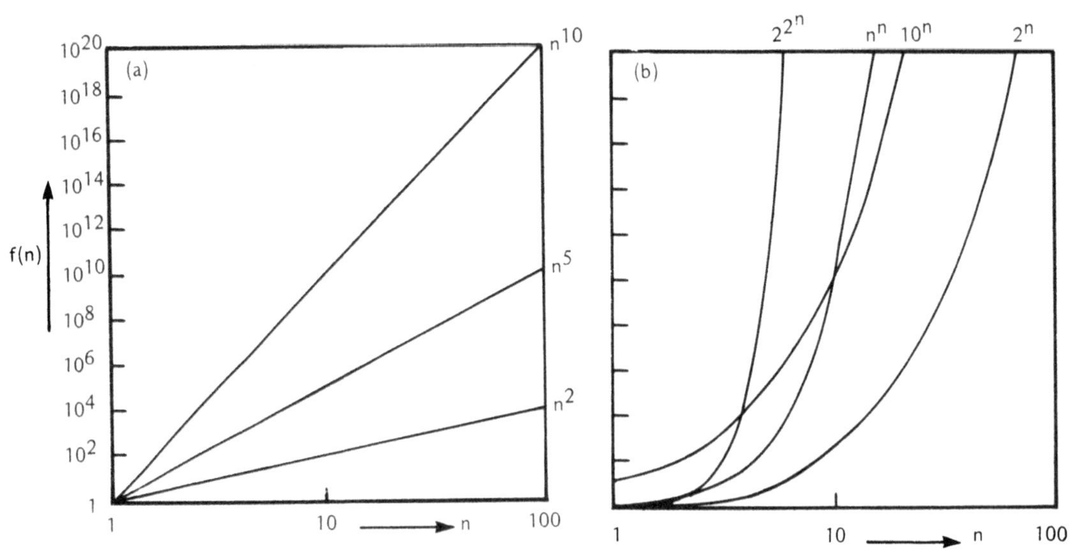

Figure 6.2. Plots of some typical time complexity functions: (a) polynomial, (b) exponential.

TABLE 6.2
Effects of Increases in Computing Speed on Problem-Solving Capabilities for Some Time Complexity Functions

	Size of largest problem instance solvable in some unit time with				
Time complexity function	current computer technology	technology hundred times faster	technology thousand times faster	technology million times faster	technology X times faster
n	n_1	$100n_1$	$1{,}000n_1$	$1{,}000{,}000n_1$	Xn_1
n^2	n_2	$10n_2$	$31.6n_2$	$1{,}000n_2$	$\sqrt{X}n_2$
n^5	n_3	$2.5n_3$	$3.98n_3$	$15.8n_3$	$\sqrt[5]{X}n_3$
n^{10}	n_4	$1.58n_4$	$2n_4$	$3.98n_4$	$\sqrt[10]{X}n_4$
2^n	n_5	$n_5+6.64$	$n_5+9.97$	$n_5+19.93$	$n_5+\log X/\log 2$
3^n	n_6	$n_6+4.19$	$n_6+6.29$	$n_6+12.58$	$n_6+\log X/\log 3$
10^n	n_7	n_7+2	n_7+3	n_7+6	$n_7+\log X$

polynomial time algorithm known to solve them, nor have they been proven intractable. A common trait oof such problems is that they can be "solved" in polynomial time by nondeterministic computers such as nondeterministic Turing machines. Such problems are called *NP-problems* (nondeterministic polynomial time problems) and form a set called the *problem class NP*. The term "solve" is used here in the sense that if the machine guesses the solution, it can verify its correctness in polynomial time. The notion of a nondeterministic polynomial time algorithm is thus used solely as a convenient definitional device for capturing the notion of polynomial time verifiability of a proposed (guessed) solution of the actual problem. It is known that any NP problem can be solved by a deterministic algorithm with time complexity $O(2^{p(n)})$, where p is a polynomial function.

The class NP contains the class P because any problem that is solvable in polynomial time on a deterministic Turing machine is also solvable (i.e., verifiable) in polynomial time on a nondeterministic Turing machine. A considerable number of NP-problems have been proven to have the property that every other NP-problem can be converted to them in polynomial time. Such problems are distinguished as *NP-complete problems.*

Since the class NP consists of many practically important problems, it is highly desirable to resolve its status. The question of whether or not NP-problems are intractable is therefore one of the most important questions in mathematics, computer science, and systems science. Its implications for systems problem solving are quite profound. The question is often stated in the form "is NP = P?". It can be answered by proving for any of the NP-complete problems that it is either a P-problem (i.e., solvable in polynomial time) or a problem inherently intractable (i.e., solvable only in exponential time). If any one of the NP-complete problems is proven intractable, then NP $\neq$ P. If, on the other hand, such a problem is proven tractable, then NP = P. Since there are strong indications that NP $\neq$ P under the usual rules of inference, the question becomes primarily one of discovering some unorthodox rules of inference under which any one of the NP-complete problems could be proven tractable.

The classification of problems from the standpoint of their solvability and computational complexity is summarized in Figure 6.3. The class denoted as coNP consists of problems that are complementary to the NP-problems in the sense that their answers are complements of the answers obtained for the corresponding NP-problems. It is not known whether NP = coNP, but it is known that the intersection NP $\cap$ coNP is not empty and contains all P-problems as well some other problems.

Although computational complexity has been predominantly studied in terms of the time it takes to perform a computation, the amount of computer memory required is frequently just as important. This requirement is usually referred to as the *space requirement*. It can be studied in terms of a *space complexity function*, analogous to the time complexity function. It is known, however, that any problem solvable in polynomial time can be solved in polynomial space as well. Indeed, the number of cells operated on by the read–write head of a Turing machine in a particular computation (which represents the space requirement) cannot exceed the number of steps involved in the computation (which represents the time requirement). It is not certain, however, whether all problems that are solvable in polynomial space are solvable in polynomial

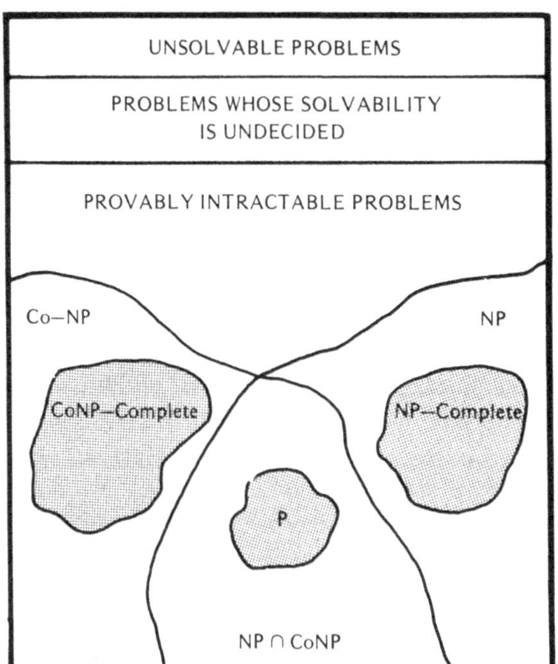

Figure 6.3. Classification of problems from the standpoint of their solvability and tractability.

time. It is for this reason that the time complexity is used to classify problems as either tractable or intractable. In practice, however, both of these requirements are equally important.

6.6. COMPLEXITY WITHIN GSPS

> *Associated with the learning process is the* complexity of describing, *the* d-complexity. *It is measured by the difficulty associated with extracting the description of a system.... associated with the interpretation process is the* complexity of interpreting, *the* i-complexity. *It is measured by the difficulty associated with extracting the interpretation (meaning) of a description.*
>
> —LARS LOFGREN

In architectural considerations regarding the GSPS, various issues involving complexity of systems or problems arise in several contexts. As far as the notion of systems complexity is concerned, it has two distinct roles within the GSPS. First, it represents a requirement type that is involved in some systems problems identifiable within the GSPS framework. Second, it is used to express the size of the systems, which are involved in the various problems, for the purpose of estimating the computational complexities of specific problem instances.

Systems complexity as a requirement type in systems problems can be viewed, generally, as a basis for defining a preference relation on the set of systems under consideration. As such, it is primarily user-oriented. It is thus always desirable that a complexity measure, if applicable in a systems problem, be defined by the user. However, in some situations, users may not have their own complexity measures and the GSPS should be able to offer some options. Furthermore, if the user is unable to select one of the suggested options or does not care which one is used, GSPS should proceed with one of the options that is declared for the given problem as a *default measure of systems complexity.*

In general, different epistemological systems types require different measures of complexity. Additional variations may then be required, at least in some instances, for methodological distinctions within each epistemological systems type. Some obvious complexity measures, which seem reasonable as default measures (or measures to be offered on the "menu"), are introduced in the context of representative problems in Chapters 3–5.

When systems complexity is intended to express the size of problem instances, on the basis of which the time complexity function of a problem is determined, it is usually defined in terms of the length of a description of the systems involved. Since the description length depends on the encoding scheme employed, it must be based in each case on the encoding scheme actually used in the GSPS implementation for describing systems of the given type.

Problem complexity arises in the development of GSPS as well as in its use. During its development, various issues of problem complexity are encountered in connection with the research on methodological tools. They include, for example: determination of time and space complexity functions for particular algorithms and encodings, delimitation of boundaries of practical solvability for the individual problems, comparison of competing algorithms from the standpoint of their computational complexities, research into new algorithms for problems whose current algorithms are unsatisfactory from the standpoint of complexity, development of efficient heuristic algorithms for intractable problems, and investigation of suitable simplifying assumptions under which various intractable problems become tractable.

One of the functions of the GSPS should be a routine analysis of the complexity status of each requested problem. A brief report of its results and, if appropriate, a list of desirable options should be presented to the user before any further processing is initiated. Based on this report, the user may decide to confirm his original request, accept one of the proposed options, modify his problem in some fashion, or cancel his request altogether.

Although the complexity analysis may be implemented in a variety of ways, it should cover the following fundamental questions. First, it should address the question of solvability. If the problem is unsolvable, it must be rejected and, if feasible, some meaningful and solvable modifications to it should be suggested to the user. Second, the problem should be classified with respect to three fixed values of computational complexity; the Bremermann limit (or some less conservative variant of it), a value that characterizes the limit of contemporary computer technology (this value must be periodically adjusted according to advances in computer technology), and a value that represents the actual limit of the specific GSPS realization employed. This leads to four rough classes of systems problems that are described in Figure 6.4. They are introduced for the purpose of making the user aware of the tractability status of his problem. If the problem is fundamentally intractable, he should abandon it and concentrate on some restricted and computationally less demanding reformulations. If it is potentially tractable, but beyond the capabilities of current computer technology, a reformulation

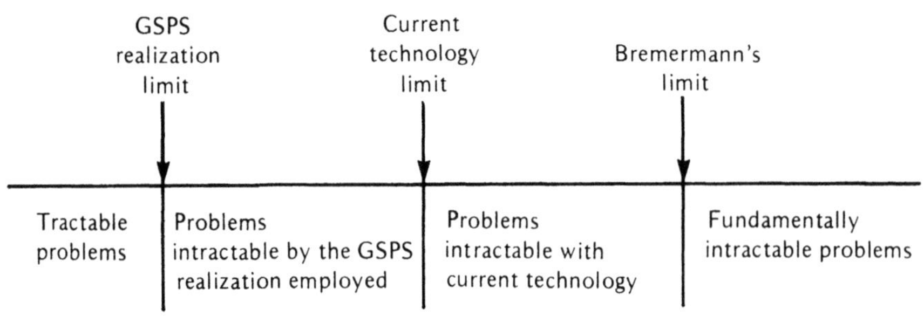

Figure 6.4. Rough classification of systems problems by their complexities.

is also necessary. However, the original problem need not be disregarded. If sufficiently significant, it should be filed and reconsidered again at some point in the future.

When the requested problem is within the capabilities of current technology, the complexity analysis should be more detailed. If the problem can be solved by the available GSPS realization, the user should be provided with the estimated computing cost. If not solvable, the user should be given a rough characterization of the required computer resources (speed, space), as well as approximate computing cost.

In some application areas, such as weather forecasting, production control, engineering testing, or management decision making, a particular value of the largest acceptable solution time is often included among the requirements of a systems problem. In other contexts, a value of the largest acceptable computer cost is specified. Such restrictions must of course be appropriately incorporated in the complexity analysis and the resulting report presented to the user.

It should be emphasized that the size of a problem instance is not the only determinant of its computational complexity. That is, problem instances of the same type and size may have very different complexities. Most studies in the area of computational complexity are oriented primarily to the characterization of the worst case problem instances. Although this orientation is theoretically sound, it usually results in estimates that are rarely reached in practice and are therefore too pesimistic. To ameliorate this situation, the worst case estimates are sometimes supplemented with average case estimates. However, such estimates are based on the assumption that all problem instances are equally likely, which does not necessarily reflect the actual probability distribution of problem instances encountered in practice. The problem of determining the actual distributions for various problem types is predominantly an empirical problem. One can hope to study this problem by monitoring and analyzing problem instances requested by users of the GSPS and other systems problem solving packages.

NOTES

6.1. The axioms of systems complexity formulated in Section 6.3 are supposed to be as general as possible. Although it is possible, in principle, to further generalize them by replacing the sum function in axiom (C5) with a general aggregation function, such a generalization does not seem to have sufficient intuitive appeal. In any event, the various axiomatic formulations of systems complexity that can be found in the current literature are considerably less general than that given in Section 6.3 [CA1, GO4]. One exception are three axioms of systems complexity by Conant [CO2], which resemble axioms (C3)–(C5).

6.2. Systems complexity has been studied from various points of view. The theory of finite state automata (or, more generally, finite semigroups) is one special area in which the notion of systems complexity has been well developed, basically in terms of the Krohn–Rhodes decomposition theorem [AR3, KR4, 5]. Another special kind of complexity, which is well covered in the literature, is complexity of sequences of symbols taken from some finite alphabets. Within the GSPS, this kind of complexity is clearly applicable to data systems with totally ordered support sets. The complexity of a sequence is usually defined in terms of the number of bits needed to describe a minimal computer program (for some fixed computer, say the Turing machine) by which the sequence can be reproduced. Various versions of this general approach to complexity of sequences are described in the literature by Kolmogorov [KO4], Chaitin [CH1, 2], Loveland [LO3], and others [MA5, SI2]. In all these variants, the notion of complexity is intimately linked to the notions of information and randomness. A good survey of the prominent issues associated with this sort of systems complexity is covered in Chapter 5 of a book by Fine [FI2].

The complexity of sequences is not the only area in which there is a relationship between complexity and information. One area of research on systems complexity focuses on the so-called maximum-entropy complexity measures, which are based on the Shannon measure of information. They were introduced by Ferdinand [FE3] on the basis of the maximum entropy principle, which he applied to obtain the prior probability distribution associated with the expected number of defects in a system under a given state of knowledge of that system. He also investigated the effect of systems modularity (i.e., structure refinement in the sense discussed in Chapter 4) on the number of expected defects in the system [FE2] and applied some of these results to computer software. He found, using some reasonable assumptions and relevant empirical information, that a computer program is subject to the minimum number of defects if it is organized hierarchically and the size of the subsystems (subroutines) at each level of the hierarchy is equal to $(2n)^{1/3}$, where n expresses the size of the whole system for which the subsystems are defined.

Ferdinand's defect complexity has been further investigated by Cornacchio [CO11], George [GE2], and Kapur [KA1].

There are also some other ways in which the Shannon entropy is used to define systems complexity. For example, Van Emden uses it to define complexity as "the way in which a whole is different from the composition of its parts" [VA2].

Graphs are useful means for describing systems of certain types (e.g., ST-systems, C-structure systems). Some efforts have therefore been made to define the complexity of graphs [AL1, MO4], but it seems that this subject area has not been developed to its potential as yet.

Systems complexity has also been studied with respect to the design of engineering systems, particularly large-scale systems such as telephone exchanges and digital computers. When the components available for constructing a system and the task to be performed by the system are precisely defined, one of the complexity issues is to find the minimum number of components required. An interesting exposition of this subject area was written by Pippenger [PI1].

Although this brief survey is representative of the main developments related to systems complexity, it is by no means complete. For further information, the reader may consult a bibliography on systems complexity [CO12].

6.3. The initial ideas of computability were introduced in the 1930s by Turing [TU1], Kleene [KL1], Post [PO5], and Church [CH10]. Two formalisms—Turing machines and recursive functions—are usually used in current literature. The notion of Markov algorithms [MA4] is also used, but considerably less frequently. The area of computability is thoroughly covered in a number of books, for example [DA1, RO2].

Although the significant difference between polynomial and exponential time algorithms was already recognized in the mid-1960s, the foundations of the current theory of NP-completeness were established only in the early 1970s, primarily due to contributions by Cook [CO10] and Karp [KA2]. For further details regarding computational complexity, an excellent book by Garey and Johnson is recommended [GA7]; in addition to comprehensive coverage of the area of computational complexity, it also contains an extensive bibliography and a well-documented catalog of over 300 NP-complete problems and some open problems in the NP class.

6.4. A fundamental study of the notion of complexity, which covers both systems complexity and problem complexity, was performed by Lars Löfgren [LO1]. He classifies complexities into description complexities and interpretation complexities and shows that some of the familiar complexity measures fit into one of these two classes. In terms of the GSPS framework, this classification reflects the ordering of epistemological types of systems. Any search for a higher epistemological type is associated with the notion of description complexity. On the other hand, any problem whose aim is to determine a lower epistemological type involves the notion of interpretation complexity.

EXERCISES

6.1. Propose some intuitively reasonable measures of systems complexity for behavior systems, ST-systems, and structure systems, and check for each of them if the axioms (C1)–(C5) formulated in Section 6.3 are satisfied.

6.2. Consider some problems in your area of interest and check for each of them whether or not it is transcomputational.

6.3. Calculate some entries in Table 6.1.

6.4. Verify the mathematical expressions in the last column of Table 6.2.

6.5. Extend Table 6.2 for time complexity functions n^n, $n!$, and 2^{2^n}.

6.6. Show that for any positive number N there exists a number n_0 such that

$$n! > N^n$$

for all $n \geq n_0$.

6.7. Prove that each of the following propositions is either true or false:

(a) $3n^5 + 10n^3 + n^2 + 25$ is of complexity $O(n^5)$;

(b) $2^n + 3^n$ is of complexity $O(2^n)$;

(c) $2^n + 3^n$ is of complexity $O(3^n)$;

(d) $n!$ is of complexity $O(n^n)$;

(e) n^n is of complexity $O(n!)$;

(f) 10^{3^n} is of complexity $O(2^{2^n})$.

7

GOAL-ORIENTED SYSTEMS

All life is a purposeful struggle and your only choice is the choice of a goal.

—AYN RAND

7.1. PRIMITIVE, BASIC, AND SUPPLEMENTARY CONCEPTS

> *The only justification for our concepts is that they serve to represent the complex of our experiences; beyond this, they have no legitimacy.*
>
> —ALBERT EINSTEIN

To achieve a global comprehension of the whole GSPS conceptual framework, it is desirable to look at the framework from a distance, with the aim of recognizing some significant categories of the concepts. Such an inspection of the conceptual framework is likely to reveal three main categories of concepts. Let us refer to concepts in these categories as primitive, basic, and supplementary systems concepts.

Primitive systems concepts are characterized by their independence from any other concepts in the framework. They represent thus a starting point in the development of a conceptual frameworks. Once selected, they basically determine the possible range of conceptual frameworks that can be built upon them. It is appropriate to say that the richness of the chosen primitive concepts determines the richness of the conceptual frameworks that are derivable from them. If any of the primitive concepts is excluded, the range of derivable conceptual frameworks may be considerably reduced or may even become pragmatically worthless.

In the GSPS framework, primitive systems concepts are those associated with the source system: attributes (or input and output attributes), appearance sets, backdrops, backdrop sets, specific and general variables (or input and output variables) and their state sets, specific and general supports and their support sets, observation channels (crisp or fuzzy), and exemplification/abstraction channels. Also included are, of course, the various methodological distinctions associated with these concepts.

All other concepts of the GSPS framework are defined in terms of the primitive concepts above. Upon careful inspection of these other, derived concepts, two categories emerge naturally. One of them consists of concepts that are connected with various forms by which constraints among variables are characterized. These are concepts involved in defining all the epistemological systems types except the source system, namely, data, translation rules in support sets, sampling variables (generated, generating, and input), masks, behavior and ST-functions (basic and generative), environment (external and internal), subsystems and supersystems, elements of structure systems, coupling variables, couplings (neutral or directed), elements of metasystems, and replacement procedures. Since these concepts are all connected with the basic issue of systems problem solving—the characterization, determination, and use of constraints among various kinds of entities—it is terminologically appropriate to call them *basic systems concepts.*

Let us refer to the remaining GSPS concepts as *supplementary systems concepts.* These concepts are neither primitive nor basic GSPS concepts, but can be defined in terms of them and, in some cases, also in terms of other supplementary concepts. They usually have different meanings depending on the systems types to which they are applied. One such concept is *systems complexity*, which is discussed in Chapter 6. It is a general concept, defined by its general axioms, which subsumes a large spectrum of more specific notions of systems complexity. These can be ordered by their degrees of specificity. To be of any practical use in systems problem solving situations, the notion of systems complexity must be sufficiently specific at least to the extent that it is defined in terms of some specific systems type.

Two important classes of concepts, which belong to the category of supplementary systems concepts, are those of the *goal* and the *performance* of a system. There are general concepts that, like the concept of systems complexity, subsume a variety of special cases. It is the purpose of this chapter to introduce these concepts at a general level, discuss some of their specific meanings, and outline their role in systems problem solving.

7.2. GOAL AND PERFORMANCE

> *There is nothing insignificant in the world. It all depends on how one looks at it.*
>
> —Johann Wolfgang Goethe

The concept of a *goal* of a system can be defined in many different ways. A general view is adopted for the GSPS according to which the goal of a system is "in the eyes of the user." That is, given a system of some epistemological type, identified by its primary traits, a goal associated with the system is a specific restriction of its primary or secondary traits, which the user considers desirable under given circumstances.

A given system may thus be viewed from the standpoint of different goals. It satisfies each of them to some degree. This degree, which is called the *performance of the system with respect to the goal*, should measure (in some manner) the closeness between the actual and desirable manifestations of those traits of the system that are involved in the goal. It is often expressed in terms of an appropriate function, which is called a *performance function.*

Let $\mathscr{X}$ denote a set of systems that differ in those traits that are assigned, in a given situation, to the notion of a goal, and are equal in all other traits. Then, a performance function, say function ω, has the form

$$\omega: \mathscr{X} \times \mathscr{X} \to [0, 1], \tag{7.1}$$

where $\omega(\mathbf{x}, \mathbf{x}^*)$ represents the degree to which a particular system $\mathbf{x} \in \mathscr{X}$ approximates a specified goal system (desirable, ideal system) $\mathbf{x}^* \in \mathscr{X}$. The performance function can be conveniently expressed in terms of an appropriate distance function

$$\delta: \mathscr{X} \times \mathscr{X} \to \mathbb{R}^+ \tag{7.2}$$

by the formula

$$\omega(\mathbf{x}, \mathbf{x}^*) = \frac{\delta_m(\mathbf{x}, \mathbf{y}) - \delta(\mathbf{x}, \mathbf{x}^*)}{\delta_m(\mathbf{x}, \mathbf{y})} = 1 - \frac{\delta(\mathbf{x}, \mathbf{x}^*)}{\delta_m(\mathbf{x}, \mathbf{y})}, \tag{7.3}$$

where

$$\delta_m(\mathbf{x}, \mathbf{y}) = \max_{\mathbf{x}, \mathbf{y} \in \mathscr{X}} \delta(\mathbf{x}, \mathbf{y}).$$

Example 7.1. Assume that a goal is defined in terms of a desirable behavior function f_B^* within the set of behavior systems characterized by the same source system and mask M. Set $\mathscr{X}$ is then represented in this example by a set of behavior systems

$$\mathbf{F}_B = (\mathbf{S}, M, f_B)$$

that differ only in behavior functions f_B. One possible way of expressing distance between systems in this set, and which seems appropriate in this case, is the city-block (Hamming) distance

$$\delta(\mathbf{x}, \mathbf{y}) = \sum_{\mathbf{c} \in \mathbf{C}} |{}^x f_B(\mathbf{c}) - {}^y f_B(\mathbf{c})|. \tag{7.4}$$

When the behavior functions are probabilistic, then $\delta_m(\mathbf{x}, \mathbf{y}) = 2$ and

$$\omega(\mathbf{x}, \mathbf{x}^*) = 1 - \delta(\mathbf{x}, \mathbf{x}^*)/2; \tag{7.5}$$

when they are possibilistic, then $\delta_m(\mathbf{x}, \mathbf{y}^*) = |\mathbf{C}|$ and

$$\omega(\mathbf{x}, \mathbf{x}^*) = 1 - \delta(\mathbf{x}, \mathbf{x}^*)/|\mathbf{C}|, \tag{7.6}$$

It is obvious that different goals and performance functions are applicable to different types of systems. However, different types of goals, each of which requires a special performance function, can be defined even for the same systems types. For behavior systems, for example, goals can be defined in terms of desirable behavior functions, ranges of behavior functions, sets of local behavior functions for specific subsets of the support set, sets of behavior functions representing desirable subsystems, etc. A special performance function is obviously needed for each of these types of goals.

The concepts of goal and performance provide a basis for defining the notion of goal-oriented systems, to which the rest of this chapter is devoted.

7.3. GOAL-ORIENTED SYSTEMS

> *A system that tends to improve its performance while pursuing its task or goal and does so without outside help is called self-organizing.*
>
> —HANS J. BREMERMANN

Assume that a goal type and a relevant performance function are defined for a set of systems of some epistemological type. As explained in Section 7.2, with each system in the set is associated the value of the performance function, which indicates the degree to which the system satisfies the goal. This fact suggests a trivial way of defining goal-oriented systems: a system is viewed as goal-oriented if and only if its performance with respect to the given goal is greater that some specified threshold value (typically 0.5 or larger).

Another way of defining goal-oriented systems, which seems operationally more meaningful, is to view the notion of goal-orientation in relative terms. That is, one system is viewed as goal-oriented with respect to another system of the same type and a specified goal if and only if it performs better (according to some performance function) with respect to the goal. Formally, given two systems $\mathbf{x}, \mathbf{y} \in \mathscr{X}$ of the same type, a specific goal $\mathbf{x}^* \in \mathscr{X}$, and a relevant performance function ω, system $\mathbf{x}$ is goal-oriented with respect to system $\mathbf{y}$ and goal $\mathbf{x}^*$, and under performance function ω, if and only if

$$\omega(\mathbf{x}, \mathbf{x}^*) > \omega(\mathbf{y}, \mathbf{x}^*).$$

Let the difference

$$\Delta\omega(\mathbf{x}, \mathbf{y} \mid \mathbf{x}^*) = \omega(\mathbf{x}, \mathbf{x}^*) - \omega(\mathbf{y}, \mathbf{x}^*) \tag{7.7}$$

be called the *degree of goal-orientation* of $\mathbf{x}$ with respect to $\mathbf{y}$, given goal $\mathbf{x}^*$.

A system that has a positive degree of goal-orientation with respect to another system must contain some traits, other than those included in the latter system or associated with the goal, that are responsible for its improved performance. Let us call these *goal-seeking traits*. Such traits are, for instance, some additional variables or states in generative systems, additional elements or couplings in structure systems, additional elements or replacement procedures in meta-systems, and the like.

To illustrate these general notions associated with goal-orientation, let us apply them in the rest of this section to systems of a specific epistemological type—neutral behavior systems. In this case, the goal-seeking traits are such variables whose inclusion into the system improves its performance. Such variables are called *goal-seeking variables*.

Consider a set of neutral behavior systems of the usual form

$$\mathbf{F}_B = (\mathbf{S}, M, f_B),$$

which differ only in their behavior functions f_B. Since the rest of this section deals solely with this type of system, no confusion will be caused if we simplify the notation by excluding the subscripts B. In addition, we may conveniently identify each system in the set either by its behavior function f or the associated distribution **f**.

Assume that the systems considered are probabilistic. Consistent with the notation introduced in Chapter 3, let **c** denote overall states of the sampling variables associated with mask M and let $\mathbf{c} \in \mathbf{C}$. Assume that the system identified by the probability distribution

$$\mathbf{f}^* = (f^*(\mathbf{c}) \mid \mathbf{c} \in \mathbf{C})$$

is viewed as a goal. Then, for each system in the set, identified by the probability distribution

$$\mathbf{f} = (f(\mathbf{c}) \mid \mathbf{c} \in \mathbf{C}),$$

its distance $\delta(\mathbf{f}, \mathbf{f}^*)$ from the goal can be expressed, for example, by formula (7.4). The performance $\omega(\mathbf{f}, \mathbf{f}^*)$ of each system in the considered set is then determined by substituting its distance into formula (7.5).

Consider now a behavior system

$$\mathbf{F}' = (\mathbf{S}', M', f')$$

whose source system $\mathbf{S}'$ contains all entities included in $\mathbf{S}$, but contains in addition some variables through which mask M is extended into mask M'. Let states of sampling variables associated with the set difference $M' - M$ be denoted by **z**, let $\mathbf{z} \in \mathbf{Z}$, and let

$$\mathbf{f}' = (f'(\mathbf{c}, \mathbf{z}) \mid \mathbf{c} \in \mathbf{C}, \mathbf{z} \in \mathbf{Z})$$

denote the probability distribution of system $\mathbf{F}'$.

To calculate the distance between systems identified by the distributions $\mathbf{f}'$ and $\mathbf{f}^*$, the extended distribution $\mathbf{f}'$ must be converted to a form comparable with $\mathbf{f}^*$. This can be done by means of the formula

$$f''(\mathbf{c}) = \sum_{\mathbf{z} \in \mathbf{Z}} f'(\mathbf{c}, \mathbf{z}). \tag{7.8}$$

Let the distribution

$$\mathbf{f}'' = (f''(\mathbf{c}) \mid \mathbf{c} \in \mathbf{C})$$

now be used to identify system $\mathbf{F}'$. Formulas (7.4) and (7.5) are then applicable for calculating the distance $\delta(\mathbf{f}'', \mathbf{f}^*)$ and performance $\omega(\mathbf{f}'', \mathbf{f}^*)$, respectively. If

$$\Delta\omega(\mathbf{f}'', \mathbf{f} \mid \mathbf{f}^*) > 0,$$

calculated by formula (7.7), the system **F**′ is a *goal-oriented system* with respect to system **F**, given goal **f***; variables associated with $M' - M$ are then called *goal-seeking variables.*

It is important to realize that a necessary condition for system **F**′ to be goal-oriented with respect to system **F** is that **f** is not a projection of **f**′. Indeed, if **f** were a projection of **f**′, then **f**″ = **f** and, consequently,

$$\Delta\omega(\mathbf{f}'', \mathbf{f} \mid \mathbf{f}^*) = 0.$$

This condition means that variables of **S**′ that are not contained in **S** must represent attributes that are not available on the object for which system **F** is defined.

Example 7.2. Consider a computer system in which the utilization of three expensive units is of particular interest. Three variables v_1, v_2, v_3 are defined, one for each of these units, by which activities of the units are described in time. Each of the variables has two states: 0, which indicates that the unit is not active at the time of observation, and 1, which indicates that the unit is active. The goal is to keep all the units active all the time.

Suppose that an extensive hardware monitoring of the variables is performed (see Example 3.8) and the probability distribution **f** specified in Table 7.1a is obtained for the memoryless mask. Also shown in Table 7.1a is the probability distribution **f*** that represents the goal. Using formulas (7.4) and (7.5), we obtain $\delta(\mathbf{f}, \mathbf{f}^*) = 1.4$ and $\omega(\mathbf{f}, \mathbf{f}^*) = 0.3$.

Assume now that a new unit, say a communication channel, is added to the computer system in such a manner that it affects the activities of the three units under consideration. Assume further that the new unit is relatively inexpensive, when compared with the other units, so that its own utilization is not important. The unit is introduced only for the purpose of enhancing the utilization of the other three units. The goal thus remains the same.

TABLE 7.1

Illustration of a Goal-Oriented System and a Goal-Seeking Variable (Example 7.2)

		(a)					(b)					(c)	
v_1	v_2	v_3	**f**	**f***	v_1	v_2	v_3	v_4	**f**′	v_1	v_2	v_3	**f**″
0	0	1	0.15	0	0	1	1	1	0.10	0	1	1	0.10
0	1	0	0.20	0	1	0	0	0	0.02	1	0	0	0.02
1	0	0	0.10	0	1	0	1	0	0.03	1	0	1	0.03
1	1	0	0.25	0	1	1	0	0	0.04	1	1	0	0.05
1	1	1	0.30	1	1	1	0	1	0.01	1	1	1	0.80
					1	1	1	0	0.25				
					1	1	1	1	0.55				

Let variable v_4 be defined for the new unit in the same way as the other variables are defined for their units. Assume that hardware monitoring is performed again for the new system, which includes variable v_4, and results in the probability distribution given in Table 5.1b. To make it comparable with $\mathbf{f}^*$, we use formula (7.8) to calculate distribution $\mathbf{f}''$ (Table 7.1c). Then, $\delta(\mathbf{f}'', \mathbf{f}^*) = 0.4$ and $\omega(\mathbf{f}'', \mathbf{f}^*) = 0.8$. Hence,

$$\Delta\omega(\mathbf{f}'', \mathbf{f} | \mathbf{f}^*) = 0.8 - 0.3 = 0.5,$$

i.e., the extended system is goal-oriented and variable v_4 is a goal-seeking variable in this case; the degree of goal-orientation (improvement in performance due to the additional unit) is 0.5.

Consider a different goal, in which $f^*(c) = 0.5$ for the last two states listed in Table 7.1a. We obtain

$$\delta(\mathbf{f}, \mathbf{f}^*) = \delta(\mathbf{f}'', \mathbf{f}^*) = 0.9$$

and

$$\omega(\mathbf{f}, \mathbf{f}^*) = \omega(\mathbf{f}'', \mathbf{f}^*) = 0.55.$$

The new system is thus not goal-oriented in this case. Although the new unit, represented by variable v_4, influences activities of the other units considerably, it does not bring the system closer to the goal. That means that variable v_4 is not a goal-seeking variable with respect to this alternative goal.

Consider still another goal, in which $f^*(c) = 0.2$ for each state listed in Table 7.1c. Then,

$$\omega(\mathbf{f}, \mathbf{f}^*) = 0.85, \; \omega(\mathbf{f}'', \mathbf{f}^*) = 0.4,$$

and

$$\Delta\omega(\mathbf{f}'', \mathbf{f} | \mathbf{f}^*) = -0.45.$$

In this case, variable v_4 is undesirable and may be called a goal-evading variable; it diverts the system from its goal and thus reduces its performance.

7.4. STRUCTURE SYSTEMS AS PARADIGMS OF GOAL-ORIENTED BEHAVIOR SYSTEMS

> *... a contemporary scientist does not try at all cost to construct a single global model of reality. ... He has learned over time that it may be preferable to construct a network of local models, perhaps of comparable complexity, but almost certainly of entirely different structure.*
>
> —RICHARD BELLMAN AND CHARLENE PAUL SMITH

Goal-oriented systems, as defined in the previous section, are characterized by the separation of goal-seeking variables from other variables involved, and by the requirement that the goal-seeking variables positively contribute toward achieving the considered goal. An exploration of possible ways in which states of goal-seeking variables can be generated is obviously of considerable importance for comprehending the notion of goal-oriented systems and, particularly, for developing methods by which an appropriate goal-oriented system can be designed. Such an exploration leads necessarily to some specific types of structure systems. Each of them can be viewed as a paradigm that describes a principle (scheme, form) in terms of which states of goal-seeking variables are generated. Let these paradigms be called *structure paradigms of goal-oriented systems.*

First, let us consider goal-oriented behavior systems of the neutral type. As explained in the previous section, their variables are partitioned into the goal-seeking variables and the remaining variables. Since the goal is defined in terms of the latter variables, it is reasonable to call them *goal-implementing variables.*

The goal-seeking variables affect the goal-implementing variables, while, at the same time, they may be affected by them. It is thus natural to view the goal-oriented behavior system as a structure system with two elements. One of them generates states of the goal-implementing variables, while the other one generates states of the goal-seeking variables. Let these elements be called a *goal-implementing element* and a *goal-seeking element*, and let them be described formally as behavior systems

$$^1\mathbf{F} = (^1\mathbf{S}, {}^1M, {}^1f)$$

and

$$^2\mathbf{F} = (^2\mathbf{S}, {}^2M, {}^2f)$$

respectively. Since no confusion can arise, subscripts B are not used here.

Assume that ^1V, ^2V are sets of variables in source systems $^1\mathbf{S}$, $^2\mathbf{S}$, respectively. In general, $^2V \subseteq {}^1V$ since the goal-seeking variables are required (by definition) to influence the goal-implementing variables, while the opposite influence is not required. To describe the manner in which the individual variables are generated, the two elements must be viewed as directed (as explained in Sections 4.3 and 4.4), even though no external environment is recognized. The directed couplings between the elements are

$$\hat{C}_{2,1} = {}^2V \quad \text{and} \quad \hat{C}_{1,2} = {}^1V',$$

where $^1V' \subseteq {}^1V$. It is reasonable to recognize three structure paradigms, which are distinguished from each other by the coupling $\hat{C}_{1,2}$:

$$\hat{C}_{1,2} = \varnothing;$$
$$\hat{C}_{1,2} \subset {}^1V;$$
$$\hat{C}_{1,2} = {}^1V.$$

These paradigms differ in the extent to which information about the goal-implementing variables is utilized for generating the goal-oriented variables. Let us call them an

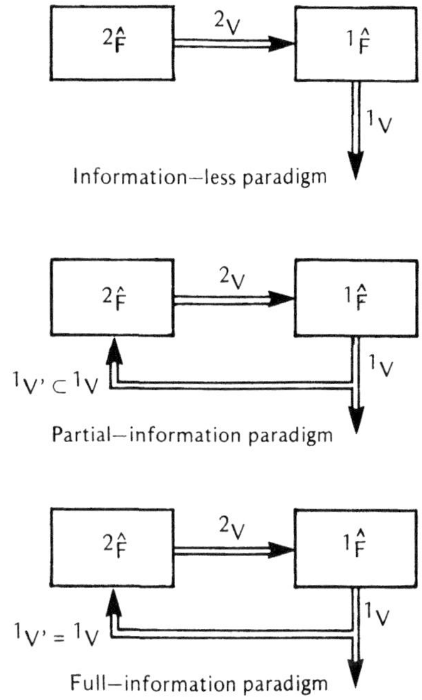

Figure 7.1. Structure paradigms of a goal-oriented behavior system of the neutral type (without input variables).

information-less paradigm, a *partial-information paradigm*, and a *full-information paradigm*, respectively. Block diagrams of these three paradigms are illustrated in Figure 7.1.

Let us now consider structure paradigms of goal-oriented behavior systems of the directed type. In addition to the sets of variables included in their neutral counterparts, they contain a set of input variables, say set X. Possible directed couplings between the goal-implementing element, goal-seeking element, and environment (identified as element 0) can be summarized by the matrix

	0	1	2
0	$\varnothing$	X	X'
1	${}^1V$	$\varnothing$	${}^1V'$
2	$\varnothing$	${}^2V$	$\varnothing$

where $X' \subseteq X$ and ${}^1V' \subseteq {}^1V$. Three characteristic cases can be distinguished for X':

$$X' = \varnothing;$$
$$X' \subset X;$$
$$X' = X.$$

Similarly, three cases can be distinguished for ${}^1V'$:

$$
\begin{aligned}
{}^1V' &= \varnothing; \\
{}^1V' &\subset {}^1V; \\
{}^1V' &= {}^1V.
\end{aligned}
$$

Each of the cases for X' can be combined with any of those distinguished for ${}^1V'$. This leads to nine paradigms. They are listed in Figure 7.2 and classified into four categories, each of which is represented by one of the block diagrams and given a name that is common in the literature.

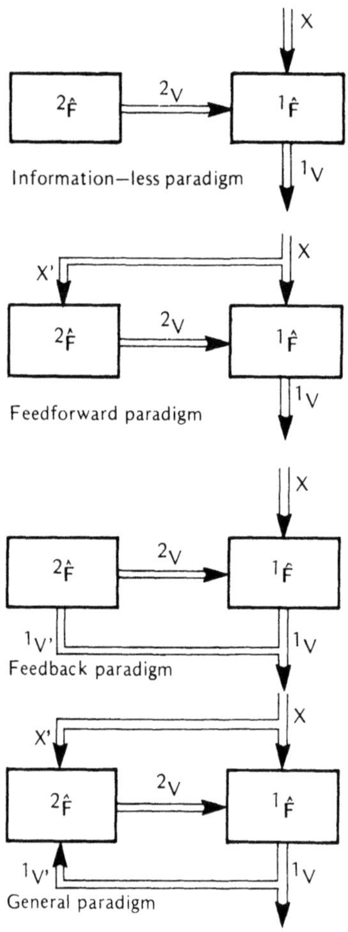

Label	$C_{0,2}$	$C_{1,2}$	Type
1	$\varnothing$	$\varnothing$	Information–less
2	$X' \subset X$	$\varnothing$	Partial input information
3	$X' = X$	$\varnothing$	Full input information
4	$\varnothing$	${}^1V' \subset {}^1V$	Partial output information
5	$\varnothing$	${}^1V' = {}^1V$	Full output information
6	$X' \subset X$	${}^1V' \subset {}^1V$	Partial input/output information
7	$X' = X$	${}^1V' \subset {}^1V$	Full input/partial output information
8	$X' \subset X$	${}^1V' = {}^1V$	Partial input/full output information
9	$X' = X$	${}^1V' = {}^1V$	Full input/output information

Figure 7.2. Structure paradigms of a goal-oriented behavior system of the directed type.

The concept of a goal-oriented behavior system, in general, and its various structure paradigms summarized in Figures 7.1 and 7.2, in particular, enjoy a broad range of applicability under a large variety of interpretations. One of the interpretations, which has been studied quite extensively, is *regulation*. Any goal-oriented system that performs some sort of regulation is called a *regulator*. Its goal is to keep variables of ${}^1V$ in a particular state or a particular subset of states in spite of disturbances, which are represented by the input variables. The goal-seeking variables assume the role of regulating variables; elements ${}^1\mathbf{F}$, ${}^2\mathbf{F}$ are regulated and regulating elements, respectively. A regulator can also be defined in terms of a ST-system. The goal is then to keep the system in a subset of state transitions that involve the regulated variables.

Another interpretation of the goal-oriented behavior system is to view it as a learning system. Elements ${}^1\mathbf{F}$ and ${}^2\mathbf{F}$ are a learning element and teaching element, respectively. The goal is to produce responses (states of variables in ${}^1V$) to individual stimuli (states of variables in X) that are considered (defined) as "correct." States of the goal-seeking variables represent in this case some sort of reinforcement.

Still another interpretation is to view the goal-oriented behavior system as a *decision-making system*. Elements ${}^1\mathbf{F}$, ${}^2\mathbf{F}$ become decision-implementing and decision-making elements, respectively. States of input variables represent so-called "states of nature" (e.g., relevant external circumstances, possible moves of an opponent, recognized characteristics of some sort, and the like). States of the variables in ${}^1V$ represent outcomes, on which a *utility function* is defined. The goal of the system is to maximize the utility function. The role of the goal-seeking variables is to make selections from a set of decision alternatives that affect the outcomes in a positive way with respect to the goal. Given this role, they may be called, e.g., decision-making, selection-making, or utility-seeking variables.

Some additional interpretations of the goal-oriented behavior system could be described, such as error-correcting systems or self-organizing systems. However, the purpose of this book is not to cover goal-oriented systems comprehensively and in detail, but only to indicate their role in systems problem solving. Some of the interpretations of goal-oriented systems (especially regulation and various kinds of decision making) are covered quite extensively in the literature. For more details, see the Notes to Chapter 7.

7.5. DESIGN OF GOAL-ORIENTED SYSTEMS

> *Every good regulator of a system must be a model of that system.*
>
> —Roger Conant and W. Ross Ashby

In systems inquiries, the notion of structure paradigms of goal-oriented behavior systems provides the investigator with a useful systems description, according to which some of the investigated variables are viewed as contributing toward a desired goal. The basic problem is to determine which of the variables under investigation exhibit a high degree of this goal-seeking capability.

In designing goal-oriented systems, on the other hand, each structure paradigm specifies a frame within which the designer is required to operate. In other words, each structure paradigm represents a set of assumptions (restrictions) regarding the system to be designed that the designer must not violate.

Since designed systems are always directed, only the structure paradigms specified in Figure 7.2 are applicable for discussing the various issues involved in the design of goal-oriented systems. These paradigms can be partially ordered by the severity of their restrictions. The more severe is the restriction, the less freedom is left to the designer to perform his task and, consequently, the less general is the paradigm. Let a structure paradigm be called *less general* than another structure paradigm if and only if it contains more restrictions (assumptions) than the latter. When applied to the set of structure paradigms specified in Figure 7.2, this ordering by the degree of generality forms a lattice that is described by its Hasse diagram in Figure 7.3. For specific sets X and ${}^1V$, a more refined lattice can be defined by allowing X' and ${}^1V'$ to represent any subsets of X and ${}^1V$, respectively.

In a typical problem of designing a goal-oriented system, the following particulars are given:

i. a directed behavior system ${}^0\mathbf{F} = ({}^0\mathbf{S}, {}^0M, {}^0f)$ that represents the goal-implementing element without the goal-seeking variables;
ii. a goal or a performance function that is compatible with the system ${}^0\mathbf{F}$;

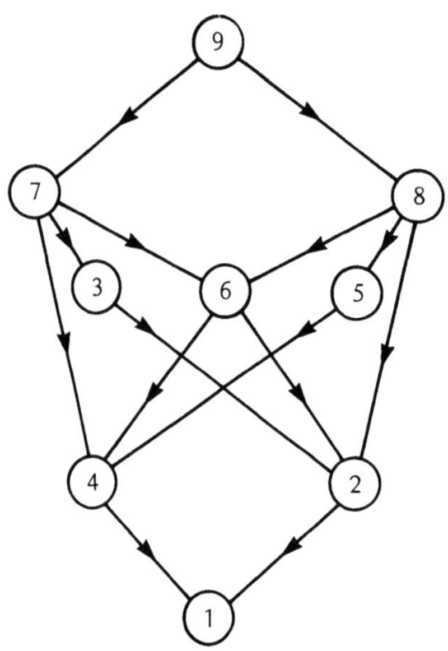

Figure 7.3. Lattice of the structure paradigms specified in Figure 7.2 under the ordering by generality.

iii. an inventory of available elements;
iv. an objective criterion (function);
v. a structure paradigm of goal-oriented systems and, possibly, other constraints regarding the designed system.

The following are the main issues involved in the design of goal-oriented systems:

1. The goal and performance function must be made explicit. If only the goal is given in the problem statement, it is up to the designer to choose an appropriate performance function. If only the performance function is given, it is normally assumed that the goal is represented by a behavior function for which the performance function reaches its maximum. This may, of course, lead to several behavior functions. The designer must either select one of them as the goal or, alternatively, select more than one goal and view the designed system as a goal-oriented metasystem.

2. A suitable set of goal-seeking variables must be selected. These variables cannot be arbitrary; they must exert some influence upon the output variables of the given system ${}^0\mathbf{F}$ and produce thereby a nontrivial extension of system ${}^0\mathbf{F}$ into a new behavior system

$${}^1\mathbf{F} = ({}^1\mathrm{S}, {}^1M, {}^1f).$$

This extended system represents the goal-implementing element of any of the structure paradigms of goal-oriented behavior systems. The selection of proper goal-seeking variables is crucial. Once selected, they determine the best performance that can be achieved under their influence. If this is not adequate, then other possible sets of goal-seeking variables must be explored.

3. The main difficulty in selecting proper goal-seeking variables is that the influences of the various sets of variables under consideration upon the goal variables (behavior functions ${}^1f$) are usually not known. These must therefore be determined as part of the design problem. Procedures described in Chapter 3 are relevant for dealing with this issue.

4. Once a set of goal-seeking variables is accepted and the corresponding goal-implementing system ${}^1\mathbf{F}$ is determined, the next step is to determine some particular way in which the goal-seeking variables are generated, i.e., to determine the goal-seeking system ${}^2\mathbf{F}$. The objective is to determine a behavior function ${}^2f$ of system ${}^2\mathbf{F}$ for which the performance function reaches its maximum within the constraints of the required structure paradigm. Various optimization methods can be employed for solving this problem. Their choice depends, primarily, on the methodological type of the systems involved as well as the nature of the performance function.

5. A structure system that implements the goal-seeking behavior system ${}^2\mathbf{F}$ by the available element types must be designed. This is a standard problem of systems design which is discussed in Section 4.5.

7.6. ADAPTIVE SYSTEMS

> *. . . all systems are adaptive, and the real question is what they are adaptive to and to what extent.*
>
> —LOTFI A. ZADEH

A goal-oriented system designed by the procedure outlined in the previous section is optimal (or close to optimal) only under the assumption that neither the behavior function ${}^1f$ of the goal-implementing element nor the goal f^* change. If this assumption is not satisfied, the actual increase in performance effected by the goal-seeking element may be far below the calculated increase. The goal-seeking element may even have an adverse effect on the goal-implementing element and cause a decrease in performance. In order to maintain a high level of performance, despite changes in ${}^1f$ or f^*, the goal-oriented system must be capable of adaptation.

There are various reasons why ${}^1f$ and f^* may change. In general, a change in ${}^1f$ means that the goal-implementing system ${}^1f$ was not properly conceptualized during the process of design to account for the change. This means, ultimately, that the goal-implementing variables are affected by some input variables that are not recognized in ${}^1f$. Once a goal-oriented system is designed and implemented for a particular function ${}^1f$, changes in ${}^1f$ are not under the control of the user. Changes in the goal, on the other hand, are fully determined by the user. In general, he may consider it desirable to change the goal because the circumstances for which the goal-oriented system was designed have changed.

In order to enable a goal-oriented system to adapt to changes in the goal, its goal-seeking element must be designed for a set of alternative goals and supplemented with a special input variable whose states represent the goals. Let us call this kind of goal-oriented system, in which the adaptation is restricted to a specific set of goals, a *multigoal-oriented system.* A block diagram of its structure paradigm (of the general type) is shown in Figure 7.4, where v^* denotes the variable that defines the current goal ($v^* \in V^*$) and ${}^2F^+$ denotes the goal-seeking system extended by the input variable v^*. The goal is determined in this case either by the user or by another system to which the input variable v^* is coupled.

Changes in the goal may be determined by a *goal-generating* system that is included as an element in the multigoal-oriented system itself. This possibility is illustrated by the block diagram in Figure 7.5a. Elements ${}^1\mathbf{F}$ and ${}^2\mathbf{F}^+$ form a basic multigoal-oriented system (as in Figure 7.4); element ${}^3\mathbf{F}$ is a goal-generating element. An overall goal is in this case replaced by sequences of subgoals, represented by states of variable v^*. These sequences are determined by a behavior function ${}^3f$ in terms of variables in sets X, ${}^1V$, ${}^2V$ or, depending on the structure paradigm employed, some subsets of them. Let goal-oriented systems of this type be called *autonomous multigoal-oriented systems.*

One obvious motivation for considering autonomous multigoal-oriented systems

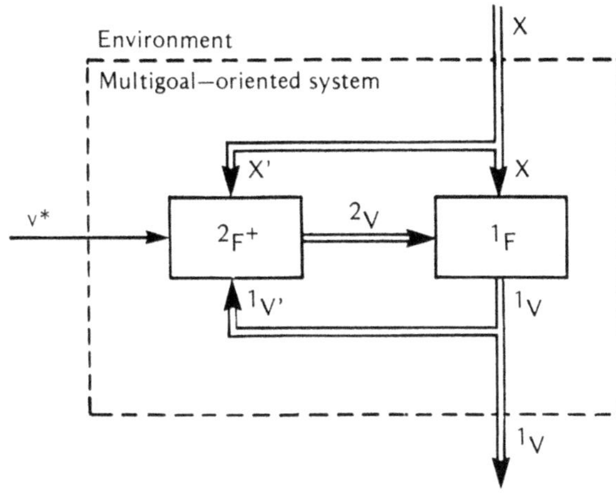

Figure 7.4. General structure paradigm of a multigoal-oriented system.

is to simplify the overall optimization problem involved in the design of a goal-seeking element. The optimization problem is decomposed into several simpler optimization problems, each of which is demarcated by specific conditions expressed in terms of the variables involved. From this point of view, it is more appropriate to consider the system as an ordinary goal-oriented system with a two-level goal-seeking element, as illustrated in Figure 7.5b.

Observe that both block diagrams in Figure 7.5 are the same as far as couplings between the three elements and environment are concerned. They differ solely in the way in which the elements are conceptually combined into larger elements. This subtle difference is an illustration of a basic phenomenon associated with systems of all kinds: the same system can be considered from different viewpoints when subjected to some operation of coarsening or refinement.

Goal-oriented systems with a k-level goal-seeking element ($k > 2$) can be defined recursively on the basis of the block diagram in Figure 7.5b. Any such system represents a hierarchical decomposition of the overall goal into subgoals, and consists of k levels of decomposition.

Let us now consider goal-oriented systems that adapt to changes in the behavior function ${}^1f$. The goal-seeking element of any such system must be able to perform the following two tasks:

i. to process data associated with variables in the sets X, ${}^1V$, ${}^2V$ and form a model of system ${}^1\mathbf{F}$; and
ii. to employ the model of ${}^1\mathbf{F}$ for generating the goal-seeking variables in such a manner that their positive effect toward the goal reaches its maximum or, at least, is close to the maximum.

Task (i) can be performed in many different ways. Methods discussed in Chapter 3 are directly relevant for this purpose. However, it is more appropriate to view the goal-

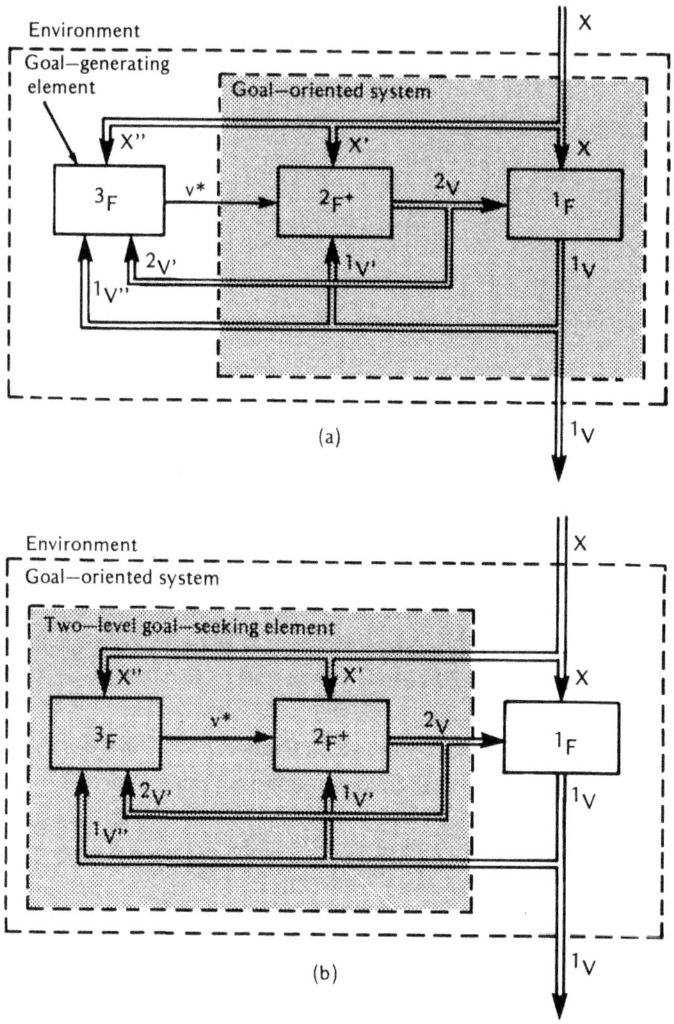

Figure 7.5. General structure paradigm of a goal-oriented system viewed as either (a) an autonomous multigoal-oriented system, or (b) a goal-oriented system with two-level goal-seeking element.

implementing element and its model as a metasystem. Then, the goal-seeking element must be capable of identifying change in the sense discussed in Section 5.6.

Example 7.3. To illustrate the notion of goal-oriented systems that are adaptive to changes in the goal-implementation system, a sophisticated adaptive system is described in this example. The object on which the system is defined is a computer equipped with a mechanism that allows it to move within a square area that is divided into smaller squares in chessboard fashion. This area is called the *operating area* of the computer. Assume that 10 rows and 10 columns are distinguished in the operating area which are labeled by identifiers x and y, respectively ($x, y \in N_{0,9}$). Assume further that each square, which represents a particular location of the computer, is labelled by a single identifier

$$l = 10x + y \, (l \in N_{0,99}).$$

The computer also has extensive *error-correcting capabilities*. During its execution of computing jobs, it is able to identify malfunctions in hardware, regardless of the kind of disturbances in the environment that caused them. It can also eliminate, within certain limits, the effect of malfunctions on the proper execution of the computing jobs. When the number of hardware malfunctions becomes too large, beyond the error-correcting capabilities of the computer, normal operation of the computer is threatened. It is thus desirable to counter the unknown disturbances by integrating the computer into an adaptive goal-oriented system whose goal is to minimize the number of hardware malfunctions in the long run. Since the computer can move, a natural way of countering external disturbances is to move selectively within the operating area according to some strategy designed to seek this goal.

A general block diagram of the proposed goal-oriented system is shown in Figure 7.6. Two variables, which are observed in time, are involved at this level:

m—goal-implementing variable, which represents the number of malfunctions that occurred in the computer during each defined period of time ($m \in \mathbb{N}_0$);
c—goal-seeking variable, which specifies (controls) the location of the computer ($c \in N_{0,99}$).

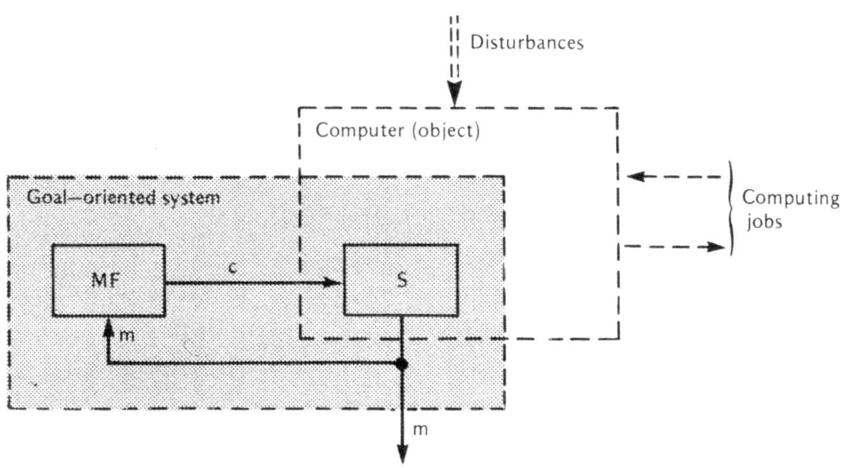

Figure 7.6. General block diagram of an adaptive goal-oriented system described in Example 7.3.

Since the nature of disturbances and their variability is not known, the goal-implementing element, in itself, can be viewed only as a source system **S**. It accepts at each time a state of variable c and generates a state of variable m. The manner in which m is generated is not known and may change.

The goal-seeking element monitors variable m and generates the goal-seeking variable c, which controls the movement of the computer. It is viewed as a metasystem

$$\mathbf{MF} = (T, \mathscr{F}, r),$$

where

$$\mathscr{F} = \{{}^l\mathbf{F} = ({}^l\mathbf{S}, {}^lM, {}^lf) \mid l \in N_{0,99}\},$$

T is defined in terms of regular time intervals during each of which the computer does not move, and r specifies that each state l of variable c determines uniquely a particular behavior element ${}^l\mathbf{F}$ of the metasystem. Elements ${}^l\mathbf{F}$ of the metasystem **MF** are thus replaced according to the replacement of states of the goal-seeking variable c.

A particular element ${}^l\mathbf{F}$ of the metasystem **MF** is based on the mask specified in Figure 7.7. It defines the input variable m, output variable c, and six internal variables. Variable v_l represents the total number of visits of the computer at location (square) l; variable m_l specifies the mean number of identified hardware malfunctions in the computer based on all visits at location l, and the remaining variables specify the same mean number for the four adjacent squares of square l; variables c', v_l', m_l' represent the next states of variables c, v_l, m_l respectively, and are the only generated variables.

Variables m_l' and v_l' are generated in a deterministic manner by the formulas

$$\begin{aligned} m_l' &= \frac{v_l m_l + m}{v_l + 1}, \\ v_l' &= v_l + 1; \end{aligned} \tag{7.9}$$

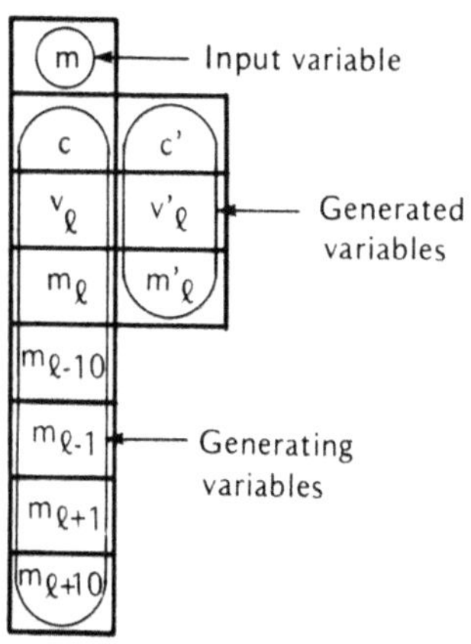

Figure 7.7. Masks ${}^lM$ of the behavior systems ${}^l\mathbf{F}(l \in N_{0,99})$ described in Example 7.3.

assume that m_l' is calculated to an accuracy of 0.01. Variable c' in system ${}^l\mathbf{F}$ can take states only from the set

$$L = \{l-10,\ l-1,\ l,\ l+1,\ l+10\}.$$

It is generated probabilistically in such a way that probabilities of the individual states are indirectly proportional to the mean number of malfunctions at the corresponding locations. However, to make the system adaptive to unexpected changes in the spatial distribution of disturbances, all of the states must be assigned nonzero probabilities. Hence, if $m_c\,(c \in L)$ is too large so that $1/m_c < 0.01$, then $1/m_c$ is set to α, where $\alpha > 0$ is a chosen constant; assume, e.g., $\alpha = 0.01$.

Behavior function ${}^l f$ of system ${}^l\mathbf{F}$ is determined in the following way. First, we define

$$q_c = \begin{cases} 1/m_c & \text{if } m_c \geq 0.01 \\ 0.01 & \text{if } m_c < 0.01 \end{cases}$$

for $c \in L - \{l\}$, and

$$q_l = \begin{cases} 1/m_l' & \text{if } m_l' \geq 0.01 \\ 0.01 & \text{if } m_l' < 0.01 \end{cases}$$

Then,

$${}^l f(c' \mid m_l', m_{l-10}, m_{l-1}, m_{l+1}, m_{l+10}) = q_c \cdot k$$

where $c' \in L$ and k is a normalization constant calculated by the formula

$$k = 1/(q_{l-10} + q_{l-1} + q_l + q_{l+1} + q_{l+10}).$$

For locations at the boundary of the operating area, obvious adjustments must be made in calculating the probabilities.

A block diagram indicating the main components involved in the goal-seeking metasystem **MF** is shown in Figure 7.8. Block 1 is a memory in which states of sampling variables of all behavior systems in F are stored. Its input l represents the replacement procedure: it reads from the memory states of those sampling variables that are pertinent to system ${}^l\mathbf{F}$ and guarantees that states of variables m_l' and v_l' are stored at a proper location in the memory. Block 2 represents Eqs. (7.9), which define the generation of variables m_l' and v_l'. Block 3 represents the determination of the auxiliary variables q_c $(c \in L)$, in terms of which the probabilities

$$P_{c'} = {}^l f(c' \mid m_l', m_{l-10}, m_{l-1}, m_{l+1}, m_{l+10}),\ c' \in L,$$

are calculated in block 4. Block 5 is a random generator by which a state of variable c' $(c' \in L)$ is generated according to the probability distribution. Block 6 is a delay that represents the intervals of time during which the computer does not move.

Before the described goal-oriented system is put into operation, its memory (block 1) may be filled with any information about the special distribution of

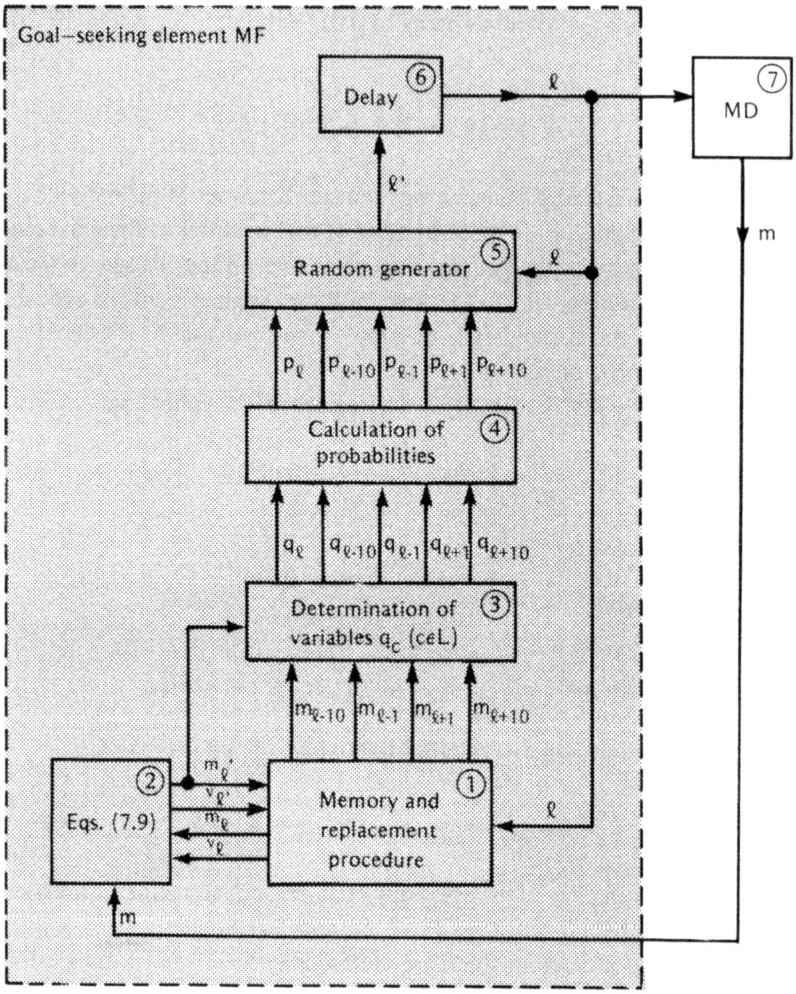

Figure 7.8. Block diagram of the goal-seeking element of the adaptive system described in Example 7.3.

malfunctions that is available at that time. If no information is available, then states of all variables in the memory are set to zero. Once put into operation, the goal-oriented system controls the movements of the computer according to its model of the computer environment (operating area). The model is represented by the content of the memory of the goal-seeking metasystem **MF** (block 1 in Figure 7.8). It is continuously updated by monitoring variables *l* and *m*. The system acts according to its anticipation of the effect of the environment on the proper operation of the computer. Systems of this kind, which are able to develop a model of their environment and use it in an anticipatory manner, are the most sophisticated adaptive goal-oriented systems. They are usually referred to in the literature as *anticipatory systems.*

Let us reflect now on the three aspects of the described adaptive system. First, the hardware malfunctions can be divided into three categories:

i. the malfunctions that originate from within the computer hardware itself (hardware defects);
ii. the malfunctions that are caused by disturbances distributed evenly within the operating area;
iii. the malfunctions that are caused by local disturbances within the operating area.

It is obvious that the goal-seeking variable (location in the operating area) can influence only malfunctions in category (iii).

Second, it should be emphasized that the described system reacts to any event which endangers its normal activity (correct execution of requested computing jobs). It does not require that the nature of such events be predetermined: everything that threatens its ability to operate normally evokes a reaction tending to preserve this ability. It is thus reasonable to call the system a *self-preserving system* (i.e., a system that attempts to preserve its ability to operate normally).

Third, the minimal value α of q_c ($c \in L$), which is fixed for a particular system, has an important effect on the way in which the system adapts to changes in the environment. If α is too small compared to the normal values of q_c, as determined by the malfunctions in categories (ii) and (iii), the system is slow in recognizing changes in the environment. If α is too large, the system is fast in recognizing changes, but its model of the environment is underutilized when the changes are substantially slower than the rate at which the computer can move.

7.7. AUTOPOIETIC SYSTEMS

> *An autopoietic system is organized (defined as a unity) as a network of processes of production (transformation and destruction) of components that produces the components that: (1) through their interactions and transformations continuously regenerate and realize the network of processes (relations) that produced them; and (2) constitute it (the machine) as a concrete unity in the space in which they exist by specifying the topological domain of its realization as such a network.*
>
> —FRANCISCO J. VARELA

The aim of this section is to describe a GSPS formulation of a rather unorthodox class of goal-oriented systems that are usually referred to as *autopoietic systems.* The term "autopoiesis" is of Greek origin and literally means "self-production." However, the self-production involved in autopoietic systems is not arbitrary, but must satisfy certain requirements.

In general, autopoietic systems function within a finite and discrete space-time support set. The space is usually two dimensionsal or three dimensional, but a k-dimensional space ($k \geq 1$, finite) can also be used. An overall characterization of autopoietic systems is that they form and maintain in time a spatially distinguished unit by a set of movement and production rules defined for entities of several different types distributed within the space.

Autopoietic systems are thus goal oriented. The goal is some kind of a boundary, usually called a *topological boundary*, that allows the observer to recognize a part of the space as a unit. Some of the movement or production rules that are essential for achieving this goal in a particular autopoietic system may thus be viewed as goal-seeking traits of the system.

The idea of autopoietic systems originated in biology, where it is exemplified by a great variety of instances. Some of the simplest biological objects whose formation and maintenance as spatial units can be well characterized in terms of autopoietic systems are biological cells, as well depicted by Milan Zeleny [ZE4]:

> We observe self-production phenomena intuitively in living systems. The cell, for example, is a complex production system, producing and synthesizing macromolecules of proteins, lipids, and enzymes, among others; it consists of about 10^5 macromolecules on the average. The entire macromolecular population of a given cell is renewed about 10^4 times during its lifetime. Throughout this staggering turnover of matter, the cell maintains its distinctiveness, cohesiveness, and relative autonomy. It produces myriads of components, yet it does not produce only something else—*it produces itself.* A cell maintains its identity and distinctiveness even though it incorporates at least 10^9 different constitutive molecules during its life span. This maintenance of unity and wholeness, while the components themselves are being continuously or periodically disassembled and rebuilt, created and decimated, produced and consumed, is called "autopoiesis."

An autopoietic system is usually described in terms of components of certain types, which are often given such suggestive names as "substrates," "catalysts," "holes," "links," and the like. At every instant of the defined time set, each recognized location of the defined space is occupied by exactly one component of a particular type. The components undergo spatial transformations in time according to specific rules (movement and production rules). If these rules are chosen properly, the system is capable of forming and maintaining a topological boundary of some kind (its goal) and may be thus considered as an autopoietic system.

One way of describing autopoietic systems in the GSPS language is to view them as metasystems of the form

$$\mathbf{MD} = (T, \mathscr{D}, r).$$

Set $\mathscr{D}$ in a particular autopoietic system consists of all data systems that are definable in terms of a specific space (as a support) and a single variable whose states represent components of the autopoietic system. Thus, for example, states 0, 1, 2, 3, etc., may represent holes, catalysts, substrates, links, etc., respectively. Support set T of the metasystem is the time set of the autopoietic system; it is totally ordered and its elements represent appropriate time intervals. The replacement procedure consists of all the movement and production rules of the autopoietic system. A replacement of one data system by another occurs always when one time in T is replaced by the next one.

Example 7.4. Let us describe a simple autopoietic system as an example of a metasystem. It consists of

- a time set T, which may conveniently be represented by the set of non-negative integers;
- a set D of data systems based on the same image system whose support is a two-dimensional space defined by two Cartesian coordinates $x, y \in N_{0,9}$ and which contains a single variable with five states, which are given the following suggestive names:
  0—hole,
  1—catalyst,
  2—substrate,
  3—link,
  4—bonded link.
- a replacement procedure r that is defined by the following rules (see Note 7.5):
  1. Two neighboring substrates either of which is in the neighborhood of a catalyst are joined to form a link.
  2. Neighboring links are joined to form bonded links. (A closed bonded link constitutes a boundary the maintenance of which is the goal of the autopoietic metasystem.)
  3. Randomly selected links, whether free or bonded, disintegrate, yielding two substrates or two links, respectively. (The resulting components may later rebond.)
  4. Substrates may move into any neighboring empty space, passing through a single chain of bonded links if necessary. Links may move into empty spaces and may also displace substrates, either pushing them into adjacent holes or trading positions with them. Catalysts have all the freedom of movement of links, and may displace them. However, unlike substrates, neither links nor catalysts may pass through bonded-link segments. Bonded links do not move.

Each time instant thus defines a distinct data system $\mathbf{D} \in \mathscr{D}$. The individual data systems are replaced in time by applying the rules of the replacement procedure. The procedure begins with an initial condition (data system) in which each location contains

2	2	2	2	2	2	2	2	2	2
2	2	2	2	2	2	2	2	2	2
2	2	2	2	2	2	2	2	2	2
2	2	2	2	2	2	2	2	2	2
2	2	2	2	2	2	2	2	2	2
2	2	2	2	1	2	2	2	2	2
2	2	2	2	2	2	2	2	2	2
2	2	2	2	2	2	2	2	2	2
2	2	2	2	2	2	2	2	2	2
2	2	2	2	2	2	2	2	2	2

t = 0

2	2	2	2	2	2	2	2	2	2
2	2	2	2	2	2	2	2	2	2
2	2	3	3	3	2	2	2	2	2
2	2	3	0	0	0	3	2	2	2
2	2	3	0	1	0	2	2	2	2
2	2	2	0	0	0	2	2	2	2
2	2	2	2	2	2	0	2	2	2
2	2	2	2	2	2	2	2	2	2
2	2	2	2	2	2	2	2	2	2
2	2	2	2	2	2	2	2	2	2

t = 1

2	2	3	2	2	2	2	2	2	2
2	0	2	2	4	2	2	2	2	2
2	2	3	2	4	4	4	2	2	2
2	2	4	0	3	0	3	2	2	2
2	2	4	0	1	0	4	2	2	2
2	2	4	4	4	4	4	2	2	2
2	2	2	2	2	2	3	2	2	2
2	2	2	2	2	2	2	2	2	2
2	2	2	2	2	2	2	2	2	2
2	2	2	2	2	2	2	2	2	2

t = 2

2	2	3	2	2	2	2	2	2	2
2	0	2	2	4	2	2	2	2	2
2	2	3	2	4	4	4	2	2	2
2	2	4	0	3	0	3	2	2	2
2	2	4	0	1	0	4	2	2	2
2	2	4	4	4	4	4	2	2	2
2	2	2	2	2	2	3	2	2	2
2	2	2	2	2	2	2	2	2	2
2	2	2	2	2	2	2	2	2	2
2	2	2	2	2	2	2	2	2	2

t = 3

2	2	3	2	2	2	2	2	2	0
2	2	2	2	4	2	2	2	2	2
2	2	4	3	4	4	4	2	2	2
2	2	4	0	3	0	4	2	2	2
2	2	4	2	1	3	4	2	2	2
2	2	4	4	4	4	4	2	2	2
2	2	2	2	2	2	0	2	2	2
2	2	2	2	2	2	2	2	2	2
2	2	2	2	2	2	2	2	2	2
2	2	2	2	2	2	2	2	2	2

t = 4

2	2	2	2	0	2	2	2	2	2
2	2	3	4	4	2	2	2	2	2
2	2	4	0	4	4	4	2	2	2
2	2	4	0	3	2	4	2	2	2
2	2	4	3	1	3	4	2	2	2
2	2	4	4	4	4	4	2	2	2
2	2	2	2	2	2	3	2	2	2
2	2	2	2	2	2	2	2	2	2
2	2	2	2	2	2	2	2	2	2
2	2	2	2	2	2	2	2	2	2

t = 5

2	2	2	2	2	0	2	2	2	2
2	2	4	4	4	2	2	2	2	2
2	2	4	2	4	4	4	2	2	2
2	2	4	3	3	2	4	2	2	2
2	2	4	0	1	3	4	2	2	2
2	2	4	4	4	4	4	2	2	2
2	2	2	2	2	2	3	2	2	2
2	2	2	2	2	2	2	2	2	2
2	2	2	2	2	2	2	2	2	2
2	2	0	2	2	2	2	2	2	2

t = 6

Figure 7.9 Illustration to Example 7.4 (autopoietic system).

either a substrate or a catalyst. The first seven time instants from one series of applications of *r* are given in Figure 7.9. The goal of this system is to form a closed space defined by bonded links (topological boundary). We can see in Figure 7.9 that the system is indeed goal oriented. The topological boundary begins to emerge at $t = 2$ and is completed at $t = 6$.

NOTES

7.1. The goal-oriented systems that have been studied most thoroughly are *regulators*. General principles of regulation are primarily due to Ross Ashby and Roger Conant [AS1, 2, CO2, 3, CO9]. *Ashby's law of requisite variety* is of particular significance [AS2, 3, PO4]. It states that the capacity of any physical device as a regulator cannot exceed its capacity as a channel of communication. It follows from this law that the variety in the regulating variables can be reduced to a desirable level (which is the goal of the regulator) only by an increase in the variety of the regulating variables to at least the appropriate minimum. The law is often paraphrased by the simple assertion: "only variety can destroy variety."

Hierarchically organized multilevel regulators were investigated by Aulin-Ahmavaara [AU1–2]. His results are expressed in its most general form by the following statement: "The weaker in average are the regulatory abilities and the larger the uncertainties of available regulators, the more hierarchy is needed in the organization of regulation and control to attain the same result of regulation, if possible at all." This statement is usually called the *law of requisite hierarchy*. It follows from this law that the lack of regulatory capability can be compensated for, to some degree, by conceptualizing the regulator as a hierarchical multigoal structure system.

In addition to these general principles of regulation, many results regarding regulation have been published under the label "*control theory*" [DI1, HS1, SA1, 2]. These are primarily results regarding feedback regulators and devoted to special classes of systems (continuous, linear).

7.2. An excellent survey of the various meanings given to the term "*adaptive system*" is in a paper by Brian Gaines [GA1]. For further study of adaptive systems, three books are recommended, written by Bellman [BE3], Holland [HO1], and Tsypkin [TS1]. Anticipatory systems are thoroughly covered in a large monograph by Robert Rosen [RO8].

7.3. The idea of *autopoietic systems* was proposed by three Chilean biologists, Humberto Maturana, Francisco Varela, and Ricardo Uribe in the early 1970s. Its first exposition in English was published in 1974 [VA5]. For further study of autopoietic systems, a book carefully edited by Milan Zeleny is recommended [ZE4], where the reader can find additional bibliographical information.

7.4. The self-preserving anticipatory system described in Example 7.3 was proposed by Antonin Svoboda in 1960 [SV3]. The idea is further developed, primarily through computer simulation, by several contributors; a summary can be found in Ref. [WI2].

7.5. The autopoietic system described in Example 7.4 is adopted from the original paper by Maturana, Varela, and Uribe (Note 7.3). For a more specific formulation of the replacement procedure, the reader is advised to consult this paper.

7.6. The literature on self-organizing systems is quite extensive. A survey of this subject area can be found in one of my papers [KL6]. Decision making viewed as a goal-oriented system is well discussed in Ref. [WH3].

EXERCISES

7.1. Consider a behavior system in some area of interest and describe it as a goal-oriented system, i.e., define a goal and a performance function, and identify possible goal-seeking variables.

7.2. Repeat example 7.2 for different goals and different performance functions.

7.3. Compare advantages and disadvantages of the feedback and feedforward structure paradigms of goal-oriented systems (Figure 7.2). Under which conditions is either of them preferable?

7.4. Simulate the anticipatory self-preserving system described in Example 7.3 on a computer and generate some scenarios.

7.5. Simulate the autopoietic system described in Example 7.4 on a computer and operate it for different initial data systems (consult also Refs. [VA5] and [ZE4]).

8

SYSTEMS SIMILARITY

> *Who has taught us the true analogies, the profound analogies which the eyes do not see, but which reason can divine? It is the mathematical mind, which scorns content and clings to pure form.*
>
> —Henri Poincare

8.1. SIMILARITY

> *If the parts composing an individual become greater or less, but in such proportion that they will preserve the same mutual relations of motion and rest, the individual will still preserve its original nature, and its actuality will not be changed.*
>
> —BENEDICT DE SPINOZA

In a common dictionary, the term *similarity* is typically defined as a quality of "having characteristics in common" or being "alike in substance or essentials" (*Webster's Third New International Dictionary*). According to this definition, two entities are considered similar if they are equal or, at least, comparable in some of their properties, but not necessarily in all of them. In addition, it is assumed that the properties in which the two entities are equal have some significance in a given context. Different kinds of similarities can thus be defined for a set of entities, depending upon the properties that are considered significant for a particular purpose.

Geometric similarity seems to be the first kind of similarity that was formulated and developed in a rigorous manner. It was defined by Euclid (third century B.C., in Volume VI of his *Elements*) as follows: "Those straight-sided geometric figures are called similar which have equal angles, and whose sides subtending equal angles are proportional." According to this definition, the two geometric figures shown in Figure 8.1 are similar because

i. their angles are equal, i.e.,

$$\alpha' = \alpha, \quad \beta' = \beta, \quad \gamma' = \gamma, \quad \delta' = \delta; \quad \text{and}$$

ii. their sides subtending equal angles are proportional, i.e.,

$$a' = ka, \; b' = kb, \ldots, e' = ke,$$

where k is a constant of proportionality. One figure is thus obtained from the other one by a simple linear transformation, which enlarges or reduces the latter, but does not distort it. Such a transformation is usually called a *distortionless linear transformation.*

The notion of geometric similarity of planar figures can be considerably generalized by allowing their angles to be subject to linear transformation as well. This

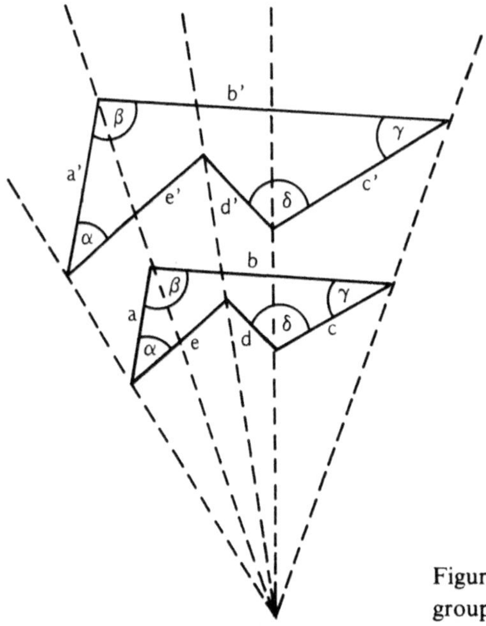

Figure 8.1. Simple geometric similarity based on a group of distortionless linear transformations.

can be expressed conveniently by representing planar figures by sets of points in a two-dimensional Cartesian space with coordinates x and y. A generalized geometric similarity of planar figures, based on a general linear transformation of the coordinates, is then represented by equations

$$\begin{aligned} x' &= k_{1,x}x + k_{2,x}y + k_{3,x}, \\ y' &= k_{1,y}x + k_{2,y}y + k_{3,y}, \end{aligned} \tag{8.1}$$

where the subscripted k's are constant coefficients; these coefficients must be such that Eqs. (8.1) have a unique solution for x and y, given x' and y'. Transformation (8.1), which can be extended to three-dimensional Cartesian space in an obvious way, is usually called a *general affine transformation*. It includes various special cases of transformations (and similarities), such as the distortionless transformation, symmetric reflection, one-dimensional enlargement or reduction, rotation, and the like.

Example 8.1 Consider a discrete Cartesian space $N_{0,7}^2$. Several planar figures that are all similar under the general affine transformation are shown in Figure 8.2. They represent various characteristic special cases: (a) the identity transformation (original figure); (b) symmetric reflection with respect to y and shift; (c) rotation and shift;

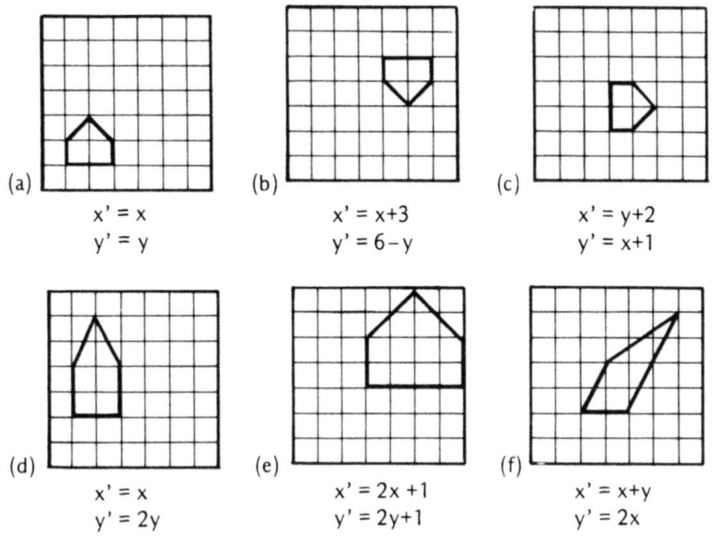

Figure 8.2. Examples of planar figures that are similar under the group of general affine transformations.

(d) enlargement along y; (e) enlargement and shift (geometric similarity); (f) a general type. An example of equations of form (8.1) that are not acceptable as an affine transformation are equations

$$x' = x + y,$$
$$y' = x + y.$$

Indeed, these equations do not provide a unique solution for x, y: any solution in which $x' = y'$ is acceptable independently of x, y.

The geometric tranformation (equal proportionality of all dimensions) preserves geometric objects completely except for their enlargement or reduction. It is thus reasonable to say that it defines a strong similarity on a set of geometric objects such as planar figures or solid bodies. The general affine transformation preserves geometric objects only approximately since it allows them to undergo a wide range of distortions; it may thus be said to define a weak similarity of geometric objects. The distortions permitted in the general affine transformation can be restricted in various ways. For example, we may require that all the coefficients in Eqs. (8.1) be equal to zero except $k_{1,x}$ and $k_{2,y}$. This type of transformation generalizes the notion of geometric similarity by allowing different constants of proportionality for different dimensions, but represents a less general kind of similarity than does the general affine transformation.

For purposes of replacing one element of a set with another element, similarity among these elements must be viewed as an *equivalence relation* defined on the set. As such, it can be expressed as a *partition* of the set, a *function* defined on it, or a *group of transformations* (one-to-one and onto functions) of the set to itself (automorphisms). In the case of geometric (distortionless) similarity, for instance, the set of all equilateral triangles, all squares, all circles, or all oblongs whose corresponding sides stand in a particular ratio are examples of equivalence classes of planar figures and, at the same time, blocks of the partition based on this kind of similarity. On the other hand, each particular constant of proportionality represents one element of the group of transformations that characterizes the similarity.

Although it is not necessary, it is usually by means of a group of transformations that an intuitive idea of similarity on a given set of entities is first formalized. Once this group is defined, an attempt is usually made to determine an operationally useful set whose elements represent the individual equivalence classes (particular blocks) of the equivalence relation imposed upon the set by the group of transformations. Such a set is usually called a *set of invariants* of the group. The equivalence relation can be then represented by an appropriate function from the set of entities on which the similarity is defined onto the set of invariants.

Consider, for example, the set of all structure systems that can be defined for a particular set of variables in the sense discussed in Section 4.7. The group of permutations of the variables imposes an equivalence relation on the set. An example of a set of invariants of this group is the set of all *unlabeled* block diagrams that have the same number of entries as the number of variables under consideration.

Different kinds of similarity defined on the same set can be partially ordered by their degrees of generality. Let one kind of similarity be a generalization of another kind if and only if the set partition corresponding to the former is a coarsening of the partition corresponding to the latter. For example, the geometric similarity that involves different constants of proportionality is less general than the similarity based on the group of general affine transformations; at the same time, it is more general than the geometric similarity based on the group of distortionless transformations.

If one entity is viewed as similar to another entity, then each preserves some properties of the latter under some transformation. In fact, the properties that for a given purpose it is desirable to preserve usually form the intuitive basis for defining an appropriate group of transformations. For example, if we want to preserve the weight of geometrically similar bodies made of different materials, the constant of proportionality c must satisfy the equation

$$c = (w/w')^{1/3},$$

where w, w' are weights per unit volume of the materials of any two bodies that qualify as similar.

8.2. SIMILARITY AND MODELS OF SYSTEMS

> *... there is no invariable one-way relationship between the model and the modelled, the symbol and the symbolized; each can be either. ... Nor is there a need for any model to be literally like the thing modelled.*
>
> —G. SPENCER BROWN

When a similarity relation is defined on a set of systems, it is usually referred to as a *modelling relation.* Two systems are similar if they preserve some common traits and can be converted to each other by appropriate transformations applied to other traits.

In systems problem solving, it is often an advantage (sometimes even a necessity) to deal with a problem in terms of a substitute system of some sort rather than the actual system for which the problem is formulated. The use of a suitable substitute system may be, for example, cheaper, faster, less dangerous, more convenient, easier to understand or control, more precise, less controversial, or better adjusted to the human scale. The two systems—the actual system and its substitute—must be similar in an appropriate and sufficiently strong sense with respect to the problem of concern.

Consider two systems, say **x** and **y**, that are similar under a set of transformations applied to some of their traits. Assume that **x** is the system under investigation and **y** is a desirable substitute. Then, **x** is called the *original system* (or just the *original*), **y** is called a *modelling system*, and **y** together with the relevant transformations is called a *model* of **x**. Since the similarity relation is symmetric (as any equivalence relation), we can just as well view system, **y** as the original system. Which of the two systems is viewed as the original system depends on the circumstances. Whether or not the other system is suitable as a model of the original system is decided solely on pragmatic grounds. It is a decision made by the user. He is likely to accept the model as a substitute for the original if, in his opinion, it has clear advantages over the original and, at the same time, it is not worse than any of the available competing models.

The term "model" is thus used in this book in connection with a particular relationship of one system to another system. It indicates that the two systems are similar in some sense and that one of them can replace the other one, under a suitable transformation, for some purpose. A modelling system attains its meaning as a model only if it is supplemented with a transformation that connects it in a desirable way with an original. In other words, each model requires an original. There may, of course, be different models of the same original.

In addition to this specific meaning of the term "model", which conforms well to our common-sense understanding of it and is almost universally accepted, the term is used in the literature to express several other concepts. The fact that the term has become so highly overworked, and liable to careless misuse, is unfortunate. The term enjoys at least three different meanings within the context of systems problem solving, in addition to the one reserved for it in this book.

First, the term "model" is frequently used for all entities that in this book are called systems, while the term "system" is used for our concept of an object. According to this terminology, a system is thus a part of the world that is a subject of some investigation and any representation (image) of it is viewed as its model. As a terminological consequence, the process of systems inquiry is referred to as systems modelling. In our terminology, all systems are viewed as abstractions, some of which are descriptions (images) of real-world phenomena, and we reserve the term "modelling" for a similarity relationship between systems.

The second meaning of the term "model" is used for a set of assumptions within which a problem is solved. Such assumptions are, for example, axioms of a mathematical theory that is employed for solving problems of some type. In our terminology, the term "paradigm" is used for this purpose.

The term "model" is also used in the literature for a system that is a simplified version of another system. Since the relationship of simplification is antisymmetric while the relationship of similarity is symmetric it is certainly desirable to distinguish these two concepts terminologically.

If we distinguish purely abstract systems, which have no physical interpretation (no observation channel), from interpreted systems (let us call them physical systems), we obtain four categories of modelling relationships, depending on the nature of the original system and modelling system:

Original system	*Modelling system*	*Category*
Physical	Abstract	I
Abstract	Physical	II
Physical	Physical	III
Abstract	Abstract	IV

Category I consists of mathematical models of all kinds. These models are based on accepted physical and other laws of nature. They make it possible to answer questions regarding physical systems by mathematical reasoning (symbol manipulation or numerical calculation) rather than by experimentation with their physical originals. For example, they allow to answer questions concerning artificial physical systems before they are actually implemented. In general, models in this category allow one to perform *gedanken experiments* (thought experiments) on mathematical models of hypothetical physical originals. An example: magnitudes of electric currents and voltages in a hypothetical electric circuits are calculated by solving appropriate algebraic or differential equations rather than by implementing the actual circuit and making relevant measurements on it.

Category II is best exemplified by computers of all kinds, whether analog, discrete, or hybrid. Also included are various special physical systems, each of which is designed as a universal model for mathematical systems of some particular class. Examples are linear analyzers (to deal with linear algebraic equations), polynomial analyzers (to deal with polynomial functions), or electrolytic tanks (to deal with partial differential equations). The use of models in this category consists in manipulating some variables of suitable physical systems and making measurements of other of their variables that represent solutions of some specific mathematical equations or answers to other mathematical questions.

Category III has an important role in engineering. The easiest to comprehend are scale models, which are systems that are simply enlarged or reduced at a certain scale with respect to the original. They are used, for instance, for investigating the dynamic properties of new designs of airplanes, helicopters, or rockets in wind tunnels, for testing new types of ships in special water tanks, for assessing various dam, bridge, and many other kinds of construction projects. Even though these cases are characterized by simple geometric similarity, results obtained for a scaled system must be subjected to appropriate transformations to be meaningful for its original. Indeed, when reducing (or enlarging) the linear dimensions of the original at a ratio of, say c, the areas (e.g., the area of the transverse or longitudial cross section of an aircraft wing) are reduced (or enlarged) at the ratio of c^2, and the volumes at the ratio of c^3. Problems associated with formulating proper transformations for models of this sort and their various generalizations are treated by the *theory of similarity or similitude.* Other examples of models in this category include computers simulated on other computers, aircraft simulators for training pilots, and medical simulators such as an artificial heart or artificial kidneys.

Category IV consists of models that are of great importance to applied mathematics. They are associated with various kinds of mathematical transformations (such as Laplace or Fourier transformations) with the aid of which mathematical systems of one kind (e.g., differential equations) are modeled by other mathematical systems (e.g., algebraic equations). Instead of dealing with the original, we can use a modelling system, which is usually considerably simpler, and apply the results obtained back to the original. In some cases (e.g., the Laplace transformation), detailed vocabularies have been compiled which specify the correspondence between the originals and the modelling systems.

A general introduction to the modelling relationship between systems has been given; the rest of this chapter is devoted to the formulation of basic types of models within the GSPS conceptual framework.

8.3. MODELS OF SOURCE SYSTEMS

> *Both the poet's metaphors and the scientist's abstractions discuss something in terms of something else. And the course of analogical extension is determined by the particular kind of interest uppermost at the time.*
>
> —KENNETH BURKE

Source systems are primitive in the sense that they contain no information about the relationship (constraint) among their variables. Given two neutral source systems, say **S** and **S′**, the only meaningful basis on which their similarity can be defined is to require that the properties recognized in each individual state set and support set (ordering, distance, etc.) be preserved under a transformation represented by one-to-one correspondences between

i. sets of variables of the two systems;
ii. sets of supports of the two systems;
iii. state sets of variables assigned by (i);
iv. support sets of supports assigned by (ii).

This basically means that systems **S** and **S′** are isomorphic in their general image systems, while they may be totally different in their semantic aspects (exemplification and observation channels).

► Formally, two neutral source systems

$$\mathbf{S} = (\mathbf{O}, \dot{\mathbf{I}}, \mathbf{I}, \mathcal{O}, \mathcal{E}),$$
$$\mathbf{S}' = (\mathbf{O}', \dot{\mathbf{I}}', \mathbf{I}', \mathcal{O}', \mathcal{E}'),$$

where

$$\mathbf{I} = (\{v_i, V_i) | i \in N_n\}, \{(w_j, W_j) | j \in N_m\}),$$
$$\mathbf{I}' = (\{v_i', V_i') | i \in N_n\}, \{(w_j', W_j') | j \in N_m\}),$$

are similar if and only if

i. v_i corresponds to $v_{p(i)}$ $(i \in N_n)$, where p denotes a permutation of N_n, i.e., p: $N_n \leftrightarrow N_n$;
ii. w_j corresponds to $w_{q(j)}$ $(j \in N_m)$, where q denotes a permutation of N_m, i.e., q: $N_m \leftrightarrow N_m$;
iii. for each $i \in N_n$, $V_i \leftrightarrow V_{p(i)}$ is a one-to-one correspondence under which all properties recognized in V_i are preserved in $V_{p(i)}$;
iv. for each $j \in N_j$, $W_j \leftrightarrow W_{q(j)}$ is a one-to-one correspondence under which all properties recognized in W_j are preserved in $W_{q(j)}$. ◄

If neutral source systems **S**, **S′** are similar in the sense that their image systems are isomorphic, then either of them can be viewed as the original system and the other as the modelling system. The modelling system together with the transformation expressed by the one-to-one correspondences (i)–(iv) is then viewed as a model of the other system, the original. It must be emphasized, however, that the modelling relationship at the level of source systems is of little practical use since it does not imply any similarity between the original and modelling system as far as the ways in which their variables are constrained. It has a pragmatic significance only when the source systems under consideration are components of some epistemologically higher types of systems. The notion of similarity is in such cases sharpened by additional requirements. Similarity of source systems may for some purposes be required as a necessary condition for a similarity of epistemologically higher types of systems, in which they are included, but it is never a sufficient condition for such similarity.

Example 8.2. Transpositions in music from one key to another are examples of similarities in source systems. Consider, for instance, the source system **S** defined in Example 2.6 (Figure 2.8b) and another source system, say **S**′, in which each note of V_1 is replaced with the note one tone lower, each chord in V_3 is replaced with an equivalent chord one tone lower, and everything else is exactly the same as in system **S**. Assume further that the variables representing pitch, rhythm, and harmony in one of the systems are assigned to the variables representing the same attributes in the other system and that all one-to-one correspondences between the respective state sets and support sets are identity mappings. Then, system **S** ad **S**′ can be viewed as similar under the identity transformation of their image systems. Given any musical score that can be described within **S**, the identity transformation transposes it into a similar score described in terms of **S**′. These correspondences are meaningful for the source systems since the musical scores are their secondary traits. If, however, **S** is a part of some data system **D** and **S**′ is a part of another data system **D**′, then the similarity between **S** and **S**′ is not sufficient for a meaningful similarity between the data systems **D**, **D**′. Indeed, data **d**, **d**′ of systems **D**, **D**′, respectively, may be any scores describable in the respective source system with no recognizable similarity at all.

Consider now two directed source systems, say $\hat{\mathbf{S}}$ and $\hat{\mathbf{S}}'$, as an original system and a modelling system, respectively. Since each variable of these systems is declared as an input variable or an output variable, its source of control (either the system itself or its environment) is uniquely determined. As a consequence, the one-to-one correspondences between the sets of variables and state sets of $\hat{\mathbf{S}}$ and $\hat{\mathbf{S}}'$, which are required for a similarity between neutral source systems, can be replaced by input and output mappings (not necessarily one-to-one) such that it is guaranteed that each variable is controlled from one source only. This condition of control uniqueness (similar to the one required for structure systems, as discussed in Chapter 4) implies that the input mappings between sets of variables are oriented from $\hat{\mathbf{S}}'$ to $\hat{\mathbf{S}}$ and the output mappings between sets of variables are oriented from $\hat{\mathbf{S}}$ to $\hat{\mathbf{S}}'$. Mappings between the state sets are oriented exactly the other way around.

Since the mappings between elements of $\hat{\mathbf{S}}$ and $\hat{\mathbf{S}}'$ are not required to be one-to-one, the modelling relationship between directed source systems is not symmetric; it is a quasiordering (reflexive and transitive relation) defined on any set of directed source systems. This means that the capability of $\hat{\mathbf{S}}'$ to serve as a modelling system of $\hat{\mathbf{S}}$ does not imply that $\hat{\mathbf{S}}$ can serve as a modelling system of $\hat{\mathbf{S}}'$ as well. This means, in turn, that instead of being isomorphic, the image systems of $\hat{\mathbf{S}}$ and $\hat{\mathbf{S}}'$ are connected to one another by two homomorphic relations, one associated with input variables and the other with output variables, and an isomorphic relation between their supports. $\hat{\mathbf{S}}$ is a homomorphic image of $\hat{\mathbf{S}}'$ with respect to input variables, while $\hat{\mathbf{S}}'$ is a homomorphic image of $\hat{\mathbf{S}}$ with respect to output variables; they are isomorphic with respect to their supports.

► Formally, a directed source system

$$\hat{\mathbf{S}}' = (\hat{\mathbf{O}}', \hat{\hat{\mathbf{I}}}, \hat{\mathbf{I}}', \hat{O}', \hat{E}')$$

is a modelling system of another directed source system

$$\hat{\mathbf{S}} = (\hat{\mathbf{O}}\ \hat{\hat{\mathbf{I}}}, \hat{\mathbf{I}}, \hat{O}, \hat{E}),$$

where

$$\hat{\mathbf{I}} = (\{(v_i, V_i) \mid i \in N_n\}, u, \{(w_j, W_j) \mid j \in N_m\}),$$
$$\hat{\mathbf{I}}' = (\{(v'_k, V'_k) \mid k \in N_{n'}, u', \{(w'_j, W'_j) \mid j \in N_m\}),$$

if and only if

i. for all $k \in N_n$, such that $u'(k) = 0$, variable v'_k represents variable $v_{p_0(k)}$, where

$$p_0 : Y_0 \to X_0$$

is a function defined on the sets

$$X_0 = \{i \mid i \in N_n, u(i) = 0\}$$

and

$$Y_0 = \{k \mid k \in N_{n'}, u'(k) = 0\}$$

of input variables of $\hat{\mathbf{S}}$ and $\hat{\mathbf{S}}'$, respectively;

ii. for each $k \in N_{n'}$, mapping $v_{p_0(k)} \to V_k$ is homomorphic with respect to the recognized properties in $v_{p_0(k)}$;

iii. for all $i \in N_n$ such that $u(i) = 1$, variable v_i represents variable $v_{p_0(i)}$, where

$$p_1 : X_1 \to Y_1$$

is a function defined on the sets

$$X_1 = \{i \mid i \in N_n, u(i) = 1\}$$

and

$$Y_1 = \{k \mid k \in N_{n'}, u'(k) = 1\}$$

of output variables of $\hat{\mathbf{S}}$ and $\hat{\mathbf{S}}'$ respectively;

iv. for each $i \in N_n$, mapping $V_{p_1(i)} \to V_i$ is homomorphic with regard to properties recognized in V_i;

v. w_j corresponds to $w'_{q(j)}$ $(j \in N_m)$, where q denotes a permutation of N_m;

vi. for each $j \in N_m$, the one-to-one correspondence $W_j \to W'_{q(j)}$ is isomorphic with respect to properties recognized in W_j.

System $\hat{\mathbf{S}}'$, together with the mappings (i)–(vi) is a model of $\hat{\mathbf{S}}$.Let the pairs of mappings be given the following names:

input mappings: (i), (ii);

output mappings: (iii), (iv);

support mappings: (v), (vi).

Some examples of these mappings are discussed in the context of models of directed systems of higher epistemological types. ◄

8.4. MODELS OF DATA SYSTEMS

> *The essence of modelling lies in establishing relations between pairs of systems.*
>
> —BERNARD P. ZEIGLER

It is easy to extend the notion of models of source systems to models of data systems. Consider two neutral data systems

$$\mathbf{D} = (\mathbf{S}, \mathbf{d}),$$
$$\mathbf{D}' = (\mathbf{S}', \mathbf{d}'),$$

and let **D** be viewed as an original system. Then, **D**′ qualifies as a modelling system of **D** if:

i. there exists a transformation (a set of one-to-one correspondences) between **S** and **S**′ under which **S**′ becomes a modelling system of **S** (Section 8.3); and
ii. **d** is preserved in **D**′ under the modeling relationship between **S** and **S**′.

A model of data systems is thus a model of its source system that, in addition, preserves its data. In other words, if **D**, **D**′ are an original data system and a modelling data system, then they are isomorphic (under appropriate one-to-one correspondences) in their image systems and with respect to data.

Example 8.3. Consider a melody whose score is given in Figure 8.3a. As in Example 2.6 (Figure 2.8), let the melody be described as a data system **D**′. Assume that the time set and variable representing the rhythm, say variable v'_1, are defined exactly as in Example 2.6. Assume further that variable v'_2 represents the pitch as defined in Figure 8.3b. Let **D** denote the data system defined in Example 2.6 without variable v_3 (harmony). Then, it can be easily verified that **D**′ is a modeling system of **D** (and vice versa) under the following one-to-one correspondences:

i. v'_1 corresponds to v_2 and v'_2 corresponds to v_1;
ii. time t' corresponds to time t;

Figure 8.3. Illustration of a modelling data system (Example 8.3).

iii. $V'_1 \leftrightarrow V_2$ is an identity mapping and $V'_2 \leftrightarrow V_1$ is defined by the equations

$$v'_2 = \begin{cases} 19 - v_1 + 1 & \text{for } v_1 \neq 0 \\ 0 & \text{for } v_1 = 0 \end{cases}$$

iv. $T' \leftrightarrow T$ is an identity mapping between the two time sets.

In musical terminology, the relationship between the two melodies would be described as an inversion combined with a transposition. The melody of **D**′ is obtained from the one of **D** by inverting it and transposing the result by one full tone down and, similarly, the melody of **D** is obtained from the one represented by **D**′ by inverting it and transposing the results by one full tone up.

For some purposes, it may be sufficient to define models of data systems in a looser sense by withdrawing the requirement of a modelling relationship between **S** and **S**′. In such cases, the modeling relationship between **D** and **D**′ is expressed in terms of one-to-one correspondences $\mathbf{V} \leftrightarrow \mathbf{V}'$ and $\mathbf{W} \leftrightarrow \mathbf{W}'$ between sets of overall states and overall

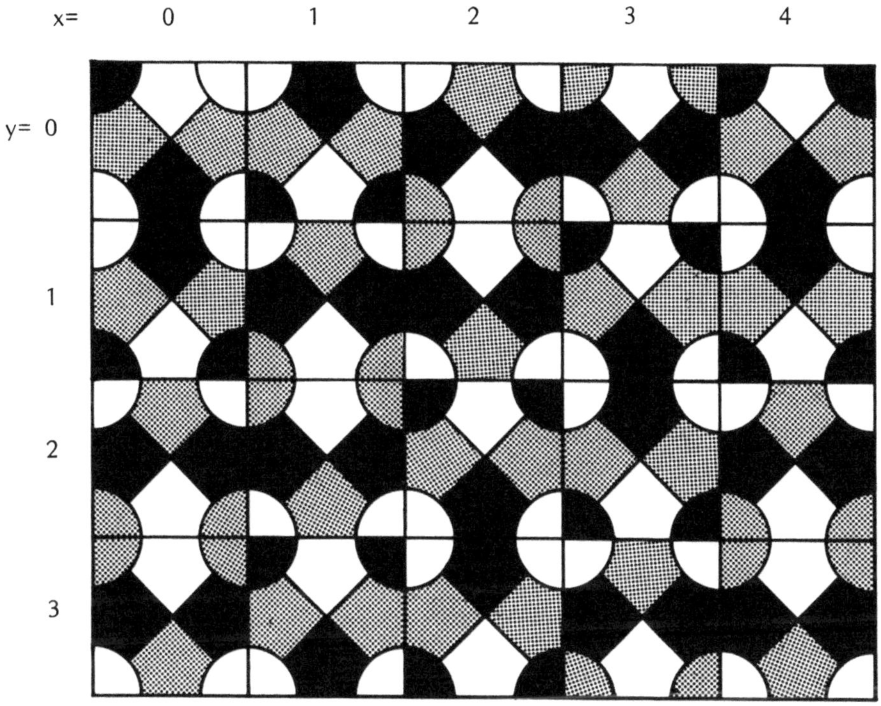

(a)

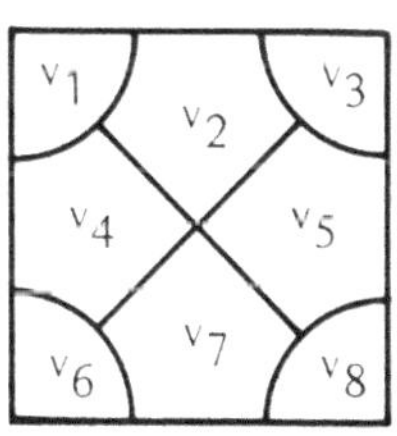

(b)

t=	0	1	2	3	
$v'_{1,t}=$	0	0	1	1	....
$v'_{2,t}=$	0	1	1	0	....

1 period

(c)

v_1	v_2	v_3	v_4	v_5	v_6	v_7	v_8	v'_1	v'_2
B	W	B	G	G	W	B	W	0	0
W	B	W	G	G	B	W	B	0	1
W	G	W	B	B	G	W	G	1	1
G	W	G	B	B	W	G	W	1	0

(d)

Figure 8.4. Illustration of a global data modelling (Example 8.4).

support instances of the two data systems, respectively, under which the data are preserved.

It should be emphasized that the applicability of models of this sort is somewhat restricted since they do not involve specific relationships between individual variables and supports of the two systems. To characterize their global relationship, let these models be called *global data models*.

Example 8.4. Consider the mosaic shown in Figure 8.4a. It can be described as a data system **D** with the following components:

- a support that consists of a two-dimensional space x, y ($x \in N_{0,4}$, $y \in N_{0,3}$) represented by a rectangular area divided regularly into 20 squares in the chessboard fashion as indicated in Figure 8.4a;
- eight variables v_i ($i \in N_8$), each of which describes the color in a particular subarea of each square as specified in Figure 8.4b;
- state sets of all the variables are equal and consist of three colors: black—B, white—W and grey—G;
- data **d**, which represent formally the mosaic: a 4×5 data matrix whose entries are 8-tuples of symbols B, W, G that represent color combinations in the individual squares of the support set.

Consider another data system, say **D**′, which consists of two-state variables v_1, v_2 observed in time t. Its data **d**′ are periodic and one period is shown in Figure 8.4c. The reader can easily verify that data **d** are preserved in data **d**′ under the transformation

$$t = x + 5y \qquad (t \leq 19)$$

between the support sets and the transformation specified in Figure 8.4d for the state sets of the two systems.

8.5. MODELS OF GENERATIVE SYSTEMS

> *The great danger of analogy is that a similarity is taken as evidence of an identiy.*
>
> —KENNETH BURKE

As data are required to be preserved in models of data systems, it is required that behavior or ST-functions be preserved in models of generative systems. However, there is no reason to require that the masks of behavior systems be preserved in their models. Indeed, the same results can often be obtained either by support sequences of states of one variable or by simultaneous states of several variables. This fact is best exemplified by the various units of serial and parallel computers (e.g., serial and parallel adders).

If transformations in masks between original and modeling generative systems are permitted, they replace transformations between their source systems. This implies that source systems of the original and modeling generative systems are related only by their supports and support sets. There are, of course, special cases, in which transformations between masks are the same as transformations between variables of source systems (e.g., memoryless behavior systems).

As shown in Chapter 3, each ST-system can be uniquely converted to an isomorphic behavior system. The discussion of models of generative systems may thus be restricted, without any loss of generality, to behavior systems. This also enables us to simplify the notation by excluding subscripts B and S, which are normally used to distinguish between behavior systems and ST-systems.

Consider now two neutral behavior systems

$$\mathbf{F} = (\mathbf{S}, M, f),$$
$$\mathbf{F}' = (\mathbf{S}', M', f'),$$

and let $\mathbf{F}$ be viewed as an original system. Then, $\mathbf{F}'$ qualifies as a modelling system of $\mathbf{F}$ if

i. there exist one-to-one correspondences between supports and support sets in $\mathbf{S}$ and those in $\mathbf{S}'$ under which recognized properties of the supports are preserved, as required in the definition of a similarity between source systems (Section 8.3);
ii. there exists a one-to-one correspondence $M \leftrightarrow M'$ between the masks of the two systems;
iii. for each sampling variables s_i of $\mathbf{F}$ that is assigned to a sampling variable s_k of $\mathbf{F}'$ ($i, k \in N_{|M|}$) by (ii), there exists a one-to-one correspondence $S_i \leftrightarrow S_k$ between their state sets under which all properties recognized in S_i are preserved in S_k;
iv. f is preserved in $\mathbf{F}'$ under the transformation (one-to-one correspondences) (i)–(iii).

Let a system $\mathbf{F}'$ together with some one-to-one correspondences (i)–(iii) that preserve f be called a *strong-behavior model of* $\mathbf{F}$. This notion can be generalized for some purposes in certain specific ways, as discussed later in this section.

Example 8.5. A modelling relationship between two behavior systems (directed, probabilistic) is illustrated in Figure 8.5. It is assumed that the support sets of both systems are totally ordered, while no properties are recognized in the state sets. Source systems of the behavior systems are not described as they are irrelevant for describing the modelling relationship.

A one-to-one correspondence between sampling variables of the two behavior systems is specified by the connections with arrows on top of the tables of behavior functions. Transformations between state sets of the corresponding variables are expressed by the equations that label the connections. Also indicated in the figure is the implied transformation between overall states of the two systems. It is obvious that a necessary (but not sufficient) condition for a modelling relationship between two

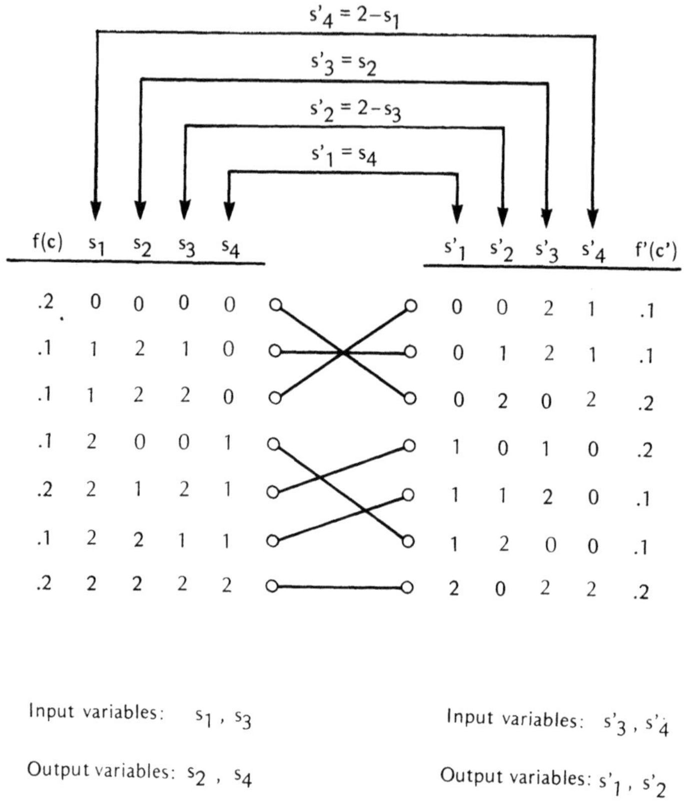

Figure 8.5. Modelling relationship of two behavior systems (Example 8.5).

behavior systems is that the distribution $f'(c')$ (probabilistic, possibilistic, or some other) is a permutation of the distribution $f(c)$. It is assumed that both systems are memoryless and that their directed behavior functions are derived from their neutral counterparts in Figure 8.5.

Example 8.6. Consider a cylindrical vessel that is connected to a conical vessel by a very thin pipe whose volume is negligible (see Figure 8.6). A memoryless and deterministic behavior system can be defined on this hydraulic object in terms of two

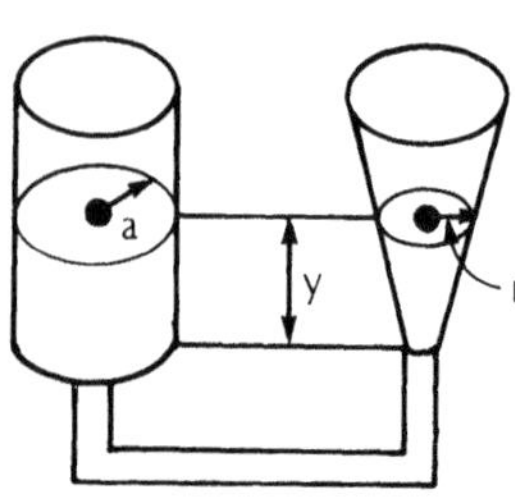

Figure 8.6. Hydraulic model of a mathematical behavior system (Example 8.6).

variables and a relationship between them that follows from the dimensions of the vessels. The variables are

x—the amount of water poured into the vessels, measured in cm^3 to an accuracy of $1\ cm^3$;

y—the height at which the water will settle, measured in cm to an accuracy of 0.1 cm.

It is obvious that the variables are observed in time.

Consider now a mathematical behavior system of two variables x' and y', whose behavior function $y' = f'(x')$ is defined by real number solutions of the equation

$$py'^3 + qy' = x',$$

where p, q are some constant coefficients. The hydraulic system can be used for appropriate dimensions of the vessels as a model of the mathematical system, i.e., it can be used for solving the cubic equation.

To show the modelling relationship between the two systems, let a and b denote the radius of the cylinder (in cm) and the ratio of the radius and altitude of the cone, respectively. Then, the amount of water in the cylinder is equal to

$$\pi a^2 y.$$

The amount of water in the cone is equal to

$$\tfrac{1}{3}\pi r_y^2 y,$$

where r_y is the radius of the cone corresponding to the altitude y (in cm). Since $b = r_y/y$, the amount of water in the cone can be expressed as

$$\tfrac{1}{3}\pi b^2 y^3.$$

The total amount of water in both vessels is equal to x so that

$$\pi a^2 y + \tfrac{1}{3}b^2 y^3 = x.$$

To convert this to the equation of the mathematical system, expressed in terms of variables x' and y', values a and b must be chosen such that they satisfy the equations

$$\pi a^2 = q$$
$$\tfrac{1}{3}\pi b^2 = p.$$

It follows immediately that the hydraulic system with

$$a = (q/\pi)^{1/2}$$

and

$$b = (3p/\pi)^{1/2}$$

is capable of solving the requested cubic equation since the relationship between x and y is expressed by exactly the same equation as the relationship between x' and y'.

When behavior systems are directed, the modelling relationship does not require one-to-one correspondences between sets of variables and state sets of the original and modelling systems. Input and output mappings can be employed instead, if desirable, as discussed in Section 8.3 in the context of directed source systems.

Example 8.7. A very simple example of the modelling relationship between two directed behavior systems is the principle of the slide rule: products of two real numbers are determined by adding appropriate distances on two rules. Input and output mappings of this well-known example are specified in Figure 8.7, which is self-explanatory.

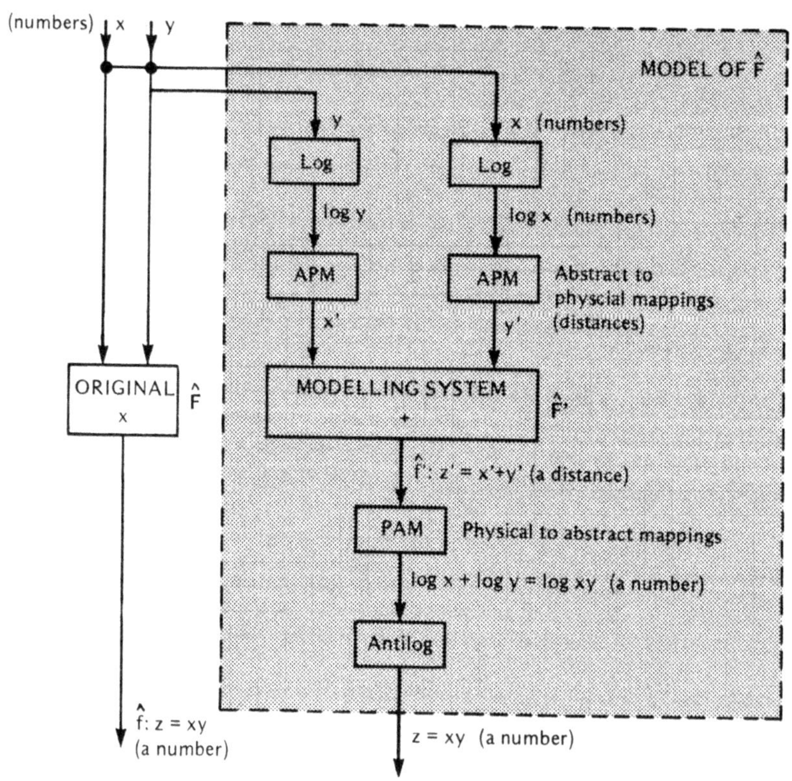

Figure 8.7. Slide rule as a model of behavior.

Examples 8.8. To illustrate a modelling relationship between directed behavior systems in which input mappings are not one-to-one, assume

- an original system that consists of input variables x, y, an output variable z, and the behavior function (deterministic, memoryless)

$$z = 3 \sin(x + y);$$

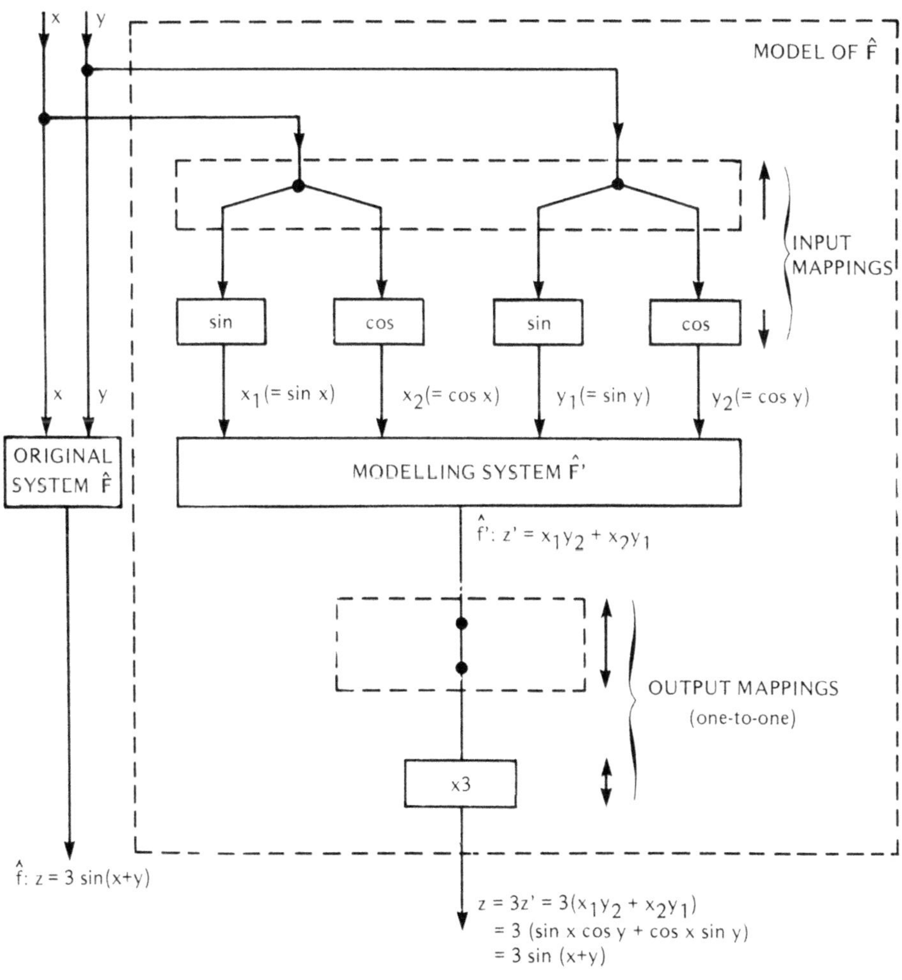

Figure 8.8. Modelling relationships between two directed behavior systems in which the input mappings are not one-to-one (Example 8.7).

- a modelling system that consists of input variables x_1, x_2, y_1, y_2, output variable z', and the behavior function

$$z' = x_1 y_2 + x_2 y_1.$$

Assume further that both systems are based on the same support.

Although the two behavior functions look quite different, the second system can be employed as a modelling system under proper input and output mappings. Such mappings are specified in Figure 8.8. Observe the inverted orientation of the mappings between the sets of variables and those between the respective state sets.

Example 8.9. Consider a simple electric circuit with two semiconductor diodes whose diagram is shown in Figure 8.9. It has two input variables x, y and one output variable z, all of which represent voltages. Assume that only two different voltage magnitudes are presented at the inputs, one of which is low and the other high with respect to the constant voltage V. Let these two magnitudes (states of the variables) be denoted by L and H, respectively. When applying each combination of these two voltages at the inputs and measuring the output variable, we would obtain the behavior function $\hat{f}$ specified in Figure 8.9b, where symbols L, H have the same meaning for the output variable and the input variables.

Assume now that we want to utilize the behavior system defined on the electric circuit for modelling logic functions. This can be done by introducing relevant propositions regarding the three physical variables and then using function $\hat{f}$ to determine the truth values of each proposition. Relevant propositions are of two kinds; either "variable x (or y or z) is in state L," or "variable x (or y or z) is in state H." The physical system may be utilized for modelling different logic functions depending on the combination of the two kinds of propositions that is applied to the three variables. For instance, if the first kind of proposition is applied to all three variables, then states L or H make the corresponding proposition true or false, respectively. Let T, F denote the truth and falsity, respectively, of the individual propositions. Then, the utilization of the physical system for modelling the logical function OR (disjunction) is illustrated in Figure 8.9c. All logic functions that can be modeled by the physical system are summarized in Figure 8.9d, where numbers 1 and 2 are labels of propositions of the first and second kind, respectively.

In a manner analogous to the notion of global data modelling we can also introduce the notion of *global behavior modelling*. It is defined in terms of a one-to-one correspondence between overall states of sampling variables of two behavior systems under which their behavior functions are preserved. That is, a global behavior modelling is an isomorphic relation between two sets of overall states.

Example 8.10. The relationship of global data modelling discussed in Example 8.4 (mosaic) can be easily reformulated as a relationship of global behavior modelling. Consider a two-column mask and let v_1'' and v_2'' be defined by the equations

$$v''_{1,t} = v'_{1,t+1} \quad \text{and} \quad v''_{2,t} = v'_{2,t+1}.$$

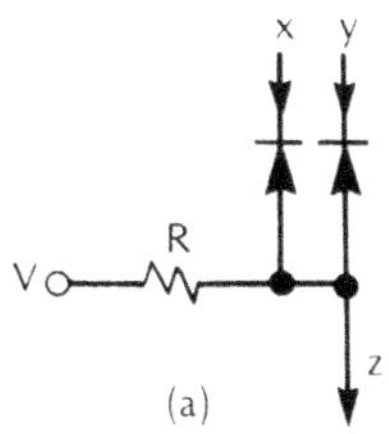

(a)

	x	y	z
	L	L	L
$\hat{f}$:	L	H	L
	H	L	L
	H	H	H

(b)

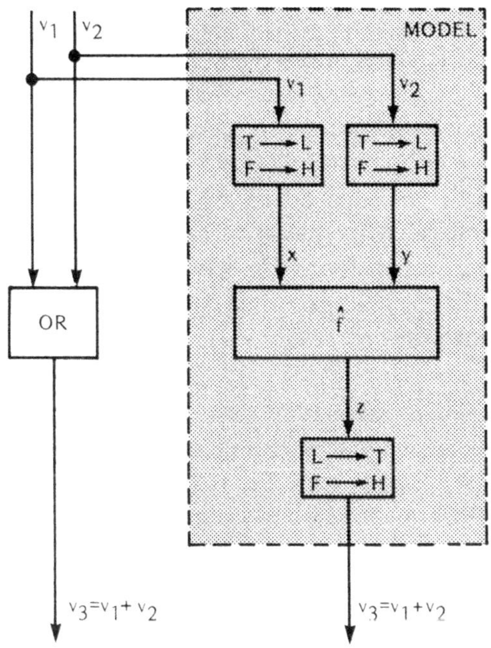

(c)

x	y	z	Logic function
1	1	1	OR
1	1	2	NOR
1	2	1	Implication ($v_2 \rightarrow v_1$)
1	2	2	Inhibition ($v_2 \nrightarrow v_1$)
2	1	1	Implication ($v_1 \nrightarrow v_2$)
2	1	2	Inhibition ($v_1 \rightarrow v_2$)
2	2	1	NAND
2	2	2	AND

(d)

Figure 8.9. Physical system capable of modelling several abstract systems (Example 8.9).

Then,

v'_1	v'_2	v''_1	v''_2
0	0	0	1
0	1	1	1
1	1	1	0
1	0	0	0

is the set of all overall states of the sampling variables (according to the data specified in Figure 8.4c). The system is deterministic: states of generated sampling variables v''_1, v''_2 are uniquely determined by states of the generating variables v'_1, v'_2. Under the one-to-one correspondence specified in Figure 8.4d, it is isomorphic to the system representing the mosaic, provided that the latter is reformulated as a behavior system in the same manner.

Modelling relationships between behavior systems can be generalized by replacing the transformation between masks by a transformation between locations (cells) of specific segments of data arrays under which the generated data are preserved for the same initial conditions. Let us illustrate this possibility by an example.

Example 8.11. Consider a modelling relationship between a serial binary adder (see Example 4.9) and a parallel binary adder. Assume that the adders are employed for adding pairs of binary numbers with 16 digits each. Then, the masks and behavior functions of the serial and parallel adders are specified in Figures 8.10a, b, respectively. Variables x, y (or x_i, y_i) represent digits of the numbers to be added, z (or z_i) is the sum digit, and c (or c_i) is the carry to the next place. The support is time in both cases.

The parallel adder is memoryless. It consists of 64 variables and adds two 16-digit binary numbers in one discrete time. The serial adder needs 16 successive discrete times to perform the same addition. The two numbers are represented in the serial adder by a time sequence of 16 states of the input variables x, y, and the sum is represented by a time sequence of 16 states of the output variable z, together with the last state of variable c, as illustrated in Figure 8.10c; the last state of c has the meaning of the most significant digit of the sum. Assume that immediately after one operation of addition is completed, another one begins with an appropriate initial condition (the carry is reset to zero).

To characterize a modelling relationship between the two adders, it is obviously not sufficient to define it solely in terms of one-to-one correspondences between their sampling variables, state sets, and support sets. A *combined transformation* that involves the sampling variables together with the support set must be introduced. Such a transformation is in this case a one-to-one correspondence between states of variables x_i, y_i, z_i, c_i at times t_p and states of variables x, y, z, c at times $t_s = i + 16(t_p - 1)$, respectively ($i \in N_{16}$), where t_p and t_s denote the times of the parallel and serial adders, respectively. We can see that if

$$x_{i,t_p} = x_{t_s} \quad \text{and} \quad y_{i,t_p} = x_{t_s}$$

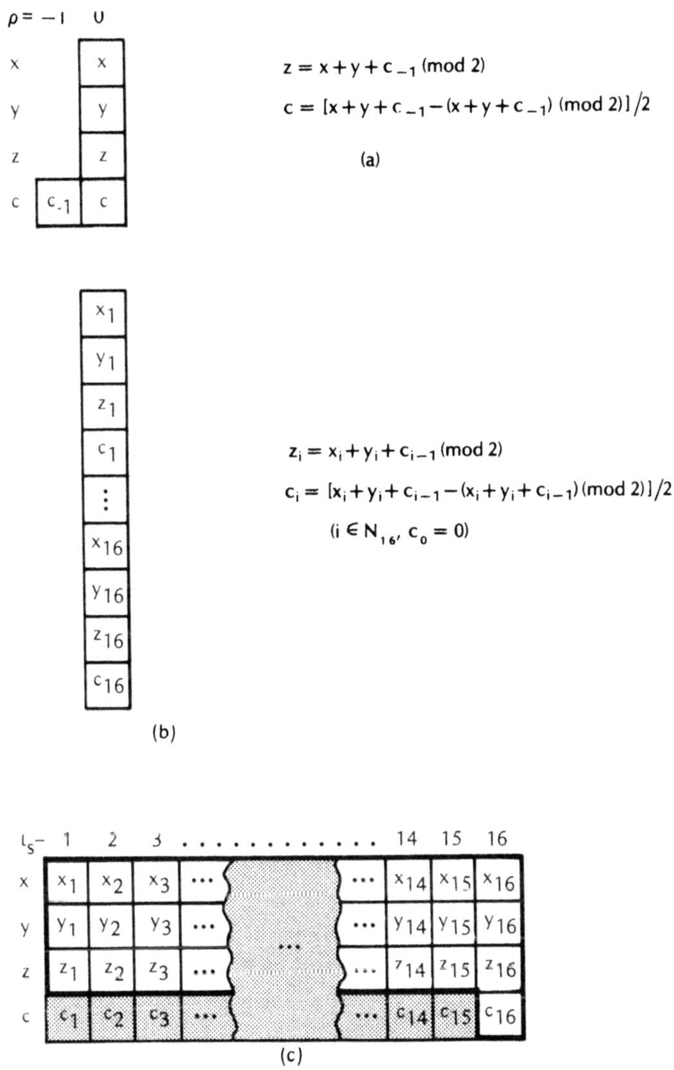

Figure 8.10. Modelling relationship between serial and parallel binary adders (Example 8.11).

for all $i \in N_{16}$, then

$$z_{i,t_p} = z_{t_s} \quad \text{and} \quad c_{i,t_p} = c_{t_s}$$

for all $i \in N_{16}$ under this combined transformation, i.e., the adders can replace each other to perform the same task. In fact, it is sufficient to satisfy the second equation only for $i = 16$ (the last carry).

The combined transformation, illustrated by the serial and parallel adders, involves a translation in the support set of one of the systems whose modelling relationship is under consideration. It introduces new variables in a manner analogous to the definition of sampling variables. Since these variables are introduced for the purpose of establishing a modelling relationship with another system, it is appropriate to call them *modelling variables.*

There are various ways in which a general model of behavior can be formally defined in terms of the combined transformation between modelling variables, along the lines suggested in Example 8.11. The reader should be able at this stage to develop his own favorite formulation.

8.6. MODELS OF STRUCTURE SYSTEMS

> *. . . although the modelling relation appears to be dyadic, it is in effect triadic; anything can be taken as a model of anything else if and only if we can sort out the relevant respects in which one entity is like another, the relevant properties which have both in common.*
>
> —MARX W. WARTOFSKY

As emphasized previously, the purpose of replacing one system (an original system) by another (a modelling system) is to gain some advantage in dealing with a problem regarding the original system. When one structure system is used for modelling another structure system, it is often sufficient that the constraint among some set of variables be preserved (under appropriate mappings between variables, supports, state sets, and support sets), and it is totally irrelevant whether or not the structure of the original system (its elements and couplings among them) is preserved in its model. The constraint of interest is obtained for both the original structure system and its model by a specific composing procedure, as explained in Chapter 4. Models of this sort are actually models of generative systems, even though the generative systems involved are derived from the corresponding structure systems.

Any two structure systems that have the same unbiased reconstruction in the reconstruction problem (Section 4.7) are examples of a modelling relationship of this kind. Another example is the modelling relationship between serial and parallel adders, which is discussed in Example 8.10.

A genuine model of a structure system must preserve not only the constraint among variables involved, but also the structure (i.e., elements and couplings) of the original. It is important to realize, however, that elements of the modelling system need not be the same as those of the original system. They are only required to satisfy the modelling relationship pertaining to the epistemological level at which they are defined.

Consider two structure systems

$$\mathbf{SX} = \{({}^iV, {}^i\mathbf{X}) \mid i \in N_q\},$$
$$\mathbf{SX}' = \{({}^jV', {}^j\mathbf{X}') \mid j \in N_q\},$$

where ${}^i\mathbf{X}$, ${}^j\mathbf{X}'$ denote source, data, or generative systems, either neutral or directed, and **X**, **X**′ are symbols of systems types. Let **SX** be viewed as an original system. Then, **SX**′ qualifies as a modelling system of **SX** if there exists a one-to-one correspondence between the elements of **SX** and **SX**′, say

$$z: N_q \leftrightarrow N_{q'},$$

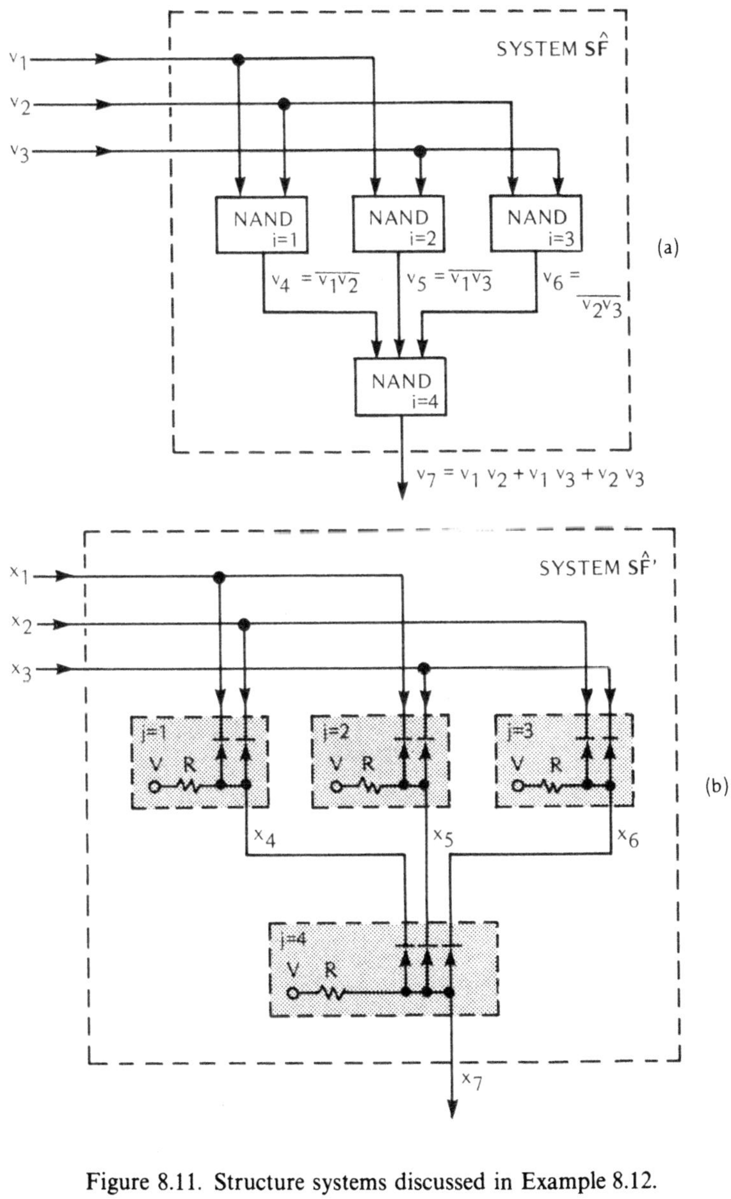

Figure 8.11. Structure systems discussed in Example 8.12.

under which

i. if $j = z(i)$, then ${}^{j}\mathbf{X}'$ together with appropriate mappings is a model of ${}^{i}\mathbf{X}$;
ii. all couplings of **SX** are preserved in **SX**′ under the mappings involved in the modelling relationships between elements ${}^{i}\mathbf{X}$ and ${}^{j}\mathbf{X}'$ $(i,j \in N_q)$.

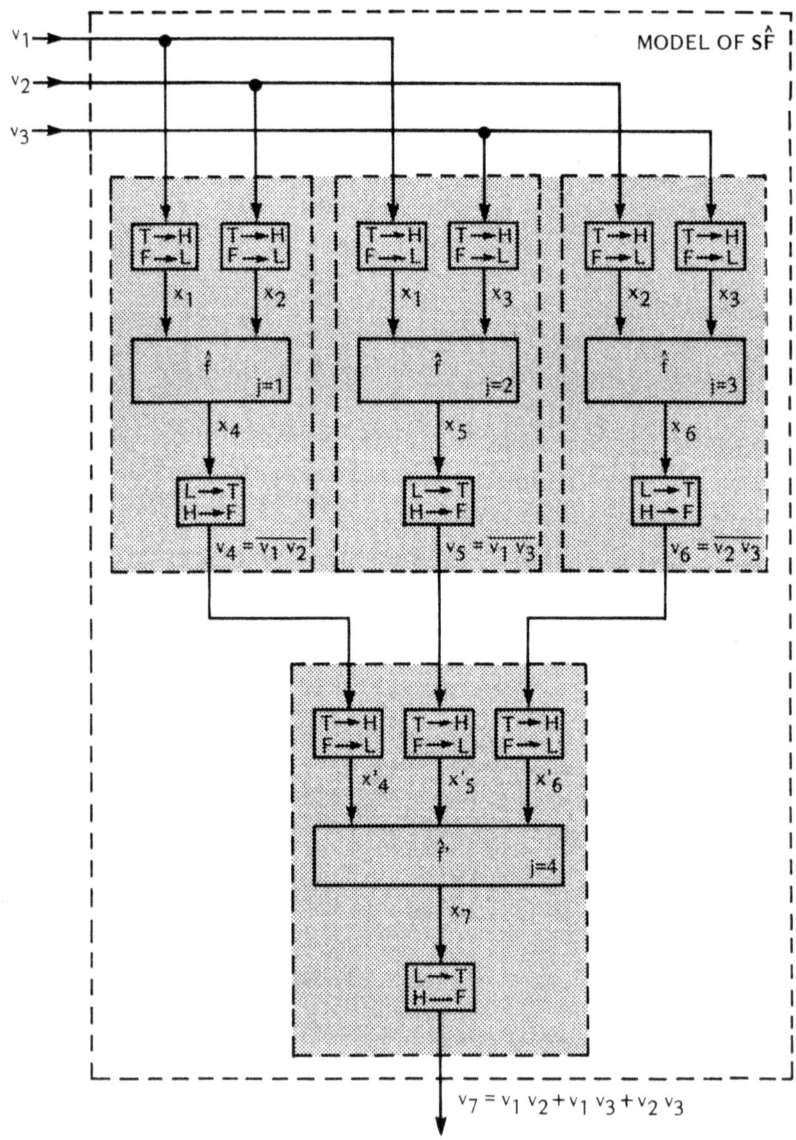

Figure 8.12. A model of the structure system specified in Figure 8.11a (Example 8.12).

Example 8.12. Consider a structure system $\hat{\mathbf{SF}}$ described by the block diagram in Figure 8.11a. Variables $v_1, v_2, \ldots, v_7$ represent propositions. Each of them has two states: true (T) and false (F). The system implements majority function (see Example 4.8): its output variable v_7 is in state T if and only if two or three of the input variables v_1, v_2, v_3 are in state T.

Consider now a structure system $\hat{\mathbf{SF}}'$ described by the block diagram in Figure 8.11b. Its elements are electric circuits with semiconductor diodes whose behavior is described in Example 8.9. This system can be used to model the structure system under the identity mapping

$$z: j = i,$$

provided that appropriate mappings are introduced between variables and state sets of the two systems by which elements of system $\hat{\mathbf{SF}}'$ become models of the corresponding elements of system $\hat{\mathbf{SF}}$, the full model of $\hat{\mathbf{SF}}$, which is based on system $\hat{\mathbf{SF}}'$ is described in Figure 8.12; $\hat{f}'$ denotes behavior function defined in Figure 8.9b and $\hat{f}$ denotes an extension of the same function for the three input variables (i.e., the output variable is in state H only if all input variables are in state H). It can easily be seen that

i. each element of $\hat{\mathbf{SF}}$ is represented by a model in the overall model of the structure system $\hat{\mathbf{SF}}$;
ii. all couplings between elements of $\hat{\mathbf{SF}}$ are preserved in terms of couplings between their models in the overall model of $\hat{\mathbf{SF}}$.

8.7. MODELS OF METASYSTEMS

> *. . . the use of models and so-called "analogies" in science is simply a change of language: one configuration is used to represent another.*
>
> —AARON SLOMAN

Although metasystems are suitable for modelling purposes in some situations, it is normally sufficient that the modelling metasystem preserve only the constraint among some variables (under appropriate mappings between variables, supports, state sets, and support sets) that is represented by the original metasystem. It is usually irrelevant whether or not the elements and replacement procedure of the original metasystem are also preserved in its model. If they are not preserved, the model is actually not a metasystem model, even though metasystems are involved.

A genuine model of a metasystem must preserve not only the constraint among the variables involved, but also the elements and replacement procedure of the original. Given two metasystems,

$$\mathbf{MX} = (\mathbf{W}, \mathscr{X}, r).$$
$$\mathbf{MX}' = (\mathbf{W}', \mathscr{X}', r'),$$

where $\mathscr{X}$, $\mathscr{X}'$ are sets of their elements, let **MX** be viewed as an original metasystem. Then, **MX**$'$ qualifies as a modelling metasystem of **MX** if there exists a one-to-one correspondence between $\mathscr{X}$ and $\mathscr{X}'$, say

$$z: \mathscr{X}' \leftrightarrow \mathscr{X}$$

under which

i. if $\mathbf{x}' = z(\mathbf{x})$, then $\mathbf{x}'$ together with appropriate mappings is a model of $\mathbf{x}$;
ii. the replacement procedure r is preserved in **MX**$'$ under a one-to-one correspondence between **W** and **W**$'$, and the mappings involved in the modelling relationships between elements in $\mathscr{X}$ and $\mathscr{X}'$ assigned to each other by z.

It is left to the reader as an exercise to generalize this definition to metasystems of higher order and to construct some examples of models of metasystems.

NOTES

8.1. For an introduction to the theory of similarity, a book by Gukhman [GU3] is recommended. More extensive coverage, with applications to heat conduction and diffusion, fluid dynamics, elastic deformation, and chemical reactions, can be found in a book by Szücz [SZ1]. Some additional references may also be useful [LA2, SK1].

8.2. It seems that the most natural mathematical formalism for dealing with the various kinds of systems similarities is offered by category theory [AR2, GO2, HE3]. This formalism is not pursued in this chapter since category theory has not been incorporated into general education as yet and, consequently, it would be unreasonable to assume that the reader is familiar with it. Interested readers may consult a book by Rosen [RO6] in which a category-theoretic formulation of systems similarity is outlined.

EXERCISES

8.1. Determine the affine transformation between
(a) two circles of different size;
(b) a circle and an elipse;
(c) a square and an oblong.

8.2. The position of the center of a metal ball suspended by a spring and subjected to external force (Figure 8.13a) is characterized by a function $x(t)$ of time that satisfies the differential equation

$$m\ddot{x} + r\dot{x} + ex = f(t),$$

where m, r, e denote the mass of the ball, coefficient of mechanical resistance, and elasticity of the spring, respectively, and $f(t)$ represents an input (externally impressed) force, acting

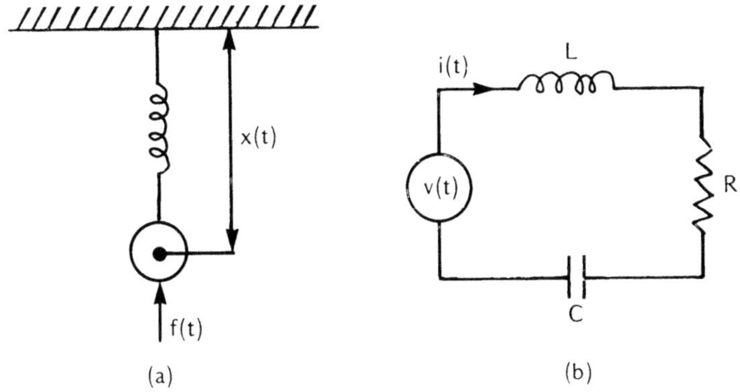

Figure 8.13. An example of similar systems (Exercise 8.2).

vertically upon the ball in time. Electric current in the circuit shown in Figure 8.13b is described by a function $i(t)$ of time that satisfies the differential equation

$$L\ddot{i} + R\dot{i} + i/C = v(t),$$

where L, R, C denote the inductance, resistance, and capacitance of the circuit, respectively; $v(t)$ represents an ideal source of voltage. Discuss the similarity between the two systems, in which f, v and x, i are viewed as input and output variables, respectively.

8.3. Explain the principle of the Laplace transform in terms of the model of behavior.

8.4. Formulate a modelling relationship between an appropriate source system for describing some sort of mosaic (e.g., Example 8.4) and a source system for describing musical scores (e.g., Example 2.8). Apply the modelling relationship for converting a specific mosaic into the corresponding musical score and vice versa.

8.5. Assume that the behavior functions of two directed, deterministic, and memoryless behavior systems $\hat{\mathbf{F}}_B$ and $\hat{\mathbf{F}}'_B$ are expressed by the tables

	v_1	v_2	v_3
	0	0	1
$\hat{f}_B$:	0	1	0
	1	0	0
	1	1	1

	x_1	x_2	x_3
	0	0	0
$\hat{f}'_B$:	0	1	1
	1	0	1
	1	1	0

respectively. Specify the input and output mappings under which

(a) system $\hat{\mathbf{F}}'_B$ becomes a modelling system of $\hat{\mathbf{F}}_B$;

(b) system $\hat{\mathbf{F}}_B$ becomes a modelling system of $\hat{\mathbf{F}}'_B$.

8.6. Change states of x_3 in Exercise 8.5 to obtain a system that cannot be used as a modelling system of $\hat{\mathbf{F}}_B$. How many such changes are there?

8.7. Let each of the following pairs of equations represent behavior functions of two directed behavior systems (deterministic and memoryless). Assume that output variables are on the left sides of the equations. For each of the pairs, find input and output mappings such that the latter system, together with the mappings, can be used to model the former:

(a) $v_3 = v_1 v_2$, $x_3 = (x_1 + x_2)^2 - (x_1 - x_2)^2$;

(b) $v_2 = 2\cos^2 v_1$, $x_2 = 1/(1 + x_1)$;

(c) $v_3 = av_1 + bv_2$, $x_3 = cx_1 + dx_2$;

8.8. A set of ST-functions is defined by the directed graphs in Figure 8.14. The edges shown represent all *possible* transitions in each case. Determine pairs of these ST-functions that can be used to model each other.

8.9. Two directed behavior systems, $\hat{\mathbf{F}}_B$ and $\hat{\mathbf{F}}'_B$, are defined on simple electric circuits as shown in Figure 8.15.

(a) Considering the resistors as elements, convert the behavior systems to the corresponding structure systems $\mathbf{SF}_B$ and $\mathbf{SF}'_B$.

(b) Show that $\mathbf{SF}'_B$ can model $\mathbf{SF}_B$ and vice versa.

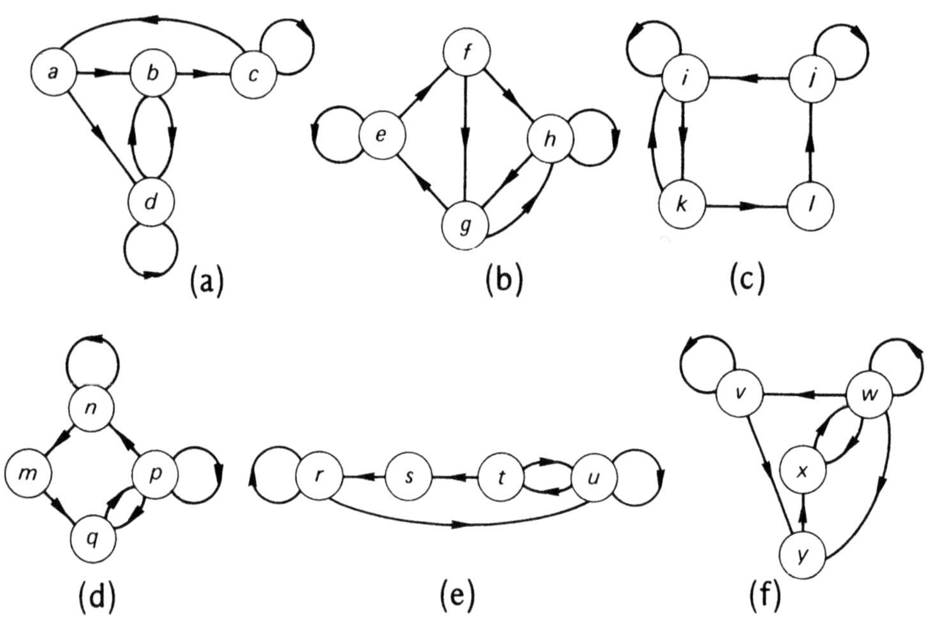

Figure 8.14. Illustration to Exercise 8.8.

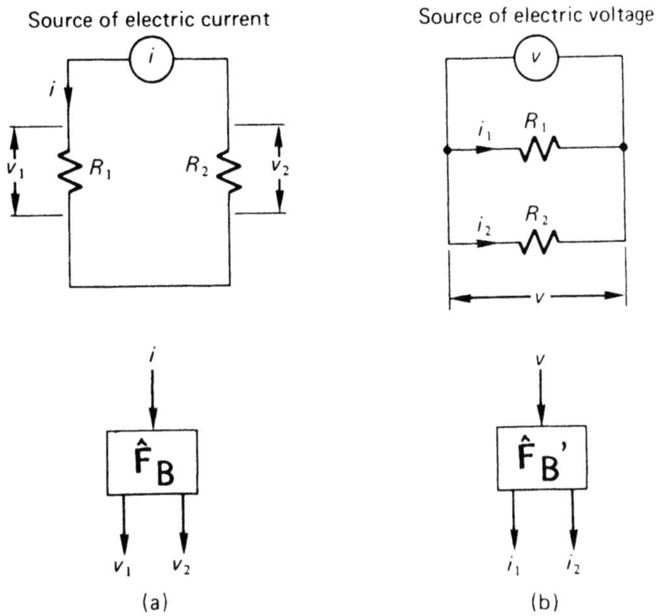

Figure 8.15. Illustrations to Exercise 8.9.

9

GSPS: ARCHITECTURE, USE, EVOLUTION

Perhaps it is true that nothing worth knowing can be taught—all the teacher can do is to show that there are paths.

—RICHARD ALDINGTON

9.1. EPISTEMOLOGICAL HIERARCHY OF SYSTEMS: FORMAL DEFINITION

> *Philosophy may be ignored but not escaped; and those who most ignore least escape.*
>
> —DAVID HAWKINS

▶ Epistemological types of systems that are recognized within the GSPS framework include source systems and their components (object systems, specific and general image systems), data systems, generative systems (behavior systems or ST-systems), structure systems of various types and levels, and metasystems of various types and levels. In addition, each of these systems types can be either neutral or directed.

The GSPS epistemological types of systems can be partially ordered. Given two of them, say types **x** and **x**′, we say that **x** is an *epistemologically lower systems type* than **x**′ if and only if

i. given any particular system of type **x**′, there exists a unique procedure based solely on this system and using all information available in it by which a single system of type **x** is derived for appropriate initial and other relevant conditions; and
ii. given a particular system of type **x**, there is no procedure by which a single system of type **x**′ can be derived solely from it, i.e., there is always some ambiguity and, consequently, some degree of arbitrariness in the determination of a system of type **x**′ from a given system of type **x**.

Given two epistemological systems type, **x** and **x**′, let the symbol

$$\mathbf{x} \precsim \mathbf{x}'$$

denote that type **x** is epistemologically lower than type **x**′. Then, the pair

$$\mathscr{H}_l = (\mathscr{E}_l, \precsim),$$

where $\mathscr{E}_l$ denotes the set of all GSPS epistemological systems types (either neutral or directed) such that the combined number of levels of structure systems and metasystems in any systems type does not exceed $l\,(l \in \mathbb{N})$, defines the *GSPS epistemological hierarchy of systems.*

For each $l \in \mathbb{N}$, $\mathscr{H}_l$ is a semilattice. If only homogeneous structure systems or metasystems are considered, $\mathscr{H}_l$ consists of

$$|\mathscr{E}_l| = 3\,(2^{l+1} - 1) \tag{9.1}$$

elements. An example for $l = 2$ is given in Figure 9.1. For an arbitrary finite l, $\mathscr{H}_l$ (restricted to homogeneous structure systems and metasystems) is defined by Table 9.1. Symbols

$$(\mathbf{S}^{s_i}\mathbf{M}^{m_i})^j\mathbf{X}$$

and

$$(\mathbf{M}^{m_i}\mathbf{S}^{s_i})^j\mathbf{X}$$

stand for the sequences

$$\mathbf{S}^{s_1}\,\mathbf{M}^{m_1}\,\mathbf{S}^{s_2}\,\mathbf{M}^{m_2}\cdots\mathbf{S}^{s_j}\,\mathbf{M}^{m_j}\,\mathbf{X}$$

and

$$\mathbf{M}^{m_1}\,\mathbf{S}^{s_1}\,\mathbf{M}^{m_2}\,\mathbf{S}^{s_2}\cdots\mathbf{M}^{m_j}\,\mathbf{S}^{s_j}\,\mathbf{X},$$

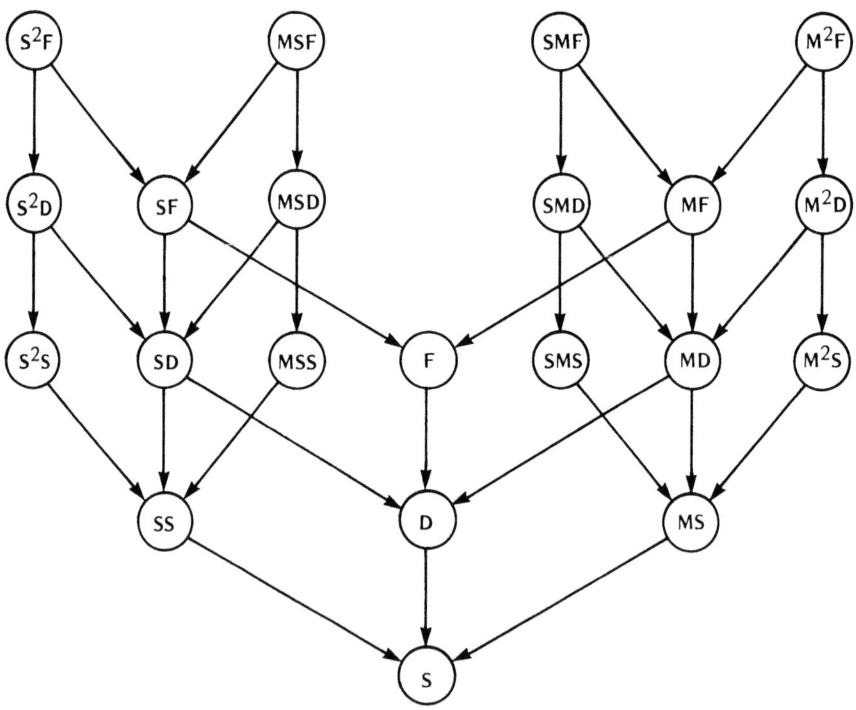

Figure 9.1. Semilattice of the GSPS epistemological systems types in which the combined number of structure systems and metasystems in any systems type is two or less: $\mathscr{H}_2 = (\mathscr{E}_2, \leq)$.

TABLE 9.1
Epistemological Hierarchy (Semilattice) of the GSPS Systems Types: $\mathscr{H}_l = (\mathscr{E}_l, \preceq)$.

Systems type	Immediate successor in the semilattice
$\mathbf{S}$	None
$\mathbf{D}$	$\mathbf{S}$
$\mathbf{F}$	$\mathbf{D}$
$(\mathbf{S}^{s_i}\mathbf{M}^{m_i})^j\mathbf{S}$	$(\mathbf{S}^{s_i-1}\mathbf{M}^{m_i})^j\mathbf{S}$ for a particular i, $(\mathbf{S}^{s_i}\mathbf{M}^{m_i-1})^j\mathbf{S}$ for a particular i
$(\mathbf{S}^{s_i}\mathbf{M}^{m_i})^j\mathbf{D}$	$(\mathbf{S}^{s_i-1}\mathbf{M}^{m_i})^j\mathbf{D}$ for a particular i, $(\mathbf{S}^{s_i}\mathbf{M}^{m_i-1})^j\mathbf{D}$ for a particular i, $(\mathbf{S}^{s_i}\mathbf{M}^{m_i})^j\mathbf{S}$
$(\mathbf{S}^{s_i}\mathbf{M}^{m_i})^j\mathbf{F}$	$(\mathbf{S}^{s_i-1}\mathbf{M}^{m_i})^j\mathbf{F}$ for a particular i, $(\mathbf{S}^{s_i}\mathbf{M}^{m_i-1})^j\mathbf{F}$ for a particular i, $(\mathbf{S}^{s_i}\mathbf{M}^{m_i})^j\mathbf{D}$
$(\mathbf{M}^{m_i}\mathbf{S}^{s_i})^j\mathbf{S}$	$(\mathbf{M}^{m_i-1}\mathbf{S}^{s_i})^j\mathbf{S}$ for a particular i, $(\mathbf{M}^{m_i}\mathbf{S}^{s_i-1})^j\mathbf{S}$ for a particular i,
$(\mathbf{M}^{m_i}\mathbf{S}^{s_i})^j\mathbf{D}$	$(\mathbf{M}^{m_i-1}\mathbf{S}^{s_i})^j\mathbf{D}$ for a particular i, $(\mathbf{M}^{m_i}\mathbf{S}^{s_i-1})^j\mathbf{D}$ for a particular i, $(\mathbf{M}^{m_i}\mathbf{S}^{s_i})^j\mathbf{S}$
$(\mathbf{M}^{m_i}\mathbf{S}^{s_i})^j\mathbf{F}$	$(\mathbf{M}^{m_i-1}\mathbf{S}^{s_i})^j\mathbf{F}$ for a particular i, $(\mathbf{M}^{m_i}\mathbf{S}^{s_i-1})^j\mathbf{F}$ for a particular i, $(\mathbf{M}^{m_i}\mathbf{S}^{s_i})^j\mathbf{D}$

respectively, where $\mathbf{X}$ denotes a systems type, and j, s_i, m_i $(i \in N_q)$ are some natural numbers; it is allowed that $m_j = 0$ in the former sequence and $s_j = 0$ in the latter sequence, and $\mathbf{S}^0\mathbf{X}$, $\mathbf{M}^0\mathbf{X}$ is interpreted as $\mathbf{X}$. It is required that

$$\sum_{i \in N_j} (s_i + m_i) \leq l. \blacktriangleleft$$

9.2. METHODOLOGICAL DISTINCTIONS: A SUMMARY

> *Too large a generalization leads to mere barrenness. It is the large generalization, limited by a happy particularity, which is the fruitful conception.*
>
> —ALFRED NORTH WHITEHEAD

Each class of systems characterized by an epistemological type is further classified by the methodological distinctions that are applicable to it. As explained previously (Chapters 2, 3), the aim of methodological distinctions is to distinguish systems that require different methods when involved in systems problems.

TABLE 9.2
A Summary of Essential Methodological Distinctions Recognized at the Architectural Level of the GSPS†

System trait	Methodological distinctions	Applicable to epistem. types of systems	Examples of possible extensions	Cross-references
Variables	Neutral Directed	All	Mixed	Sec. 2.5
Observation channels	Crisp Fuzzy	All	Mixed	Eqs. (2.2), (2.3) Eqs. (2.8)–(2.10)
State sets and support sets	No property Meaningful combinations of properties of ordering, distance and continuity	All	Interval scales Ratio scales Other scales	Secs. 2.3, 2.4 (Fig. 2.2, Table 2.1)
Data	Completely specified Incompletely specified With "don't care" entries	**D** or **XD**	—	Sec. 2.6
Data	Crisp Fuzzy	**D** or **XD**	Mixed	Eq. (2.21) Eq. (2.27)
Data	Periodic Aperiodic	**D** or **XD**	Mixed	Sec. 2.6
Mask	Memoryless Memory-dependent	**F** or **XF**	Compact	Sec. 3.2
Constraint function	Behavior function ST-function	**F** or **XF**	—	Sec. 3.8
Behavior or ST function	Deterministic Probabilistic Possibilistic	**F** or **XF**	Based on other subsets of fuzzy measures	Eqs. (3.14), (3.16) Eqs. (3.18), (3.19) Eqs. (3.20), (3.21)
Elements of structure systems	Consistent Inconsistent	**SXF**	—	Secs. 4.3, 4.4

† **X** denotes a sequence of operators **S** and **M**. **F** denotes generative systems of all types.

Selection of appropriate methodological distinctions for the various epistemological types of systems is an important issue in the GSPS architecture. The selection should begin with the most general methodological distinctions, applicable to the various systems traits, and proceed in increasing order of specificity. It is desirable to order the resulting methodological distinctions by their specificities. Since several categories are always combined, the resulting ordering is only partial.

At the architectural level, it is desirable to identify basic categories of methodological distinctions and commit only to a small set of important and rather general distinctions in each category. At the same time, however, the GSPS architecture should allow for desirable extensions of these sets in the various GSPS implementations.

Methodological distinctions that are considered sufficiently significant to be recognized in the GSPS architecture are introduced at various places in this book in connection with the individual epistemological types of systems. A list of categories of these essential distinctions is given in Table 9.2. Each category is characterized by the system trait for which it is defined, a list of its individual methodological distinctions, epistemological types of systems to which it applies, examples of possible extensions, and appropriate cross-references.

9.3. PROBLEM REQUIREMENTS

> *As soon as a problem is clearly defined, its solution is often simple.*
>
> —ROBERT ROSEN

Problem requirements define problems on a set of recognized systems. Each requirement involves either a single system or a pair of systems. In a manner similar to that in which systems are classified into systems types, problem requirements are classified into requirement types. When properly defined, each requirement type should be such that variations of specific requirements within it do not demand methodological variations.

Although specific problem requirements as well as their types are meaningful only in connection with the individual system types, four broad categories of problem requirements can be recognized, in general. Each requirement can be either a requirement to answer a *question*, a requirement to satisfy a *request*, a requirement to achieve an *objective*, or a requirement to satisfy some *restriction*. Several requirements may be combined in a single problem, e.g., one or more objectives may be combined with one or more restrictions.

As emphasized on a number of occasions, the GSPS should allow the user to define his own choice of specific requirements within each of the recognized requirement types. Once a requirement type is identified, which can be done only in the context of one or two systems types identified prior to the requirement type, the user should be invited to *define his own choice* of a specific variant of this type. If he does not take this opportunity, the GSPS should offer him a *"menu" of available options*, one of which is declared as a default option. If he does not care to select one of them, the GSPS should use the *default option*.

At the architectural level, it is desirable to commit for each systems type or a pair of systems types only to a limited set of essential requirement types. It is assumed that this initial set will be gradually extended in the process of the evolution of GSPS (Section 9.8).

Types of problem requirements cannot be defined in isolation from the types of systems to which they are applied. Each of them together with the systems types involved forms a problem type. Requirement types are thus definable only as parts of problem types. The most significant requirement types, which are introduced in various contexts within the book, are summarized in the next section.

9.4. SYSTEMS PROBLEMS

> *It must, in all justice, be admitted that never again will scientific life be as satisfying and serene as in the days when determinism reigned supreme. In partial recompense for the tears we must shed and the toil we must endure is the satisfaction of knowing that we are treating significant problems in a more realistic and productive fashion.*
>
> —RICHARD BELLMAN

Systems problems are formed by systems and requirements that are relevant to the systems involved. They are naturally classified into problem types according to the underlying classification of systems and requirements into types.

Some problems may involve a single system. Let these problems, which have the form of a question or request regarding some properties of the given system, be called *elementary problems*. For example, given a particular ST-system, we may ask: "is any state reachable from any other state?"; or, alternatively, we may request for each pair of states a list of the shortest sequences of transitions from one of the states to another.

For each systems type recognized, the GSPS should be equipped to deal with elementary problems as comprehensively as possible. An important subset of elementary problems consists of *validity (correctness) tests* regarding systems specified by users. They include, for example, tests of required properties of probability or possibility distributions involved, various consistency conditions, control uniqueness, and the like. These tests should be executed by the GSPS authomatically for each particular system specified by the user in the process of his problem formulation. If any of the tests fails, the system must be rejected as invalid and the user informed of the reasons.

All systems that do not belong to the category of elementary problems involve two or more systems. Of these, problems that involve two systems are explicitly defined and methodologically supported in the GSPS; let them be referred to as *basic problems*. The remaining problems, each of which involves more than two systems, are then expressed and dealt with in terms of appropriate sequences of basic and elementary problems.

Basic systems problems are classified into problem types. Each of them consists of an ordered pair of systems types and a set of requirement types that are applicable to the systems types involved. Since a problem solution is always oriented from some given entities to some unknown entities, the order of systems types is essential for describing this orientation. The first of the two systems types is viewed as a general characterization of an *initial system*, i.e., a system that is given in a particular problem of the given type. The second systems type characterizes the class of systems in terms of which the solution is expressed; let them be called *terminal systems*. According to the role of terminal systems, systems problems of two kinds can be recognized.

Problems of the first kind have the following canonical formulation. Given a particular initial system of type z, determine those terminal systems of type z' for which given requirements (relevant to systems types z and z') are satisfied. The solution of a problem of the first kind is thus a set of particular systems of type z'.

Problems of the second kind have a different canonical formulation. Given a particular initial system of type z and a particular terminal system of type z', determine some property, specified by given requirements, of the terminal system in relation to the initial system. The solution of a problem of the second kind is thus some property by which the two given systems are related.

An example of a problem type of the first kind is the problem of deriving from a given data system all generative systems that satisfy the requirements of minimal misfit, complexity, and generative uncertainty, and whose masks are submasks of a specified largest acceptable mask. This problem type is formulated in Section 3.4 and discussed in Section 3.6; it is summarized in Figure 9.2.

An example of a problem type of the second kind is the problem of determining the change in the performance of a behavior system due to some specified variables. In this problem type, which is discussed in Section 7.3, some property of the initial system represents a goal, and the performances of the terminal system with and without the specified variables are compared in terms of a specified performance function. A summary of this problem type is given in Figure 9.3.

Problem categories that represent the GSPS kernel are depicted in Figure 9.4. Each problem category is characterized by the epistemological types of systems involved; it is represented in the figure by an arrow labeled with an integer identifier. Problem categories are clusters of problem types. Each of them contains problem types that differ from each other in methodological distinctions and requirement types, but involve the same pairs of epistemological types of systems.

Some key problem types in the categories shown in Figure 9.4 are outlined, either directly or as parts of larger problems, in previous chapters of this book. The following is a summary of these problem types, classified by categories.

Categories 1–6 obviously include all elementary problem types associated with the individual epistemological system types, but they include many basic problem types as

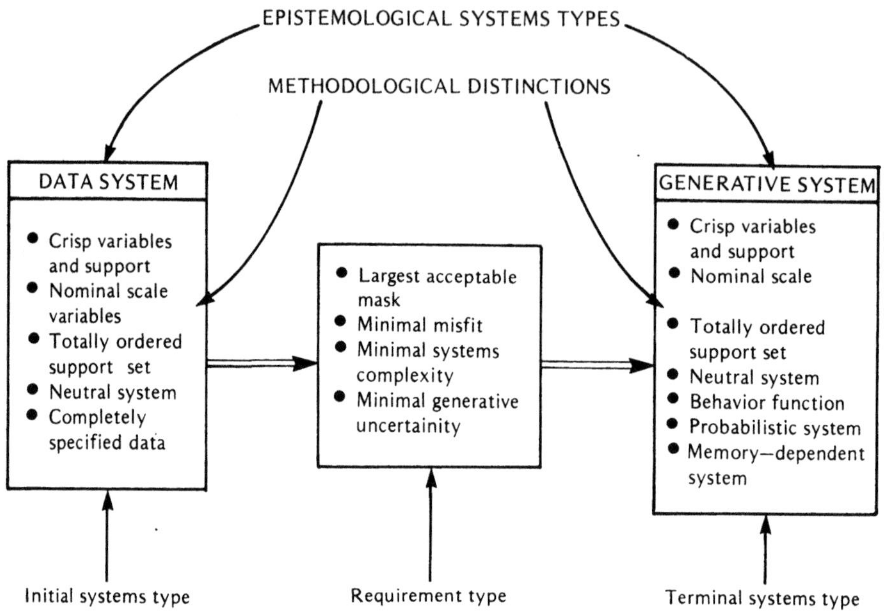

Figure 9.2. An example of a basic problem type of the first kind (see Sections 3.4 and 3.6).

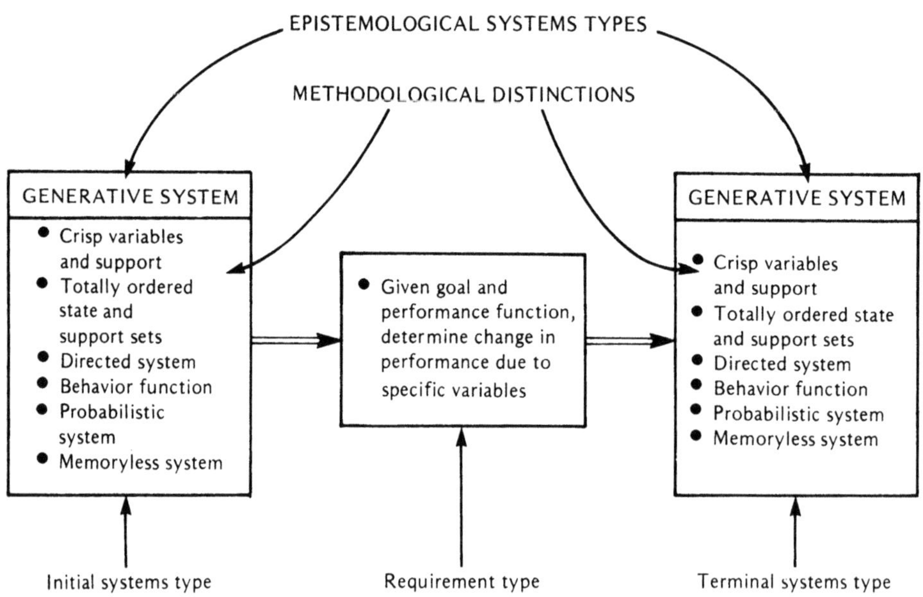

Figure 9.3. An example of a basic problem type of the second kind (see Section 7.3).

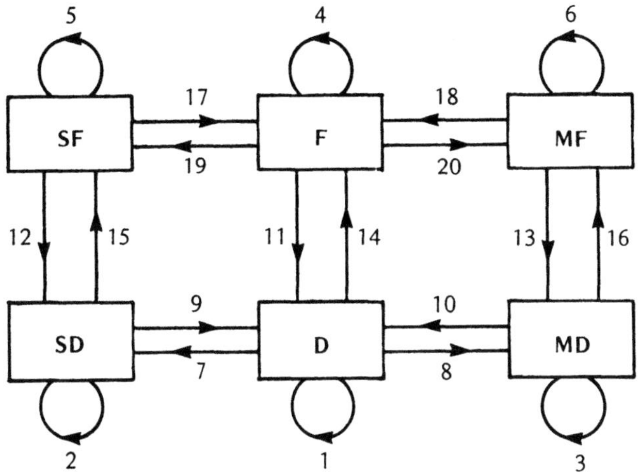

Figure 9.4. GSPS kernel: a summary of key problem categories.

well. In particular, they contain the following classes of basic systems types:

- various problem types of systems simplification, as exemplified by the problems discussed in Section 3.9;
- various problem types associated with the modelling relationship between systems, as discussed for the individual epistemological levels in Chapter 8;
- various problem types in which systems at the individual epistemological levels are compared according to some criteria, e.g., by their complexities (Chapter 6), their performance with respect to various goals (Chapter 7), their information loss (Sections 4.6–4.8), etc.

Categories 7 and 8 include problem types in which data are partitioned, according to specified criteria, with respect to the variables or support set involved.

Categories 9 and 10 consist of problem types in which an overall data system is determined from elements of a structure system or a metasystem, respectively. They include, for example, problem types in which resolution of data inconsistencies is required, such as those mentioned in Section 4.3 for structure data systems.

Categories 11–13 contain various problem type in which data are generated by epistemologically higher types of systems for specified initial and other conditions. In all problems of these categories, the GSPS is used as a tool for computer simulation.

Category 14 consists of problem types in which generative systems are derived from given data systems under specified requirement types. As discussed in Sections 3.4, 3.6, and 3.10, it is usually required that the resulting generative systems be based only on masks that do not exceed a specified largest acceptable mask, that the data they generate be in perfect fit with the given data, and that their generative uncertainties and complexities be minimized. Additional requirements may be added to these problems, such as further restrictions on the resulting generative systems or optimization criteria expressed in terms of special preference relations defined on the relevant set of generative systems. Also included are the problems associated with the introduction of internal variables, as discussed in Section 3.10.

Categories 15 and 16 consist basically of the same problem types as those in category 14, but each individual problem must be repeated for all elements of the given structure data system or data metasystem, respectively.

Category 17 contains problem types associated with the determination of reconstruction families, unique reconstructions based on various requirement types, and related issues, as discussed in Sections 4.6–4.8.

Category 18 represents problem types associated with the simulation of generative systems by metasystems of various methodological types.

Category 19 includes various types of decomposition problems involved in systems design (Section 4.5), as well as some aspects of the reconstruction problem (Sections 4.7, 4.8).

Category 20 subsumes all problem types associated with the identification of support instances that are associated with changes in the constraint among variables of a given generative system. These problem types are exemplified by the problem type discussed in Section 5.6.

Most problem types of practical significance require sequences of basic or elementary problem types. For example, the reconstruction problem type introduced in Section 4.7 involves a sequence of problem types from categories 4, 5, 17, 19; some types of design problems require problem types from categories 11, 14, 17, 19; certain kinds of computer simulation may involve problem types from categories 11 and 17 or 11 and 18, etc.

9.5. GSPS CONCEPTUAL FRAMEWORK: FORMAL DEFINTION

> *When a philosopher invents a new approach to reality, he promptly finds that his predecessors saw something as a unit which he can subdivide, or that they accepted distinctions which his system can name as unities. The universe would appear to be something like a cheese; it can be sliced in an infinite number of ways—and when one has chosen his own pattern of slicing, he finds that other men's cuts fall at the wrong places.*
>
> —Kenneth Burke

► The following definitions relate the adjectives "identifiable" and "admissible" with the terms "system," "requirement," and "problem." We say that a system, requirement, or problem is *identifiable* if it can be formulated in terms of the GSPS language; we say that it is *admissible* if it is identifiable and can be dealt with in the context of a particular GSPS implementation. In addition, we say that a problem is *solvable* if it is admissible and can be solved by the methodological tools available in a particular GSPS implementation. These adjectives apply to particular systems, requirements and problems as well as to their types.

Let $\mathscr{X}$ denote the set of all *admissible epistemological types of systems*. Then,

$$\mathscr{X} \subset \mathscr{E},$$

where $\mathscr{E}$ denotes the set of all identifiable epistemological systems type. Typically,

$$\mathscr{X} \subseteq \mathscr{E}_l$$

for some particular $l \geq 1$, where $\mathscr{E}_l$ ($l \in \mathbb{N}$) is the set of all identifiable epistemological systems types in which the combined number of structure systems and metasystems is l or less.

Let $\mathscr{Y}$ denote the set of *admissible types of methodological distinctions* and let $\mathscr{Z}$ denote the set of all *admissible types of systems*. Then,

$$\mathscr{Z} \subset \mathscr{X} \times \mathscr{Y},$$

$\mathscr{Z}$ is a proper subset of $\mathscr{X} \times \mathscr{Y}$ since some methodological distinctions are not applicable to all epistemological types of systems. Let $\mathscr{S}$ be the set of all *admissible systems* (i.e., particular systems, not systems types). Then, $\mathscr{Z}$ imposes a partition

$$\mathscr{S}/\mathscr{Z} = \{\mathscr{S}_{\mathbf{z}} \mid \mathbf{z} \in \mathscr{Z}\}$$

on $\mathscr{S}$, where $\mathscr{S}_{\mathbf{z}}$ denotes the set of all admissible systems of type $\mathbf{z}$. Let

$$\mathbf{s}_{\mathbf{z},i} \in \mathscr{S}_{\mathbf{z}} \; (\mathbf{z} \in \mathscr{Z}).$$

Then $\mathbf{s}_{\mathbf{z},i}$ is an admissible system of type $\mathbf{z}$ identified (distinguished from other systems of type $\mathbf{z}$) by identifier $\mathbf{i}$.

Let $\mathscr{Q}_{\mathbf{z}}$ denote the set of all admissible requirement types that are applicable to a single system of type $\mathbf{z}$ and let $\mathscr{Q}_{\mathbf{z},\mathbf{z}'}$ denote the set of all admissible requirement types that are applicable to a pair of systems of types $\mathbf{z}, \mathbf{z}'$. Then, we can define

$$^{1}\mathscr{Q} = \bigcup_{\mathbf{z} \in \mathscr{Z}} \mathscr{Q}_{\mathbf{z}}$$

and

$$^{2}\mathscr{Q} = \bigcup_{\mathbf{z},\mathbf{z}' \in \mathscr{Z}} \mathscr{Q}_{\mathbf{z},\mathbf{z}'}.$$

The set

$$\mathscr{Q} = {}^{1}\mathscr{Q} \cup {}^{2}\mathscr{Q}$$

obviously consists of all *admissible requirement types*.

Let $\mathscr{R}_z$ be the set of all admissible requirements (i.e., particular requirements, not requirement types) that are applicable to a single system of type $\mathbf{z}$. Then, $\mathscr{Q}_z$ imposes a partition

$$\mathscr{R}_z / \mathscr{Q}_z = \{\mathscr{R}_{z,j} | \mathbf{q}_{z,j} \in \mathscr{Q}_z\}$$

on $\mathscr{R}_z$, where $\mathscr{R}_{z,j}$ denotes the set of admissible requirements of type $\mathbf{j}$ that are applicable to a single system of type $\mathbf{z}$.
Let

$$r_{z,j,u} \in \mathscr{R}_{z,j},$$

i.e., let $\mathbf{r}_{z,j,u}$ denote an admissible requirement of type $\mathbf{j}$ that is applicable to a single system of type $\mathbf{z}$ and is uniquely labeled by an identifier $\mathbf{u}$.

Let $\mathscr{R}_{z,z'}$ be the set of all admissible requirements that are applicable to a pair of systems of types $\mathbf{z}$ and $\mathbf{z}'$. Then, $\mathscr{Q}_{z,z'}$ imposes a partition

$$\mathscr{R}_{z,z'} / \mathscr{Q}_{z,z'} = \{\mathscr{R}_{z,z',k} | \mathbf{q}_{z,z',k} \in \mathscr{Q}_{z,z'}\}$$

on $\mathscr{R}_{z,z'}$, where $\mathscr{R}_{z,z',k}$ denotes the set of all admissible requirements of type $\mathbf{k}$ that are applicable to a pair of systems of types $\mathbf{z}$ and $\mathbf{z}'$.
Let

$$\mathbf{r}_{z,z',k,u} \in \mathscr{R}_{z,z',k},$$

i.e., $\mathbf{r}_{z,z',k,u}$ denotes an admissible requirement of type $\mathbf{k}$ that is applicable to a pair of systems of types $\mathbf{z}$ and $\mathbf{z}'$, and is uniquely labeled by an identifier $\mathbf{u}$.

The notions introduced of admissible systems, requirements, and their types make it possible now to define admissible problems and their types. Let ${}^1\mathscr{P}$ and ${}^2\mathscr{P}$ denote, respectively, the set of all *admissible types of elementary problems* and the set of all *admissible types of basic problems*. Then,

$${}^1\mathscr{P} = \{(\mathbf{z}, \mathbf{q}_{z,j}) | \mathbf{z} \in \mathscr{Z}, \mathbf{q}_{z,j} \in \mathscr{Q}_z\},$$
$${}^2\mathscr{P} = \{(\mathbf{z}, \mathbf{z}', \mathbf{q}_{z,z',k}) | \mathbf{z}, \mathbf{z}' \in \mathscr{Z}, \mathbf{q}_{z,z',k} \in \mathscr{Q}_{z,z'}\},$$

and

$$\mathscr{P} = {}^1\mathscr{P} \cup {}^2\mathscr{P}$$

is the set of all admissible problem types. Although $\mathbf{z}$ may be the same systems type as $\mathbf{z}'$ in the formulation of ${}^2\mathscr{P}$, these two types of systems stand for two different systems in each particular basic problem. This distinguishes basic problems from elementary problems, each of which involves only one particular system.

On the basis of the characteristics of requirement types in $\mathscr{Q}_{z,z'}$ ($\mathbf{z}, \mathbf{z}' \in \mathscr{Z}$), set ${}^2\mathscr{P}$ is naturally partitioned into sets ${}^{21}\mathscr{P}$ and ${}^{22}\mathscr{P}$ of basic problem types of the first and second kind, respectively.
Hence,

$$\mathscr{P} = {}^1\mathscr{P} \cup {}^{21}\mathscr{P} \cup {}^{22}\mathscr{P}.$$

Set $\mathscr{Z}$ imposes a partition

$$^{1}\mathscr{P}/\mathscr{Z} = \{^{1}\mathscr{P}_{\mathbf{z}} \mid \mathbf{z} \in \mathscr{Z}\}$$

on $^{1}\mathscr{P}$, where $^{1}\mathscr{P}_{\mathbf{z}}$ denotes the set of all admissible types of elementary problems that are applicable to systems type $\mathbf{z}$. Similarly, set $\mathscr{Z}^2$ imposes a partition

$$^{2}\mathscr{P}/\mathscr{Z}^{2} = \{^{2}\mathscr{P}_{\mathbf{z},\mathbf{z}'} \mid \mathbf{z}, \mathbf{z}' \in \mathscr{Z}\}$$

on set $^{2}\mathscr{P}$, where $^{2}\mathscr{P}_{\mathbf{z},\mathbf{z}'}$ denotes the set of all admissible types of basic problems that are applicable to a pair of systems of types $\mathbf{z}, \mathbf{z}'$. Similar partitions can also be defined on sets $^{21}\mathscr{P}$ and $^{22}\mathscr{P}$. Observe that sets $^{1}\mathscr{P}_{\mathbf{z}}$ and $^{2}\mathscr{P}_{\mathbf{z},\mathbf{z}'}$ are categories of problems in the sense discussed in Sections 9.4 and illustrated in Figure 9.4.

Let $^{1}\mathscr{P}\mathscr{P}$ denote the set of all admissible elementary problems. Then, set $^{1}\mathscr{P}$ imposes a partition

$$^{1}\mathscr{P}\mathscr{P}/^{1}\mathscr{P} = \{^{1}\mathscr{P}\mathscr{P}_{\mathbf{z},j} \mid \mathbf{z} \in \mathscr{Z}, \mathbf{q}_{\mathbf{z},j} \in \mathscr{Q}_{\mathbf{z}}\}$$

on $^{1}\mathscr{P}\mathscr{P}$, where $^{1}\mathscr{P}\mathscr{P}_{\mathbf{z},j}$ denotes the set of all admissible elementary problems that are defined in terms of a single system of type $\mathbf{z}$ and a requirement of type $\mathbf{q}_{\mathbf{z},j}$.

Let $^{2}\mathscr{P}\mathscr{P}$ denote the set of all admissible basic problems. Then, set $^{2}\mathscr{P}$ imposes a partition

$$^{2}\mathscr{P}\mathscr{P}/^{2}\mathscr{P} = \{^{2}\mathscr{P}\mathscr{P}_{\mathbf{z},\mathbf{z}',k} \mid \mathbf{z}, \mathbf{z}' \in \mathscr{Z}, \mathbf{q}_{\mathbf{z},\mathbf{z}',k} \in \mathscr{Q}_{\mathbf{z},\mathbf{z}'}\}$$

on $^{2}\mathscr{P}\mathscr{P}$, where $^{2}\mathscr{P}\mathscr{P}_{\mathbf{z},\mathbf{z}',k}$ denotes the set of all admissible basic problems that are defined in terms of a pair of systems of types $\mathbf{z}$ and $\mathbf{z}'$, and a requirement of type $\mathbf{q}_{\mathbf{z},\mathbf{z}',k}$. Each set $^{2}\mathscr{P}\mathscr{P}_{\mathbf{z},\mathbf{z}',k}$ is further partitioned into subsets of basic problems of the first and second kind, say subsets $^{21}\mathscr{P}\mathscr{P}_{\mathbf{z},\mathbf{z}',k}$ and $^{22}\mathscr{P}_{\mathbf{z},\mathbf{z}',k}$, respectively. Then,

$$^{21}\mathscr{P}\mathscr{P} = \cup\, ^{21}\mathscr{P}\mathscr{P}_{\mathbf{z},\mathbf{z}',k},$$
$$^{22}\mathscr{P}\mathscr{P} = \cup\, ^{22}\mathscr{P}\mathscr{P}_{\mathbf{z},\mathbf{z}',k},$$

where the set unions are taken over all $\mathbf{z}, \mathbf{z}' \in \mathscr{Z}$ and $\mathbf{q}_{\mathbf{z},\mathbf{z}',k} \in \mathscr{Q}_{\mathbf{z},\mathbf{z}'}$, are the sets of all basic problems of the first and second kind, respectively.

Three fundamental kinds of admissible problems can now be described formally:

Elementary problems, say $^{1}p_{\alpha}$, have the form

$$^{1}p_{\alpha} = (\mathbf{s}_{\mathbf{z},i}, \mathbf{r}_{\mathbf{z},j,u}),$$

where α denotes a unique identifier of quadruples $(\mathbf{z}, \mathbf{i}, \mathbf{j}, \mathbf{u})$.

Basic problems of the first kind, say $^{21}p_{\beta}$, have the form

$$^{21}p_{\beta} = (\mathbf{s}_{\mathbf{z},i}, \mathbf{z}', \mathbf{r}_{\mathbf{z},\mathbf{z}',k,u}),$$

where β denotes a unique identifier of quintuples $(\mathbf{z}, \mathbf{i}, \mathbf{z}', \mathbf{k}, \mathbf{u})$;

Basic problems of the second kind, say ${}^{22}p_\gamma$ have the form

$$ {}^{22}p_\gamma = (\mathbf{s}_{z,i}, \mathbf{s}_{z',t}, \mathbf{r}_{z,z',k,u}), $$

where γ denotes a unique identifier of sixtuples $(\mathbf{z}, \mathbf{i}, \mathbf{z}', \mathbf{t}, \mathbf{k}, \mathbf{u})$.

Let $\mathcal{PP}$ denote the set of all admissible problems definable within the GSPS. Then,

$$ \mathcal{PP} \subset ({}^{1}\mathcal{PP} \cup {}^{21}\mathcal{PP} \cup {}^{22}\mathcal{PP})^*, $$

where the asterisk denotes the set of all sequences that can be formed by elements of the set, i.e., all sequences of elementary or basic problems. Similarly, set $\mathcal{P}$ of all admissible problem types of the GSPS is defined by

$$ \mathcal{P} = ({}^{1}\mathcal{P} \cup {}^{21}\mathcal{P} \cup {}^{22}\mathcal{P})^*, $$

where ${}^{1}\mathcal{P}$, ${}^{21}\mathcal{P}$, ${}^{22}\mathcal{P}$ are finite sets of elementary and basic problem types.

We can see now that the GSPS conceptual framework can be viewed at this global level as a language for describing admissible types of systems problems whose alphabet consists of all admissible elementary and basic problem types, each of which is defined in turn by the underlying epistemological and methodological types of systems, and requirement types. ◄

9.6. OVERVIEW OF GSPS ARCHITECTURE

The key for a successful rebuilding of our environment—which is the architect's task—will be our determination to let the human element be the dominating factor. . . . Architecture needs conviction and leadership. It cannot be decided upon by clients or by Gallup Polls, which would most often only reveal a wish to continue what everybody knows best.

—Walter Gropius

The purpose of architecture is to identify and properly characterize those functions of an artifact under design, be it a building, machine, or expert system, that are necessary for achieving a given goal. The choice and characterization of basic functions of the GSPS, whose goal is to provide the user with expert's knowledge for dealing with systems problems, are discussed on previous pages of this book in various degrees of detail. The aim of this section is to present a concise and comprehensive overview of these functions. A block diagram in Figure 9.5 is employed for this purpose; the reader is advised to use it as a guide when reading the following description.

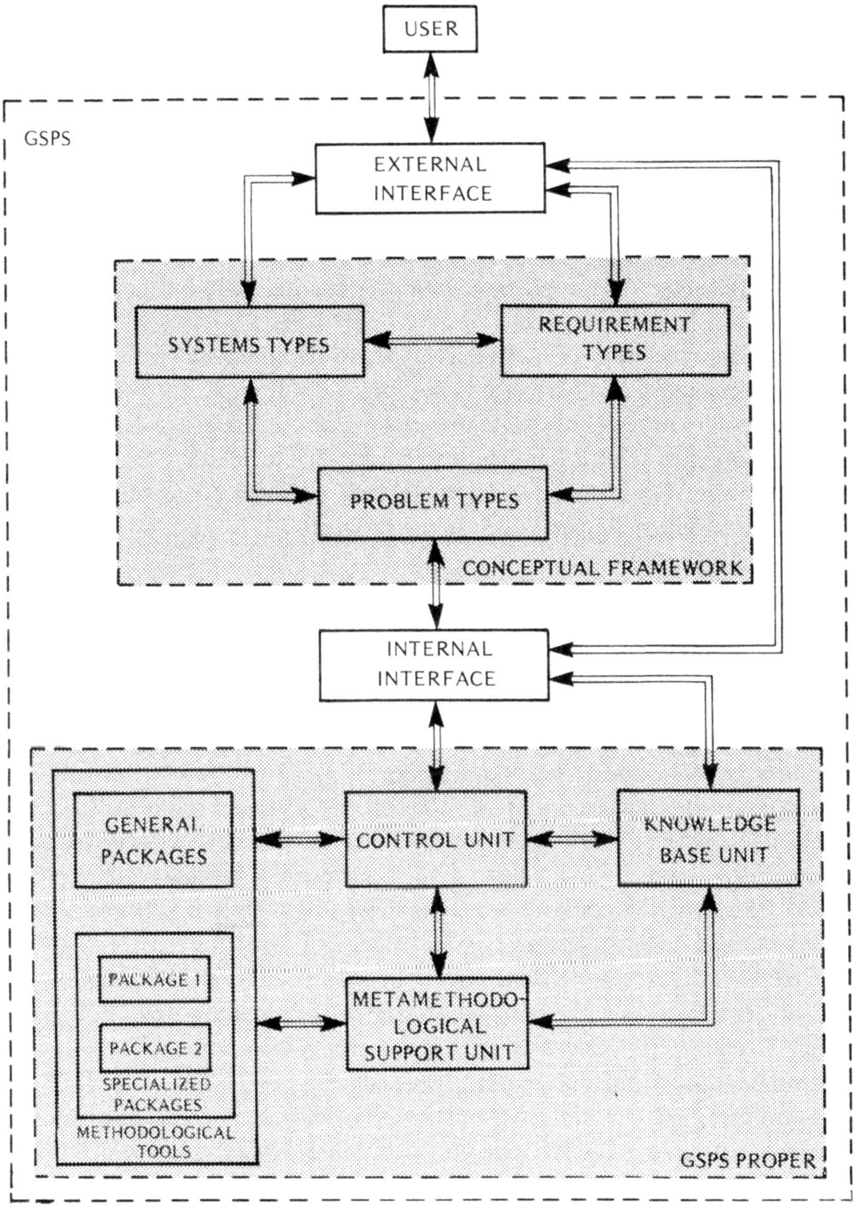

Figure 9.5. A sketch of the GSPS architecture.

The *conceptual framework* that is developed in Chapters 2–8 and summarized in Section 9.5 is the kernel of the GSPS. It is a language that is used within the GSPS to describe recognized types of systems, requirements, and problems. As far as the GSPS architecture is concerned, it commits only to the epistemological hierarchy of systems types, as defined in Section 9.1. The remaining concepts, such as the individual requirement types or methodological distinctions of systems, are characterized in the GSPS architecture only in general terms. Their specific delineation is a subject of a particular GSPS implementation. It is understood that each GSPS implementation is based on a specific set of system types, each represented by a combination of certain epistemological and methodological features, as well as specific sets of requirement types and the resulting problem types; these are referred to as *admissible types* of systems, requirements, and problems (i.e., they are admissible by the given GSPS implementation).

As explained in Section 9.5, meaningful types of systems problems are formed by sequences of elementary and basic problem types. The latter are formed, in turn, by systems types and requirement types that are compatible in the sense that they can be applied to each other. The three main categories that form the GSPS conceptual framework—system types, requirement types, and problem types—are thus interrelated. This is indicated in Figure 9.5 by the bidirectional connections between the respective blocks.

The conceptual framework is a linguistic medium in terms of which the user communicates with the GSPS proper. The latter consists of an appropriate knowledge base, a set of methodological tools, a metamethodological support, and a control unit. The user-GSPS communication, which is two-way communication, is facilitated at either end by an appropriate interface. Let us call the one at the user's end an *external interface* and the other one an *internal interface.*

Two types of the external interface should be recognized, both of which may be incorporated in a particular GSPS implementation. One of them is designed to serve a *sophisticated user*, defined as a person who is sufficiently familiar with the GSPS conceptual framework (at least at the level described in this book) and with the limitations of the GSPS implementation he intends to use. This type of interface is based on the assumption that the user does not need any assistance in formulating his systems and requirements and, consequently, the only function of the interface is to check for possible inconsistencies in the user's formulations.

The other type of external interface is designed to serve a *general user*, whose knowledge of the GSPS conceptual framework is not sufficient. Its function is not only to check inconsistencies in user's formulations, but to provide the user with a broader assistance in formulating his problem. This means that the external interface must include appropriate *inquiry procedures* for the identification of system and requirement types, as well as the identification of particular systems and requirements of given types. Such procedures may prove very difficult, if not totally unrealistic, if the user is not expected to have any knowledge of the GSPS conceptual framework. It is thus envisoned that in any GSPS implementations, at least those available in the near future, some *minimal knowledge of the GSPS conceptual framework* will be required of the general user. Such minimal knowledge may be made a standard part of every GSPS manual.

The mentioned inquiry identification procedures, which are an important part of the GSPS external interface, can be developed at various levels, depending on the minimal knowledge of the GSPS conceptual framework required of the user. The less knowledge is required, the more difficult and less developed the procedures are. At the extreme level, where no knowledge is required, the development of successful inquiry procedures is a major research challenge. The various issues involved, which are beyond the scope of this book, are studied predominantly within the area of artificial intelligence.

The *GSPS proper* consists of four functional units: a set of methodological tools, a knowledge base, a metamethodological support unit, and a control unit. All GSPS concepts are represented within the GSPS proper in some standard form that is facilitated by the internal interface. It is understood that the knowledge base unit, as well as the methodological and metamethodological units, must be equipped with appropriate reasoning strategies.

Methodological tools are packages of methods (and the corresponding computer programs) by which some of the admissible problem types can be solved. They are divided into general packages and specialized packages. The *general packages* are designed for problem types formulated in terms of the most general methodological distinctions available within a given GSPS implementation; the *specialized packages* are designed for all problem types based on any less general methodological distinctions.

Each methodological tool is a set of methods (and the corresponding computer programs) for solving some elementary or basic problem types and a procedure (computer program) that specifies the order in which the individual methods are to be employed. Methodological tools are thus formed from a common pool of methods (programs), available for the admissible elementary and basic problem types, by a procedure (a control program) that employs the required methods in appropriate sequences.

The GSPS also incorporates various metamethodological considerations. These are handled by the *metamethodological support unit*. This unit contains information about the ordering of all admissible problems by their generality and their methodological status. The ordering by generality reflects basically the ordering of methodological distinctions of systems and requirements. The methodological status of a problem has to do with its solvability, time and space computational demands, and characteristics of relevant methodological tools.

Theoretical unsolvability of a problem is distinguished from unsolvability that is solely due to the limitations of the given GSPS implementation. If the problem is unsolvable in the latter sense, the GSPS may respond in two ways. First, alternative formulations of the problem, which are based on stronger assumptions and are solvable by the GSPS, are offered to the user; we say that the methodological support unit performs appropriate shifting from the given methodological paradigm to more specific paradigms. Second, the knowledge base unit is called upon to provide the user with useful information regarding the original problem such as references to relevant program libraries, papers, or books.

For each particular problem for which a methodological tool is available in the given GSPS implementation, the metamethodological support unit should perform an adequate analysis of its computational complexity. If the problem turns out to be practically intractable, the unit should determine, if possible, alternative formulations of the problem (based on stronger assumptions) that are computationally tractable. In addition, the unit should determine for each requested problem that can be solved by the given GSPS implementation an estimated computing cost and other relevant characterizations of the method employed.

As mentioned previously, the *knowledge support unit* contains useful information regarding those problems that cannot be solved by the GSPS implementation involved. In addition, however, it may contain other relevant information regarding systems and systems problems. Examples are the various theoretical or experimental laws, principles, or rules of thumb of systems science, such as the law of requisite variety or the law of requisite hierarchy.

The user communicates with the units of the GSPS proper either through the conceptual framework or by a direct connection between the external interface and the internal interface. The former communication is involved in the formulation of problems; the latter is associated with the various metamethodological considerations and utilization of the knowledge base.

The necessary coordination of the three described units of the GSPS proper is performed by a *control unit*. It basically makes decisions, according to the user's requests and other conditions, about which unit to activate and how.

9.7. GSPS USE: SOME CASE STUDIES

> *God gave us the nuts, but he will not crack them for us.*
>
> —Johann Wolfgang Goethe

The aim of this section is to present some additional examples of systems problems and describe how the GSPS deals with them. Included are only problems that relate to the methods described in this book. The main focus is on case studies that involve more than one methodological tool and include aspects that are not sufficiently emphasized in previous text. Some of the case studies are adopted from literature; results of these are compared with the published results. Other case studies are based on data that were not previously published.

It should be mentioned that the quality of data, which in all instances were obtained indirectly from various sources, should not be the issue here. The purpose of the case studies is to illustrate the GSPS potential, and not to argue issues that are outside the GSPS domain, such as data gathering and interpretation of results. Although limited interpretation of some results is suggested, it must be emphasized that this would normally be left to the GSPS users—experts in the areas which the individual case studies represent.

For the sake of simplicity, variables are often represented by their identifiers (subscripts) and subsets of variables are separated by slashes in the following examples (see, e.g., Figures 9.6b,c).

Example 9.1. This example deals with experimental data collected in a study regarding growth of plants following different treatment. The data were collected for 960 plum trees (Note 9.1) and involve three binary variables:

v_1—mortality (0, alive; 1, dead);
v_2—time of planting (0, at once; 1, in spring);
v_3—root cutting (0, long; 1, short).

The data were collected under controlled conditions: 240 plum trees were observed for each of the controlled variables v_2 and v_3. The experiment was designed to investigate the effect of time of planting and length of cutting on the tree survival.

Since all variables are binary, there is no possibility of further reducing their

	v_1	v_2	v_3	$N(\mathbf{c})$	$f(\mathbf{c})$
$\mathbf{c} =$	0	0	0	156	0.162
	0	0	1	107	0.111
	0	1	0	84	0.087
	0	1	1	31	0.032
	1	0	0	84	0.087
	1	0	1	133	0.139
	1	1	0	156	0.162
	1	1	1	209	0.218

(a)

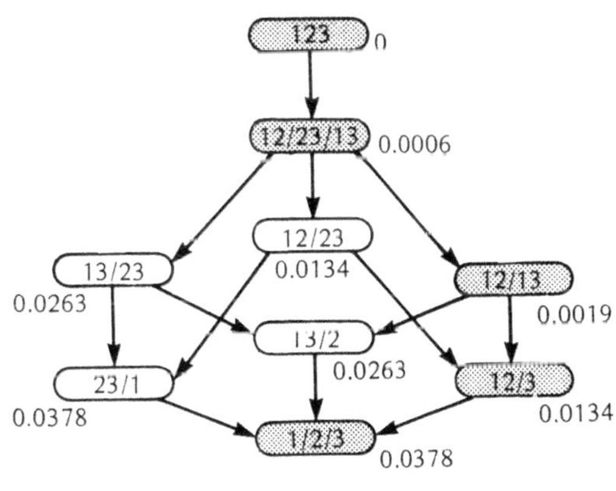

(b)

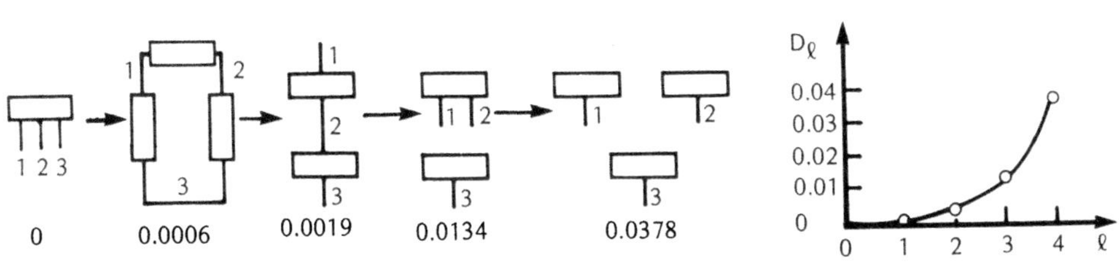

(c) (d)

Figure 9.6. A summary of Example 9.1 (tree survival).

resolution forms. The population is 30 times larger than the number of states of the variables; this is quite sufficient for probabilistic analysis. Since the support set is an unordered population, the study is restricted to the determination of the behavior function and performing reconstructability analysis of it. Frequencies of the individual states of the variables and the probability distribution derived from them in terms of relative frequencies are given in Figure 9.6a. After all reconstruction hypotheses in $\mathscr{G}_3$ are evaluated, we obtain the distances specified in Figure 9.6b. Reconstruction hypotheses that belong to the solution set are shaded in the figure (including, for the sake of completeness, is also the overall system itself).Observe that the reconstruction hypotheses 23/1 and 13/2 need not be evaluated because their distances cannot be smaller than the largest distances of their predecessors, i.e., they cannot be smaller than 0.0263. Hence, they are inferior to the hypothesis 12/3, whose actual distance is 0.0134, and must be rejected. Block diagrams of members of the solution set and their ordering are given in Figure 9.6c. The dependence of the minimum distance on the refinement level is plotted in Figure 9.6d.

We may conclude from the results in Figure 9.6 that the tree survival is affected by both time of planting and length of cutting, but time of planting is about twice as significant as length of cutting. We can also see that these controlled variables (v_2 and v_3) are independent as far as their influence on the tree mortality is concerned.

Example 9.2. Although this example is only a small variation on Example 9.1 from the methodological point of view, the two examples are totally different in their semantic and pragmatic aspects. This example deals with experimental data collected for 114 90-day-old mice in a study regarding infanticide, which means the killing of young of the same species. The experiments concern one category of infanticide—that relating to sexual competition between males. They were performed to test the so-called sexual-competition hypothesis, based on Darwin's concept of sexual selection, according to which a male that commits infanticide can increase his reproductive success at the expense of competitors by killing a competitor's offspring and then mating with the mother.† Variables involved in the experiments are:

v_1—dominance (0, dominant; 1, subordinate);
v_2—sexual experience (0, naive; 1, experienced);
v_3—attitude to competitor's offsprings (0, infanticide; 1, parental, 2, ignoring).

As in Example 9.1, we can only determine a behavior function and perform its reconstructability analysis. A summary of results, based on probabilistic characterization of constraint among the variables, is given in Figure 9.7.

When we disregard the totally different interpretation, the results are very similar to those obtained in Example 9.1. We may conclude that the attitude to the young is

† For more details regarding the study, see *Science* **25**, pp. 1270–1272, 1982.

	v_1	v_2	v_3	N(**c**)	f(**c**)
c =	0	0	0	28	0.246
	0	0	1	4	0.035
	0	0	2	2	0.018
	0	1	0	5	0.044
	0	1	1	25	0.219
	0	1	2	3	0.026
	1	0	0	5	0.044
	1	0	1	9	0.079
	1	0	2	8	0.070
	1	1	0	7	0.061
	1	1	1	15	0.132
	1	1	2	3	0.026

(a)

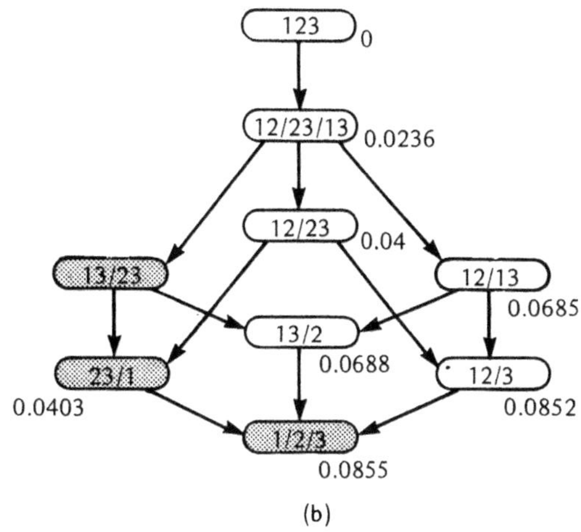

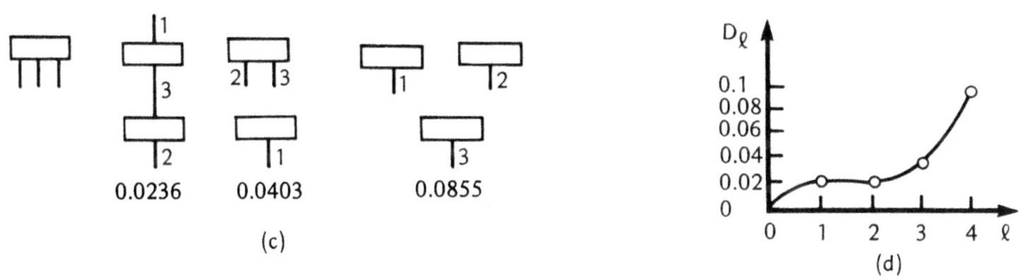

Figure 9.7. A summary of Example 9.2 (infanticide).

affected by both dominance and sexual experience (hypothesis 13/23), but sexual experience is almost twice as significant as dominance (compare distances of 23/1 and 13/2). We can also see that whenever the subsystem 12 is excluded, the distance virtually does not change. Hence, dominance and sexual experience are independent of each other.

Example 9.3. This example is based on data that were collected as part of an investigation into satisfaction with housing conditions in Copenhagen (Note 9.2). A population of 1,681 residents from selected areas living in rented homes built between 1960 and 1968 were questioned on their satisfaction, the degree of contact with other residents, and their feeling of influence on apartment management. At the same time, the type of housing of each resident was recorded. Four variables are thus involved:

v_1—housing type (0, tower blocks; 1, apartments; 2, atrium houses; 3, terraced houses);
v_2—feeling of influence (0, low; 1, medium; 2, high);
v_3—degree of contact (0, low; 1, high);
v_4—satisfaction with housing conditions (0, low; 1, medium; 2, high).

Since all 72 states of the variables were observed, we can define a lexicographic ordering of the states and represent the data by the sequence of frequencies $N(\mathbf{c})$ that is based on this ordering. Assuming that the ordering of variables is v_1, v_2, v_3, v_4, the lexicographic ordering is unique and, hence, the data are uniquely defined by the following sequence of frequencies $N(\mathbf{c})$: 21, 21, 28, 14, 19, 37, 34, 22, 36, 17, 23, 40, 10, 11, 36, 3, 5, 23, 61, 23, 17, 78, 46, 43, 43, 35, 40, 48, 45, 86, 26, 18, 54, 15, 25, 62, 13, 9, 10, 20, 23, 20, 8, 8, 12, 10, 22, 24, 6, 7, 9, 7, 10, 21, 18, 6, 7, 57, 23, 13, 15, 13, 13, 31, 21, 13, 7, 5, 11, 5, 6, 13.

As in previous examples, the data are based on an unordered population and, consequently, the mask evaluation is meaningless. The population is 23 times larger than the number of overall states of the variables. Hence, no coarsening of resolution forms of the variables is needed.

Since this example is methodologically similar to the previous examples, we provide the reader with results obtained for two options of reconstructability analysis (Table 9.3) and leave their discussion and interpretation to the reader. The two options are based on probabilistic and possibilistic behavior functions, respectively. In either case, only C-structures are used and, at each level of refinement, only those C-structures are refined whose distance is not greater than the minimum distance at that level plus 20% of the minimum distance. The refined structures are indicated in Table 9.3 by asterisks; those with minimum distances are identified by two asterisks.

Example 9.4. This example describes a simple study of the relationship between the political situation and stock market performance in the United States in this century. The political situation is characterized by three binary variables:

v_1—party affiliation of the President (0, Democrat; 1, Republican);
v_2—house control (0, Democratic; 1, Republican);
v_3—senate control (0, Democratic; 1, Republican).

The stock market performance is represented by a single variable:

v_4—stock market (0, down; 1, up).

These variables are observed in time during the period 1897–1981. The whole period is divided into 21 equal intervals associated with the regular presidential period of four years. Variables v_2 and v_3 are determined by the house and senate control, respectively, at the beginning of each of the periods.

It is obvious that the source system represents a highly simplified characterization of the attributes of concern. It is used here only to illustrate how the GSPS could be utilized for the investigation of systems of this sort. For instance, it may be desirable to distinguish more states for variable v_4 or to divide the time period involved into intervals (not necessarily equal) that are distinguished by changes in the variables themselves. Some additional political or economic variables may also be included, such

TABLE 9.3
Summary of Reconstructability Analysis in Example 9.3 (Housing Conditions in Copenhagen)

(a) Probabilistic option

l	Structure	Distance
1	134/234	0.0030
	124/234	0.0045
	123/234	0.0069
	124/134	0.0023**
	123/134	0.0094
	123/124	0.0023**
2	13/14/23/24	0.0051
	124/23	0.0053
	123/24	0.0077
	124/13	0.0038**
	123/14	0.0109
	134/24	0.0047
	124/34	0.0062
	12/13/24/34	0.0089
	12/134	0.0110
3	14/13/24	0.0062**
	124/3	0.0065*
	12/13/24	0.0093
	12/13/14	0.0124
4	14/24/3	0.0089**
	13/24	0.0104*
	14/13/2	0.0136
	12/3/24	0.0120
	12/3/14	0.0151
5	1/24/3	0.0131**
	14/2/3	0.0163
	13/2/4	0.0178
6	1/2/3/4	0.0205**

(b) Possibilistic option

l	Structure	Distance[a]
1	134/234	0.0197
	124/234	0.0235
	123/234	0.0312
	124/134	0.0182*
	123/134	0.0845
	123/124	0.0162**
2	13/14/23/24	0.0360*
	124/23	0.0335*
	123/24	0.0408
	124/13	0.0325**
	123/14	0.0957
	134/24	0.0329*
	124/34	0.0350*
	12/13/24/34	0.0478
	12/134	0.0973
3	14/13/24	0.0487*
	124/3	0.0454**
	12/13/24	0.0585
	12/13/14	0.1128
	14/24/34	0.0560
	13/24/34	0.0693
	134/2	0.1095
	14/23/24	0.0531*
	12/23/24	0.0662
	12/23/14	0.1119
	12/34/24	0.0660
	12/34/14	0.1141
	13/23/24	0.0678
	13/14/23	0.1128
4	14/3/24	0.0658**
	12/3/24	0.0779*
	12/3/14	0.1237
	13/24	0.0785*
	14/13/2	0.1241
	1/23/24	0.0943
	14/23	0.1296
5	1/3/24	0.1087**
	14/3/2	0.1427
	12/3/4	0.1492
	13/2/4	0.1530
6	1/2/3/4	0.1832**

[a] Refined structures are indicated by asterisks; those with minimum distances are identified by two asterisks.

as a global characterization of the political situation in the states, inflation, unemployment, change in GNP, etc.

Data representing the described source system are given in Table 9.4. There are many different ways in which the GSPS can be used in processing the data and deriving systems at higher epistemological levels. The following is one possible scenario.

First, the user wanted to determine admissible masks for $\Delta\mathbf{M} = 2$ and both probabilistic and possibilistic distinctions. Since he had no special requirements, the GSPS provided him with the results in Figure 9.8; they are based on the usual objective criteria—the generative uncertainty and mask size—which are employed in the default option. The figure is self-contained; the shaded areas in the tables identify sampling variables associated with the individual masks.

We can see that both options result in almost the same set of admissible masks, but the possibilistic option includes one additional mask for four sampling variables (the memoryless one). The user decided tentatively to accept the mask with five variables, but he wanted to be sure that there are no better masks with five variables within a larger acceptable mask. He thus requested admissible masks for $\Delta\mathbf{M} = 3$ and $|{}^{i}M| = 5$. He obtained the same mask as before, which represents sampling variables s_2, s_3, s_5, s_7, s_8. This mask is thus well supported and the user accepted it as a basis for further processing. Since only 21 observations are available and there are 32 states of the sampling variables, the user decided to perform reconstructability analysis of the behavior system based on the chosen mask only for the possibilistic option. Admissible reconstruction hypotheses are listed in Table 9.5, where the same notation is used as in Example 9.3; the sampling variables are conveniently relabeled as indicated in Figure 9.9a.

When inspecting the dependence of D_l on l (Figure 9.9b), it is clear that D_l is quite small for $l \leq 6$ and increases considerably for $l > 6$. Hence, the structure 12/35/23/24 at level 6 emerges as the most informative reconstruction hypothesis, i.e., the most refined one in a cluster characterized by small distances. Its block diagram with appropriate directions of variables is given in Figure 9.10. Variable 2, which is the only generating variable in the overall system, is taken as input variable of the individual subsystems; the fact that it is determined by the previous state of variable 5 and not arbitrarily by an

TABLE 9.4
Data Matrix in Example 9.4

t	1	2	3	4	5	6	7	8	9	10	11	12	13	14	15	16	17	18	19	20	21
v^1	1	1	1	1	0	0	1	1	0	0	0	0	0	0	1	1	0	0	1	1	0
v_2	1	1	1	1	0	0	1	1	1	0	0	0	0	0	1	1	0	0	0	0	0
v_3	1	1	1	1	0	0	1	1	1	0	0	0	0	0	1	0	0	0	0	0	0
v_4	1	1	1	1	0	0	0	1	1	1	0	0	1	0	1	1	1	1	0	0	0

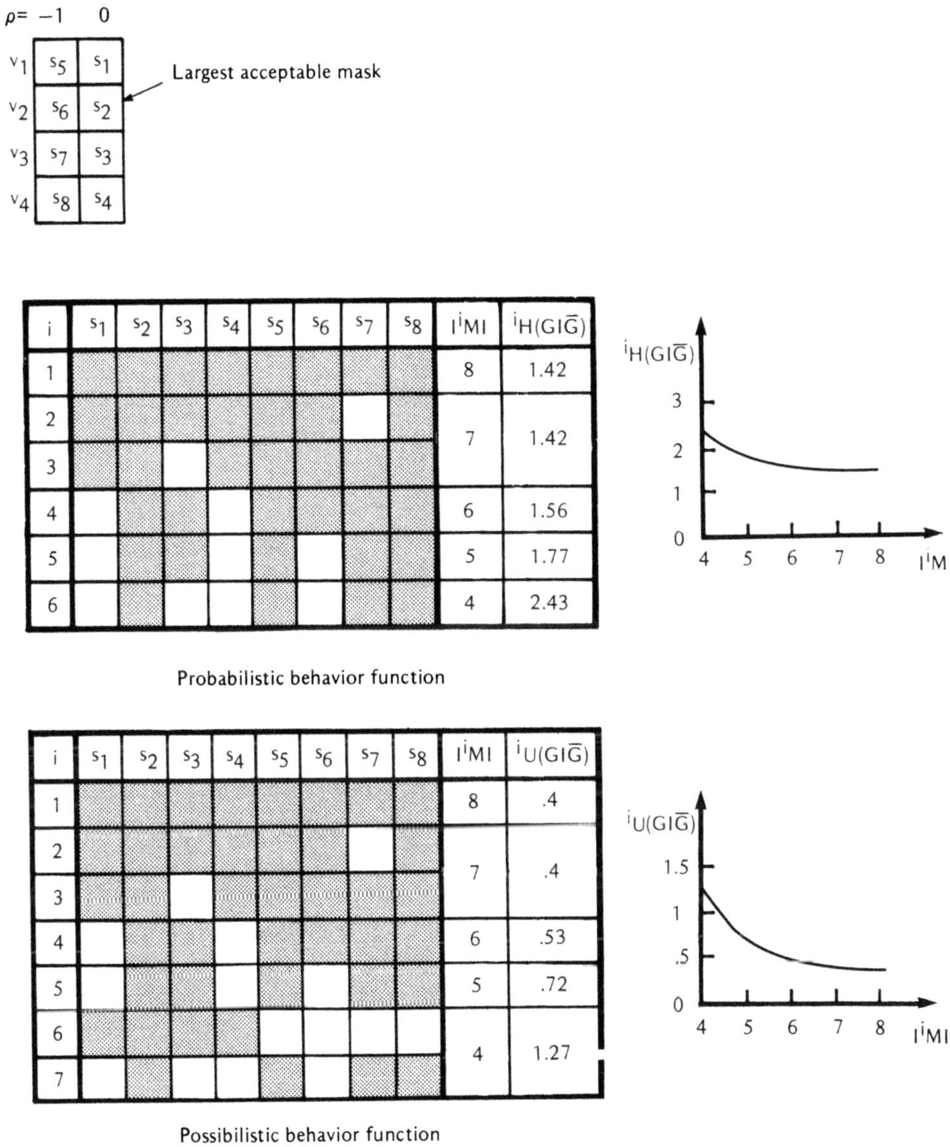

Figure 9.8. Mask evaluation in Example 9.4 for $M = 2$ (stock market and U.S. Federal Government).

environment is indicated by the block denoted as DELAY through which the two variables are connected. Directions of the remaining variables then become unique.

Several interpreted global conclusions can be derived from the admissible

TABLE 9.5
Admissible Reconstruction Hypotheses in Example 9.4 Based on the Mask Specified in Figure 9.9a

l	Structure	D_l
1	1234/1345	0.0
2	123/1345	0.0097
	123/135/124/145	
	1234/135	
3	123/135/124	0.0138
4	123/124/35	0.0277
5	124/35/23	0.0333
6	12/35/23/24	0.0579
7	1/35/23/24	0.1667
	12/35/23/4	
8	1/35/23/4	0.2805
9	1/23/4/5	0.4138
10	1/2/3/4	0.5610

reconstruction hypotheses in Table 9.5 and, particularly, from the key hypothesis in Figure 9.10:

i. senate control is strongly related to stock market performance at the same time period (observe that variables 2 and 3 belong to the same subsystem in all admissible hypotheses except the last one);
ii. stock market performance during one time interval strongly determines senate control during the next interval (variables 3 and 5 become disconned only at refinement level 9);
iii. senate control is the most important variable in the investigated system (variable 2 is involved in three subsystems of the key reconstruction

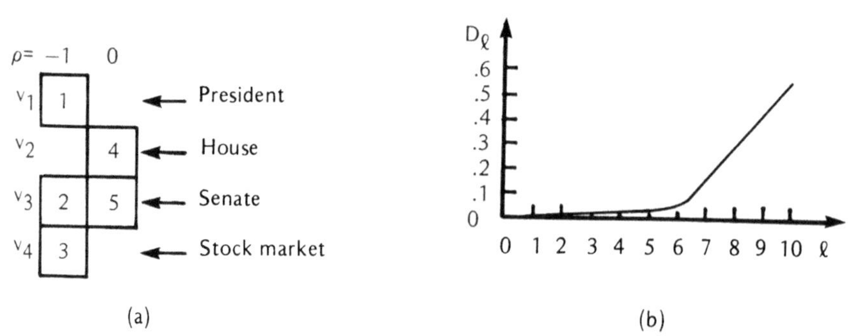

Figure 9.9. Illustration to Table 9.5 (Example 9.4).

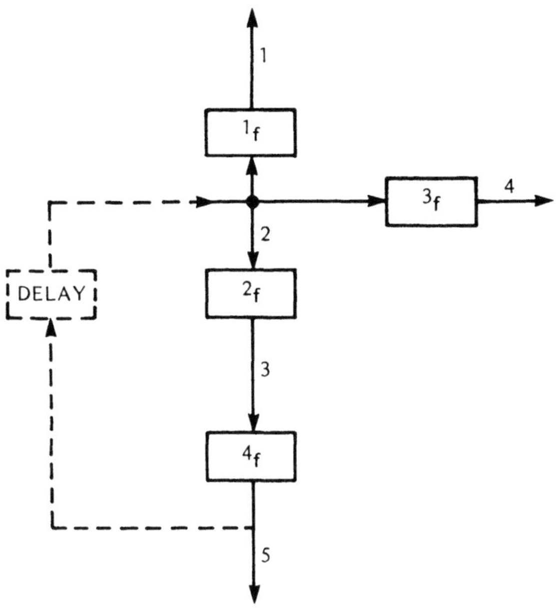

1	2	^1f
0	0	1
0	1	.25
1	0	.5
1	1	1

2	3	^2f
0	0	.75
0	1	1
1	0	.25
1	1	1

2	4	^3f
0	0	1
0	1	.5
1	0	.25
1	1	1

3	5	^4f
0	0	.75
0	1	.5
1	0	1
1	1	1

Figure 9.10. Key reconstruction hypothesis in Example 9.4 (stock market and U.S. Federal Government).

hypothesis): it is related, with the same degree of significance (compare admissible hypotheses at levels 6 and 7), to the President's party affiliation during the same time interval and house control during the next interval.

More specific interpreted conclusions can be derived from the possibility distributions of the subsystems of the key hypothesis in Figure 9.10 and, possibly, from the overall system reconstructed from the key hypothesis (employing the various reconstruction characteristics discussed in Sections 4.9 and 4.10), which the GSPS would make available upon request. We leave these more specific interpretations of the results to the interested reader.

Example 9.5. This example describes a segment of a typical interaction between an ecologist and the GSPS. It illustrates a possible use of the GSPS in dealing with a combination of ecological and climatological data. The data involve Oneida Lake, the largest lake wholly within New York State (33.6 km long and 8.8 km wide on the average). The following ten variables are considered:

v_1—total zooplankton biomass (ng/l);
v_2—total phytoplankton biomass (ng/l);
v_3—chlorophyll *a* (ng/l);

v_4—nitrate nitrogen (ng/l);
v_5—soluble reactive silicon (ng/l);
v_6—soluble reactive phosphorus (ng/l);
v_7—water temperature (°C);
v_8—solar radiation (langleys/day);
v_9—precipitation, water equivalent (in./day);
v_{10}—wind, average speed (miles/h).

When consulting the GSPS, the ecologist has a 10×193 data matrix that represents states of these variables observed daily for 193 days during the period from April 12, 1977 through October 21, 1977 (see Note 9.3).

Since the variables were measured with a rather high precision, but only 193 observations are available, resolution forms of the variables must be coarsened drastically to make it possible to derive meaningful results from the data. From the various options, the ecologist decides to use the criterion of equal frequency (a partitioning into equally populated blocks). He also decides to reduce the state set of each of the variables except v_9 into three states and to reduce the state set of v_9 into two states. Based on these decisions, the GSPS determines the resolution forms specified in Table 9.6 and employs these forms to convert the original data into their new form. These new data (a 10×193 matrix of integers), which are the basis for further processing, are not given here to save space.

It is clear that even after the drastic coarsening of the resolution forms, the number of observations is still too small when compared with the number of all overall states defined for the variables (204 times less). Hence, the ecologist decides to use solely the possibilistic option, which is considerably less demanding on data size, and to explore some meaningful subsets of the variables.

TABLE 9.6
Resolution Forms in Example 9.5 (Oneida Lake)

	State identifiers		
Variables	0	1	2
v_1 (zooplankton)	[1.5–147.7)	[147.7–215.2)	[215.2–338.9]
v_2 (phytoplankton)	[202.3–2170.4)	[2170.4–5122.6)	[5122.6–14963.2]
v_3 (chlorophyll)	[1.3–8.3)	[8.3–12.7)	[12.7–27.5]
v_4 (nitrogen)	[0.0–54.5)	[54.5–253.8)	[253.8–543.8]
v_5 (silicon)	[25.0–337.7)	[337.7–605.6)	[605.6–1364.9]
v_6 (phosphorus)	[1.4–2.1)	[2.1–4.4)	[4.4–24.0]
v_7 (temperature)	[2.2–15.2)	[15.2–20.2)	[20.2–23.4]
v_8 (solar radiation)	[0.0–221.2)	[221.2–442.3)	[442.3–663.5]
v_9 (precipitation)	[0.0]	[0.0–1.75]	
v_{10} (wind)	[3.7–7.6)	[7.6–10.2)	[10.2–18.6]

TABLE 9.7
Admissible Reconstruction Hypotheses Based on Variables v_1–v_6 of Example 9.5 (Oneida Lake)

l	Structure	D_l
1	12345/12356	0.0021
2	12345/1356	0.0049
3	1345/1356/1235	0.0098
4	1356/1235/345	0.0242
5	1356/123/345	0.0386
6	136/123/345/356	0.0544

First, reconstruction properties of a system containing only variables v_1–v_6 are determined. Admissible reconstruction hypotheses at refinement levels 1–6 are listed in the usual form in Table 9.7; distances of those at higher refinement levels are unacceptably large according to criteria specified by the ecologist. A block diagram of the most refined admissible hypothesis, which we may refer to as hypothesis **SF**, is given in Figure 9.11.

A number of interpreted results can be derived from the solution set in Table 9.7. One of them is the significance of variable 3 (chlorophyll), which is strongly interrelated with all the other variables. Hypothesis **SF** (Figure 9.11) can also be used as a guideline for choosing appropriate subsets of variables for further, more specific studies. Since variable 1 (zooplankton) is of primary interest in this investigation, the ecologist decides to exclude variables 4 and 5, which are not directly related to variable 1 in **SF**, from further studies. In addition, he excludes variable 2 (phytoplankton) on the grounds that it is well represented by variable 3 (chlorophyll); this is due to a strong relationship between the two variables as well as their ecological meaning. The remaining variables are 1, 3, 6.

To investigate variables 1, 3, 6 more thoroughly, the ecologist requests the determination of all admissible masks with three generating sampling variables that are

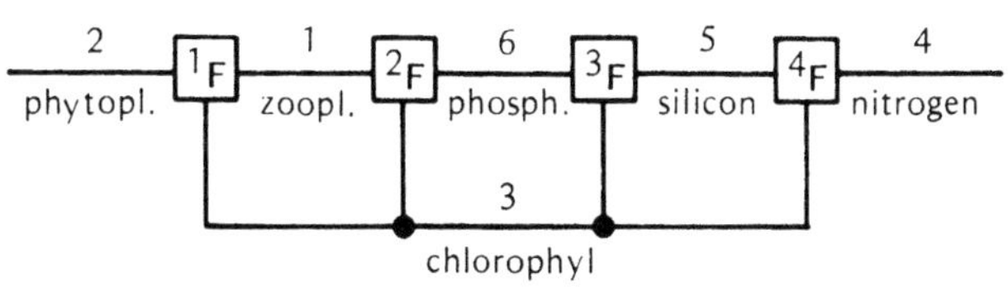

Figure 9.11. Key reconstruction hypothesis SF based on variables v_1–v_6 and memoryless mask in Example 9.5 (Oneida Lake).

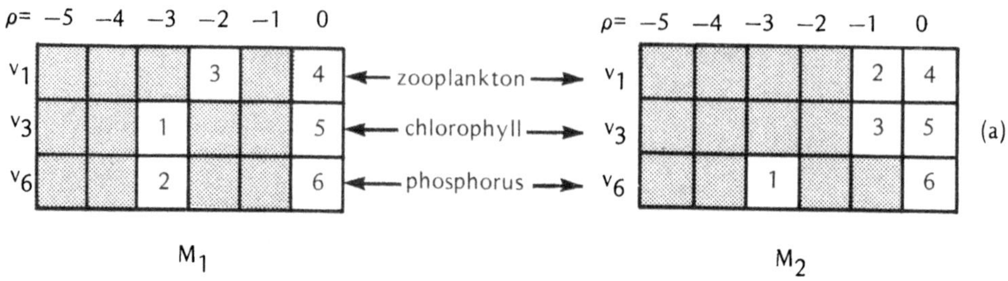

Figure 9.12. Admissible masks in Example 9.5 (Oneida Lake).

defined within the largest acceptable mask with $\Delta \mathbf{M} = 6$. This request results in two admissible masks, M_1 and M_2, which are specified in Figure 9.12. When reconstructability analysis is applied to behavior systems derived from the data for these masks, the admissible reconstruction hypotheses listed in Table 9.8 are obtained. Refinements beyond level 10 are not included since the distance is not acceptable according to the ecologist's criteria. Observe that identifiers 1–6 of sampling variables have different meanings in the two masks.

Table 9.8 is a rich source of interpreted conclusions, including those regarding directions of variables, but they may involve other than systems considerations and, hence, I omit them here. I describe, however, one additional interaction between the ecologist and GSPS. The ecologist decides to supplement the three ecological variables—v_1 (zooplankton), v_3 (chlorophyll), v_6 (phosphorus)—with four climatological

TABLE 9.8
Admissible Reconstruction Hypotheses Based on Masks M_1 and M_2 in Example 9.5: Oneida Lake (see Figure 9.12).

Mask M_1			Mask M_2		
l	Structure	D_l	l	Structure	D_l
1	13456/23456	0.0006	1	12346/12356	0.0003
2	1356/23456	0.0019	2	1246/1356/2346/2356	0.0010
3	1356/2456/3456	0.0032	3	1356/2346/2356	0.0017
4	156/2456/3456	0.0094	4	1356/234/2356	0.0033
5	156/2456/346	0.0137	5	1356/2356/34	0.0046
6	156/2456/34	0.0159	6	1356/256/34	0.0087
7	156/245/34/256	0.0232	7	1356/25/34	0.0148
8	15/245/34/256	0.0339	8	156/25/34/356	0.0231
9	15/245/34/26	0.0401	9	16/25/34/356	0.0341
10	15/24/34/26/25	0.0549	10	16/25/34/36/56	0.0542

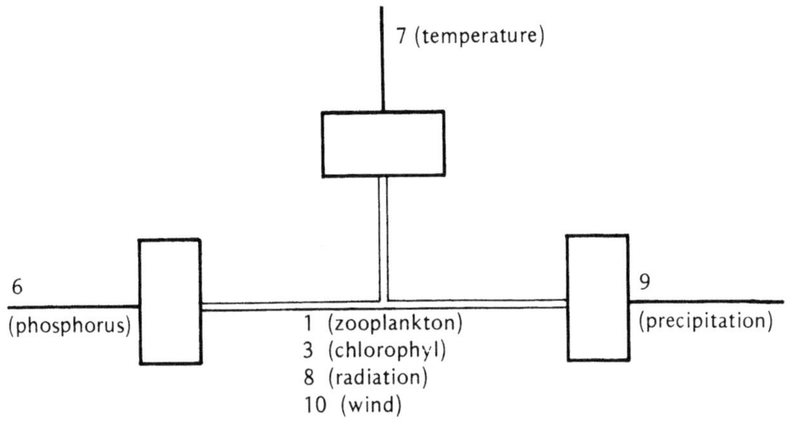

Figure 9.13. The most refined admissible reconstruction hypothesis for variables 1, 3, 6, 7, 8, 9, 10 and memoryless mask in Example 9.5 (Oneida Lake).

variables—v_7 (water temperature), v_8 (solar radiation), v_9 (precipitation), v_{10} (wind)—and perform reconstructability analysis on the behavior system based on these seven variables and memoryless mask. A block diagram of the most refined reconstruction hypothesis is shown in Figure 9.13. Among possible interpreted results, it is particularly interesting for the ecologist that variable 10 (wind) has the same significance as variable 8 (solar radiation). This seems to be a consequence of the morphological fact that Oneida Lake is unusually shallow.

The ecologist may now use the GSPS to further study the central variables 1, 3, 8, 10 in the last successful hypothesis (Figure 9.13), but the previous description seems sufficient to illustrate the role of the GSPS in the active and creative process of scientific inquiry.

Example 9.6. A case study, in which one of the existing implementations of the GSPS was applied to open heart surgery, is outlined in this example (Note 9.4). The following six physiological variables, which are monitored at intervals of 30 sec (i.e., at time instances 0, 30 sec, 60 sec, 90 sec, . . .) on a patient during an open heart surgery, form the source system of interest:

v_1—systolic blood pressure (SBP);
v_2—mean blood pressure (MBP);
v_3—central venous pressure (CVP);
v_4—cardiac output (CO);
v_5—heart rate (HR);
v_6—left atrial pressure (LAP).

Five states are defined for each variable; they are labeled 1, 2, 3, 4, 5. State 3 represents the range of appearances of the corresponding attribute that is medically established as normal. States 1 and 5 represent critical ranges, which require that appropriate actions be taken immediately to induce desirable changes or, else, the patient would die. States 2 and 4 are undesirable (dangerous), but not critical. The aim of this study is to determine reconstruction properties of the system whose knowledge would help the anesthesiologist to properly deal with undesirable states of the variables during the surgery.

A probabilistic behavior function associated with this case study and based on the mask defined in Figure 9.14a is given in Table 9.9. It characterizes an average male patient, about 45 years old, and was derived from data collected during 100 successful operations. Observe that (i) the normal overall state, in which all variables are in state 3, is by far the most probable state, and (ii) some undesirable states do not occur for some variables at all (e.g., states 1, 2, and 5 of variable v_1) for this category of patients. It is likely that different behavior functions would be obtained for different categories of patients (female patients, different ages, etc.). The mask for which the behavior function is defined was determined by the GSPS as the best mask in the following sense: it is the only mask, within $\Delta M = 2$ and with six sampling variables, that results in the most deterministic ST-system. The ST-function, which is of considerable interest to the anesthesiologist, was also determined by the GSPS, but it is not given here due to its large size (a 27×27 matrix or a list with 729 entries).

The GSPS can be used in many other ways to process the data. As an illustration, let a particular option of reconstructability analysis be applied to the behavior function in Table 9.9. The option is characterized by the following requirements:

- analyze C-structures first and, then, analyze G-structures in the r-equivalence classes of the most refined admissible C-structures;
- largest acceptable increase in distance between refinement levels of C-structures is 0.016;

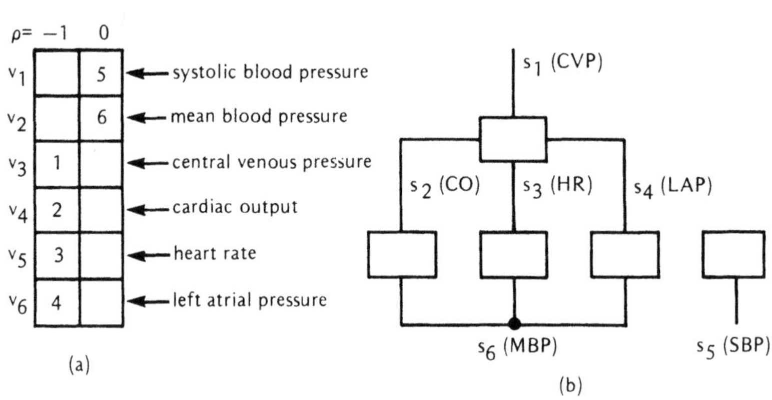

Figure 9.14. The most refined admissible G-structure in Example 9.6 (open heart surgery).

TABLE 9.9
Behavior Function in Example 9.6 (Open Heart Surgery)

s_1	s_2	s_3	s_4	s_5	s_6	$f(\mathbf{c})$
c = 3	1	3	3	4	5	0.005
3	2	3	1	3	3	0.025
3	2	3	3	3	3	0.171
3	2	3	3	3	4	0.005
3	2	3	3	3	5	0.005
3	2	3	3	4	3	0.005
3	2	3	3	4	4	0.005
3	2	3	3	4	5	0.005
3	2	3	3	5	5	0.030
3	2	3	4	3	3	0.010
3	2	5	3	4	5	0.005
3	3	3	1	3	3	0.005
3	3	3	3	3	3	0.442
3	3	3	3	3	4	0.015
3	3	3	3	3	5	0.005
3	3	3	3	4	3	0.025
3	3	3	3	4	4	0.005
3	3	3	3	4	5	0.010
3	3	3	3	5	5	0.025
3	3	3	4	3	3	0.035
4	2	3	3	3	3	0.025
4	3	3	3	3	3	0.101
4	3	3	3	3	4	0.005
4	3	3	3	3	5	0.005
4	3	3	3	4	4	0.010
4	3	3	3	4	5	0.010
4	3	3	3	5	5	0.005

- *G*-structures in the analyzed *r*-equivalence classes are acceptable only if their distance is the same as the distance of the corresponding *C*-structure;
- the city-block (Hamming) distance is required;
- $\Delta = 0.0001$ for the iterative join procedure.

According to these requirements, admissible *C*-structures at the individual levels of refinement are listed in Table 9.10a. The only admissible *G*-structure is 1234/26/36/46/5; its best predecessors in the analyzed *r*-equivalence class are listed in Table 9.10b and its block diagram is shown in Figure 9.14b. This structure should be of considerable help to the anesthesiologist. For example, it indicates that the systolic blood pressure is highly independent of the other variables, particularly heart rate and mean blood pressure (see Table 9.10a for $l = 1, 2$) and that no more than four of the six variables have to be considered simultaneously. Although a number of additional interpreted results can be

TABLE 9.10
Admissible Reconstruction Hypotheses of the Behavior Function in Table 9.9 (Example 9.6)—Open Heart Surgery

(a) *C*-structures			(b) *G*-structures		
l	Structure	D_l	l	Structure	D_l
1	12346/12456	0.0	1	1234/236/246/346/5	0.0127
	12345/12346	0.0	2	1234/236/346/5	0.0127
2	12346/1245	0.0	3	1234/346/26/5	0.0127
3	12346/145	0.005	4	1234/26/36/46/5	0.0127
4	12346/15	0.0089			
5	12346/5	0.0099			
6	1234/2346/5	0.0127			

derived from the resulting structure system, it is better to leave further interpretation to the appropriate medical experts.

Example 9.7. This example illustrates the use of the GSPS in archaeology. It describes a small part of a large study performed during the period 1978–1980 (Note 9.5). The object of study is Brown Knoll, a prehistoric settlement in East-Central New York State. The settlement is atop a large gravel knoll jutting out from the valley wall in Colliersville, as the juncture of the Susquehanna River and Schenevus Creek. From previous archaeological work done at the site, it is known that Brown Knoll consists of a number of loci representing the material remains of various kinds of activities of prehistoric hunter–gatherers—stone toolmaking, bone and woodworking, nut gathering, fire and house building, etc.—ranging over a span of time from approximately 3,000 B.C. to 500 B.C. The site itself is not commonly excavated by archaeologists because it lacks vertical stratigraphy within its soil, which is necessary to separate the various prehistoric occupations in time. It has also been partly disturbed by plowing in recent times.

In excavating the site, two kinds of units were used: (i) 1,068 shovel test pits (30 cm in diameter, to a depth of 30 cm—the maximum depth of archaeological material in this case), spaced 5 m apart; and (ii) 291 larger excavation units (1 × 1 m square). The shovel test pits were used to locate high artifact concentrations, which were then excavated in the larger, square units. The set of space locations of the individual excavation units (either test pits or squares) represents the support set in this example. Although the regular distribution of test pits in space can be employed for exploring spatial relationship by analyzing a set of masks, this example is restricted to memoryless masks.

Four attributes are defined by the number of artifacts found at the individual excavation units in each of the following four categories of waste material: decortification flakes, blocks, shatter, and flakes. Two to five states are distinguished for each of the corresponding variables. They were determined by the GSPS on the basis of the equal frequency requirement. Examples of two sets of the resulting resolution forms are specified in Table 9.11. Several other sets of resolution forms, all based on the frequency requirement, were determined in the original study and reconstructability analysis of the corresponding memoryless behavior systems performed for each of them. It was found that these changes in resolution forms of the variables involved had little effect on the solution set. This fact makes the admissible reconstruction hypotheses well supported.

In this example, only set II of resolution forms (defined in Table 9.11) is considered. It is applied to three related data sets that are based on

- unplowed shovel test pits;
- plowed shovel test pits;
- all shovel test pits in the site.

Probabilistic behavior functions derived from these three data sets, which are denoted f_1, f_2, and f_3, respectively, are specified in Table 9.12. Admissible sets of reconstruction hypotheses (based only on C-structures) for each of them are listed in Table 9.13; the corresponding dependencies of D_l on l are shown in Figure 9.15.

TABLE 9.11
Examples of Two Sets of Resolution Forms in Example 9.7
(Archaeological Excavations)

Set I

Variables	State identifiers 0	1	2	3
v_1 (decort. flakes)	0	1,2	≥3	—
v_2 (blocks)	0	1,2	≥3	—
v_3 (shatter)	0	1,2	≥3	—
v_4 (flakes)	0–2	3–8	9–14	≥15

Set II

Variables	State identifiers 0	1	2	3	4
v_1 (decort. flakes)	0	≥1	—	—	—
v_2 (blocks)	0,1	≥2	—	—	—
v_3 (shatter)	0	≥1	—	—	—
v_4 (flakes)	0	1	2	3,4	≥5

TABLE 9.12
Probabilistic Behavior Functions in Example 9.7
(Archaeological Excavations)

v_1	v_2	v_3	v_4	$f_1(\mathbf{c})$	$f_2(\mathbf{c})$	$f_3(\mathbf{c})$
c=0	0	0	0	0.628	0.211	0.437
0	0	0	1	0.103	0.137	0.119
0	0	0	2	0.050	0.114	0.081
0	0	0	3	0.042	0.123	0.079
0	0	0	4	0.056	0.142	0.095
0	0	1	0	0.012	0.016	0.014
0	0	1	1	0	0.014	0.007
0	0	1	2	0.004	0.010	0.007
0	0	1	3	0.002	0.012	0.007
0	0	1	4	0.010	0.027	0.018
0	1	0	0	0.004	0.012	0.008
0	1	0	1	0.005	0.006	0.006
0	1	0	2	0.002	0.010	0.006
0	1	0	3	0.004	0.014	0.009
0	1	0	4	0.016	0.020	0.018
0	1	1	0	0.002	0.004	0.003
0	1	1	1	0	0	0
0	1	1	2	0	0.004	0.002
0	1	1	3	0	0	0
0	1	1	4	0.004	0.008	0.006
1	0	0	0	0.002	0.023	0.011
1	0	0	1	0.007	0.012	0.009
1	0	0	2	0.004	0.008	0.006
1	0	0	3	0.005	0.025	0.014
1	0	0	4	0.016	0.027	0.020
1	0	1	0	0.002	0.002	0.002
1	0	1	1	0.004	0.002	0.003
1	0	1	2	0.002	0	0.001
1	0	1	3	0	0.008	0.004
1	0	1	4	0.009	0	0.005
1	1	0	0	0	0	0
1	1	0	1	0.002	0	0.001
1	1	0	2	0	0.002	0.001
1	1	0	3	0	0	0
1	1	0	4	0.007	0.002	0.005
1	1	1	0	0	0	0
1	1	1	1	0	0	0
1	1	1	2	0	0	0
1	1	1	3	0	0	0
1	1	1	4	0.002	0.002	0.002

TABLE 9.13
Admissible Sets of Reconstruction Hypotheses in Example 9.7 (Archaeological Excavations)

(a) Unplowed shovel test pits (f_1)			
1	134/234	0.0006	(b), (c)
2	134/24	0.0016	neither
3	14/24/34	0.0059	(c)
4	14/24/3	0.0126	(b), (c)
5	24/1/3	0.0286	(b)
6	1/2/3/4	0.0400	(b), (c)
(b) Plowed shovel test pits (f_2)			
1	134/234	0.0023	(a), (c)
2	13/234	0.0039	neither
3	1/234	0.0058	neither
4	14/24/3	0.0075	(a), (c)
5	24/1/3	0.0095	(a)
6	1/2/3/4	0.0120	(a), (c)
(c) All shovel test pits (f_3)			
1	134/234	0.0007	(a), (b)
2	14/234	0.0019	neither
3	14/24/34	0.0036	(a)
4	14/24/3	0.0077	(a), (b)
5	14/2/3	0.0140	neither
6	1/2/3/4	0.0208	(a), (b)

Example 9.8. This example is based on a source system defined by one of my assistants, Michael Pittarelli, according to the instructions in Exercise 2.7. He defined the following five variables on himself, each of which recognizes four states (low—0, medium—1, high—2, very high—3):

v_1—energy level;
v_2—amount of protein consumed;
v_3—amount of simple carbohydrate consumed;
v_4—amount of exercise;
v_5—amount of sleep (previous night).

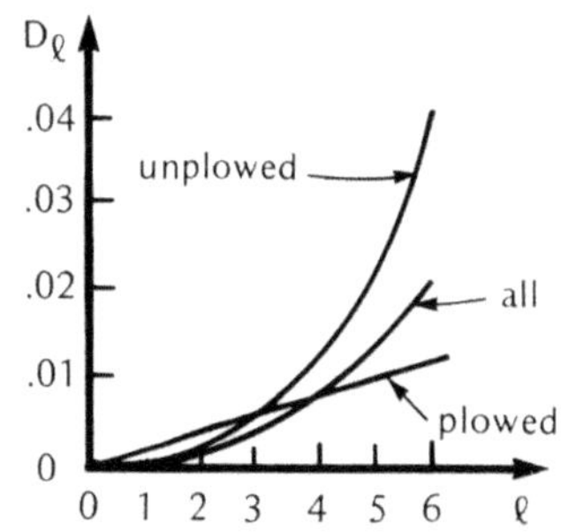

Figure 9.15. A summary of reconstructability analysis in Example 9.7 (archaeological excavations).

Each morning, the states of variables v_1–v_4 are recorded for the previous day, and the state of variable v_5 is recorded for the current day. The following is a sample of data collected for the initial period of 50 days:

$$v_{1,t} = 21101101022112221021022121021212212133321012222102$$
$$v_{2,t} = 22113110121022321102111211211021212221311011121212$$
$$v_{3,t} = 21111312211231110133311222322221132201232323121331$$
$$v_{4,t} = 30100102023002010021022010002013001202030011101200$$
$$v_{5,t} = 01211222133132200221223331321301003211001121120011$$

When the data are analyzed with the aim of determining admissible behavior systems for $\Delta\mathbf{M} = 2$ and the usual requirement (uncertainty, complexity), the set of admissible masks is the same for both possibilistic and probabilistic options. They are specified in Figure 9.16a, where the sampling variables have the same meaning as defined in Figure 9.16b. The dependence of the two kinds of generative uncertainties on mask size is depicted graphically in Figure 9.16c. Reconstructability analysis can now be performed for behavior systems based on some of these admissible masks. As an illustration, it is described here only for the mask with six sampling variables ($i = 6$, Figure 9.17a), possibilistic option (the number of observations is not sufficient for probabilistic analysis), and C-structures. The dependence of D_l on l for this mask is shown in Figure 9.17b. We can see that the refinement levels are naturally clustered into $l = 1$, 2, 3 and the rest. For the first three levels, the admissible reconstruction hypotheses are unique:

13456/23456	($l = 1$),
1345/2345/1456/2456	($l = 2$),
1345/1456/2456	($l = 3$).

Each of these hypotheses is also well discriminated (by the information distance) from competing hypotheses at the same refinement level. For $l \geq 4$, on the contrary, large clusters of admissible reconstruction hypotheses occur at each level. Hence, the hypothesis 1345/1456/2456 emerges clearly as the most informative one. Its block diagram is shown in Figure 9.17c.

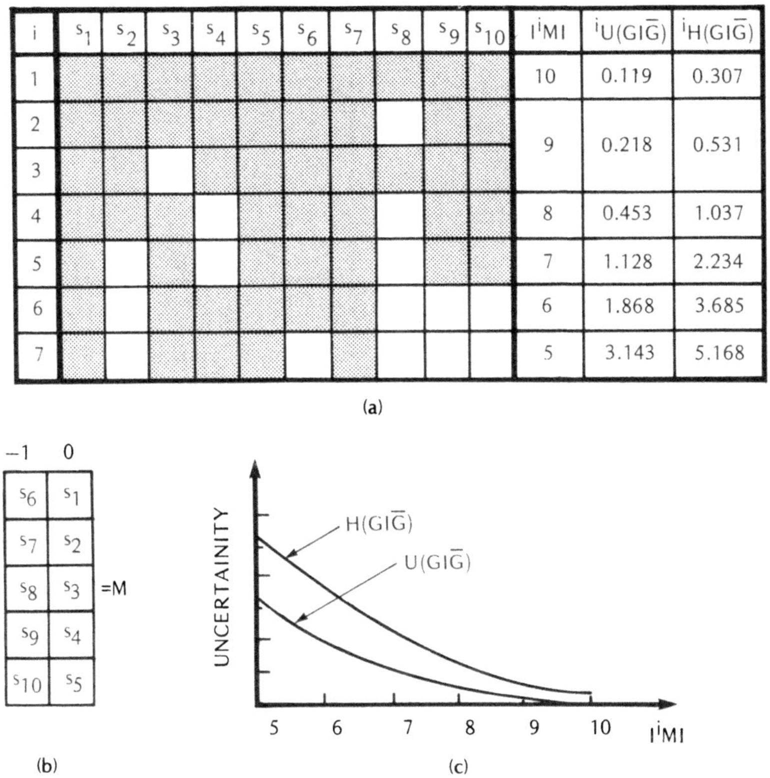

Figure 9.16. Admissible masks within the largest acceptable mask M in Example 9.8 (self-observation).

To express the effect of variables 1, 2, 4, and 5 upon variables 3 and 6, which is of primary interest in the investigated system, appropriate directions must be introduced for the variables. Variable 6 is determined either from ^2f or ^3f (but not both). This dilemma can be resolved by determining the generative uncertainties of both these alternatives and selecting the one with smaller uncertainty. One of the two functions (^2f or ^3f) is then of no use, and the corresponding subsystem can be excluded. Another possibility of resolving the dilemma associated with the control of variable 6 is to combine subsystems 1456 and 2456 into one larger subsystem 12456, as illustrated in Figure 9.8d.

The source system defined in this example (or its variations) can also be studied from the standpoint of a metasystem. In such a case, data would be partitioned according to periods characterized by special features such as special diets, excessive exercise (before and during athletic competitions), and the like. Each characteristic subset of data would be analyzed independently of the other subsets, and the resulting behavior or structure systems would then be integrated into one metasystem by an appropriate replacement procedure.

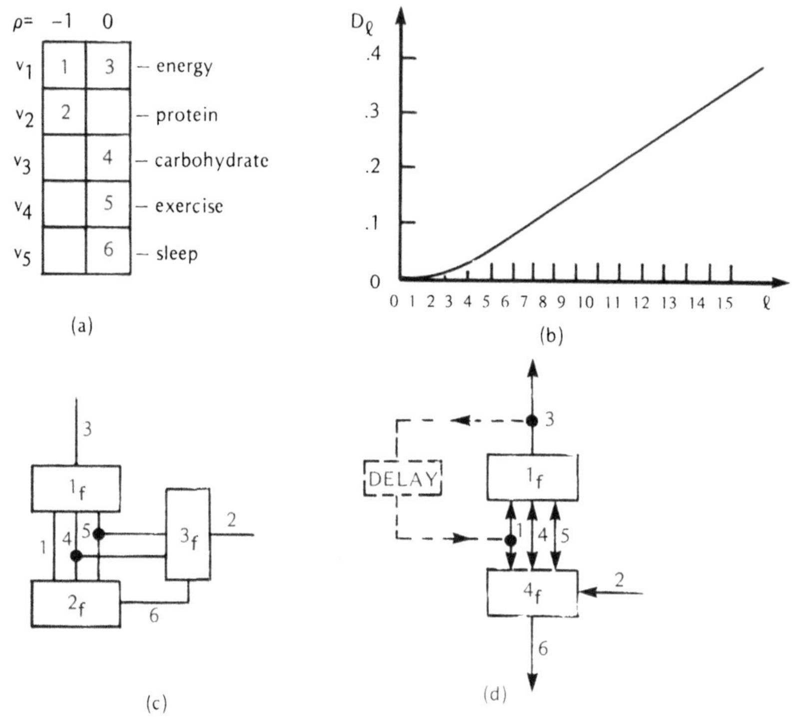

Figure 9.17. Illustration to reconstructability analysis in Example 9.8 (self-observation).

Example 9.9. Consider a simple switching circuit consisting of two elements representing logic functions AND and OR that are connected as shown in Figure 9.18a. For all states of the input variables v_1, v_2, v_3, v_4, the output variables v_5, v_6 can assume states specified in Figure 9.18b. Due to states *a* and *b*, indicated in the figure, the system is clearly non-deterministic. It has a memory ability. When $v_1 = v_2 = 1$ and $v_3 = v_4 = 0$, the actual state of the output variables v_5 and v_6 contains information about the last change of the input variables. For example, if the last change involved only variables v_1 and v_2, then $v_5 = v_6 = 0$; if it involved only variables v_3 and v_4, then $v_5 = v_6 = 1$.

Assume now that, due to some defect in the couplings of the system, state *a* does not occur anymore, but everything else remains the same. Assume further that we have no direct access to the system to identify the defect. Then, one way of identifying it indirectly is to analyze reconstruction properties of the new behavior system (the one without state *a*) in the neighborhood of the structure representing the correct system given in Figure 9.18a. When considering all states listed in Figure 9.18b except *a* as possible states of the system and the remaining ones as its impossible states, and after analyzing (in the possibilistic fashion) *C*-refinements of the correct structure 1256/3456, we obtain two structures with zero distance: 125/3456/156 and 125/3456/256. When further refining both of them, we find that only 125/3456 has zero distance. Refining it

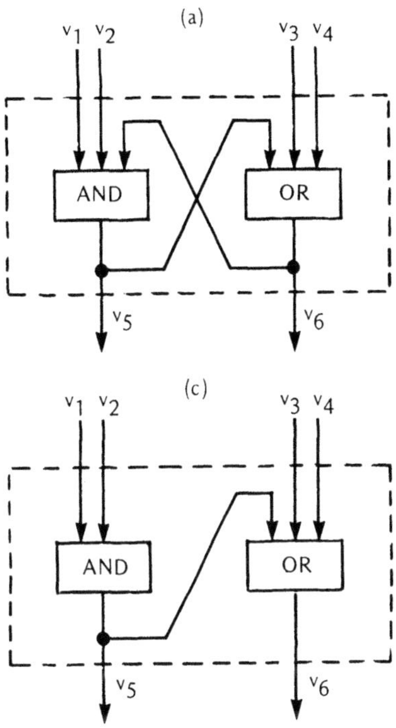

(b)

	v_1	v_2	v_3	v_4	v_5	v_6
	0	0	0	0	0	0
	0	0	0	1	0	1
	0	0	1	0	0	1
	0	0	1	1	0	1
	0	1	0	0	0	0
	0	1	0	1	0	1
	0	1	1	0	0	1
	0	1	1	1	0	1
	1	0	0	0	0	0
	1	0	0	1	0	1
	1	0	1	0	0	1
	1	0	1	1	0	1
a =	1	1	0	0	0	0
b =	1	1	0	0	1	1
	1	1	0	1	1	1
	1	1	1	0	1	1
	1	1	1	1	1	1

Figure 9.18. Identification of a coupling defect (Example 9.9).

further, we find that none of its immediate refinements has zero distance. Structure 125/3456 represents thus the new system; its block diagram is shown in Figure 9.18c. The defect is now obvious: variable v_6 was decoupled from the AND element.

The RC-procedure was sufficient in this case since we expected changes in the couplings, but no disintegration of the elements. The user could also have employed the GSPS in an interactive mode and requested the evaluation of only those structures that correspond to coupling defects, such as structures 1256/346, 125/3456, 256/3456, 156/3456, etc.

I must add that this example does not take into account the way in which the system is physically implemented. Each particular implementation may have some additional properties that must be considered. Nevertheless, the GSPS can always be used in assisting the user with this kind of problems.

Example 9.10. Hardware monitoring, as one approach to computer performance evaluation, is briefly introduced in Example 3.8. An alternative use of the GSPS in this application area is illustrated here by describing a particular case study.

It was observed that the CPU utilization of a computer system had periodically dropped from almost 100% to 80% or less for periods of about 20 min on average.

Hardware monitoring of nine carefully chosen variables of the computer system was arranged in order to find conditions which contribute to this peculiar performance.

The monitoring was done in a time period during which one of the drops in the CPU utilization occurred. The total period of monitoring was 49 min long, with a time resolution level of 30 sec. That means that 98 observations were made. In each of the observations, percentages of activities of all the monitored units during the respective interval of 30 sec were recorded. The GSPS was then used to convert these raw data into data based on Boolean variables. States of these new variables were required to discriminate between "high" and "low" activities of the corresponding computer units on the basis of their average utilizations determined from the raw data. All the Boolean variables $v_i (i \in N_9)$ were defined in the same way as

$$v_i = \begin{cases} 0, & \text{if the utilization is less than average } a_i \\ 1, & \text{otherwise.} \end{cases}$$

The following list describes the assignment between the variables and computer units, and specifies for each variable v_i the average utilization a_i:

v_1—CPU ($a_1 = 90\%$);
v_2—supervisor ($a_2 = 43\%$);
v_3—problem activity ($a_3 = 45\%$);
v_4—channel 1 ($a_4 = 10\%$);
v_5—channel 3 ($a_5 = 10\%$);
v_6—channel 5 ($a_6 = 10\%$);
v_7—unit 160, connected to CPU through channel 1 ($a_7 = 3\%$);
v_8—unit 162, connected to CPU through channel 1 ($a_8 = 56\%$);
v_9—unit 163, connected to CPU through channel 1 ($a_9 = 44\%$).

The determination of the values $a_i (i \in N_9)$ from the raw data was the first use of the GSPS in this study. Its second use was the transformation of the raw data into the Boolean form. Since the data matrix is rather large (9×98), it is not reproduced here.

The Boolean data were then processed by the GSPS for the memoryless mask and a probabilistic behavior function was determined. The reconstruction properties of the function were then analyzed for such C-structures which do not require the use of the iterative join procedure. The city-block (Hamming) distance was requested by the user. All admissible reconstruction hypotheses for the given requirements are listed in Table 9.14; the dependence of the distance on the refinement level is illustrated graphically in Figure 9.19. By following the refinement path summarized in the table, the computer analyst can determine which projections of the overall behavior function are significant, to which degree, and where in the refinement lattice are they located. His overall aim is to utilize this information for developing strategies through which to minimize underutilization of the computer system.

TABLE 9.14
Admissible Reconstruction Hypotheses in Example 9.10 Computer Performance Evaluation

l	Structure	$\delta_{1,l}$
1	12345678/12346789	0.0000
	12345678/12456789	0.0000
	12345678/13456789	0.0000
	12456789/23456789	0.0000
2	12345678/1246789	0.0000
	12345678/1346789	0.0000
	2345678/12456789	0.0000
	12345678/1456789	0.0000
3	12345678/146789	0.0000
4	12345678/14789	0.00825
5	12345678/1789	0.01515
6	1235678/1245678/1789	0.02025
7	235678/1245678/1789	0.02026
8	235678/1245678/179	0.02715
9	235678/1245678/19	0.0391
10	234678/125678/245678/19	0.0501
11	235678/12567/245678/19	0.0551
12	235678/12567/24578/19	0.06655
13	235678/1257/24578/19	0.08285
14	235678/1257/2478/19	0.09945
15	235678/1257/478/19	0.1113
16	235678/1257/47/19	0.12215
17	23578/25678/1257/47/19	0.1469
18	23578/2567/1257/47/19	0.1594
19	23578/267/1257/47/19	0.16855
20	23578/26/1257/47/19	0.17775
21	23578/26/1257/4/19	0.1923
22	23578/26/127/4/19	0.20675
23	2357/3578/26/127/4/19	0.21265
24	2357/578/26/127/4/19	0.24105
25	235/578/26/127/4/19	0.2707
26	235/578/6/127/4/19	0.2883
27	235/57/78/6/127/4/19	0.31105
28	235/78/6/127/4/19	0.34045
29	23/35/78/6/127/4/19	0.34355
30	23/35/78/6/127/4/9	0.3507
31	23/35/78/6/17/27/4/9	0.36715
32	23/5/78/6/17/27/4/9	0.3723
33	3/4/5/6/78/17/27/9	0.39955
34	3/4/5/6/8/9/17/27	0.41165
35	2/3/4/5/6/8/9/17	0.4209
36	1/2/3/4/5/6/7/8/9	0.66085

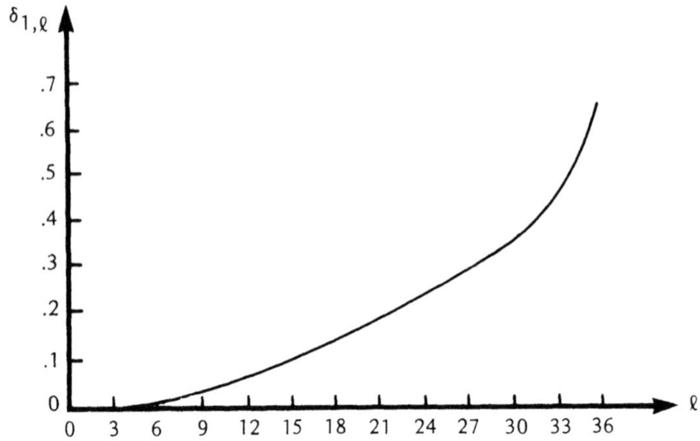

Figure 9.19. Distance versus refinement level in Example 9.10 (computer performance evaluation).

Example 9.11. Numerous functions have been proposed to evaluate offensive performance is baseball, each defined in terms of several variables that characterize offensive actions of individual players. Ten of these functions, referred to as offensive performance estimators, are defined in the paper "An evaluation of major league baseball offensive performance models" by J. M. Bennett and J. A. Flueck (*The American Statistician*, February 1983, pp. 76–81). Data consisting of five of these estimators are given in this paper for 33 leading players in both National League and American League; the estimators are called batting average (BA), slugging percentage (SP), offensive average (OA), offensive performance average (OPA), and expected run production average (ERPA).

One way of evaluating these estimators and determining their relative significance is to perform reconstructability analysis of the data. The data system consists of five variables (the five estimators), one support (the population of 33 leading baseball players), and 5×33 data matrix, which is not reproduced here. Before performing reconstructability analysis, the user decided to coarsen resolution forms of the variables by the equal frequency option to three states for each variable. The result is shown in Table 9.15a. The data were then sampled and a possibilistic behavior function based on Eq. (3.33) determined. Finally, admissible reconstruction hypotheses were determined within the set $\mathscr{C}_5$; they are listed in Table 9.15b. Variables are denoted by integers and have the following meaning: 1—BA; 2—SP; 3—OA; 4—OPA; 5—ERPA. The increase in distance (loss of information) with increasing refinement is shown in Figure 9.20. It is clear that the rate of change is small for $l \leq 6$ and increases drastically for $l > 6$. Hence, the most informative reconstruction hypothesis is the one at level 6: 14/24/34/45. This clearly indicates that variable 4 (OPA) represents the most significant performance estimator since each of the other estimators is directly derivable from it. Additional conclusions can be obtained by inspecting the whole set of admissible reconstruction hypotheses.

TABLE 9.15
Illustration to Example 9.11 (Baseball)

(a) Resolution forms

Batting average (BA):
0—[0, 0.277); 1—(0.277, 0.307); 2—[0.307, 0.8]
Slugging percentage (SP):
0—[0, 0.443); 1—[0.443, 0.461); 2—[0.461, 0.8]
Offensive average (OA):
0—[0, 0.515); 1—[0.515, 0.533); 2—[0.533, 0.8]
Offensive performance average (OPA):
0—[0, 0.472); 1—[0.472, 0.491); 2—[0.491, 0.8]
Expected run production average (ERPA):
0—[0, 0.147); 1—[0.147, 0.16); 2—[0.16, 0.8].

(b) Admissible reconstruction hypotheses

l	Structure	Distance
1	1234/1345	0.0069
2	1234/145	0.0221
3	1234/45	0.0357
4	123/234/45	0.0580
5	124/34/45	0.0771
6	14/24/34/45	0.1030
7	1/24/34/45	0.1778
8	1/2/34/45	0.2777
9	1/2/3/45	0.3975
10	1/2/3/4/5	0.5309

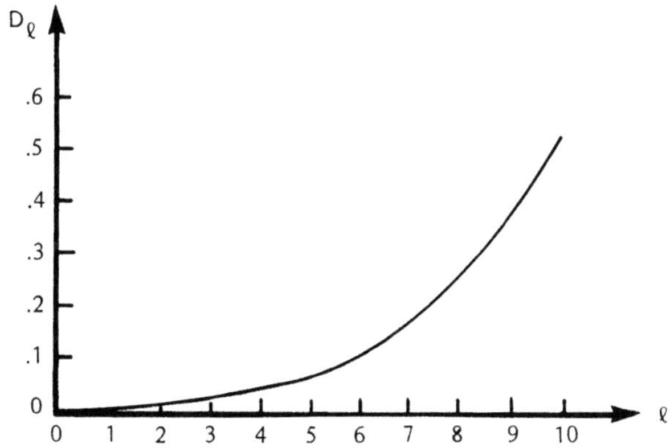

Fig. 9.20. Distance versus refinement level in Example 9.11 (baseball).

9.8. GSPS EVOLUTION

> *As our ability to solve problems expands, the scale of the problems attacked themselves seem to expand at a similar rate. As a result there always exist over the horizon new categories of problems of greater size to tackle.*
>
> —David M. Himmelbau

According to current terminology, the GSPS is an expert system. Brian Gaines and Mildred Shaw give an excellent characterization of expert systems†:

> Expert systems on larger computers allow the recorded mind of an expert, or the composite "mind" of many, to guide others less skilled in complex tasks. In using them you are discussing with a "colleague" problems of medical diagnosis or oil exploration, or the inventive processes of mathematical discovery. You tell them your problem and the information you have. You discuss it with them, query their judgement and ultimately come to a decision based on a collaboration with someone who may be long dead.
>
> Computer technology affects the core of our being because conversation is at the heart of human civilization. We are a species remarkable for our adaptive and learning capabilities, the effect of which is immensely amplified by our capability for conversation. Only one person need learn from experience. Others can be told and media allow the telling to transcend space and time. New media *do* change our world and the computer provides the first one for encoding not just a conversation but the capability to converse itself, not just a picture but the capability to be in the world portrayed.

According to John Sowa, "an expert system is a knowledge-based system that incorporates enough knowledge to reach expert levels of performance" [SO1]. While most expert systems described in the literature are designed to provide the user with the expertise in a traditional discipline (such as a specific subject area of medicine or law), the role of the GSPS is to assist the user in dealing with systems problems. Its expertise is thus systems knowledge and methodology, and, consequently, its utility transcends boundaries between the traditional disciplines.

When compared with expert systems of another kind, those which aim at solving problems in general (general problem solvers), the GSPS is restricted by focusing solely on systems aspects of problems. This restriction is intentional and reflects my philosophical view that a man–computer symbiosis represents the best arrangement for dealing with problems in general, and that the problems for which the computer symbiont has the best potential for being well utilized are precisely those recognizable as systems problems. The aim of the GSPS is to make most of this potential.

†*Nature*, **30**, April 28, 1983, p. 772.

One of the main characteristics of the GSPS architecture, which is not sufficiently emphasized in the previous text, is its evolutionary nature. To explain this aspect properly, I must first make it clear that the GSPS conceptual framework as well as its knowledge and methodological bases have not developed in isolation from the various traditional disciplines. In fact, the traditional disciplines were viewed as the only indigenous source of systems ideas from which the GSPS architecture should emerge.

Virtually all traditional disciplines have been involved, in one way or another, with systems problems of certain types and, consequently, have developed some methods for dealing with these problems. Although some of the disciplines have been more successful in this respect than others, each of them has been advancing the systems knowledge and methodology within its own context-dependent boundaries and with virtually no contribution to other disciplines.

I must emphasize that, to a large extent, the GSPS conceptual framework, as described in this book, has evolved (in the course of many years, say two decades or so) by the processes of distilling the notions of systems and associated problems from many distinct disciplines, abstracting them from their narrow contexts, categorizing them, and, finally, integrating them into a coherent whole. Although the current framework is rather stable, these processes are still on-going and may contribute to further extensions and other evolutionary changes in the framework in the future. Similar processes have also been involved in the evolution of the GSPS knowledge base and its methodological and metamethodological bases.

This gathering of systems concepts, methods, and knowledge from the various traditional disciplines is obviously only a part of the overall process by which the GSPS architecture has been evolving. In fact, it has been only a basis for the actual research on the GSPS, whose aim has been to make the conceptual framework as complete as possible by identifying and filling gaps in it, and to advance the knowledge and methodological bases in a comprehensive way, particularly in important problem areas that are not adequately developed.

One additional aspect of the evolutionary nature of the GSPS architecture is quite important: it is required that the GSPS be adaptive to the needs of its users. This means, in practical terms, that the GSPS is expected to keep records of all user–GSPS interactions and this information is then utilized to guide further research on the GSPS. Records of unsuccessful interactions are particularly important for this purpose, since they may indicate specific weaknesses in the user–GSPS interface, methodologically underdeveloped problem types, or even the need for extending the conceptual framework.

The outlined processes by which the GSPS architecture has been evolving are summarized by the diagram in Figure 9.21. Numbers attached to the individual connections in the diagram have the following meaning:

1—processes of distilling, abstracting, categorizing, and integrating relevant concepts, knowledge, and methods from the traditional disciplines;

2—four areas of research on the GSPS;

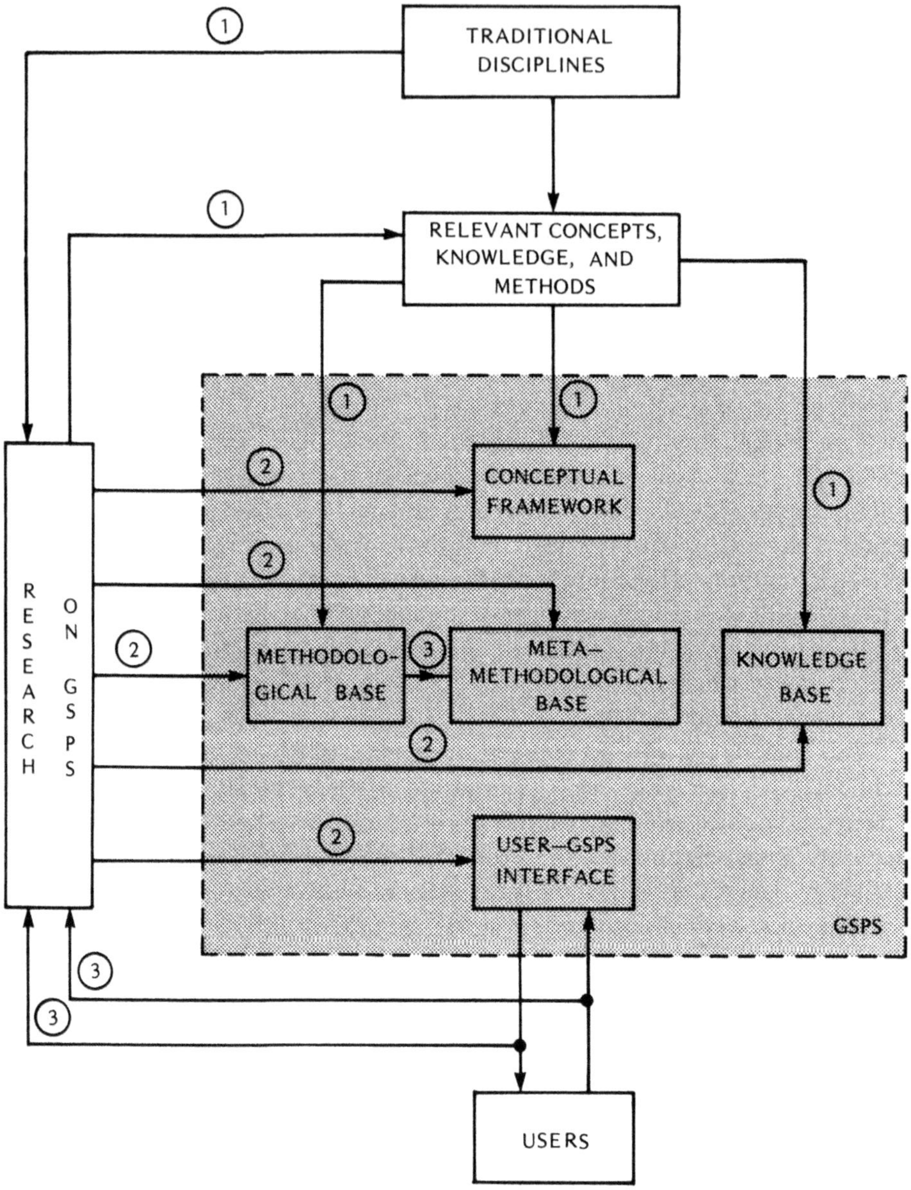

Figure 9.21. A summary of evolution of the GSPS architecture.

3—information about the user–GSPS interactions that is utilized in guiding the research on the GSPS.

Since the environment within which the GSPS has evolved—the traditional disciplines and the GSPS users who represent these disciplines—has been playing a crucial role in this evolution, it seems appropriate to characterize the GSPS architecture as an *indigenous architecture.*

Let me end this section and this book by an appropriate quote:

> Completeness without completion is useful.
> Fulfillment without being fulfilled is desirable.

NOTES

9.1. Example 9.1 is often discussed in statistical literature. For instance, it is included in the book by Bishop *et al.* [BI1, p. 87], where information regarding the original source of data is given.

9.2. Example 9.3 is analyzed by statistical techniques in the book *Applied Statistics* by D. R. Cox and E. J. Snell (Chapman and Hall, New York, 1981, p. 155), where information regarding the original source of data is given.

9.3. Example 9.5 is adopted from the Ph.D. dissertation by Masahiko Higashi at SUNY-Binghamton [HI1]. The ecological data were collected by the Cornell University Biological Field Station; the climatological data were taken from *Local Climatological Data* (Syracuse, New York, 1977), published by the U.S. Department of Commerce. Further details regarding the data can be found in the original source [HI1].

9.4. The case study outlined in Example 9.6 is described in a paper by H. J. J. Uyttenhove [UY3]. It was performed at the Academic Hospital of Leiden University in co-operation with the Technical University of Eindhoven in the Netherlands in 1979. The processing was done on SAPS, which is a small software system designed in the spirit of the GSPS [UY2].

9.5. Example 9.7 is based on a small part of a workshop project performed by J. C. Wanser on the GSPS at SUNY-Binghamton during the period 1978–1980. The whole project is summarized in a report "Systems Modelling of an Archaeological Site" (Systems Science Department, SUNY-Binghamton, New York, 1980). Although the information distance is employed in Example 9.7, the original project was performed in terms of the city-block distance.

9.6. The method of converting data collected on a computer system by a hardware monitor into their Boolean form, which is used in Example 9.10, was proposed by Robert Orchard of Bell Laboratories. He also provided me with the data and problem statement of the example.

9.7. A good overview of expert systems, including useful bibliographical and historical remarks, was prepared by Weiss and Kulikowski [WE4].

9.8. The closing quote of this book is from *Tao Te Ching* by Lao Tsu.

EXERCISES

9.1. Derive formula (9.1).

9.2. Considering only the essential methodological distinctions listed in Table 9.2, determine the total number of types of systems that are based on one support and three variables provided they are:

(a) source systems;
(b) data systems;
(c) generative systems;
(d) structure generative systems.

9.3. Repeat Exercise 9.2 for
(a) two supports and four variables;
(b) one support and n variables;
(c) two supports and n variables.

9.4. Decide for each type of subproblems involved in reconstructability analysis whether it is a problem of the first kind or second kind.

9.5. Considering only systems in which all variables are of the same methodological type and, also, all supports are of the same methodological type, determine the number of all admissible types of systems when epistemological types specified in Figure 9.1 and types of methodological distinctions listed in Table 9.1 are admissible.

9.6. Extend the epistemological hierarchy in Figure 9.1 for $l = 3$.

APPENDICES

APPENDIX A

LIST OF SYMBOLS

GENERAL SYMBOLS

$\{x, y, \ldots\}$	set of elements $x, y, \ldots$
$\{x \mid p(x)\}$	set determined by property p
$(x_1, x_2, \ldots, x_n)$	n-tuple
$[x_{i,j}]$	matrix
$[a, b]$	closed interval of real numbers
$[a, b)$	interval of real numbers closed in a and open in b
$[a, \infty)$	set of real numbers greater than or equal to a
$x \in X$	set membership
$X = Y$	set equality
$X \neq Y$	set inequality
$X - Y$	set difference
$X \subseteq Y$	set inclusion
$X \subset Y$	proper set inclusion ($X \neq Y$)
$\varnothing$	empty set
$\mathcal{P}(X)$	power set (set of subsets) of X
$\lvert X \rvert$	cardinality of set X
$X \cap Y$	set intersections
$X \cup Y$	set union
$X \times Y$	Cartesian product of sets X and Y
X^2	Cartesian product $X \times X$
$X \to Y$	function from X into Y
$f \circ g$	composition of functions f and g
$f * g$	join of functions or relations f and g
f^{-1}	inverse of function f
$[f \downarrow X]$	projection of behavior function with respect to variables in set X
$\overset{f}{\equiv}$	equivalence relation associated with function f
$X / \overset{f}{\equiv}$	set of equivalence classes (or a partition) of set X based on equivalence relation $\overset{f}{\equiv}$
$<$	less than
$\leq$	either less than or equal to or partial ordering
$\preceq$	epistemological ordering of systems
$\prec$	substate relation
$x \wedge y$	meet (greatest lower bound) in a lattice
$x \vee y$	join (least upper bound) in a lattice
$x \mid y$	x given y

$x \Rightarrow y$	x implies y
$x \Leftrightarrow y$	x if and only if y
$\forall$	for all
$\exists$	there exists at least one
$\sum$	sum
$\prod$	product
$\max(x_1, x_2, \ldots, x_n)$	maximum of $x_1, x_2, \ldots, x_n$
$\min(x_1, x_2, \ldots, x_n)$	minimum of $x_1, x_2, \ldots, x_n$
iff	if and only if
$\mathbb{N}$	set of positive integers
$\mathbb{N}_0$	set of nonnegative integers
N_n	set $\{1, 2, \ldots, n\}$
$N_{n,m}$	set $\{n, n+1, \ldots, m\}$
$\mathbb{R}$	set of all real numbers
$n!$	n factorial $[= n(n-1) \ldots 1]$
$\binom{n}{r}$	combinatorial number $n!/(n-r)!r!$

SPECIAL SYMBOLS

a_i, A_i	attribute and its set of appearances
b_j, B_j	backdrop and its set of instances
c	l-cut function—Eq. (3.53)
$\mathbf{c}$	overall state of a system ($\mathbf{c} \in \mathbf{C}$)
$\mathbf{C}$	set of all overall states of a system
$C_{x,y}, \hat{C}_{x,y}$	neutral or directed coupling between elements $\mathbf{x}$ and $\mathbf{y}$ of a structure system Eqs. (4.5), (4.8)
$\mathscr{C}_n$	set of all C-structures with n variables
$d, \tilde{d}$	crisp or fuzzy data—Eqs. (2.21), (2.27)
$\mathbf{d}, \tilde{\mathbf{d}}$	crisp data matrix and fuzzy data array
$\mathbf{D}, \hat{\mathbf{D}}$	neutral or directed data system—Eqs. (2.22), (2.24)
${}^s\mathbf{D}, {}^s\hat{\mathbf{D}}$	neutral or directed data system with semantics—Eqs. (2.23), (2.25)
D	information distance—Eqs. (4.40), (4.42)
e_i	exemplification of general variable v_i—Eq. (2.4)
$\mathbf{e}$	overall state of input variables
$\bar{\mathbf{e}}$	overall state of all variables of a system except input variables
$\mathbf{E}$	set of all input states of a system
$\bar{\mathbf{E}}$	set of all overall states of a system that do not involve input variables
$\mathscr{E}$	overall exemplification channel—Eq. (2.14)
$f_B, \hat{f}_B$	neutral or directed behavior function—Eqs. (3.18), (3.26)
$f_{GB}, \hat{f}_{GB}$	neutral or directed generative behavior function—Eqs. (3.22), (3.29)
$f_s, \hat{f}_s$	neutral or directed ST-function—Eqs. (3.74), (3.90)
$f_{GS}, \hat{f}_{GS}$	neutral or directed generative ST-function—Eqs. (3.75), (3.91)
$\mathbf{F}_B, \hat{\mathbf{F}}_B$	neutral or directed behavior system—Eqs. (3.10), (3.27)
$\mathbf{F}_{GB}, \hat{\mathbf{F}}_{GB}$	neutral or directed generative behavior system—Eqs. (3.15), (3.30)
$\mathbf{F}_S, \hat{\mathbf{F}}_S$	neutral or directed ST-system—Eqs. (3.77), (3.93)

$\mathbf{F}_{GS}$, $\hat{\mathbf{F}}_{GS}$	neutral or directed generative ST-system—Eqs. (3.78), (3.94)
$\mathscr{F}_{SF}$	reconstruction family of structure system **SF**
$\mathbf{g}$, $\bar{\mathbf{g}}$	state of generated or generating sampling variables
$\mathbf{G}$, $\bar{\mathbf{G}}$	set of all overall states of generated or generating sampling variables
$\mathscr{G}_n$	set of all G-structures with n variables
H	Shannon entropy—Eq. (3.37)
$\mathbf{I}$, $\dot{\mathbf{I}}$	general or specific neutral image system—Eqs. (2.11), (2.12)
$\hat{\mathbf{I}}$, $\hat{\dot{\mathbf{I}}}$	general or specific directed image system—Eqs. (2.17), (2.18)
L_f	level set of possibility distribution **f**—Eq. (3.55)
$\mathscr{L}_V$	lattice of variables (V-lattice)
$\mathscr{L}_{V_i}$	resolution lattice
M	mask—Eq. (3.5)
M_i	submask associated with variable v_i—Eq. (3.8)
M_G	generative mask—Eq. (3.11)
ΔM	depth of mask M
M^+	extended mask
MX	metasystem based on systems of type **X**
$N(c)$	frequency (number of observations) of state **c**
o_i, $\hat{o}_i$	crisp and fuzzy observation channels of variable v_i—Eq. (2.2) and (2.8) or (2.9)
$\mathbf{O}$, $\hat{\mathbf{O}}$	neutral or directed object system—Eqs. (2.1), (2.16)
$\mathbf{O}$, $\hat{\mathbf{O}}$	crisp or fuzzy overall observation channel—Eq. (2.13)
$\mathscr{P}_n$	set of all P-structures with n variables
r	replacement function of a metasystem—Eq. (5.2)
r_j	translation rule—Eq. (3.1)
R	set of translation rules
s_k, S_k	sampling variable—Eq. (3.3), and its state set
$\mathbf{S}$, $\hat{\mathbf{S}}$	neutral or directed source system—Eqs. (2.15), (2.19)
SX	structure system based on systems of type **X**
u	input/output identifier
U	U-uncertainty (possibilistic measure of uncertainty/information)—Eq. (3.56)
v_i, $\dot{v}_i$	general or specific variable
V_i, $\dot{V}_i$	general or specific state set of variable v_i, $\dot{v}_i$, respectively
V_i	overall state set of variable $v_i (i \in N_n)$
w_j, $\dot{w}_j$	general or specific support
W_j, $\dot{W}_j$	general or specific support set of w_j, $\dot{w}_j$, respectively
W	overall support set
ε_j	exemplification of general support w_j—Eq. (2.5)
ω_j, $\tilde{\omega}_j$	crisp and fuzzy support observation channel of b_j—Eqs. (2.3), (2.10)

APPENDIX B

GLOSSARY OF RELEVANT MATHEMATICAL TERMS

The purpose of this glossary is to define those mathematical terms for which some readers are expected to have the need for brief reminders. The glossary is, of course, not sufficient for learning the corresponding concepts. It should be adequate, however, for refreshing one's memory of mathematical subjects that have been mastered at some point. Readers with insufficient mathematical background, who may have difficulties in comprehending some of the definitions, are advised to consult any of the following books:

Bavel, Z., *Math Companion for Computer Science*, Reston, Virginia, 1982.

Korfhage, R. R., *Discrete Computational Structures*, Academic Press, New York, 1974.

Levy, L. S., *Discrete Structures of Computer Science*, Wiley, New York, 1980.

Preparata, F. P., and R. T. Yeh, *Introduction to Discrete Structures for Computer Science and Engineering*, Addison-Wesley, Reading, MA, 1973.

Stanat, D. F., and D. F. McAllister, *Discrete Mathematics in Computer Science*, Prentice-Hall, Englewood Cliffs, NJ, 1977.

For mathematical concepts in the relatively new area of fuzzy sets, the following book is recommended:

Dubois, D., and H. Prade, *Fuzzy Sets and Systems: Theory and Applications*, Academic Press, New York, 1980.

GLOSSARY

Algebraic structure is a tuple

$$(X, Y, \ldots, R_1, R_2, \ldots, a, b, \ldots)$$

whose elements are drawn from among sets $X, Y, \ldots$, relations $R_1, R_2, \ldots$, some of which may be functions or operations defined on various Cartesian products of the sets, and distinguished members $a, b, \ldots$ of the sets.

Antisymmetric relation is a binary relation R on set X such that $(x, y) \in R$ and $(y, x) \in R$ implies $x = y$.

Binary relation R on set X is a subset of the Cartesian product $X \times X$. Formally, $R \subset X \times X$.

Cartesian product $X_1 \times X_2 \times \cdots \times X_n$ is the set of all ordered n-tuples such that the first element in each n-tuple is a member of X_1, the second element is a member of X_2, etc. Formally,

$$X_1 \times X_2 \times \cdots \times X_n = \{(x_1, x_2, \ldots, x_n) | x_1 \in X_1, x_2 \in X_2, \ldots, x_n \in X_n\}.$$

Alternative name: set product.

Compatibility relation on set X is a relation that is reflexive and symmetric. Alternative name: tolerance relation.

Composition of two binary relations (or functions) $R \subset X \times Y$ and $S \subset Y \times Z$ is a binary relation $R \circ S \subset X \times Z$ that consists of all pairs (x, z) such that $(x, y) \in R$ and $(y, z) \in S$ for some $y \in Y$. Formally,

$$R \circ S = \{(x, z) | (x, y) \in R \text{ and } (y, z) \in S \text{ for some } y \in Y\}.$$

Connected relation is a binary relation R on X such that $x \neq y$ implies either $(x, y) \in R$ or $(y, x) \in R$.

Distance defined on set X is a function d: $X \times X + \mathbb{R}$ such that

i. $d(x, y) \geq 0$;

ii. $d(x, y) = d(y, x) = 0$ iff $x = y$;

iii. $d(x, z) \leq d(x, y) + d(y, z)$.

Equivalence relation is a binary relation that is relexive, symmetric, and transitive.

Function $f: X \to Y$ is a subset of $X \times Y$ such that $(x, y) \in f$ and $(x, z) \in f$ implies $y = z$. Alternative names: correspondence, map, mapping.

Fuzzy set X defined within a *crisp* universal set U is the set of ordered pairs $X = \{(u, m_x(u)) | u \in U\}$, where $m_x(u)$ is the grade of membership of u in X.

Graph G is a pair (X, R), where X is a nonempty set and R is a binary relation on X. Elements of X are called vertices and elements of R are called edges of the graph. When R is a symmetric relation, G is called an undirected graph.

Group is an algebraic structure $(X, \circ, e)$, where X is a nonempty set, $\circ$ is an associative binary operation, and e is an identity element of X that possesses the following properties:

i. $e \circ a = a$;

ii. $a \circ e = a$;

iii. for each $x \in X$ there exists $x^{-1} \in X$ such that $x \circ x^{-1} = e$. Element x^{-1} is called an inverse of x.

Homomorphism from an algebraic structure (X, R) into another algebraic structure (Y, S) is a function h: $X \to Y$ such that $(x_1, x_2) \in R$ implies $(h(x_1), h(x_2)) \in S$. When function h is such that $(x_1, x_2) \in R$ iff $(h(x_1), h(x_2)) \in S$, the homomorphism is called strong.

Hypergraph H is a pair (X, S), where X is a finite set and S is a family of nonempty subsets of X whose union is set X. Formally, $H = (X, S)$, where X is a finite set and $S = \{S_i | S_i \neq \varnothing \text{ and } \cup_i S_i = X \text{ for } i \in I\}$.

Inverse R^{-1} of a binary relation (or a function) R is obtained by reversing each of the pairs on R. Formally, $R^{-1} = \{(x, y) | (y, x) \in R\}$.

Isomorphism between algebraic structures (X, R) and (Y, R) is a one-to-one correspondence $h: X \leftrightarrow Y$ such that

$$(x_1, x_2) \in R \text{ iff } (h(x_1), h(x_2)) \in S.$$

Join (or **least upper bound**) of a subset Y of a partially ordered set X is an element x of X such that x is an upper bound of Y and $x \leq y$ for every upper bound y of Y.

Join of two binary relations (or functions) $R \subset X \times Y$ and $S \subset Y \times Z$ is a ternary relation $R * S \subset X \times Y \times Z$ that consists of all triples (x, y, z) such that $(x, y) \in R$ and $(y, z) \in S$. Formally, $R * S = \{(x, y, z) | (x, y) \in R \text{ and } (y, z) \in S\}$.

Lattice is a partially ordered set each of which subsets has a meet and join.

Lexicographic ordering on set $X = X_1 \times X_2 \times \cdots \times X_n$ with a linear ordering $\leq_i$ defined on each set $X_i (i \in N_n)$ is a partial ordering on X such that $(x_1, x_2, \ldots, x_n) \leq (y_1, y_2, \ldots, y_n)$ iff $x_k \leq_i y_k$ for the smallest integer k such that $x_k \neq y_k$.

Linear ordering is a partial ordering that is connected. Alternative names: total ordering, simple ordering, complete ordering.

Lower bound of a subset Y of a partially ordered set X is an element x of X such that $x \leq y$ for all $y \in Y$.

Meet (or **greatest lower bound**) of a subset Y of a partially ordered set X is an element x of X such that x is a lower bound of Y and $y \leq x$ for every lower bound of Y.

Metric distance defined on set X is a function $d: X \times X \to \mathbb{R}$ such that

i. $d(x, y) \geq 0$;
ii. $d(x, y) = d(y, x)$;
iii. $d(x, y) = 0$ iff $x = y$;
iv. $d(x, z) \leq d(x, y) + d(y, z)$.

Alternative name: metric.

One-to-one correspondence between sets X and Y is an onto function $f: X \to Y$ such that $x \neq y$ implies $f(x) \neq f(y)$. Alternative name: bijection.

Onto function $f: X \to Y$ is a function such that $y \subset Y$ implies $(x, y) \in f$ for at least one $x \in X$. Alternative names: epic function, surjection.

Operation is a function $o: X^n \to X$. When $n = 1, 2, \ldots,$ it is called unary operation, binary operation, etc., respectively.

Partially ordered set is a set with a partial ordering defined on it. Alternative name: poset.

Partial ordering is a binary relation on a set that is reflexive, antisymmetric, and transitive.

Partition $\pi(X)$ of a set X is a family of non-empty subsets of X such that each element of X belongs to exactly one of these subsets. Formally,

$$\pi(X) = \{X_i | X_i \in \mathscr{P}(X), X_i \neq \varnothing, \bigcup_{i \in I} X_i = X, \text{ and } X_i \cap X_j = \varnothing \text{ for all } i, j \in I\}.$$

Power set of set X is the set of all subsets of X.

Quasiordering is a binary relation on a set that is reflexive and transitive. Alternative name: preordering.

Reflexive relation is a binary relation R on set X such that $(x, x) \in R$ for all $x \in X$.

Symmetric relation is a binary relation R on set X such that $(x, y) \in R$ implies $(y, x) \in R$.

Transitive closure R_T of a binary relation R on a set is the smallest relation that is transitive and contains R (i.e., $R \subset R_T$).

Transitive relation is a binary relation R on set X such that $(x, y) \in R$ and $(y, z) \in R$ implies $(x, z) \in R$.

Upper bound of a subset Y of a partially ordered set X is an element x of X such that $y \leq x$ for all $y \in Y$.

APPENDIX C

SOME RELEVANT THEOREMS

The purpose of this appendix is to state some theorems that are relevant to various subject areas covered in this book and give their proofs. Although these theorems and proofs are not necessary for general understanding of the subject areas, their comprehension should result in much deeper understanding of the material.

The theorems are presented in three sections, each devoted to one subject area. When desirable, the theorems are supplemented by appropriate definitions, lemmas, and connecting remarks.

C.1. MEASURE OF UNCERTAINTY BASED ON POSSIBILITY DISTRIBUTIONS*

Let ${}^n\mathcal{F}$ denote the set of possibility distributions with at least one nonzero element that can be defined on any finite set with n elements and let

$$\mathcal{F} = \bigcup_{n \in \mathbb{N}} {}^n\mathcal{F}.$$

Let $B(N_n)$ denote the set of all permutations on N_n. For each

$$\mathbf{f} = (\rho_1, \rho_2, \ldots, \rho_n) \in \mathcal{F}$$

and for each $b \in B(N_n)$, let

$$b[\mathbf{f}] = (\rho_{b(1)}, \rho_{b(2)}, \ldots, \rho_{b(n)}) \in \mathcal{F}.$$

Definition 1. A possibility distribution $\mathbf{f} = (\rho_1, \rho_2, \ldots, \rho_n) \in {}^n\mathcal{F}$ is called a normalized possibility distribution if and only if

$$\max_i \rho_i = 1.$$

* This material is related primarily to Section 3.5; it is adopted from one of my papers [H12].

Definition 2. Given a possibility distribution

$$\mathbf{f} = (\rho_1, \rho_2, \ldots, \rho_n),$$

let

$$\hat{\mathbf{f}} = (\hat{\rho}_1, \hat{\rho}_2, \ldots, \hat{\rho}_n),$$

where $\hat{\rho}_j = \rho_{b(j)}$ for some permutation $b \in B(N_n)$ such that $\hat{\rho}_j \geqq \hat{\rho}_k$ when $j < k (j, k \in N_n)$. Then, $\hat{\mathbf{f}}$ is called an ordered possibility distribution of **f**.

Definition 3. Given a possibility distribution **f** and a permutation $b \in B(N_n)$, if $\hat{\mathbf{f}} = b[\mathbf{f}]$, then b is called an ordering permutation of **f**.

Definition 4. For each $\mathbf{f} = (\rho_1, \rho_2, \ldots, \rho_n) \in \mathscr{F}$ and each $l \in [0, 1]$, let

$$c: \mathscr{F} \times 0, 1 \to \mathscr{P}(N_n)$$

be a function such that

$$c(\mathbf{f}, l) = \{i \in N_n | \rho_i \geqq l\}.$$

This function is called an l-cut function and the set $c(\mathbf{f}, l)$ is called an l-cut of **f**.

Defintion 5. Let $\mathbf{f} = (\rho_1, \rho_2, \ldots, \rho_n) \in \mathscr{F}$. Then,

$$L_f = \{l | (\exists\, i \in N_n)(\rho_i = l) \text{ or } l = 0\}$$

is called a level set of **f**.

Definition 6. For every $n \in \mathbb{N}$, let

$${}^1\mathbf{f} = ({}^1\rho_1, {}^1\rho_2, \ldots, {}^1\rho_n) \in {}^n\mathscr{F}$$

and

$${}^2\mathbf{f} = ({}^2\rho_1, {}^2\rho_2, \ldots, {}^2\rho_n) \in {}^n\mathscr{F}.$$

be two possibility distributions. Then, ${}^1\mathbf{f}$ is called a subdistribution of ${}^2\mathbf{f}$ if and only if

$${}^1\rho_i \leqq {}^2\rho_i \text{ for all } i \in N_n;$$

let ${}^1\mathbf{f} \leqq {}^2\mathbf{f}$ be used to indicate that ${}^1\mathbf{f}$ is a subdistribution of ${}^2\mathbf{f}$.

Theorem 1. Function U defined by (3.56) or, alternatively, (3.57) possesses all of the following properties:

(U1) $U(\rho_1, \rho_2, \ldots, \rho_n) = U(\rho_{b(1)}, \rho_{b(2)}, \ldots, \rho_{b(n)})$ for all possibility distributions $\mathbf{f} \in \mathscr{F}$ and all permutations $b \in B(N_n)$ (symmetry);

(U2) $U(\rho_1, \rho_2, \ldots, \rho_n, 0) = U(\rho_1, \rho_2, \ldots, \rho_n)$ for all $\mathbf{f} \in \mathscr{F}$ (expansibility);

(U3) $U(\mathbf{f}) \leq U(^1\mathbf{f}) + U(^2\mathbf{f})$, where $\mathbf{f} = (\rho_{11}, \rho_{12}, \ldots, \rho_{1n}, \rho_{21}, \rho_{22}, \ldots, \rho_{2n} \ldots, \rho_{m1}, \rho_{m2}, \ldots, \rho_{mn})$, $^1\mathbf{f} = (^1\rho_1, {}^1\rho_2, \ldots, {}^1\rho_m) \in \mathscr{F}$ such that $^1\rho_i = \max_j \rho_{ij}$, and $^2f = (^2\rho_1, {}^2\rho_2, \ldots, {}^2\rho_n) \in \mathscr{F}$ such that $^2\rho_j = \max_i \rho_{ij}$ (subadditivity);

(U4) $U(\mathbf{f}) = U(^1\mathbf{f}) + U(^2\mathbf{f})$ if $^1\mathbf{f}$ and $^2\mathbf{f}$ in (U3) are noninteractive, i.e., $\rho_{ij} = \min(^1\rho_i, {}^2\rho_j)$ for all $i \in N_m$ and all $j \in N_n$ (additivity);

(U5) $U(1, 1) = 1$ (normalization);

(U6) U is continuous in all its arguments (continuity);

(U7) $U(^1\mathbf{f}) \leq U(^2\mathbf{f})$ if $^1\mathbf{f} \leq {}^2\mathbf{f}$ and $\max_i {}^1\rho_i = \max_i {}^2\rho_i$ for all $^1\mathbf{f}, {}^2\mathbf{f} \in {}^n\mathscr{F}$ and each particular $n \in \mathbb{N}$ (general monotonicity);

(U8) for all $\mathbf{f} \in \mathscr{F}$, $U(\mathbf{f}) = 0$ iff $\rho_i \neq 0$ for exactly one $i \in N_n$ (minimum property);

(U9) for all $\mathbf{f} \in \mathscr{F}$, $U(\mathbf{f})$ attains its maximum within $^n\mathscr{F}$ iff ρ_i has the same value for all $i \in N_n$; the maximum of $U(\mathbf{f})$ within ^nF is equal to $U(1, 1, \ldots, 1) = \log_2 n$ (maximum property);

(U10) for all $\mathbf{f} \in \mathscr{F}$, $U(\mathbf{f})$ decreases if $U(\mathbf{f}) \neq 0$ and only one maximum element of $\mathbf{f}$, say element ρ_{i_0} such that $\rho_{i_0} = \max_i \rho_i$, increases (special monotonicity).

Proof. (i) (U1) and (U7): Let us consider, for any $n \in \mathbb{N}$, $^1\mathbf{f}, {}^2\mathbf{f} \in {}^n\mathscr{F}$ such that

$$|c(^1\mathbf{f}, l)| \leqq |c(^2\mathbf{f}, l)| \qquad \text{for all } l \in [0, 1].$$

Then, clearly,

$$\log_2 |c(^1\mathbf{f}, l) \leqq \log_2 |c(^2\mathbf{f}, l)| \qquad \text{for all } l \in [0, 1]$$

and, consequently, $U(^1\mathbf{f}) \leqq U(^2\mathbf{f})$. Hence, U satisfies (U7) and, by Lemma 2 (p. 490), it also satisfies (U1) and (U7).

(ii) (U2): Let $\mathbf{f} = (\rho_1, \rho_2, \ldots, \rho_n) \in \mathscr{F}$ and $\mathbf{f}' = (\rho_1, \rho_2, \ldots, \rho_n, 0) \in \mathscr{F}$. Then, $L_f = L_{f'}$ and $c(\mathbf{f}, l) = c(\mathbf{f}', l)$ for all $l \neq 0$, but $c(f, 0)$ is not included in formula (3.56). Hence, U satisfies (U2).

(iii) (U3): Let $\mathbf{f}$, $^1\mathbf{f}$, and $^2\mathbf{f}$ be defined as in (U3). Since for each $l \in [0, 1]$

$$\rho_{ij} \geqq l \text{ implies } {}^1\rho_i = \max_j \rho_{ij} \geqq l$$

and

$$^2\rho_j = \max_i \rho_{ij} \geqq l,$$

we get

$$\begin{aligned} c(\mathbf{f}, l) &= \{(i, j) \in N_m \times N_n \mid \rho_{ij} \geqq l\} \\ &\subseteq \{(i, j) \in N_m \times N_n \mid {}^1\rho_i \geqq l \text{ and } {}^2\rho_j \geqq l\} \\ &= \{i \in N_m \mid {}^1\rho_i \geqq l\} \times \{j \in N_n \mid {}^2\rho_j \geqq l\} \\ &= c(^1\mathbf{f}, l) \times c(^2\mathbf{f}, l). \end{aligned}$$

Hence,

$$|c(\mathbf{f}, l)| \leqq |c(^1\mathbf{f}, l)| \times |c(^2\mathbf{f}, l)| \qquad \text{for all } l \in [0, 1].$$

By definition,

$$l_f = \max_{i,j} \rho_{ij} = \max_i \left(\max_j \rho_{ij} \right) = \max_i {}^1\rho_i$$
$$= l_{^1f}.$$

Similarly, $l_f = l_{^2f}$. Hence,

$$\begin{aligned} U(\mathbf{f}) &= \frac{1}{l_f} \int_0^{l_f} \log_2 |c(\mathbf{f}, l)| \, dl \\ &\leqq \frac{1}{l_f} \int_0^{l_f} \log_2 |c(^1\mathbf{f}, l)| \times |c(^2\mathbf{f}, l)| \, dl \\ &= \frac{1}{l_f} \int_0^{l_f} \log_2 |c(^1\mathbf{f}, l)| \, dl + \frac{1}{l_f} \int_0^{l_f} \log_2 |c(^2\mathbf{f}, l)| \, dl \\ &= U(^1\mathbf{f}) + U(^2\mathbf{f}). \end{aligned}$$

This concludes the proof that U satisfies (U3).

(iv) (U4): Let $\mathbf{f}$, $^1\mathbf{f}$ and $^2\mathbf{f}$ be defined as in (U4). Then, clearly, $\rho_{ij} \geqq l \Leftrightarrow {}^1\rho_i \geqq l$ and ${}^2\rho_j \geqq l$ for all $l \in [0, 1]$. Hence, as in (iii), we get $c(\mathbf{f}, l) = c(^1\mathbf{f}, l) \times c(^2\mathbf{f}, l)$ and, consequently, $|c(\mathbf{f}, l)| = |c(^1\mathbf{f}, l)| \times |c(^2\mathbf{f}, l)|$ for all $l \in [0, 1]$. Following the scheme of reasoning in (iii), we can show that $l_f = l_{^1f} = l_{^2f}$ and, eventually, $U(\mathbf{f}) = U(^1\mathbf{f}) + U(^2\mathbf{f})$. Therefore, U satisfies (U4).

(v) (U5): $U(1, 1) = \log_2 2 = 1$.

(vi) (U6): Let $\mathbf{f} = (\rho_1, \rho_2, \ldots, \rho_n) \in \mathcal{F}$, let $\alpha \in N_n$ be a particular integer, and let $L_f = \{l_1, l_2, \ldots, l_r\}$, where $0 = l_1 < l_2 < \cdots < l_r (= l_f)$. For the sake of clarity, let us distinguish three cases.

Case I. Assume $\rho_\alpha = l_\beta$ for some β such that $2 \leqq \beta \leqq r - 1$.

(A) Let Δl denote a real number such that $0 < \Delta l < l_{\beta+1} - l_\beta$ and let $\mathbf{f}' = (\rho'_1, \rho'_2, \ldots, \rho'_r)$ denote a possibility distribution such that $\rho'_i = \rho_i$ for all $i \neq \alpha$ and $\rho'_\alpha = \rho_\alpha + \Delta l$. Then, for all $l \in [0, 1]$, we get

$$\begin{aligned} c(\mathbf{f}', l) - \{\alpha\} &= \{i \in N_n | \rho'_i \geqq l\} - \{\alpha\} \\ &= \{i \in N_n | i \neq \alpha, \rho'_i \geqq l\} \\ &= \{i \in N_n | i \neq \alpha, \rho_i \geqq l\} \\ &= \{i \in N_n | \rho_i \geqq l\} - \{\alpha\} \\ &= c(\mathbf{f}, l) - \{\alpha\} \end{aligned}$$

For $l \leqq \rho_\alpha$, we also get $l \leqq \rho'_\alpha$ (since $\rho'_\alpha \geqq \rho_\alpha$) and, consequently, both $\alpha \in c(\mathbf{f}, l)$ and $\alpha \in c(\mathbf{f}', l)$; hence, $c(\mathbf{f}', l) = c(\mathbf{f}, l)$. For $\rho_\alpha < l \leqq \rho_\alpha + \Delta l (= \rho'_\alpha)$, it is clear that $\alpha \notin c(\mathbf{f}, l)$ and $\alpha \in c(\mathbf{f}', l)$; hence, $|c(\mathbf{f}', l)| = |c(\mathbf{f}, l)| + 1$. For $l > \rho_\alpha + \Delta l$, both $\alpha \notin c(\mathbf{f}, l)$ and $\alpha \notin c(\mathbf{f}', l)$: hence, again, $c(\mathbf{f}', l) = c(\mathbf{f}, l)$. We may conclude now that

$$U(\mathbf{f}') - U(\mathbf{f}) = \frac{1}{l_f} \int_{\rho_\alpha}^{\rho_\alpha + \Delta l} [\log_2(|c(\mathbf{f}, l)| + 1) - \log_2 |c(\mathbf{f}, l)|] dl,$$

where $c(\mathbf{f}, l) = c(\mathbf{f}, l_\beta)$ for all $l \in [\rho_\alpha, \rho_\alpha + \Delta l]$ (since $l_\beta = \rho_\alpha$ and $\rho_\alpha + \Delta l < l_{\beta+1}$). Hence,

$$\begin{aligned} U(\mathbf{f}') - U(\mathbf{f}) &= \frac{1}{l_f} [\log_2(|c(\mathbf{f}, l_\beta)| + 1) - \log_2 |c(\mathbf{f}, l_\beta)|] \Delta l \\ &= \frac{1}{l_f} \log_2 \frac{|c(\mathbf{f}, l_\beta)| + 1}{|c(\mathbf{f}, l_\beta)|} \Delta l. \end{aligned} \tag{*}$$

Since

$$\frac{1}{l_f} \log_2 \frac{|c(\mathbf{f}, l_\beta)| + 1}{|c(\mathbf{f}, l_\beta)|}$$

is a constant, equation (*) implies that U is continuous at $\mathbf{f}$ from the right with respect to its argument ρ_α.

(B) Let $l_{\beta-1} - l_\beta < \Delta l < 0$, and let $\mathbf{f}'$ be defined in the same way as in (A). Then, following the same reasoning used in (A), we obtain

$$U(\mathbf{f}') - U(\mathbf{f}) = \frac{1}{l_f} \log_2 \frac{|c(\mathbf{f}, l_{\beta-1})|}{|c(\mathbf{f}, l_{\beta-1})| - 1} \Delta l, \tag{**}$$

which implies that U is continuous from the left with respect to ρ_α.

From (A) and (B), we conclude that U is continuous with respect to ρ_α.

Case II. Assume $\rho_\alpha = l_r$.

(A) Set $l_r < 1$. First, let $0 < \Delta l < 1 - l_r$. Following the same reasoning used in Case I(A), we get

$$\begin{aligned} |U(\mathbf{f}') - U(\mathbf{f})| &= \left(\frac{1}{l_{f'}} - \frac{1}{l_f}\right) \cdot C \\ &= \frac{C}{l_r \cdot (l_r + \Delta l)} \Delta l < \frac{C}{l_r^2} \Delta l. \end{aligned}$$

where $C = \int_0^{l_r} \log_2 |c(\mathbf{f}, l)| dl$ and l_r^2 are constant. Hence, U is continuous from the right. Next, let $l_r - l_{r-1} < \Delta l < 0$. Suppose $|c(\mathbf{f}, l_r)| = 1$. Then, by the same reasoning used in the first part of this case [i.e., Case II(A)], we get

$$|U(\mathbf{f}') - U(\mathbf{f})| \leqq \frac{C'}{l_r l_{r-1}} |\Delta l|$$

where, $C' = \int_0^{l_{r-1}} \log_2 |c(\mathbf{f}, l)| dl$ and $l_r l_{r-1}$ are constant. Hence, U is continuous from the left. If $|c(\mathbf{f}, l_r)| \neq 1$, then all the arguments in Case I(B) are valid, and we get (**), which implies that U is continuous from the left. Hence, U is continuous with respect to ρ_α.

(B) Set $l_r = 1$. All the arguments in the second part of (A) are valid and, thus, U is continuous from the left with respect to ρ_α.

Case III. Assume $\rho_\alpha = l_1 = 0$. Then, all arguments in Case I(A) are valid and U is thus continuous from the right with respect to ρ_α.

It follows from Cases I, II, III that U is continuous for any $\mathbf{f} \in \mathscr{F}$ with respect to any of its arguments

(vii) (U8): Let $\mathbf{f} = (\rho_1, \rho_2, \ldots, \rho_n) \in \mathscr{F}$ and let $L_f = \{l_1, l_2, \ldots, l_r\}$. Then, by the definition of U, we get

$$\begin{aligned} U(\mathbf{f}) = 0 &\Leftrightarrow |c(f, l_2)| = 1 \quad \text{and} \quad r = 2 \\ &\Leftrightarrow \rho_i \neq 0 \text{ for exactly one } i \in N_n. \end{aligned}$$

Hence, U satisfies (U8).

(viii) (U9): Let $n \in \mathbb{N}$, $\mathbf{f} = (\rho_1, \rho_2, \ldots, \rho_n)$ and $L_f = \{l_1, l_2, \ldots, l_r = l_f\}$. Since $|c(\mathbf{f}, l_{i+1})| \leqq n$ for all $i \in N_{r-1}$, we get

$$\begin{aligned} U(\mathbf{f}) &= \frac{1}{l_r} \sum_{i=1}^{r-1} (l_{i+1} - l_i) \log_2 |c(\mathbf{f}, l_{i+1})| \\ &\leqq \frac{1}{l_r} \sum_{i=1}^{r-1} (l_{i+1} - l_i) \log_2 n \\ &= \frac{1}{l_r} \log_2 n \sum_{i=1}^{r-1} (l_{i+1} - l_i) \\ &= \log_2 n. \end{aligned}$$

Hence, $U(\mathbf{f}) \leqq \log_2 n$, where the equality holds if and only if $r = 2$ and $|c(\mathbf{f}, l_2)| = n$, i.e., if and only if $\rho_i = l_2$ for some constant $l_2 > 0$ and all $i \in N_n$. We can thus conclude that U satisfies (U9).

(ix) (U10): Let us consider $\mathbf{f} = (\rho_1, \rho_2, \ldots, \rho_n) \in \mathscr{F}$ such that $U(\mathbf{f}) \neq 0$. Let $I_f = \{i \in N_n | \rho_i \geqq \rho_j \text{ for all } j \in N_n\}$. Since $U(\mathbf{f}) \neq 0$, it follows from (vii) that $|\{i \in N_n | \rho_i > 0\}| \geqq 2$ and, consequently, $\rho_i > 0$ for all $i \in I_f$.

Let $\lambda \in I_f$. Consider $f^\lambda = (\rho_1^\lambda, \rho_2^\lambda, \ldots, \rho_n^\lambda) \in \mathscr{F}$ such that

$$\rho_i^\lambda \begin{cases} = \rho_i, & \text{for} \quad i \neq \lambda, \\ > \rho_i, & \text{for} \quad i = \lambda. \end{cases}$$

Then, clearly,

$$c(\mathbf{f}^\lambda, l) = \begin{cases} c(\mathbf{f}, l), & \text{for} \quad 0 \leqq l \leqq \rho_\lambda, \\ \{\lambda\}, & \text{for} \quad \rho_\lambda \leqq l \leqq \rho_\lambda^\lambda. \end{cases}$$

Hence,

$$\begin{aligned} U(\mathbf{f}^{\lambda}) &= \frac{1}{\rho_{\lambda}^{\lambda}} \int_{0}^{\rho_{\lambda}^{\lambda}} \log|c(\mathbf{f}^{\lambda}, l)|\,dl \\ &= \frac{1}{\rho_{\lambda}^{\lambda}} \left[\int_{0}^{\rho_{\lambda}} \log|c(\mathbf{f}^{\lambda}, l)|\,dl + \int_{\rho_{\lambda}}^{\rho_{\lambda}^{\lambda}} \log|c(\mathbf{f}^{\lambda}, l)|\,dl \right] \\ &= \frac{1}{\rho_{\lambda}^{\lambda}} \left[\int_{0}^{\rho_{\lambda}} \log|c(\mathbf{f}^{\lambda}, l)|\,dl + \int_{\rho_{\lambda}}^{\rho_{\lambda}^{\lambda}} (\log 1)\,dl \right] \\ &= \frac{1}{\rho_{\lambda}^{\lambda}} [\rho_{\lambda} U(\mathbf{f})] \\ &= \frac{\rho_{\lambda}}{\rho_{\lambda}^{\lambda}} U(\mathbf{f}). \end{aligned}$$

Since

$$\rho_{\lambda} < \rho_{\lambda}^{\lambda}, \qquad \frac{\rho_{\lambda}}{\rho_{\lambda}^{\lambda}} < 1 \qquad \text{and } U(\mathbf{f}) \neq 0,$$

we get $U(\mathbf{f}^{\lambda}) < U(\mathbf{f})$, which proves that U satisfies (U10).

The theorem is proved by (i)–(vix). Q.E.D.

Lemma 1. Let ${}^1\mathbf{f}, {}^2\mathbf{f} \in {}^n\mathscr{F}$ and let $|c({}^1\mathbf{f}, l)|$, $|c({}^2\mathbf{f}, l)|$ denote the cardinalities of l-cuts of ${}^1\mathbf{f}$, ${}^2\mathbf{f}$, respectively. Then,

$$(\forall l \in [0, 1])|[c({}^1\mathbf{f}, l)| \leqq |c({}^2\mathbf{f}, l)|] \tag{1}$$

$$\Leftrightarrow {}^1\hat{\mathbf{f}} \leqq {}^2\hat{\mathbf{f}} \tag{2}$$

$$[\exists\, b \subset B(N_n)]({}^1\mathbf{f} \leqq b[{}^2\mathbf{f}]) \tag{3}$$

Proof. Let ${}^j\mathbf{f} = ({}^j\rho_1, {}^j\rho_2, \ldots, {}^j\rho_n)$, let ${}^j\hat{\mathbf{f}} = ({}^j\hat{\rho}_1, {}^j\hat{\rho}_2, \ldots, {}^j\hat{\rho}_n)$, and let ${}^j\hat{\mathbf{f}} = {}^jb[{}^j\mathbf{f}]$, where $j = 1, 2$.

(i) *Proof of* (1) $\Rightarrow$ (2). Assume (1). Let $l_i = {}^1\hat{\rho}_i$ for each $i \in N_n$. Since

$${}^1\rho_{1_{b(k)}} = {}^1\hat{\rho}_k \geqq {}^1\hat{\rho}_i = l_i \qquad \text{for all } k \leqq i_i$$

we get

$$c({}^1\mathbf{f}, l_i) \supseteq \{{}^1b(1), {}^1b(2), \ldots, {}^1b(i)\}.$$

Thus, $|c({}^1\mathbf{f}, l_i)| \geqq i$ and, from (1), we get $|c({}^2\mathbf{f}, l_i)| \geqq i$ for all $i \in N_n$. This, together with ${}^2\rho_{{}^2b(k)} \geqq {}^2\rho_{{}^2b(j)}$ for $j > k$, implies

$$c({}^2\mathbf{f}, l_i) \supseteq \{{}^2b(1), {}^2b(2), \ldots, b^2(i)\}.$$

Thus, ${}^2\rho_{{}^2b(k)} \geqq l_i$ for all $k \leqq i$. Hence, ${}^2\hat{\rho}_i = {}^2\rho_{{}^2b(j)} \geqq l_i = {}^1\hat{\rho}_i$ for all $i \in N_n$. This, by Definition 6, implies ${}^2\hat{\mathbf{f}} \geqq {}^1\hat{\mathbf{f}}$.

(ii) *Proof of* (2) $\Rightarrow$ (3). Assume (2). Then, since ${}^j\hat{\mathbf{f}} = {}^jb[{}^j\mathbf{f}]$ $(j = 1, 2)$, we get ${}^2b[{}^2\mathbf{f}] \geqq {}^1b[{}^1\mathbf{f}]$ and, clearly,

$$ {}^1b^{-1}[{}^2b[{}^2\mathbf{f}]] \geqq {}^1b^{-1}[{}^1b[{}^1\mathbf{f}]] = {}^1\mathbf{f}, $$

where ${}^1b^{-1} \in B(N_n)$ denotes the inverse of the bijection ${}^1b$. Hence, $b[{}^2\mathbf{f}] \geqq {}^1\mathbf{f}$, where $b = {}^2b \circ {}^1b^{-1} \in B(N_n)$. This implies (3).

(iii) *Proof of* (3) $\Rightarrow$ (1). Assume (3). Let $l \in [0, 1]$ and let $c({}^1\mathbf{f}, l) = \{{}^1b(i_1), {}^1b(i_2) \ldots, {}^1b(i_k)\}$ for some $k \in N_n$. Then, by the assumption of (3), we get

$$ {}^2\rho_{{}^2b(i_j)} \geqq {}^1\rho_{{}^1b(i_j)} \geqq l $$

for $j = 1, 2, \ldots, k$. Hence,

$$ c({}^2\mathbf{f}, l) \supseteq \{{}^2b(i_1), {}^2b(i_2), \ldots, {}^2b(i_k)\} $$

and, consequently, $|c({}^2f, l)| \geqq k = |c({}^1f, l)|$.

From (i), (ii), and (iii) we may conclude that (1) $\Leftrightarrow$ (2) $\Leftrightarrow$ (3), and the lemma is proved. Q.E.D.

Lemma 2. Let ${}^1\mathbf{f}$, ${}^2\mathbf{f} \in {}^n\mathcal{F}$. Then, (U1) and (U7), taken together, are equivalent to

$$ \text{(U7')}\ \mathrm{U}({}^1\mathbf{f}) \leqq \mathrm{U}({}^2\mathbf{f}) \quad \text{if } \max_i {}^1\rho_i = \max_i {}^2\rho_i \quad \text{and} \quad (\exists b \in B(N_n))({}^1\mathbf{f} \leqq b[{}^2\mathbf{f}]) $$

as well as to

$$ \text{(U7'')}\ \mathrm{U}({}^1\mathbf{f}) \leqq \mathrm{U}({}^2\mathbf{f}) \quad \text{if } \max_i {}^1\rho_i = \max_i {}^2\rho_i \text{ and } (\forall l \in [0, 1])[|c({}^1\mathbf{f}, l)| \leqq |c({}^2\mathbf{f}, l)|]. $$

Proof: Let ${}^1\mathbf{f}$, ${}^2\mathbf{f} \in {}^n\mathcal{F}$ and let $\max_i {}^1\rho_i = \max_i 2\rho_i$.

(i) Assume that the function $V: \mathcal{F} \to [0, \infty)$ under consideration satisfies the properties (U1) and (U7). Assume further that the statement

$$ (\exists b \in B(N_n))({}^1\mathbf{f} \leqq b[{}^2\mathbf{f}]) $$

is true. Then, by (U7), we obtain

$$ U({}^1\mathbf{f}) \leqq U(b[{}^2\mathbf{f}]), $$

and, from (U1), we obtain

$$U(b[{}^2\mathbf{f}]) = U({}^2\mathbf{f}).$$

Hence, $U({}^1\mathbf{f}) \leqq U({}^2\mathbf{f})$.

(ii) Assume that U satisfies (U7′) and ${}^1\mathbf{f} \leqq {}^2\mathbf{f}$. Then, Clearly ${}^1\mathbf{f} = b[{}^2\mathbf{f}]$, where $b \in B(N_n)$ is the identity function on N_n, and, according to (U7′), $U({}^1\mathbf{f}) \leqq U({}^2\mathbf{f})$. This implies (U7). Assume now that the statement

$$[\exists b \in B(N_n)]({}^1\mathbf{f} = b[{}^2\mathbf{f}])$$

is true. Then, ${}^1\mathbf{f} \leqq b[{}^2\mathbf{f}]$ and ${}^1\mathbf{f} \geqq b[{}^2\mathbf{f}]$, and from (U7′) we get

$$U({}^1\mathbf{f}) \leqq U(b[{}^2\mathbf{f}])$$

and

$$U({}^1\mathbf{f}) \geqq U(b[{}^2\mathbf{f}]).$$

Hence,

$$U({}^1\mathbf{f}) = U(b[{}^2\mathbf{f}]),$$

which implies (U1).

It follows from (i) and (ii) that (U1) and (U7) are equivalent to (U7′). It follows directly from Lemma 1 that (U7′) is equivalent to (U7″). Q.E.D.

C.2. JOIN PROCEDURE AND THE UNBIASED (MAXIMUM ENTROPY) RECONSTRUCTION FOR PROBABILISTIC SYSTEMS*

Theorem 2. Given a probabilistic and consistent behavior structure system **SF** whose elements are characterized by behavior functions ${}^kf\,(k \in N_q)$, if **SF** is such that no inconsistencies occur during the basic join procedure, then the procedure determines the unbiased (maximum entropy) reconstruction from **SF**.

Proof (*Lewis* [*LE*2]). Let $f' = *_k {}^kf$ and f'' be behavior functions of two reconstructions from S. We can start with the inequality

$$-\sum_{\alpha} f''(\alpha) \log f''(\alpha) \leq -\sum_{\alpha} f''(\alpha) \log f'(\alpha),$$

which is well known in information theory as Gibbs' theorem. Now, $f'(\alpha)$ can be

* This material is related to Sections 4.6 and 4.7; it is adopted from one of my papers [CA6].

expressed as

$$f'(\alpha) = \prod_k {}^k f({}^k\beta).$$

where ${}^k\beta \prec \alpha$, and ${}^k f({}^k\beta)$ denotes in this proof either basic or conditional probability, as required by the individual joins. Hence,

$$\begin{aligned} -\sum_\alpha f''(\alpha) \log f'(\alpha) &= -\sum_\alpha f''(\alpha) \log \prod_k {}^k f({}^k\beta) \\ &= -\sum_\alpha f''(\alpha) \sum_k \log {}^k f({}^k\beta), \end{aligned}$$

where ${}^k\beta \prec \alpha$. Since f'' represents a reconstruction from **SF**, the terms in the last expression can be grouped together for each ${}^k\beta \prec \alpha$ in a way described by the formula

$$-\sum_{\alpha > {}^k\beta} f''(\alpha) \log {}^k f({}^k\beta),$$

which can be written as

$$-{}^k f({}^k\beta) \log {}^k f({}^k\beta).$$

When all these expressions are added, we obtain

$$-\sum_k \sum_{{}^k\beta} {}^k f({}^k\beta) \log {}^k f({}^k\beta),$$

which is exactly the same as the entropy of $f'(\alpha)$; indeed, by repeating the previous arguments, we obtain

$$-\sum_\alpha f'(\alpha) \log f'(\alpha) = -\sum_\alpha f'(\alpha) \sum_k \log {}^k f({}^k\beta),$$

where ${}^k\beta \succ \alpha$, and since $f'(\alpha)$ is a reconstruction from **SF**, we can group the terms in the last expression in the same way as before and derive, eventually, the same final expression. Hence,

$$-\sum_\alpha f''(\alpha) \log f'(\alpha) = -\sum_\alpha f'(\alpha) \log f'(\alpha),$$

and the original inequality becomes

$$-\sum_\alpha f''(\alpha) \log f''(\alpha) \leq -\sum_\alpha f'(\alpha) \log f'(\alpha),$$

which proves the theorem. Q.E.D.

In order to justify the iterative join procedure, we observe that ${}^jf * f_{i-1}$ is a degenerate join which can be expressed as

$$f_i(\alpha) = {}^jf({}^j\beta)\frac{f_{i-1}(\alpha)}{[f_{i-1} \downarrow {}^jV]({}^j\beta)}$$

where ${}^j\beta = \alpha$. This iterative scheme is exactly the same as the one proposed by Brown [BR7] and based on the idea of proportional fitting:

$$f_i(\alpha) = f_{i-1}(\alpha)\frac{{}^jf({}^j\beta)}{[f_{i-1} \downarrow {}^jV]({}^j\beta)}$$

Hence, we can take advantage of Brown's proofs (which are also well covered in a book by Bishop *et al.* [BI1]) of the following propositions to justify the procedure:

i. the iterative join procedure converges to a behavior function which conforms to the structure systems **SF**;
ii. the join procedure converges to the behavior function of the unbiased reconstruction from **SF**.

C.3. JOIN PROCEDURE AND THE UNBIASED (MAXIMUM UNCERTAINTY) RECOGNITION FOR POSSIBILISTIC SYSTEMS*

Theorem 3. $\min[{}^1f(\alpha, \beta), {}^2f(\gamma | \beta)] = \min[{}^1f(\alpha, \beta), {}^2f(\beta, \gamma)]$.

Proof. Clearly, the marginal possibility $f(\beta)$ must satisfy the inequality $f(\beta) \geq \max[{}^1f(\alpha, \beta), {}^2f(\beta, \gamma)]$. Using Eq. (3.113), the following four cases have to be considered, each of which leads to a particular result of $\min[{}^1f(\alpha, \beta), {}^2f(\gamma | \beta)]$:

(i) $f(\beta) = {}^1f(\alpha, \beta)$ and $f(\beta) > {}^2f(\beta, \gamma)$: in this case ${}^2f(\gamma | \beta) = {}^2f(\beta, \gamma)$ and, hence, the theorem holds.

(ii) $f(\beta) > {}^1f(\alpha, \beta)$ and $f(\beta) = {}^2f(\beta, \gamma)$: in this case ${}^2f(\gamma | \beta) = [{}^2f(\beta, \gamma), 1]$ and ${}^1f(\alpha, \beta) < {}^2f(\beta, \gamma)$; hence $\min[{}^1f(\alpha, \beta), [{}^2f(\beta, \gamma), 1]] = {}^1f(\alpha, \beta)$ and the theorem holds.

(iii) $f(\beta) = {}^1f(\alpha, \beta)$ and $f(\beta) = {}^2f(\beta, \gamma)$: in this case ${}^2f(\gamma | \beta) = [{}^2f(\beta, \gamma), 1]$ and $\min[{}^1f(\alpha, \beta), [{}^2f(\beta, \gamma), 1]] = {}^1f(\alpha, \beta) = {}^2f(\beta, \gamma)$; the theorem again holds.

(iv) $f(\beta) > {}^1f(\alpha, \beta)$ and $f(\beta) > {}^2f(\beta, \gamma)$: in this case ${}^2f(\gamma | \beta) = {}^2f(\beta, \gamma)$ and the theorem holds; this concludes the proof. Q.E.D.

Theorem 4. $\min[{}^1f(\beta), {}^2f(\gamma | \beta)] = \min[{}^1f(\beta), {}^2f(\beta, \gamma)]$.

* The theorems presented here are related to Sections 4.6 and 4.7 are adopted from one of my papers [CA9].

Proof. Only two cases have to be considered. If ${}^1f(\beta) > {}^2f(\beta, \gamma)$, then ${}^2f(\gamma \mid \beta) = {}^2f(\beta, \gamma)$ and the theorem holds. If ${}^1f(\beta) = {}^2f(\beta, \gamma)$, then ${}^2f(\gamma \mid \beta) = [{}^2f(\beta, \gamma), 1]$ and $\min[{}^1f(\beta), [{}^2f(\beta, \gamma), 1]] = {}^1f(\beta) = {}^2f(\beta, \gamma)$ and the theorem again holds. Q.E.D.

Theorem 5. Given a possibilistic and consistent behavior structure system SF whose elements are characterized by behavior functions ${}^kf(k \in N_q)$, the basic join procedure determines the unbiased (maximum U-uncertainty) reconstruction from SF.

Proof. Since $[f \downarrow {}^kV]({}^k\beta) = \max_{\alpha > {}^k\beta} f(\alpha)$ for each k, the largest values of the overall possibility distribution are preserved in all projections, i.e., if $\alpha_{\max} \in \mathbf{C}$ is such that $f(\alpha_{\max}) = \max_{\alpha \in \mathbf{C}} f(\alpha)$ and ${}^k\beta \prec \alpha$, then ${}^kf({}^k\beta) = f(\alpha_{\max})$ for each k and some ${}^k\beta$. Hence, values $f(\alpha_{\max})$ are preserved in each member of the reconstruction family.

Let us consider now a reconstruction of an aggregate state represented by a concatenation of three disjoint substates $\alpha_1, \beta_1, \gamma_1$, i.e., $(\alpha_1, \beta_1, \gamma_1) \in \mathbf{C}$. Let the following list include values of the overall possibility distribution for all aggregate states which are relevant for determining the value ${}^1f(\alpha_1, \beta_1)$ of the projection ${}^1f$:

$$f(\alpha_1, \beta_1, \gamma_1) = a_1 \quad (= x),$$
$$f(\alpha_1, \beta_1, \gamma_2) = a_2,$$
$$\cdots$$
$$f(\alpha_1, \beta_1, \gamma_p) = a_p.$$

Then, ${}^1f(\alpha_1, \beta_1) = \max_i a_i$. Similarly, let the values of the overall possibility distribution for all aggregate states which are relevant for determining the value ${}^2f(\beta_1, \gamma_1)$ of the projection ${}^2f$ be

$$f(\alpha_1, \beta_1, \gamma_1) = b_1 \quad (= x),$$
$$f(\alpha_2, \beta_1, \gamma_1) = b_2,$$
$$\cdots$$
$$f(\alpha_r, \beta_1, \gamma_1) = b_r.$$

Then, ${}^2f(\beta_1, \gamma_1) = \max_j b_j$. By Theorem 3, $[{}^1f * {}^2f](\alpha_1, \beta_1, \gamma_1) = \min[{}^1f(\alpha_1, \beta_1), {}^2f(\beta_1, \gamma_1)]$. The possibilities relevant for the determination of this value of the join ${}^1f * {}^2f$ are

either (i) $\max_i a_i = x$,

or (ii) $\max_i a_i = y \neq x \quad (y > x)$,

and

either (iii) $\max_i b_i = x$

or (iv) $\max_i b_i = z \neq x \quad (z > x)$.

This leads to the following four cases:

(i) and (iii): $[{}^1f * {}^2f](\alpha_1, \beta_1, \gamma_1) = x;$

(i) and (iv): $[{}^1f * {}^2f](\alpha_1, \beta_1, \gamma_1) = \min[x, z] = x;$

(ii) and (iii): $[{}^1f * {}^2f](\alpha_1, \beta_1, \gamma_1) = \min[y, x] = x;$

(ii) and (iv): $[{}^1f * {}^2f](\alpha_1, \beta_1, \gamma_1) = \min[y, z].$

The first three cases are obvious; they all lead to the correct reconstruction value which is the same value in each member of the reconstruction family. In the case (ii) and (iv), both $y > x$ and $z > x$, and $\min[y, z]$ is the largest possibility value for state $(\alpha_1, \beta_1, \gamma_1)$ which satisfies the projections ${}^1f$ and ${}^2f$. Indeed, if there were a member in the reconstruction family with a larger possibility value, say value w ($w > y$ and $w > z$), then it would be ${}^1f(\alpha_1, \beta_1) = w$ and ${}^2f(\alpha_1, \gamma_1) = w$, which contradicts the assumption of this case.

Since state $(\alpha_1, \beta_1, \gamma_1)$ was chosen as an arbitrary state, the same conclusions hold for any state of **C**. Moreover, if $(\alpha_1, \beta_1, \gamma_1)$ were only a substate of an aggregate state, we would come to the same conclusions by extending each of the previously considered states to a set of states distinguished by some additional substances, say $\alpha_1, \alpha_2, \ldots, \alpha_m$.

The same arguments which were made for the first join operation ${}^1f * {}^2f$ can be made for a second join operation, e.g., ${}^3f * [{}^1f * {}^2f]$, etc.

Since all members of the reconstruction family have the same maxima and the join operation determines for each state the largest possibility value which does not violate the given projections, it directly follows from Property (U7) of the U-uncertainty (Section C.1) that the join operation determines the reconstruction with maximum U-uncertainty.

Finally, we have to investigate the effect of a loop in the given structure system on the result of the join operation. A loop involves at least two joins, say ${}^3f * ({}^1f * {}^2f)$, such that

$${}^1f\colon X_1 \times X_2 \to [0, 1],$$

$${}^2f\colon X_2 \times X_3 \to [0, 1],$$

$${}^3f\colon X_1 \times X_3 \to [0, 1].$$

Let us consider again the reconstruction of an aggregate state $(\alpha_1, \beta_1, \gamma_1) \in \mathbf{C}$, where $\alpha_1 \in X_1$, $\beta_1 \in X_2$, $\gamma_1 \in X_3$. Let ${}^1f$, ${}^2f$, and ${}^1f * {}^2f$ be exactly the same as described previously in this proof and let the following list include values of the overall possibility distribution for all aggregate states which are relevant for determining the value ${}^3f(\alpha_1, \gamma_1)$ of the projection ${}^3f$:

$$f(\alpha_1, \beta_1, \gamma_1) = c_1 \quad (= x),$$

$$f(\alpha_1, \beta_2, \gamma_1) = c_2,$$

$$\ldots$$

$$f(\alpha_1, \beta_s, \gamma_1) = c_s.$$

Then, $^3f(\alpha_1, \gamma_1) = \max_i c_i$ and $[^3f * [^1f * {}^2f]](\alpha_1, \beta_1, \gamma_1) = \min[^3f(\alpha_1, \gamma_1), [^1f * {}^2f](\alpha_1, \beta_1, \gamma_1)]$. As before, there are two possibilities relevant for the determination of this value:

either (i) $\max_i c_i = x,$

or (ii) $\max_i c_i = w \neq x \quad (w > x),$

and

either (iii) $[^1f * {}^2f](\alpha_1, \beta_1, \gamma_1) = x,$

or (iv) $[^1f * {}^2f](\alpha_1, \beta_1, \gamma_1) = \min[y, z] > x.$

As before, cases (i)–(iii), (i)–(iv), and (ii)–(iii) lead to the correct reconstruction value which is the same value in each member of the reconstruction family. In the case (ii)–(iv),

$$[^3f * [^1f * {}^2f]](\alpha_1, \beta_1, \gamma_1) = \min[w, \min[y, z]] = \min[w, y, z].$$

Due to the same arguments which were used previously in this proof, the value $\min[w, y, z]$ is the largest possibility value for state $(\alpha_1, \beta_1, \gamma_1)$ which satisfied all the three projections ^1f, ^2f, ^3f. Hence, the theorem holds also for structure systems with loops. This means that no iterative procedure is needed for systems with loops and the proof is concluded. Q.E.D.

C.4. GENERAL INFORMATION DISTANCE FOR POSSIBILITY DISTRIBUTIONS†

For convenience, the symbols $\vee$ and $\wedge$ are used in this section as the maximum and minimum operators, respectively.

First, let us introduce a partial ordering of possibility distributions defined on the same set, which assumes a key role in our considerations in this section.

Definition 7. For any pair $^1\mathbf{f}$, $^2\mathbf{f} \in \mathscr{F}$, of normalized possibility distributions $^1\mathbf{f} \leqq {}^2\mathbf{f}$ iff $^1f(x) \leqq {}^2f(x)$ for all $x \in X$.

It is obvious that $(\mathscr{F}, \leqq)$ is a poset. It has a unique universal upper bound, say $^*\mathbf{f}$, defined by $^*f(x) = 1$ for all $x \in X$. For any pair $^1\mathbf{f}$, $^2\mathbf{f} \in \mathscr{F}$, the join $^1\mathbf{f} \vee {}^2\mathbf{f}$ exists in the poset and is defined by the equation

$$(^1f \vee {}^2f)(x) = {}^1f(x) \vee {}^2f(x) \quad \text{for all } x \in X;$$

† Results presented here are related to Section 4.9 and are adopted from my paper [H13].

the meet ${}^1\mathbf{f} \wedge {}^2\mathbf{f}$, which exists only for some pairs ${}^1\mathbf{f}, {}^2\mathbf{f} \in \mathscr{F}$, is defined by

$$({}^1f \wedge {}^2f)(x) = {}^1f(x) \wedge {}^2f(x) \qquad \text{for all } x \in X.$$

More specifically, the meet ${}^1\mathbf{f} \wedge {}^2\mathbf{f}$ exists iff there exists at least one $x \in X$ such that ${}^1f(x) = {}^2f(x) = 1$; if no such x exists, ${}^1\mathbf{f} \wedge {}^2\mathbf{f}$ is not normalized and, hence, ${}^1\mathbf{f} \wedge {}^2\mathbf{f} \notin \mathscr{F}$.

Definition 8. For any pair ${}^1\mathbf{f}, {}^2\mathbf{f} \in \mathscr{F}$ such that ${}^1\mathbf{f} \leqq {}^2\mathbf{f}$, we define the *gain of information*, $g({}^1\mathbf{f}, {}^2\mathbf{f})$, when ${}^2\mathbf{f}$ is replaced by ${}^1\mathbf{f}$ as

$$\begin{aligned} g({}^1\mathbf{f}, {}^2\mathbf{f}) &= U({}^2\mathbf{f}) - U({}^1\mathbf{f}) \\ &= \int_0^1 \log_2 \frac{|c({}^2\mathbf{f}, l)|}{|c({}^1\mathbf{f}, l)|}\, dl. \end{aligned} \tag{C.1}$$

Remark 1. Since ${}^1\mathbf{f} \leqq {}^2\mathbf{f}$ implies $c({}^1\mathbf{f}, l) \subseteq c({}^2\mathbf{f}, l)$ for all $l \in [0, 1]$, it follows immediately from Definition 8 that $g({}^1\mathbf{f}, {}^2\mathbf{f}) \geqq 0$.

Remark 2. For any ${}^1\mathbf{f}, {}^2\mathbf{f}, {}^3\mathbf{f} \in \mathscr{F}$ such that ${}^1\mathbf{f} \leqq {}^2\mathbf{f} \leqq {}^3\mathbf{f}$, the gain of information is additive, i.e.,

$$g({}^1\mathbf{f}, {}^3\mathbf{f}) = g({}^1\mathbf{f}, {}^2\mathbf{f}) + g({}^2\mathbf{f}, {}^3\mathbf{f}).$$

Indeed,

$$\begin{aligned} g({}^1\mathbf{f}, {}^3\mathbf{f}) &= U({}^3\mathbf{f}) - U({}^1\mathbf{f}) \\ &= [U({}^3\mathbf{f}) - U({}^2\mathbf{f})] + [U({}^2\mathbf{f}) - U({}^1\mathbf{f})] \\ &= g({}^2\mathbf{f}, {}^3\mathbf{f}) + g({}^1\mathbf{f}, {}^2\mathbf{f}). \end{aligned}$$

Some additional properties of the gain of information (Definition 8), needed later in our considerations, are stated by the following two lemmas.

Lemma 3. For any ${}^1\mathbf{f}, {}^2\mathbf{f} \in \mathscr{F}$ such that ${}^1\mathbf{f} \leqq {}^2\mathbf{f}$, $g({}^1\mathbf{f}, {}^2\mathbf{f}) = 0$ iff ${}^1\mathbf{f} = {}^2\mathbf{f}$ (i.e., ${}^1f(x) = {}^2f(x)$ for all $x \in X$).

Proof. Let $K(l) \equiv \log_2 |c({}^2\mathbf{f}, l)| - \log_2 |c({}^1\mathbf{f}, l)|$ for all $l \in [0, 1]$. Then,

$$g({}^1\mathbf{f}, {}^2\mathbf{f}) = \int_0^1 K(l)\, dl.$$

Since ${}^1\mathbf{f} \leq {}^2\mathbf{f}$ implies $c({}^1\mathbf{f}, l) \subseteq c({}^2\mathbf{f}, l)$ and, consequently, $|c({}^1\mathbf{f}, l)| \leqq |c({}^2\mathbf{f}, l)|$, we get

$K(l) \geqq 0$ for all $l \in [0, 1]$. Hence,

$$g({}^1\mathbf{f}, {}^2\mathbf{f}) = 0 \Leftrightarrow K(l) = 0 \text{ for all } l \in [0, 1]$$
$$\Leftrightarrow |c({}^1\mathbf{f}, l)| = |c({}^2\mathbf{f}, l)|$$
$$\text{for all } l \in [0, 1].$$

Since $c({}^1\mathbf{f}, l) \subseteq c({}^2\mathbf{f}, l)$ for all $l \in [0, 1]$, we obtain for all $l \in [0, 1]$

$$|c({}^1\mathbf{f}, l)| = |c({}^2\mathbf{f}, l)| \Leftrightarrow c({}^1\mathbf{f}, l) = c({}^2\mathbf{f}, l)$$

If ${}^1\mathbf{f} = {}^2\mathbf{f}$, then the last equation is clearly satisfied. If ${}^1\mathbf{f} \neq {}^2\mathbf{f}$, i.e., ${}^1f(x_0) \neq {}^2f(x_0)$ for some $x_0 \in X$, then ${}^1f(x_0) < {}^2f(x_0)$ and, consequently, $x_0 \notin c({}^1\mathbf{f}, l_0)$ and $x_0 \in c({}^2\mathbf{f}, l_0)$ for some l_0 such that ${}^1f(x_0) < l_0 < {}^2f(x_0)$, say $l_0 = [{}^1f(x) + {}^2f(x)]/2$. This implies that $c({}^1\mathbf{f}, l_0) \neq c({}^2\mathbf{f}, l_0)$. Hence, $c({}^1\mathbf{f}, l) = c({}^2\mathbf{f}, l)$ for all $l \in [0, 1]$ iff ${}^1\mathbf{f} = {}^2\mathbf{f}$ and, therefore, $g({}^1\mathbf{f}, {}^2\mathbf{f}) = 0$ iff ${}^1\mathbf{f} = {}^2\mathbf{f}$. Q.E.D.

Lemma 4. $g({}^1\mathbf{f}, {}^3\mathbf{f}) \geqq g({}^1\mathbf{f}, {}^2\mathbf{f})$ for any ${}^1\mathbf{f}, {}^2\mathbf{f}, {}^3\mathbf{f} \in \mathcal{F}$ such that ${}^1\mathbf{f} \leqq {}^2\mathbf{f} \leqq {}^3\mathbf{f}$, where the equality holds iff ${}^2\mathbf{f} = {}^3\mathbf{f}$.

Proof. By Remark 2,

$$g({}^1\mathbf{f}, {}^3\mathbf{f}) - g({}^1\mathbf{f}, {}^2\mathbf{f}) = g({}^2\mathbf{f}, {}^3\mathbf{f}).$$

By Remark 1, $g({}^2\mathbf{f}, {}^3\mathbf{f}) \geqq 0$, where the equality holds iff ${}^2\mathbf{f} = {}^3\mathbf{f}$ (Lemma 3). Hence, the proposition holds. Q.E.D.

The partial ordering introduced in Definition 7 can be described as the relation

$$R = \{({}^i\mathbf{f}, {}^j\mathbf{f}) \in \mathcal{F} \times \mathcal{F} \,|\, {}^i\mathbf{f} \leq {}^j\mathbf{f}|\}.$$

The information gain, as introduced in Definition 8, can then be viewed as a function on this relation, i.e.,

$$g: R \rightarrow [0, \infty),$$

which, according to Lemma 3, satisfies the nondegeneracy requirement of metric distances. It can easily be extended to satisfy the symmetry requirement as well.

Definition 9. A *symmetric extension* of function g (Definition 8) is a function

$$\hat{g}: R \cup R^{-1} \rightarrow [0, \infty),$$

where R^{-1} denotes the inverse relation of R, that possesses the following properties:

($\hat{g}1$) $\hat{g}|R = g$, where $\hat{g}|R$ denotes the restriction of the domain of function $\hat{g}$ to set R,†

($\hat{g}2$) $\hat{g}$ is symmetric, i.e. $\hat{g}({}^i\mathbf{f}, {}^j\mathbf{f}) = \hat{g}({}^j\mathbf{f}, {}^i\mathbf{f})$ for all $({}^i\mathbf{f}, {}^j\mathbf{f}) \in R \cup R^{-1}$.

Lemma 5. The only function $\hat{g}$ that qualifies as a symmetric extension of g, as specified by Definition 9, is such that

$$\hat{g}({}^1\mathbf{f}, {}^2\mathbf{f}) = |U({}^2\mathbf{f}) - U({}^1\mathbf{f})|. \tag{C.2}$$

Proof. Let $({}^1\mathbf{f}, {}^2\mathbf{f}) \in R$. Then, from ($\hat{g}1$),

$$\hat{g}({}^1\mathbf{f}, {}^2\mathbf{f}) = g({}^1\mathbf{f}, {}^2\mathbf{f}) = U({}^2\mathbf{f}) - U({}^1\mathbf{f}).$$

Let $({}^1\mathbf{f}, {}^2\mathbf{f}) \in R^{-1}$. Then, from ($\hat{g}2$),

$$\hat{g}({}^1\mathbf{f}, {}^2\mathbf{f}) = \hat{g}({}^2\mathbf{f}, {}^1\mathbf{f}),$$

and from ($\hat{g}1$),

$$\begin{aligned}\hat{g}({}^2\mathbf{f}, {}^1\mathbf{f}) = g({}^2\mathbf{f}, {}^1\mathbf{f}) &= U({}^1\mathbf{f}) - U({}^2\mathbf{f}) \\ &= |U({}^2\mathbf{f}) - U({}^1\mathbf{f})|.\end{aligned}$$

Hence,

$$\hat{g}({}^1\mathbf{f}, {}^2\mathbf{f}) = |U({}^2\mathbf{f}) - U({}^1\mathbf{f})|$$

for all

$$({}^1\mathbf{f}, {}^2\mathbf{f}) \in R \cup R^{-1}.$$

Conversely, it is obvious that $\hat{g}$ defined by (C.2) is a function on $R \cup R^{-1}$ into $[0, \infty)$ that satisfies both ($\hat{g}1$) and ($\hat{g}2$). Q.E.D.

Since function $\hat{g}$ characterizes information difference, but does not distinguish whether information is gained or lost, it seems reasonable to refer to it as information variation. It is obvious that $\hat{g}$ has the properties of nondegeneracy and symmetry. In addition, it is also additive in the same sense as function g (Remark 2). It is thus meaningful to view the information variation $\hat{g}$ as a measure of information closeness between possibility distributions in the restricted set $R \cup R^{-1}$ of $\mathcal{F} \times \mathcal{F}$.

Definition 10. A *metric distance in $\mathcal{F}$ based on information closeness* is a function

$$d: \mathcal{F} \times \mathcal{F} \to [0, \infty)$$

† For any function, say $a: A \to B$, the symbol $a|C$ is used in this section to denote the restriction of function a to the domain C, where $C \subset A$.

such that

(d1) $d({}^1\mathbf{f}, {}^2\mathbf{f}) = 0$ iff ${}^1\mathbf{f} = {}^2\mathbf{f}$ (nondegeneracy);

(d2) $d({}^1\mathbf{f}, {}^2\mathbf{f}) = d({}^2\mathbf{f}, {}^1\mathbf{f})$ (symmetry);

(d3) $d({}^1\mathbf{f}, {}^3\mathbf{f}) \leqq d({}^1\mathbf{f}, {}^2\mathbf{f}) + d({}^2\mathbf{f}, {}^3\mathbf{f})$ (triangle property);

(d4) $d|R \cup R^{-1} = \hat{g}$ (restriction of $d|$on $R \cup R^{-1}$).

Our search for a function that would qualify as a metric distance specified in Definition 10 is expressed in terms of the following three lemmas. The main results of this search are then summarized in two theorems.

Lemma 6. Let ${}^1\mathbf{f}, {}^2\mathbf{f}, {}^3\mathbf{f} \in \mathscr{F}$ and ${}^1\mathbf{f} \leq {}^2\mathbf{f}$. Then,

$$g({}^1\mathbf{f}, {}^1\mathbf{f} \vee {}^3\mathbf{f}) \geqq g({}^2\mathbf{f}, {}^2\mathbf{f} \vee {}^3\mathbf{f}), \tag{C.3}$$

where the equality holds iff $c({}^1\mathbf{f}, l) = c({}^2\mathbf{f}, l)$ or $c({}^3\mathbf{f}, l) \subseteq c({}^1\mathbf{f}, l)$ for all $l \in [0, 1]$.

Proof. It follows from ${}^1\mathbf{f} \leq {}^2\mathbf{f}$ that $c({}^1\mathbf{f}, l) \subseteq c({}^2\mathbf{f}, l)$ for all $l \in [0, 1]$. Let $\alpha = |c({}^1\mathbf{f}, l)|$, $\beta = |c({}^2\mathbf{f}, l)|$, $\gamma = |c({}^3\mathbf{f}, l)|$ for an arbitrary $l \in [0, 1]$. Then, clearly, $\beta \geqq \alpha \geqq 0$. Let $\delta = |c({}^1\mathbf{f}, l) \cap c({}^3\mathbf{f}, l)|$ and $\theta = |c({}^2\mathbf{f}, l) \cap c({}^3\mathbf{f}, l)|$. Then, clearly, $\gamma \geqq \theta \geqq \delta \geqq 0$. Hence,

$$\frac{|c({}^1\mathbf{f} \vee {}^3f, l)|}{|c({}^1\mathbf{f}, l)|} - \frac{|c({}^2\mathbf{f} \vee {}^3\mathbf{f}, l)|}{|c({}^2\mathbf{f}, l)|} = \frac{|c({}^1\mathbf{f}, l) \cup c({}^3\mathbf{f}, l)|}{|c({}^1\mathbf{f}, l)|} - \frac{|c({}^2\mathbf{f}, l) \cup c({}^3\mathbf{f}, l)|}{|c({}^2\mathbf{f}, l)|}$$

$$= \frac{\alpha + \gamma - \delta}{\alpha} - \frac{\beta + \gamma - \theta}{\beta} = \frac{\gamma(\beta - \alpha) + \alpha\theta - \beta\delta}{\alpha\beta}$$

$$\frac{\gamma(\beta - \alpha) + \alpha\theta - \beta\theta}{\alpha\beta} = \frac{(\gamma - \theta)(\beta - \alpha)}{\alpha\beta} \geqq 0,$$

where the first inequality is due to $\beta \geqq 0$ and $\theta \geqq \delta$, while the second inequality is due to $\gamma \geqq \theta$ and $\beta \geqq \alpha$. This implies that

$$\frac{|c({}^1\mathbf{f}, l) \cup c({}^3\mathbf{f}, l)|}{|c({}^1\mathbf{f}, l)|} \geqq \frac{|c({}^2\mathbf{f}, l) \cup c({}^3\mathbf{f}, l)|}{|c({}^2\mathbf{f}, l)|} \tag{C.4}$$

is satisfied for all $l \in [0, 1]$. When applying formula (C.1) to both sides of this inequality, we directly obtain the desired inequality $g({}^1\mathbf{f}, {}^1\mathbf{f} \vee {}^3\mathbf{f}) \geq g({}^2\mathbf{f}, {}^2\mathbf{f} \vee {}^3\mathbf{f})$.

If there exists some l for which the equality in (C.4) does not hold, then neither does the equality in (C.3). Hence, the equality in (C.3) holds iff the equality in (C.4) holds for all $l \in [0, 1]$, which is equivalent to $\gamma = \theta = \delta$ or $\alpha = \beta$. These equalities are satisfied, since

$$c({}^3\mathbf{f}, l) \supseteq c({}^2\mathbf{f}, l) \cap c({}^3\mathbf{f}, l) \supseteq c({}^1\mathbf{f}, l) \cap c({}^3\mathbf{f}, l)$$

and

$$c({}^2\mathbf{f}, l) \supseteq c({}^1\mathbf{f}, l).$$

iff

$$c({}^3\mathbf{f}, l) = c({}^2\mathbf{f}, l) \cap c({}^3\mathbf{f}, l) = c({}^1\mathbf{f}, l) \cap c({}^3\mathbf{f}, l)$$

or

$$c({}^2\mathbf{f}, l) = c({}^1\mathbf{f}, l),$$

i.e.,

$$c({}^3\mathbf{f}, l) \subseteq c({}^1\mathbf{f}, l)$$

or

$$c({}^2\mathbf{f}, l) = c({}^1\mathbf{f}, l). \qquad \text{Q.E.D.}$$

Lemma 7. Let

$$G({}^1\mathbf{f}, {}^2\mathbf{f}) = g({}^1\mathbf{f}, {}^1\mathbf{f} \vee {}^2\mathbf{f}) + g({}^2\mathbf{f}, {}^1\mathbf{f} \vee {}^2\mathbf{f}) \tag{C.5}$$

for all ${}^1\mathbf{f}, {}^2\mathbf{f} \in \mathscr{F}$. Then

(*a*) for any ${}^1\mathbf{f}, {}^2\mathbf{f}, {}^3\mathbf{f} \in \mathscr{F}$

$$G({}^1\mathbf{f}, {}^3\mathbf{f}) \leqq G({}^1\mathbf{f}, {}^2\mathbf{f}) + G({}^2\mathbf{f}, {}^3\mathbf{f}),$$

where the equality holds iff ${}^2\mathbf{f} \leq {}^1\mathbf{f} \vee {}^3\mathbf{f}$ and, in addition, $c({}^1\mathbf{f}, l) \subseteq c({}^2\mathbf{f}, l)$ or $c({}^3\mathbf{f}, l) \subseteq c({}^2\mathbf{f}, l)$ for all $l \in [0, 1]$;

(*b*) for any ${}^i\mathbf{f} \in \mathscr{F}$ $(i = 1, 2, \ldots, n)$,

$$G({}^1\mathbf{f}, {}^n\mathbf{f}) \leqq \sum_{i\,1}^{n-1} G({}^i\mathbf{f}, {}^{i+1}\mathbf{f}).$$

Proof. (*a*) Since ${}^1\mathbf{f} \leq {}^1\mathbf{f} \vee {}^3\mathbf{f} \leq {}^1\mathbf{f} \vee {}^2\mathbf{f} \vee {}^3\mathbf{f}$, it follows from Lemma 4 that $g({}^1\mathbf{f}, {}^1\mathbf{f} \vee {}^3\mathbf{f}) \leqq g({}^1\mathbf{f}, {}^1\mathbf{f} \vee {}^2\mathbf{f} \vee {}^3\mathbf{f})$, where the equality holds iff ${}^1\mathbf{f} \vee {}^3\mathbf{f} = {}^1\mathbf{f} \vee {}^2\mathbf{f} \vee {}^3\mathbf{f}$, i.e., ${}^2\mathbf{f} \leqq {}^1\mathbf{f} \vee {}^3\mathbf{f}$. By additivity of g (Remark 2),

$$g({}^1\mathbf{f}, {}^1\mathbf{f} \vee {}^2\mathbf{f} \vee \mathbf{f}) = g({}^1\mathbf{f}, {}^1\mathbf{f} \vee {}^2\mathbf{f}) + g({}^1\mathbf{f} \vee {}^2\mathbf{f}, {}^1\mathbf{f} \vee {}^2\mathbf{f} \vee {}^3\mathbf{f}).$$

Since ${}^2\mathbf{f} \leq {}^1\mathbf{f} \vee {}^2\mathbf{f}$, it follows from Lemma 6 that

$$g({}^1\mathbf{f} \vee {}^2\mathbf{f}, {}^1\mathbf{f} \vee {}^2\mathbf{f} \vee {}^3\mathbf{f}) \leqq g({}^2\mathbf{f}, {}^2\mathbf{f} \vee {}^3\mathbf{f}),$$

where the equality holds iff $c({}^2\mathbf{f}, l) = c({}^1\mathbf{f} \vee {}^2\mathbf{f}, l)$, i.e., $c({}^1\mathbf{f}, l) \subseteq c({}^2\mathbf{f}, l)$, or $c({}^3\mathbf{f}, l) \subseteq c({}^2\mathbf{f}, l)$, for all $l \in [0, 1]$. Hence,

$$g({}^1\mathbf{f}, {}^1\mathbf{f} \vee {}^3\mathbf{f}) \leqq g({}^1\mathbf{f}, {}^1\mathbf{f} \vee {}^2\mathbf{f}) + g({}^2\mathbf{f}, {}^2\mathbf{f} \vee {}^3\mathbf{f}), \tag{C.6}$$

where the equality holds iff ${}^2\mathbf{f} \le {}^1\mathbf{f} \vee {}^3\mathbf{f}$ and, in addition, $c({}^1\mathbf{f}, l) \subseteq c({}^2\mathbf{f}, l)$ or $c({}^3\mathbf{f}, l) \subseteq c({}^2\mathbf{f}, l)$ for all $l \in [0, 1]$.

The same arguments by which (C.6) was derived can be repeated with ${}^1\mathbf{f}$ and ${}^3\mathbf{f}$ exchanged. This leads to the inequality

$$g({}^3\mathbf{f}, {}^1\mathbf{f} \vee {}^3\mathbf{f}) \leqq g({}^3\mathbf{f}, {}^3\mathbf{f} \vee {}^2\mathbf{f}) + g({}^2\mathbf{f}, {}^2\mathbf{f} \vee {}^1\mathbf{f}), \tag{C.7}$$

where the equality holds under exactly the same conditions as for (C.6). Proposition (a) now follows directly from (C.6) and (C.7).

Proposition (b) is immediately obtained by repeatedly applying the result of (a).

Q.E.D.

Lemma 8. For any ${}^1\mathbf{f}, {}^2\mathbf{f} \in \mathscr{F}$, if ${}^1\mathbf{f} \wedge {}^2\mathbf{f} \in \mathscr{F}$, then

$$g({}^1\mathbf{f}, {}^1\mathbf{f} \vee {}^2\mathbf{f}) + g({}^2\mathbf{f}, {}^1\mathbf{f} \vee {}^2\mathbf{f}) \leqq g({}^1\mathbf{f} \wedge {}^2\mathbf{f}, {}^1\mathbf{f}) + g({}^1\mathbf{f} \wedge {}^2\mathbf{f}, {}^2\mathbf{f}), \tag{C.8}$$

where the equality holds iff $c({}^1\mathbf{f}, l) \subseteq c({}^2\mathbf{f}, l)$ or $c({}^1\mathbf{f}, l) \supseteq c({}^2\mathbf{f}, l)$ for all $l \in [0, 1]$.

Proof. Assume ${}^1\mathbf{f} \wedge {}^2\mathbf{f} \in \mathscr{F}$ and let ${}^3\mathbf{f} = {}^1\mathbf{f} \wedge {}^2\mathbf{f}$. Then, ${}^1\mathbf{f} \vee {}^2\mathbf{f} = {}^1\mathbf{f}$ and, clearly, $g({}^1\mathbf{f}, {}^1\mathbf{f} \vee {}^2\mathbf{f}) = 0$. Hence,

$$g({}^1\mathbf{f} \wedge {}^2\mathbf{f}, {}^1\mathbf{f}) = g({}^3\mathbf{f}, {}^1\mathbf{f}) = g({}^3\mathbf{f}, {}^1\mathbf{f} \vee {}^3\mathbf{f}) + g({}^1\mathbf{f}, {}^1\mathbf{f} \vee {}^3\mathbf{f}) = G({}^1\mathbf{f}, {}^3\mathbf{f}).$$

Similarly, ${}^2\mathbf{f} \vee {}^3\mathbf{f} = {}^2\mathbf{f}$, $g({}^2\mathbf{f}, {}^2\mathbf{f} \vee {}^3\mathbf{f}) = 0$, and we get

$$g({}^1\mathbf{f} \wedge {}^2\mathbf{f}, {}^2\mathbf{f}) = g({}^3\mathbf{f}, {}^2\mathbf{f}) = g({}^3\mathbf{f}, {}^2\mathbf{f} \vee {}^3\mathbf{f}) + g({}^2\mathbf{f}, {}^2\mathbf{f} \vee {}^3\mathbf{f}) = G({}^3\mathbf{f}, {}^2\mathbf{f}).$$

We can see now that

$$g({}^1\mathbf{f} \wedge {}^2f, {}^1\mathbf{f}) + g({}^1\mathbf{f} \wedge {}^2\mathbf{f}, {}^2\mathbf{f}) = G({}^1\mathbf{f}, {}^3\mathbf{f}) + G({}^3\mathbf{f}, {}^2\mathbf{f}).$$

Since

$$G({}^1\mathbf{f}, {}^2\mathbf{f}) = g({}^1\mathbf{f}, {}^1\mathbf{f} \vee {}^2\mathbf{f}) + g({}^2\mathbf{f}, {}^1\mathbf{f} \vee {}^2\mathbf{f}),$$

the inequality (C.8) follows directly from Lemma 7. Furthermore, it also follows from Lemma 7 that the equality in (C.8) is satisfied iff ${}^3\mathbf{f} \le {}^1\mathbf{f} \vee {}^2\mathbf{f}$ (i.e., ${}^1\mathbf{f} \wedge {}^2\mathbf{f} \le {}^1\mathbf{f} \vee {}^2\mathbf{f}$, which is always true) and

$$c({}^1\mathbf{f}, l) \subseteq c({}^3\mathbf{f}, l) = c({}^1\mathbf{f} \wedge {}^2\mathbf{f}, l) = c({}^1\mathbf{f}, l) \cap c({}^2\mathbf{f}, l)$$

or

$$c({}^2\mathbf{f}, l) \subseteq c({}^3\mathbf{f}, l) = c({}^1\mathbf{f} \wedge {}^2\mathbf{f}, l) = c({}^1\mathbf{f}, l) \cap c({}^2\mathbf{f}, l),$$

i.e.

$$c({}^1\mathbf{f}, l) \subseteq c({}^2\mathbf{f}, l)$$

or

$$c({}^2\mathbf{f}, l) \subset c({}^1\mathbf{f}, l)$$

for all $l \in [0, 1]$. Q.E.D.

It is interesting to notice that ${}^1\mathbf{f} \wedge {}^2\mathbf{f} \in \mathscr{F}$ is always true when $c({}^1\mathbf{f}, l) \subseteq c({}^2\mathbf{f}, l)$, or $c({}^2\mathbf{f}, l) \subseteq c({}^1\mathbf{f}, l)$ for all $l \in [0, 1]$. Indeed, since $c({}^1\mathbf{f}, l) \subseteq c({}^2\mathbf{f}, l)$ or $c({}^2\mathbf{f}, l) \subseteq c({}^1\mathbf{f}, 1)$, $c({}^1\mathbf{f} \wedge {}^2\mathbf{f}, 1) = c(\mathbf{f}, 1) \wedge c({}^2\mathbf{f}, 1)$ is equal to $c({}^1\mathbf{f}, 1)$ or $c({}^2\mathbf{f}, 1)$, neither of which is empty. Hence, $c({}^1\mathbf{f} \wedge {}^2\mathbf{f}, 1) \neq \varnothing$, i.e., ${}^1\mathbf{f} \wedge {}^2\mathbf{f} \in \mathscr{F}$.

We are now in a position to state the main results of this section by the following two theorems. However, to simplify their formulations as well as proofs, let us first introduce some convenient notation.

Let

$$\Phi_n = \{(\mathbf{f}_1, \mathbf{f}_2)(\mathbf{f}_2, \mathbf{f}_3) \cdots (\mathbf{f}_{n-1}, \mathbf{f}_n) \mid (\forall i \in N_n)((\mathbf{f}_i, \mathbf{f}_{i+1}) \in R \cup R^{-1})\},$$

where $N_n = \{1, 2, \ldots, n\}$ for each $n \in \mathbb{N}$ denotes the set of all *paths* through n elements (possibility distributions) of $\mathscr{F}$ whose successive elements $(\mathbf{f}_i, \mathbf{f}_{i+1})$ are ordered, i.e., $\mathbf{f}_i \leq \mathbf{f}_{i+1}$ or $\mathbf{f}_i \geq \mathbf{f}_{i+1}$, and let

$$\Phi = \bigcup_{n \in \mathbb{N}} \Phi_n.$$

For convenience, elements in each path of Φ will be referred to as *nodes* of the path.

Let a function

$$e: \Phi \rightarrow \mathscr{F} \times \mathscr{F}$$

such that $e(\varphi) = (\mathbf{f}_1, \mathbf{f}_n)$ for each path $\varphi = (\mathbf{f}_1, \mathbf{f}_2)(\mathbf{f}_2, \mathbf{f}_3) \ldots (\mathbf{f}_{n-1}, \mathbf{f}_n)$ be called an *edge identifier*. Clearly, function e is unique and many-to-one; given a path $\varphi \in \Phi$, e identifies its *edges* (the first and last nodes).

Given a path $\varphi = (\mathbf{f}_1, \mathbf{f}_2)(\mathbf{f}_2, \mathbf{f}_3) \ldots (\mathbf{f}_{n-1}, \mathbf{f}_n) \in \Phi$, let $N(\varphi)$ denote the sequence of nodes in the path, i.e., $N(\varphi) = (\mathbf{f}_1, \mathbf{f}_2, \ldots, \mathbf{f}_n)$, and let

$$\hat{g}(\varphi) = \sum_{i=1}^{n-1} \hat{g}(\mathbf{f}_i, \mathbf{f}_{i+1})$$

be called the length of the path. For any ${}^1\mathbf{f}, {}^2\mathbf{f} \in \mathscr{F}$, it is clear that $G({}^1\mathbf{f}, {}^2\mathbf{f}) = \hat{g}(\varphi^*)$, where $\varphi^* = ({}^1\mathbf{f}, {}^1\mathbf{f} \vee {}^2\mathbf{f})\ ({}^1\mathbf{f} \vee {}^2\mathbf{f}, {}^2\mathbf{f}) \in e^{-1}({}^1\mathbf{f}, {}^2\mathbf{f})$.

Theorem 6. (*a*) $\min(\hat{g}(\varphi) = G({}^1\mathbf{f}, {}^2\mathbf{f})$, where $\varphi \in e^{-1}({}^1\mathbf{f}, {}^2\mathbf{f})$; (*b*) if d is a metric distance in $\mathscr{F}$ based on $\hat{g}$, then $d({}^1\mathbf{f}, {}^2\mathbf{f}) \leqq G({}^1\mathbf{f}, {}^2\mathbf{f})$ for any ${}^1\mathbf{f}, {}^2\mathbf{f} \in \mathscr{F}$.

Proof. (*a*) Let ${}^1\mathbf{f}, {}^2\mathbf{f} \in \mathscr{F}$. Consider an arbitrary φ such that $\varphi \in e^{-1}({}^1\mathbf{f}, {}^2\mathbf{f})$ and $N(\varphi) = (\mathbf{f}_1, \mathbf{f}_2, \ldots, \mathbf{f}_n)$, i.e., ${}^1\mathbf{f} = \mathbf{f}_1$ and ${}^2\mathbf{f} = \mathbf{f}_n$. Let the following set be defined on φ:

$$N_\varphi = \{\mathbf{f}_i \mid i \in \{2, 3, \ldots, n-1\},\quad \mathbf{f}_{i-1} \leq \mathbf{f}_i \geq \mathbf{f}_{i+1}$$
$$\text{or } \mathbf{f}_{i-1} \geq \mathbf{f}_i \leq \mathbf{f}_{i+1}, \text{ and } \mathbf{f}_{i-1} \neq \mathbf{f}_i \neq \mathbf{f}_{i+1}\}.$$

Assume that $N_\varphi = \{\mathbf{h}_1, \mathbf{h}_2, \ldots, \mathbf{h}_m\}$, where $\mathbf{h}_k = \mathbf{f}_{i_k}$ and $i_k < i_{k+1}$ for $k \in N_m$.

Case I. Assume $N_\varphi = \varnothing$. Then either ${}^1\mathbf{f} = \mathbf{f}_1 \leq \mathbf{f}_2 \leq \ldots \leq \mathbf{f}_n = {}^2\mathbf{f}$ or ${}^1\mathbf{f} = \mathbf{f}_1 \geq \mathbf{f}_2 \geq \cdots \geq \mathbf{f}_n = {}^2\mathbf{f}$. Due to the additivity of g (Remark 2), which also holds for $\hat{g}$, we get $\hat{g}(\varphi) = \hat{g}({}^1\mathbf{f}, {}^2\mathbf{f})$. Since either ${}^1\mathbf{f} \leq {}^2\mathbf{f}$ or ${}^1\mathbf{f} \geq {}^2\mathbf{f}$, we have either ${}^1\mathbf{f} \vee {}^2\mathbf{f} = {}^2\mathbf{f}$ or ${}^1\mathbf{f} \vee {}^2\mathbf{f} = {}^1\mathbf{f}$, respectively. Hence,

$$\hat{g}({}^1\mathbf{f}, {}^2\mathbf{f}) = g({}^1\mathbf{f}, {}^1\mathbf{f} \vee {}^2\mathbf{f}) + g({}^2\mathbf{f}, {}^1\mathbf{f} \vee {}^2\mathbf{f})$$
$$= G({}^1\mathbf{f}, {}^2\mathbf{f}),$$

i.e.,

$$\hat{g}(\varphi) = G({}^1\mathbf{f}, {}^2\mathbf{f}).$$

Case II. Assume $N_\varphi \neq \varnothing$. Let $\mathbf{h}_0 = {}^1\mathbf{f}$, $\mathbf{h}_{m+1} = {}^2\mathbf{f}$ and

$$\varphi' = (\mathbf{h}_0, \mathbf{h}_1)(\mathbf{h}_1, \mathbf{h}_2) \cdots (\mathbf{h}_m, \mathbf{h}_{m+1}) \in e^{-1}({}^1\mathbf{f}, {}^2\mathbf{f}) \subseteq \Phi.$$

Then, by the addivity of $\hat{g}$, we get $\hat{g}(\varphi) = \hat{g}(\varphi')$.

(i) Assume that $\mathbf{h}_0 \leq \mathbf{h}_1$ and $\mathbf{h}_m \geq \mathbf{h}_{m+1}$. Then m is odd and for each $k \in N_m$,

$$\mathbf{h}_{k-1} \leq \mathbf{h}_k \geq \mathbf{h}_{k+1} \quad \text{if } k \text{ is odd}$$

and

$$\mathbf{h}_{k-1} \geq \mathbf{h}_k \leq \mathbf{h}_{k+1} \quad \text{if } k \text{ is even.}$$

Hence, for $k = 2j$ ($j = 0, 1, \ldots, (m+1)/2$), $\mathbf{h}_k \vee \mathbf{h}_{k+2} \leq \mathbf{h}_{k+1}$ and, by Lemma 4, we get

$$0 \leq g(\mathbf{h}_k, \mathbf{h}_k \vee \mathbf{h}_{k+2}) \leqq g(\mathbf{h}_k, \mathbf{h}_{k+1})$$

and

$$0 \leq g(\mathbf{h}_{k+2}, \mathbf{h}_k \vee \mathbf{h}_{k+2}) \leqq g(\mathbf{h}_{k+2}, \mathbf{h}_{k+1}).$$

Therefore,

$$\hat{g}(\varphi) = \hat{g}(\varphi') = \sum_{k=1}^{m} \hat{g}(\mathbf{h}_k, \mathbf{h}_{k+1}) \geqq \sum_{j=0}^{(m-1)/2} G(\mathbf{h}_{2j}, \mathbf{h}_{2j+2}),$$

and since

$$\sum_{j=0}^{(m-1)/2} G(\mathbf{h}_{2j}, \mathbf{h}_{2j+2}) \geqq G(\mathbf{h}_0, \mathbf{h}_{m+1}),$$

according to Lemma 7, we get $\hat{g}(\varphi) \geqq G(\mathbf{h}_0, \mathbf{h}_{m+1}) = G({}^1\mathbf{f}, {}^2\mathbf{f})$.

(ii) Assume that $\mathbf{h}_0 \geq \mathbf{h}_1$ or $\mathbf{h}_m \leq \mathbf{h}_{m+1}$. Three cases have to be considered.

Assume first that $\mathbf{h}_0 \geq \mathbf{h}_1$ and $\mathbf{h}_m \geq \mathbf{h}_{m+1}$. Then, $\mathbf{h}_0 \geq \mathbf{h}_1 \leq \mathbf{h}_2$ and, consequently, $\mathbf{h}_1 \leq \mathbf{h}_0 \wedge \mathbf{h}_2$. By Lemma 4,

$$g(\mathbf{h}_1, \mathbf{h}_0) + g(\mathbf{h}_1, \mathbf{h}_2) \geqq g(\mathbf{h}_0 \wedge \mathbf{h}_2, \mathbf{h}_0) + g(\mathbf{h}_0 \wedge \mathbf{h}_2, \mathbf{h}_2),$$

and by Lemma 8,

$$g(\mathbf{h}_0 \wedge \mathbf{h}_2, \mathbf{h}_0) + g(\mathbf{h}_0 \wedge \mathbf{h}_2, \mathbf{h}_2) \geqq g(\mathbf{h}_0, \mathbf{h}_0 \vee \mathbf{h}_2) + g(\mathbf{h}_2, \mathbf{h}_0 \vee \mathbf{h}_2).$$

Hence,

$$g(\mathbf{h}_1, \mathbf{h}_0) + g(\mathbf{h}_1, \mathbf{h}_2) \geqq g(\mathbf{h}_0, \mathbf{h}_0 \vee \mathbf{h}_2) + g(\mathbf{h}_2, \mathbf{h}_0 \vee \mathbf{h}_2).$$

Therefore, for

$$\varphi'' = (\mathbf{h}_0, \mathbf{h}_0 \vee \mathbf{h}_2)(\mathbf{h}_0 \vee \mathbf{h}_2, \mathbf{h}_2)(\mathbf{h}_2, \mathbf{h}_3) \cdots (\mathbf{h}_m, \mathbf{h}_{m+1}) \in e^{-1}({}^1\mathbf{f}, {}^2\mathbf{f}),$$

we get $\hat{g}(\varphi') \geqq \hat{g}(\phi'')$. Applying the same transformation to φ'' as to φ, we obtain

$$\varphi''' = (\mathbf{h}_0, \mathbf{h}_0 \vee \mathbf{h}_2)(\mathbf{h}_0 \vee \mathbf{h}_2, \mathbf{h}_3)(\mathbf{h}_3, \mathbf{h}_4) \cdots (\mathbf{h}_m, \mathbf{h}_{m+1}) \in e^{-1}({}^1\mathbf{f}, {}^2\mathbf{f})$$

and $\hat{g}(\varphi'') = \hat{g}(\varphi''')$. Thus,

$$\hat{g}(\varphi) = \hat{g}(\varphi'') \geqq \hat{g}(\varphi'') = \hat{g}(\varphi''').$$

Let $N(\varphi''') = (\mathbf{h}'_0, \mathbf{h}'_1, \ldots, \mathbf{h}'_m)$. Then, $\mathbf{h}'_0 \leq \mathbf{h}'_1$ and $\mathbf{h}'_{m-1} \geq \mathbf{h}'_m$.

Assume now that $\mathbf{h}_0 \leq \mathbf{h}_1$ and $\mathbf{h}_m \leq \mathbf{h}_{m+1}$. By the same reasoning as in the previous case, φ can be transformed into $\varphi^1 \in e^{-1}({}^1\mathbf{f}, {}^2\mathbf{f})$ such that $\hat{g}(\varphi) \geqq \hat{g}(\varphi^1)$ and $\mathbf{h}_0^1 \leq \mathbf{h}_1^1$, $\mathbf{h}_{m-1}^1 \geq \mathbf{h}_m^1$, where $N(\varphi^1) = (\mathbf{h}_0^1, \mathbf{h}_1^1, \ldots, \mathbf{h}_m^1)$.

In the third case, i.e., $\mathbf{h}_0 \geq \mathbf{h}_1$ and $\mathbf{h}_m \geq \mathbf{h}_{m+1}$, we can first transform φ into φ''' and, then, apply to φ''' the same transformation that was applied to φ to get φ^1. As a result, we get $\varphi^2 \in e^{-1}({}^1\mathbf{f}, {}^2\mathbf{f})$ such that $\hat{g}(\varphi) \geqq \hat{g}(\varphi''') \geq g'(\varphi^2)$ and $\mathbf{h}_0^2 \leq \mathbf{h}_1^2$, $\mathbf{h}_{m-1}^2 \geq \mathbf{h}_{m-1}^2$, where $N(\varphi^2) = (\mathbf{h}_0^2, \mathbf{h}_1^2, \ldots, \mathbf{h}_{m-1}^2)$.

Hence, in any of the three cases, we can transform φ into $\varphi^3 \in e^{-1}({}^1\mathbf{f}, {}^2\mathbf{f})$ such that $\hat{g}(\varphi) \geqq \hat{g}(\varphi^3)$ and $\mathbf{h}_0^3 \leq \mathbf{h}_1^3$, $\mathbf{h}_r^3 \geq \mathbf{h}_{r+1}^3$, where $N(\varphi^3) = (\mathbf{h}_0^3, \mathbf{h}_1^3, \ldots, \mathbf{h}_{r+1}^3)$. Applying now the result of (i), we get $\hat{g}(\varphi) \geqq \hat{g}(\varphi^3) \geqq G({}^1\mathbf{f}, {}^2\mathbf{f})$.

From Case I and Case II, we may conclude that

$$\hat{g}(\varphi) \geqq G({}^1\mathbf{f}, {}^2\mathbf{f}) \tag{C.9}$$

for any $\varphi \in e^{-1}({}^1\mathbf{f}, {}^2\mathbf{f})$. Since $G({}^1\mathbf{f}, {}^2\mathbf{f}) = \hat{g}(\varphi^*)$, where $\varphi^* = ({}^1\mathbf{f}, {}^1\mathbf{f} \vee {}^2\mathbf{f})({}^1\mathbf{f} \vee {}^2\mathbf{f}, {}^2\mathbf{f}) \in e^{-1}({}^1\mathbf{f}, {}^2\mathbf{f})$, the inequality (C.9) implies that

$$\min \hat{g}(\varphi) = G({}^1\mathbf{f}, {}^2\mathbf{f}),$$

where the min operator is applied to all $\varphi \in e^{-1}({}^1\mathbf{f}, {}^2\mathbf{f})$.

(*b*) Let ${}^1\mathbf{f}, {}^2\mathbf{f} \in \mathscr{F}$. Since d satisfies the triangle inequality (Definition 10),

$$d({}^1\mathbf{f}, {}^2\mathbf{f}) \leqq d({}^1\mathbf{f}, {}^1\mathbf{f} \vee {}^2\mathbf{f}) + d({}^1\mathbf{f} \vee {}^2\mathbf{f}, {}^2\mathbf{f}).$$

Since

$$({}^1\mathbf{f}, {}^1\mathbf{f} \vee {}^2\mathbf{f}) \in R, \ ({}^1\mathbf{f} \vee {}^2\mathbf{f}, {}^2\mathbf{f}) \in R^{-1}, \text{ and } d \,|\, R \cup R^{-1} = \hat{g},$$

we have

$$d({}^1\mathbf{f}, {}^1\mathbf{f} \vee {}^2\mathbf{f}) = g({}^1\mathbf{f}, {}^1\mathbf{f} \vee {}^2\mathbf{f})$$

and

$$d({}^1\mathbf{f} \vee {}^2\mathbf{f}, {}^2\mathbf{f}) = \hat{g}({}^1\mathbf{f} \vee {}^2\mathbf{f}, {}^2\mathbf{f}) = g({}^2\mathbf{f}, {}^1\mathbf{f} \vee {}^2\mathbf{f}).$$

Hence,

$$\begin{aligned} d({}^1\mathbf{f}, {}^2\mathbf{f}) &\leqq g({}^1\mathbf{f}, {}^1\mathbf{f} \vee {}^2\mathbf{f}) + g({}^2\mathbf{f}, {}^1\mathbf{f} \vee {}^2\mathbf{f}) \\ &= G({}^1\mathbf{f}, {}^2\mathbf{f}), \end{aligned}$$

and this concludes the proof of (b). Q.E.D.

Theorem 7. Function G (introduced in Lemma 7) is a metric distance in $\mathscr{F}$ based on information closeness (Definition 10).

Proof. We have to show that G satisfies requirements (d1)–(d4) specified in Definition 10.

(d1) $G({}^1\mathbf{f}, {}^2\mathbf{f}) = g({}^1\mathbf{f}, {}^1\mathbf{f} \vee {}^2\mathbf{f}) + g({}^2\mathbf{f}, {}^1\mathbf{f} \vee {}^2\mathbf{f}) = 0$

$\Leftrightarrow g({}^1\mathbf{f}, {}^1\mathbf{f} \vee {}^2\mathbf{f}) = 0$ and $g({}^2\mathbf{f}, {}^1\mathbf{f} \vee {}^2\mathbf{f}) = 0$

$\Leftrightarrow {}^1\mathbf{f} = {}^1\mathbf{f} \vee {}^2\mathbf{f}$ and ${}^2\mathbf{f} = {}^1\mathbf{f} \vee {}^2\mathbf{f}$

(by Lemma 3)

$\Leftrightarrow {}^1\mathbf{f} = {}^2\mathbf{f}$.

(d2) $G({}^1\mathbf{f}, {}^2\mathbf{f}) = G({}^2\mathbf{f}, {}^1\mathbf{f})$ is trivial.

(d3) Proved in Lemma 7.

(d4) For $({}^1\mathbf{f}, {}^2\mathbf{f}) \in R \cup R^{-1}$, either ${}^1\mathbf{f} \leq {}^2\mathbf{f}$ or ${}^2\mathbf{f} \leq {}^1\mathbf{f}$ and, thus, either ${}^1\mathbf{f} \vee {}^2\mathbf{f} = {}^2\mathbf{f}$ or ${}^1\mathbf{f} \vee {}^2\mathbf{f} = {}^1\mathbf{f}$, respectively. Hence, either $G({}^1\mathbf{f}, {}^2\mathbf{f}) = g({}^1\mathbf{f}, {}^2\mathbf{f})$ or $G({}^1\mathbf{f}, {}^2\mathbf{f}) = g({}^2\mathbf{f}, {}^1\mathbf{f})$, respectively. Hence, $G({}^1\mathbf{f}, {}^2\mathbf{f}) = \hat{g}({}^1\mathbf{f}, {}^2\mathbf{f})$. Q.E.D.

APPENDIX D

REFINEMENT LATTICES

The three tables in this appendix contain complete descriptions of the refinement lattices $(\mathscr{G}_n/i, \leq)$ and $(\mathscr{C}_n/i, \leq)$ for $n = 3, 4, 5$, as well as a characterization of the r-equivalence classes in these lattices.

Each i-equivalence class of G-structures is represented in the tables by one particular G-structure, a representative of the respective i-equivalence class. These representatives are specified in the sixth column of each table by subsets of N_n $(n = 3, 4, 5)$ that are separated by slashes. The remaining columns in the table have the following meaning: g is an identifier of i-equivalence classes of G-structures; $l(g)$ indicates the level of refinement (l-equivalence class) of G-structure g; c is an identifier of C-structures; j is an identifier by which i-equivalence classes are distinguished in each r-equivalence class (the i-equivalence classes of P-structures are located at the end of each r-equivalence class); $l(c)$ indicates the level of refinement of C-structure c; $\#g$ denotes the number of distinct G-structures in each i-equivalence class; $s(g)$ stands for the set of i-equivalence classes that are immediate successors (refinements) of the i-equivalence class identified by g; $\#s(g)$ denotes the set of numbers of immediate successors of i-equivalence class g, one for each of the i-equivalence classes in $s(g)$: individual i-equivalence classes in each set $s(g)$ are listed (in terms of their identifiers) together with the corresponding numbers in set $\#s(g)$ as pairs separated by slashes; pairs in parentheses indicate successors that are in a different r-equivalence class than the reference G-structure g.

Let $\#s(g, g')$ denote the number of immediate successors (refinements) of each general structure of the i-equivalence class g in the i-equivalence class g' and let $\#p(g', g)$ denote the numbers of immediate predecessors (coarsenings) of each general structure of the i-equivalence class g' in the i-equivalence class g. Then, by simple considerations, we find that

$$\#g \cdot \#s(g, g') = \#g' \cdot \#p(g', g).$$

Hence,

$$\#p(g', g) = \frac{\#g \cdot \#s(g, g')}{\#g'}. \tag{D.1}$$

TABLE D.1
Lattices $(\mathscr{G}_3/i, \leq)$ and $(\mathscr{C}_3/i, \leq)$

g	$l(g)$	c	j	$l(c)$	Subsystems	#g	$s(g)/\#s(g)$
1	1	1	1	1	123	1	2/1
2	2		2		12/13/23	1	(3/3)
3	3	2	1	2	12/23	3	(4/2)
4	4	3	1	3	12/3	3	(5/1)
5	5	4	1	4	1/2/3	1	None

TABLE D.2
Lattices $(\mathscr{G}_4/i, \leq)$ and $(\mathscr{C}_4/i, \leq)$

g	$l(g)$	c	j	$l(c)$	Subsystems	#g	$s(g)/\#s(g)$
1	1	1	1	1	1234	1	2/1
2	2		2		123/124/134/234	1	3/4
3	3		3		123/124/134	4	4/3
4	4		4		123/124/34	6	5/2 (7/1)
5	5		5		123/14/24/34	4	6/1 (8/3)
6	6		6		12/13/14/23/24/34	1	9/6
7	5	2	1	2	123/124	6	8/2
8	6		2		123/14/24	12	9/1 (10/2)
9	7		3		12/13/14/23/24	6	(11/4 12/1)
10	7	3	1	3	123/14	12	11/1 (13/1)
11	8		2		12/13/14/23	12	(14/1 15/1 16/2)
12	8	4	1	3	13/14/23/24	3	(16/4)
13	8	5	1	4	123/4	4	14/1
14	9		2		12/13/23/4	4	(17/3)
15	9	6	1	4	12/13/14	4	(17/3)
16	9	7	1	4	13/14/23	12	(17/2 18/1)
17	10	8	1	5	13/14/2	12	(19/2)
18	10	9	1	5	14/23	3	(19/2)
19	11	10	1	6	14/2/3	6	(20/1)
20	12	11	1	7	1/2/3/4	1	None

TABLE D.3
Lattices ($(\mathscr{G}_5/i, \leq)$) and ($(\mathscr{C}_5/i, \leq)$)

g	$l(g)$	c	j	$l(c)$	Subsystems	#g	$s(g)$/#$s(g)$
1	1	1	1	1	12345	1	2/1
2	2		2		1234/1235/1245/1345/2345	1	3/4
3	3		3		1235/1245/1345/2345	5	4/4
4	4		4		123/1245/1345/2345	10	5/1 6/4
5	5		5		1245/1345/2345	10	7/3
6	5		6		123/124/125/1345/2345	10	7/3 8/2
7	6		7		124/125/1345/2345	30	9/2 10/2
8	6		8		123/124/125/134/135/145/2345	5	10/6 11/1
9	7		9		125/1345/2345	30	12/1 13/2
10	7		10		124/125/134/135/145/2345	30	12/4 14/1 15/1
11	7		11		123/124/125/134/135/145/234/ 235/245/345	1	15/10
12	8		12		12/1345/2345	10	16/2 (55/1)
13	8		13		125/134/135/145/2345	60	16/1 17/1 18/2 19/1
14	8		14		124/125/134/135/2345	15	18/4 20/1
15	8		15		124/125/134/135/145/234/235/ 245/345	10	19/6 20/3
16	9		16		12/134/135/145/2345	20	21/3 22/1 (56/2)
17	9		17		125/135/145/2345	20	21/3 23/1
18	9		18		125/134/145/2345	60	21/2 24/1 25/1
19	9		19		125/134/135/145/234/235/245/345	30	22/1 23/2 25/4 26/1
20	9		20		124/125/134/135/234/235/245/345	15	25/4 26/4
21	10		21		12/135/145/2345	60	27/2 28/1 (57/1)
22	10		22		12/134/135/145/234/235/245/345	10	28/6 29/1
23	10		23		125/135/145/234/235/245/345	20	28/3 30/1 31/3
24	10		24		125/134/2345	15	27/2 32/1
25	10		25		125/134/145/234/235/245/345	60	28/2 31/2 32/1 33/2
26	10		26		125/134/135/145/234/235/245	30	29/1 31/2 33/4
27	11		27		12/13/145/2345	30	34/1 35/1 (59/2)
28	11		28		12/135/145/234/235/245/345	60	35/2 36/1 37/2 38/1 (60/1)
29	11		29		12/134/135/145/234/235/245	10	38/6 (60/1)
30	11		30		125/135/145/235/245/345	5	36/6
31	11		31		125/135/145/234/245/345	60	36/1 38/3 39/2
32	11		32		125/134/234/235/245/345	15	35/2 39/4
33	11		33		125/134/145/234/235/345	60	37/1 38/2 39/2 40/1
34	12		34		12/13/14/15/2345	5	41/1 (63/4)
35	12		35		12/13/145/234/235/245/345	30	41/1 42/2 43/2 (62/2)
36	12		36		12/135/145/235/245/345	30	42/4 44/1 (65/1)
37	12		37		12/135/145/234/245/345	60	42/2 43/2 45/1 (66/1)
38	12		38		12/135/145/234/235/245	60	43/2 44/1 45/2 (64/1)
39	12		39		125/135/145/234/345	60	42/1 43/1 45/2 46/1
40	12		40		125/134/145/234/235	12	45/5
41	13		41		12/13/14/15/234/235/245/345	5	47/4 (69/4)
42	13		42		12/13/145/235/245/345	60	47/1 48/1 49/2 (68/2)
43	13		43		12/13/145/234/235/345	60	47/1 49/2 50/1 (67/1 70/1)
44	13		44		12/135/145/235/245/34	15	49/4 (71/2)
45	13		45		12/135/145/234/245	60	49/2 50/2 (72/1)
46	13		46		125/135/145/234	10	48/1 50/3
47	14		47		12/13/14/15/235/245/345	20	51/3 (74/1 77/3)
48	14		48		12/13/145/23/245/345	10	51/3 (76/3)

(continued overleaf)

TABLE D.3 *(continued)*

g	*l(g)*	*c*	*j*	*l(c)*	Subsystems	#*g*	*s(g)*/# *s(g)*
49	14		49		12/13/145/235/24/345	60	51/2 52/1 (73/2 78/1)
50	14		50		12/13/145/234/235	30	51/1 52/2 (75/2)
51	15		51		12/13/14/15/23/245/345	30	53/2 (79/1 80/2 82/2)
52	15		52		12/13/145/235/24/34	15	53/2 (81/4)
53	16		53		12/13/14/15/23/24/25/345	10	54/1 (83/6 84/1)
54	17		54		12/13/14/15/23/24/25/34/35/45	1	(85/10)
55	9	2	1	2	1234/1345	10	56/2
56	10		2		123/124/1345/234	20	57/3 58/1
57	11		3		124/1345/234	60	59/2 60/1
58	11		4		123/124/134/135/145/234/345	10	60/6 61/1
59	12		5		12/1345/234	60	62/1 63/1 (86/1)
60	12		6		124/134/135/145/234/345	60	62/2 64/1 65/1 66/2
61	12		7		123/124/135/145/234/345	10	64/6
62	13		8		12/134/135/145/234/345	60	67/1 68/2 69/1 70/1 (88/1)
63	13		9		12/1345/23/24	20	69/1 (87/3)
64	13		10		124/135/145/234/345	60	67/2 71/1 72/2
65	13		11		124/134/145/234/345	30	68/4 71/1
66	13		12		124/134/135/234/345	60	68/2 70/2 72/1
67	14		13		12/135/145/234/345	60	73/2 74/1 75/1 (91/1)
68	14		14		12/134/145/234/345	120	73/1 76/1 77/1 78/1 (90/1)
69	14		15		12/134/135/145/23/24/345	20	74/1 77/3 (89/3)
70	14		16		12/134/135/145/234	60	75/1 77/1 78/2 (91/1)
71	14		17		124/13/145/234/345	30	73/4 (106/1)
72	14		18		124/135/234/345	60	73/2 75/2
73	15		19		12/13/145/234/345	120	79/1 80/1 81/1 (94/1 107/1)
74	15		20		12/135/145/23/24/345	20	80/3 (93/3)
75	15		21		12/135/145/234	60	80/1 81/2 (96/1)
76	15		22		12/134/15/234/345	30	79/1 82/2 (95/2)
77	15		23		12/134/145/23/24/345	60	80/1 82/2 (92/2 93/1)
78	15		24		12/134/145/234/35	60	81/1 82/2 (94/2)
79	16		25		12/13/14/15/234/345	30	84/2 (101/1 108/2)
80	16		26		12/13/145/23/24/345	60	83/2 (97/2 99/1 108/1)
81	16		27		12/13/145/234/35	60	83/2 (100/2 109/1)
82	16		28		12/134/15/23/24/345	60	83/1 84/1 (97/2 98/1 101/1)
83	17		29		12/13/14/15/23/24/345	60	85/1 (102/2 103/1 104/1 110/2)
84	17		30		12/134/15/23/24/35/45	10	85/1 (103/6)
85	18		31		12/13/14/15/23/24/34/35/45	10	(105/6 111/3)
86	13	3	1	3	125/2345	30	87/1 88/1
87	14		2		12/15/2345	30	89/1 (122/2)
88	14		3		125/234/235/245/345	30	89/1 90/2 91/2
89	15		4		12/15/234/235/245/345	30	92/2 93/2 (123/2)
90	15		5		125/235/245/345	60	92/1 94/2 95/1
91	15		6		125/234/245/345	60	93/1 94/2 96/1
92	16		7		12/15/235/245/345	60	97/2 98/1 (124/1 125/1)
93	16		8		12/15/234/245/345	60	97/2 99/1 (124/2)
94	16		9		125/23/245/345	120	97/1 100/1 101/1 (112/1)
95	16		10		125/235/245/34	30	98/1 101/2 (118/1)
96	16		11		125/234/345	30	99/1 100/2
97	17		12		12/15/23/245/345	120	102/1 103/1 (113/1 126/1 127/1)

TABLE D.3 *(continued)*

g	l(g)	c	j	l(c)	Subsystems	#g	s(g)/#s(g)
98	17		13		12/15/235/245/34	30	103/2 (119/1 127/2)
99	17		14		12/15/234/25/345	30	102/2 (126/2 131/1)
100	17		15		125/23/24/345	60	102/1 104/1 (114/2)
101	17		16		125/23/245/34/35	60	103/1 104/1 (113/2 119/1)
102	18		17		12/15/23/24/25/345	60	105/1 (115/1 128/1 129/1 132/1)
103	18		18		12/15/23/245/34/35	60	105/1 (116/2 120/1 129/2)
104	18		19		125/23/24/34/35/45	30	105/1 (115/4 120/1)
105	19		20		12/15/23/24/25/34/35/45	30	(117/4 121/1 130/2 133/1)
106	15	4	1	3	123/125/145/134	15	107/4
107	16		2		125/134/145/23	60	108/2 109/2 (112/1)
108	17		3		12/134/145/23/25	60	110/2 (113/2 131/1)
109	17		4		125/134/23/45	30	110/2 (114/2)
110	18		5		12/13/145/23/25/34	60	111/1 (115/2 116/1 132/2)
111	19		6		12/13/14/15/23/25/34/45	15	(117/4 133/4)
112	17	5	1	4	125/235/345	60	113/2 114/1
113	18		2		12/15/235/345	120	115/1 116/1 (135/1 142/1)
114	18		3		125/23/345	60	115/2 (139/1)
115	19		4		12/15/23/25/345	120	117/1 (136/1 140/1 143/1 151/1)
116	19		5		12/15/235/34/45	60	117/1 (137/2 143/2)
117	20		6		12/15/23/25/34/35/45	60	(138/2 141/1 144/2 152/2)
118	17	6	1	4	125/235/245	10	119/3
119	18		2		12/15/235/245	30	120/2 (142/2)
120	19		3		12/15/23/245/35	30	121/1 (143/4)
121	20		4		12/15/23/24/25/35/45	10	(134/1 144/6)
122	15	7	1	4	12/2345	20	123/1 (145/1)
123	16		2		12/234/235/245/345	20	124/3 125/1 (146/1)
124	17		3		12/235/245/345	60	126/2 127/1 (147/1)
125	17		4		12/234/235/245	20	127/3 (147/1)
126	18		5		12/23/245/345	60	128/1 129/1 (135/1 142/1)
127	18		6		12/235/245/34	60	129/2 (142/1 148/1)
128	19		7		12/23/24/25/345	20	130/1 (136/3 149/1)
129	19		8		12/23/245/34/35	60	130/1 (137/1 143/2 149/1)
130	20		9		12/23/24/25/34/35/45	20	(138/2 144/4 150/1)
131	18	8	1	4	125/145/23/34	30	132/2 (135/2)
132	19		2		12/145/23/25/34	60	133/1 (136/1 137/1 151/2)
133	20		3		12/14/15/23/25/34/45	30	(134/1 138/2 152/4)
134	21	9	1	5	12/13/14/25/35/45	10	(153/6)
135	19	10	1	5	12/235/345	60	136/1 137/1 (156/1)
136	20		2		12/23/25/345	60	138/1 (157/1 159/2)
137	20		3		12/235/34/45	60	138/1 (154/2 157/1)
138	21		4		12/23/25/34/35/45	60	(153/1 155/1 158/1 160/2)
139	19	11	1	5	125/345	15	140/2
140	20		2		12/15/25/345	30	141/1 (159/2 161/1)
141	21		3		12/15/25/34/35/45	15	(160/4 162/2)
142	19	12	1	5	15/235/345	60	143/2 (156/1)
143	20		2		15/23/25/345	120	144/1 (154/1 157/1 161/1)
144	21		3		15/23/25/34/35/45	60	(153/1 155/2 158/1 162/2)

(continued overleaf)

TABLE D.3 *(continued)*

g	$l(g)$	c	j	$l(c)$	Subsystems	$\# g$	$s(g)/\# s(g)$
145	16	13	1	5	1/2345	5	146/1
146	17		2		1/234/235/245/345	5	147/4
147	18		3		1/235/245/345	20	148/3
148	19		4		1/23/245/345	30	149/2 (156/1)
149	20		5		1/23/24/25/345	20	150/1 (157/3)
150	21		6		1/23/24/25/34/35/45	5	(158/6)
151	20	14	1	5	12/135/24/45	60	152/1 (154/1 159/2)
152	21		2		12/24/45/13/15/35	60	(153/2 155/1 160/2 163/1)
153	22	15	1	6	12/13/15/24/45	60	(164/2 167/2 168/1)
154	21	16	1	6	124/13/45	60	(155/1 169/2)
155	22		2		12/13/14/24/45	60	(164/1 167/2 170/2)
156	20	17	1	6	1/235/345	30	157/2
157	21		2		1/23/25/345	60	158/1 (169/2)
158	22		3		1/23/25/34/35/45	30	(168/1 170/4)
159	21	18	1	6	124/13/35	60	160/1 (165/1 169/1)
160	22		2		12/13/14/24/35	60	(164/2 166/1 167/1 170/1)
161	21	19	1	6	124/13/15	30	162/1 (169/2)
162	22		2		12/13/14/15/24	30	(167/2 170/2 171/1)
163	22	20	1	6	12/13/24/35/45	12	(164/5)
164	23	21	1	7	12/13/24/45	60	(172/2 176/2)
165	22	22	1	7	124/35	10	166/1 (173/1)
166	23		2		12/14/24/35	10	(174/1 176/3)
167	23	23	1	7	12/23/24/45	60	(172/2 175/1 176/1)
168	23	24	1	7	12/15/24/3/45	15	(172/4)
169	22	25	1	7	12/245/3	60	170/1 (173/1)
170	23		2		12/24/25/3/45	60	(172/2 174/1 175/1)
171	23	26	1	7	13/23/34/35	5	(175/4)
172	24	27	1	8	12/24/3/45	60	(177/2 178/1)
173	23	28	1	8	124/3/5	10	174/1
174	24		2		12/14/24/3/5	10	(177/3)
175	24	29	1	8	12/24/25/3	20	(178/3)
176	24	30	1	8	12/24/35	30	(177/1 178/2)
177	25	31	1	9	12/13/4/5	30	(179/2)
178	25	32	1	9	12/3/45	15	(179/2)
179	26	33	1	10	12/3/4/5	10	(180/1)
180	27	34	1	11	1/2/3/4/5	1	None

APPENDIX E

CLASSES OF STRUCTURES RELEVANT TO RECONSTRUCTABILITY ANALYSIS

The term *structure* is used in reconstructability analysis (RA) for families of subsets of a given set of variables, say set V. There are 2^{2^n} structures of this sort for n variables. Various special classes of structures are illustrated by the diagram in Figure E.1. The arrows in the diagram indicate a subset relationship between the classes.

A special class of structures, which are relevant in some instances to RA, are structures that satisfy the irredundancy requirement. They are called in this book *extended general structures* or *G^+-structures.* Another special class of structures are known as *hypergraphs.* A hypergraph is a family of subsets of V that covers all elements of V (i.e., it satisfies the covering condition) and does not contain the empty set. A special class of hypergraphs are *G-structures*, which represent reconstruction hypotheses in RA. They satisfy the covering condition and irredundancy condition.

The five subclasses of G-structures shown in Figure E.1 are based on undirected and reflexive graphs (symmetric and reflexive relations) defined on V. P-structures and C-structures are introduced in Section 4.7. A *P-structure* consists of all pairs (v_i, v_j) of different variables ($i \neq j$) that correspond to edges in the associated graph and all single variables that are not redundant. A *C-structure* consists of only and all maximal compatibility classes (cliques) of the associated graph. Hence, there is one P-structure and one C-structure for each graph. For n variables, there are exactly $2^{n(n-1)/2}$ of either of them.

M-structures also consist only of maximal compatibility classes of the associated graphs, but are not required to contain all of the classes (contrary to C-structures). Hence, the class of M-structures is larger than that of C-structures. *I-structures* consist of irreducible coverings of the associated graphs by maximal compatibility classes. There is a partial overlap between the classes of C-structures and I-structures, but neither of them is a subset of the other.

Loopless structures are defined as structures that do not contain any loop of the following kind: let

$$E_{k_1}, E_{k_2}, E_{k_3}, \ldots, E_{k_m}$$

be a sequence of elements of a given structure (sequence of subsets of V) such that

$$E_{k_i} \cap E_{k_{i+1}} \neq \varnothing$$

and $m \geqq 3$; then, this sequence represents a loop if and only if there exists a variable

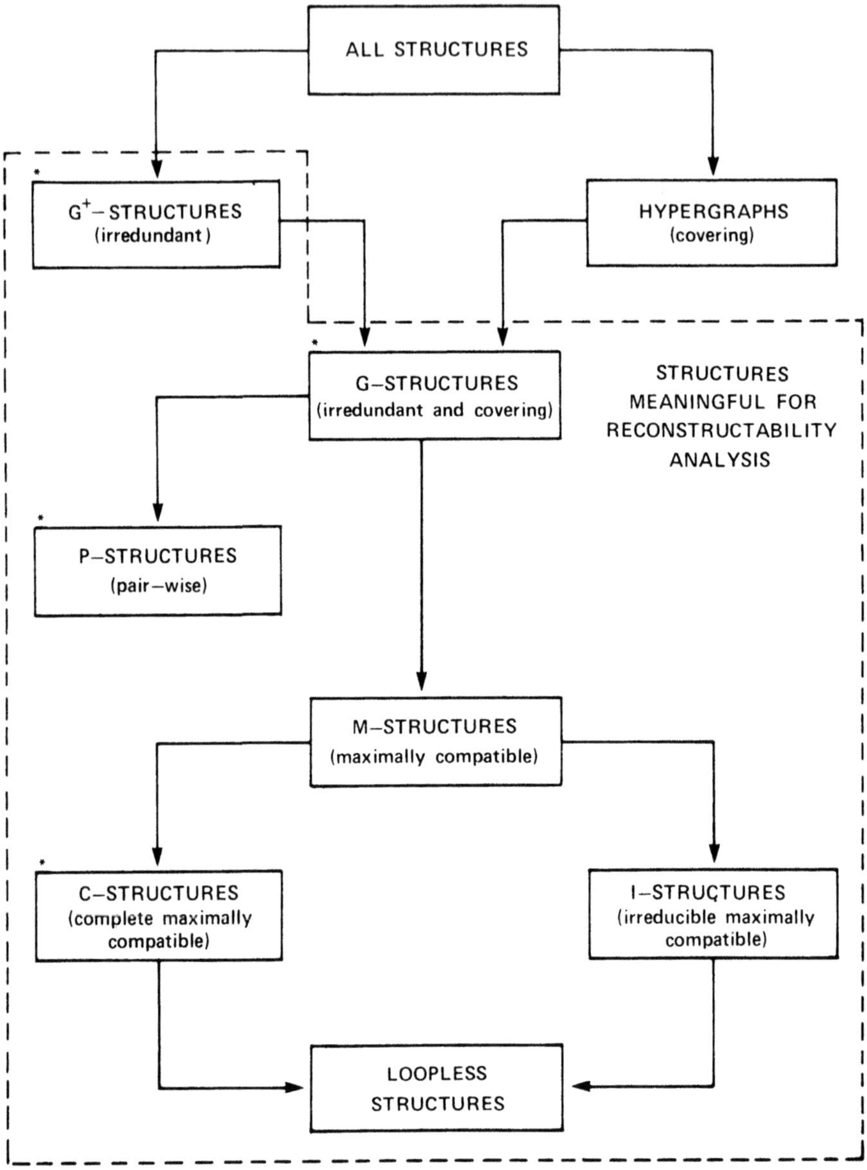

Figure E.1. Classification of structures.

$v_x \in V$ such that $v_x \in E_{k_1}$, $v_x \in E_{k_m}$, and $v_x \notin E_{k_i}$ for $i \neq 1, m$. Loopless structures form a special class of C-structures as well as I-structures.

REFERENCES

AC1 **Ackoff, R. L.**, Systems, organizations and interdisciplinary research. *General Systems Yearbook*, vol. **5**, 1960, pp. 1–8.

AC2 **Ackoff, R. L.**, *The Art of Problem Solving*. Wiley-Interscience, New York, 1978.

AC3 **Aczel, J.**, *Lectures on Functional Equations and Their Applications*. Academic Press, New York, 1966.

AC4 **Aczel, J.**, and **Z. Daroczy**, *On Measures of Information and Their Characterizations*. Academic Press, New York, 1975.

AC5 **Aczel, J.**, **B. Forte**, and **C. T. Ng**, Why the Shannon and Hartley entropies are "natural." *Advances in Applied Probability*, **6**, 1974, pp. 131–146.

AL1 **Albin, P. S.**, Calculation of a complexity parameter for graphs. *PAIS-79*, 1979, pp. 575–583.

AL2 **Alexander, C.**, *Notes on the Synthesis of Form*. Harvard University Press, Cambridge, Massachusetts, 1964.

AL3 **Allen, T. F. H.**, and **T. B. Starr**, *Hierarchy*. University of Chicago Press, Chicago, 1982.

AL4 **Allsopp, B.**, *A Modern Theory of Architecture*. Routledge & Kegan Paul, Boston, 1977.

AM1 **Amdahl, G. M.**, **G. A. Blaauw**, and **F. P. Brooks**, Architecture of the IBM System/360. *IBM Journal of Research and Development*, **8**, No. 2. 1964.

AM2 **Amoroso, S.**, Maps preserving the uniformity of neighborhood interconnection patterns in tessellation structures. *Information and Control*, **25**, 1974, pp. 1–9.

AM3 **Amoroso, S.**, and **G. Cooper**, Tessellation structures for reproduction of arbitrary patterns. *Journal of Computer and Systems Sciences*, **5**, 1971, pp. 455–464.

AR1 **Arbib, M. A.**, Automatic theory and control theory: A rapprochement. *Automatica*, **3**, 1966, pp. 161–189.

AR2 **Arbib, M. A.**, and **E. G. Manes**, *Arrows, Structures, and Functors: The Categorical Imperative*. Academic Press, New York, 1975.

AR3 **Arbib, M. A.**, **J. L. Rhodes**, and **B. R. Tils**, Complexity and group complexity of finite-state machines and finite semi-groups. In: *Algebraic Theory of Machines, Languages, and Semi-groups*, edited by M. A. Arbib, Academic Press, New York, 1968, pp. 127–145.

AS1 **Ashby, W. R.**, *Design for a Brain*. Wiley, New York, 1952.

AS2 **Ashby, W. R.**, *An Introduction to Cybernetics*. Wiley, New York, 1956.

AS3 **Ashby, W. R.**, Requisite variety and its implications for the control of complex systems. *Cybernetica*, **1**, 1958, pp. 83–99.

AS4 **Ashby, W. R.**, Constraint analysis of many dimensional relations, *General Systems Yearbook*, **9**, 1964, pp. 99–105.

AS5 **Ashby, W. R.**, Measuring the internal informational exchange in a system. *Cybernetica*, **1**, No. 1, 1965, pp. 5–22.

AS6 **Ashby, W. R.,** Mathematical models and computer analysis of the function of the central nervous system. *Annual Review of Physiology*, **28,** 1966, pp. 89–106.

AS7 **Ashby, W. R.,** Some consequences of Bremermann's limit for information-processing systems. In: *Cybernetic Problems in Bionics*, edited by H. Oestreicher and D. Moore, Gordon and Breach, New York, 1968, pp. 69–76.

AS8 **Ashby, W. R.,** Two tables of identities governing information flows within large systems. *ASC Communications*, **1,** No. 2, 1969, pp. 3–8.

AS9 **Ashby, W. R.,** Analysis of the system to be modeled. In: *The Process of Model-Building*, edited by R. M. Stogdill, Ohio Univ. Press, Columbus, 1970, pp. 94–114.

AS10 **Ashby, W. R.,** Information flows within co-ordinated systems. In: *Progress in Cybernetics*, Vol. 1, edited by J. Rose, Gordon and Breach, London, 1970, pp. 57–4.

AS11 **Ashby, W. R.,** Systems and their informational measures. In: *Trends in General Systems Theory*, edited by G. J. Klir, Wiley-Interscience, New York, 1972, pp. 78–97.

AS12 **Ashby, W. R.,** Some peculiarities of complex systems. *Cybernetic Medicine*, **9,** No. 2, 1973, pp. 1–7.

AT1 **Atkin, R. H.,** *Mathematical Structure in Human Affairs*. Heinemann, London, 1974

AU1 **Aulin, A.,** *The Cybernetic Laws of Social Progress*. Pergamon Press, New York, 1982.

AU2 **Aulin-Ahmavaara, A. Y.,** The law of requisite hierarchy. *Kybernetes*, **8,** 1979, pp. 259–266.

BA1 **Bahm, A. J.,** Wholes and parts. *The Southwestern Journal of Philosophy*, **3,** 1972, pp. 17–22.

BA2 **Balakrishnan, A. V.,** On the state space theory of linear systems. *Journal of Mathematical Analysis and Applications*, **14,** No. 3, 1966, pp. 371–391.

BA3 **Barto, A. G.,** Discrete and continuous models. *International Journal of General Systems*, **4,** No. 3, 1978, pp. 163–177.

BA4 **Bateson, G.,** *Mind and Nature*. Dutton, New York, 1979.

BE1 **Bell, D.,** *The Coming of Post-Industrial Society*. Basic Books, New York, 1973.

BE2 **Bellmann, R.,** *Dynamic Programming*. Princeton Univ. Press, Princeton, New Jersey, 1957.

BE3 **Bellmann, R.,** *Adaptive Control Processes*: A Guided Tour. Princeton Univ. Press, Princeton, New Jersey, 1972.

BE4 **Bellmann, R.,** and **C. P. Smith,** *Simulation in Human Systems: Decision-Making in Psychotherapy*. Wiley-Interscience, New York, 1973.

BE5 **Berge, C.,** *Graphs and Hypergraphs*. North-Holland/American Elsevier, Amsterdam and New York, 1973.

BE6 **Bertalanffy, L. von,** Zu einer allgemeinen Systemlehre. *Blätter für Deutsche Philosophie*, **18,** Nos. 3 and 4, 1945.

BE7 **Bertalanffy, L. von,** An outline of general systems theory. *British Journal of the Philosophy of Science*, **1,** 1950, pp. 134–164.

BE8 **Bertalanffy, L. von,** *General Systems Theory*. Braziller, New York, 1968.

BI1 **Bishop, Y. M. M., S. E. Fienberg,** and **P. W. Holland,** *Discrete Multivariate Analysis*. MIT Press, Cambridge, Massachusetts 1975.

BL1 **Blaauw, G. A., von,** Computer architecture. *Elektronische Rechenanlagen*, **14,** No. 4, 1972, pp. 154–159.

BL2 **Black, M.,** *Caveats and Critiques (Chapter II, Induction and Experience)*. Cornell Univ. Press, Ithaca, New York, 1975.

BL3 **Blauberg, I. V., V. N. Sadovsky, and E. G. Yudin,** *Systems Theory: Philosophical and Methodological Problems.* Progress, Moscow, 1977.

BO1 **Booth, T. L.,** *Sequential Machines and Automata Theory.* Wiley, New York, 1967.

BO2 **Boulding, K. L.,** General systems theory—The skeleton of science. *Management Science,* **2,** 1956, pp. 197–208. (Reprinted in *General Systems Yearbook,* **1,** 1956, pp. 11–17.)

BR1 **Bremermann, H. J.,** Optimization through evolution and recombination. In: *Self-Organizing Systems,* edited by M. C. Yovits and S. Cameron, Spartan, Washington, D.C., 1962, pp. 93–106.

BR2 **Bremermann, H. J.,** Quantifiable aspects of goal-seeking self-organizing systems. In *Progress in Theoretical Biology,* edited by M. Snell, Academic Press, New York, 1967, pp. 59–77.

BR3 **Broekstra, G.,** Constraint analysis and structure identification. *Annals of Systems Research,* I: **5,** 1976, pp. 67–80; II: **6,** 1977, pp. 1–20.

BR4 **Broekstra, G.,** On the representation and identification of structure systems. *International Journal of Systems Science,* **9,** No. 11, 1978, pp. 1271–1293.

BR5 **Broekstra, G.,** *C*-analysis of *C*-structures. *International Journal of General Systems,* **7,** No. 1, 1981, pp. 33–61.

BR6 **Brooks, F. P.,** Architectural philosophy. In: *Planning a Computer System,* edited by W. Buchholz, McGraw-Hill, New York, 1962, pp. 5–16.

BR7 **Brown, D. T.,** A note on approximations to discrete probability distributions. *Information and Control,* **2,** No. 4, 1959, pp. 386–392.

BR8 **Brown, F. M.,** Equational logic. *IEEE Transactions on Computers,* **C-23,** No. 12, 1974, pp. 1228–1237.

BR9 **Brown, G. S.,** *Probability and Scientific Inference.* Longmans, Green and Co., London, 1957.

BU1 **Buckley, W.** (ed.), *Modern Systems Research for the Behavioral Scientist.* Aldine, Chicago, 1968.

BU2 **Bunge, M.,** The GST challenge to the classical philosophies of science. *International Journal of General Systems,* **4,** No. 1, 1977, pp. 29–37.

CA1 **Casti, J. L.,** *Connectivity, Complexity and Catastrophe in Large-Scale Systems.* Wiley-Interscience, New York and London, 1979.

CA2 **Cavallo, R. E.** (ed.), Systems Research Movement: Characteristics, Accomplishments, and Current Developments. Special Issue of *General Systems Bulletin,* **9,** No. 3, 1979.

CA3 **Cavallo, R. E.,** *The Role of Systems Methodology in Social Science Research.* Martinus Nijhoff, Boston, 1979.

CA4 **Cavallo, R. E., and G. J. Klir,** A conceptual foundation for systems problem solving. *International Journal of Systems Science,* **9,** No. 2, 1978, pp. 219–236.

CA5 **Cavallo, R. E., and G. J. Klir,** Reconstructability analysis of multi-dimensional relations: A theoretical basis for computer-aided determination of acceptable systems models. *International Journal of General Systems,* **5,** No. 3, 1979, pp. 143–171.

CA6 **Cavallo, R. E., and G. J. Klir,** Reconstructability analysis: Evaluation of reconstruction hypotheses. *International Journal of General Systems,* **7,** No. 1, 1981, pp. 7–32.

CA7 **Cavallo, R. E., and G. J. Klir,** Decision making in reconstructability analysis. *International Journal of General Systems,* **8,** No. 4, 1982, pp. 243–255.

CA8 **Cavallo, R. E., and G. J. Klir,** Reconstructability analysis: Overview and bibliography. *International Journal of General Systems,* **7,** No. 1, 1981, pp. 1–6.

CA9 **Cavallo, R. E.,** and **G. J. Klir,** Reconstruction of possibilistic behavior systems. *Fuzzy Sets and Systems,* **8,** No. 2, 1982, pp. 175–197.

CE1 **Cerny, E.,** and **M. A. Marin,** A computer algorithm for the synthesis of memoryless logic circuits. *IEEE Transactions on Computers,* **C-23,** No. 5, 1974, pp. 455–465.

CE2 **Cerny, E.,** and **M. A. Marin,** An approach to unified methodology of combinational switching circuits. *IEEE Transactions on Computers,* **C-26,** No. 8, 1977, pp. 745–756.

CH1 **Chaitin, G.,** On the length of programs for computing finite binary sequences. *ACM Journal,* **16,** 1969, pp. 145–159.

CH2 **Chaitin, G.,** Information-theoretic computation complexity. *IEEE Transactions on Information Theory,* **IT-20,** No. 1, 1974, pp. 10–15.

CH3 **Chang, A. I. T.,** *The Tao of Architecture.* Princeton Univ. Press, Princeton, New Jersey, 1956.

CH4 **Checkland, P.,** *Systems Thinking, Systems Practice,* Wiley, New York, 1981.

CH5 **Christensen, R.,** *Entropy Minimax Sourcebook. Vol. I.: General Description.* Entropy, Lincoln, Massachusetts, 1981.

CH6 **Christensen, R.,** *Entropy Minimax Sourcebook. Vol. II.: Philosophical Origins.* Entropy, Lincoln, Massachusetts, 1980.

CH7 **Christensen, R.,** *Entropy Minimax Sourcebook. Vol. III.: Computer Implementation.* Entropy, Lincoln, Massachusetts, 1980.

CH8 **Christensen, R.,** *Entropy Minimax Sourcebook. Vol. IV.: Applications.* Entropy, Lincoln, Massachusetts, 1981.

CH9 **Christensen, R.,** *Foundations of Inductive Reasoning.* Entropy, Lincoln, Massachusetts, 1980.

CH10 **Church, A.,** An unsolvable problem of elementary number theory. *American Journal of Mathematics,* **58,** 1936, pp. 345–363.

CH11 **Churchman, C. W.,** *The Design of Inquiring Systems.* Basic Books, New York, 1971.

CH12 **Churchman, C. W.,** *The Systems Approach and Its Enemies.* Basic Books, New York, 1979.

CO1 **Comstock, F. L.,** and **H. J. J., Uyttenhove,** A systems approach to grading of flight simulator students. *Journal of Aircraft,* **16,** No. 11, 1979, pp. 780–786.

CO2 **Conant, R. C.,** *Information Transfer in Complex Systems, With Applications to Regulation.* Technical Report No. 13, Biological Computer Laboratory, University of Illinois, Urbana, Illinois, January 1968.

CO3 **Conant, R. C.,** The information transfer required in regulatory processes. *IEEE Transactions on Systems Science and Cybernetics,* **SSC-5,** No. 4, 1969, pp. 334–338.

CO4 **Conant, R. C.,** Laws of information which govern systems. *IEEE Transactions on Systems, Man, and Cybernetics,* **SMC-6,** No. 4, 1976, pp. 240–255.

CO5 **Conant, R. C.,** Structural modelling using a simple information measure. *International Journal of Systems Science,* **11,** No. 6, 1980, pp. 721–730.

CO6 **Conant, R. C.,** Detection and analysis of dependency structures. *International Journal of General Systems,* **7,** No. 1, 1981, pp. 81–91.

CO7 **Conant, R. C.,** Set-theoretic structure modelling. *International Journal of General Systems,* **7,** No. 1, 1981, pp. 93–107.

CO8 **Conant, R. C.,** (ed.) *Mechanisms of Intelligence: Ross Ashby's Writings on Cybernetics.* Intersystems, Seaside, California, 1981.

CO9 **Conant, R. C.,** and **W. Ross Ashby,** Every good regulator of a system must be a model of that system. *International Journal of Systems Science,* **1,** No. 2, 1970, pp. 89–97.

CO10 **Cook, S. A.,** The complexity of theorem-proving procedures. *Proc. 3rd Ann. ACM Symp. on Theory of Computing*, 1971, pp. 151–158.

CO11 **Cornacchio, J. V.,** Maximum-entropy complexity measures. *International Journal of General Systems*, **3,** No. 4, 1977, pp. 215–225.

CO12 **Cornacchio, J. V.,** System complexity—A bibliography. *International Journal of General Systems*, **3,** No. 4, 1977, pp. 267–271.

DA1 **Davis, M.,** *Computability and Unsolvability*. McGraw-Hill, New York, 1958.

DE1 **De Zurko, E. R.,** *Origins of Functionalist Theory*. Columbia University Press, 1957.

DI1 **Distefano, J. J.** and **W. Stubberud,** *Feedback and Control Systems*. Schaum, New York, 1967.

DU1 **Dubois, D.,** and **H. Prade,** *Fuzzy Sets and Systems: Theory and Applications*. Academic Press, New York, 1980.

DU2 **Dubois, D.,** and **H. Prade,** A class of fuzzy measures based on triangular norms: A general framework for the combination of uncertain information. *International Journal of General Systems*, **8,** No. 1, 1982.

DU3 **Dubois, D.,** and **H. Prade,** Unfair coins and necessity measures: Towards a possibilistic interpretation of histograms. *Fuzzy Sets and Systems*, **10,** No. 1, 1983, pp. 15–20.

DU4 **Dussauchoy, R. L.,** Generalized information theory and decomposability of systems. *International Journal of General Systems*, **9,** No. 1, 1982, pp. 13–36.

EL1 **Ellis, B.,** *Basic Concepts of Measurement*. Cambridge Univ. Press, Cambridge, 1968.

FE1 **Feinstein, A.,** *Foundations of Information Theory*. McGraw-Hill, New York, 1958.

FE2 **Ferdinand, A. E.,** Quality in programming. *IBM Technical Report* 21.485, Kingston, New York, June 1972.

FE3 **Ferdinand, A. E.,** A theory of system complexity. *International Journal of General Systems*, **1,** No. 1, 1974, pp. 19–33.

FI1 **Fienberg, S. E.,** *The Analysis of Cross-Classified Categorical Data*. MIT Press, Cambridge, Massachusetts, 1977.

FI2 **Fine, T. L.,** *Theories of Probability: An Examination of Foundations*. Academic Press, New York, 1973.

FL1 **Flagle, C. D., W. H. Huggins,** and **R. H. Roy,** (eds.), *Operations Research and Systems Engineering*. The John Hopkins Press, Baltimore, 1960.

FO1 **Foerster, H. von,** *Observing Systems*. Intersystems, Seaside, California, 1983.

FO2 **Forrester, J. W.,** *Industrial Dynamics*. MIT Press, Cambridge, Massachusetts, 1961.

FO3 **Forrester, J. W.,** *World Dynamics*. Wright-Allen, Cambridge, Massachusetts, 1971.

FO4 **Forte, B.,** Why Shannon entropy. *Symposia Mathematica*, **XV,** Academic Press, New York, 1975, pp. 137–152.

GA1 **Gaines, B. R.,** Axioms for adaptive behavior. *International Journal of Man–Machine Studies*, **4,** 1972, pp. 169–199.

GA2 **Gaines, B. R.,** Systems identification, approximation and complexity. *International Journal of General Systems*, **3,** No. 3, 1977, pp. 145–174.

GA3 **Gaines, B. R.,** Progress in general systems research. In: *Applied General Systems Research*, edited by G. J. Klir, Plenum Press, New York, 1978, pp. 3–28.

GA4 **Gaines, B. R.,** General Systems research: Quo vadis? *General Systems Yearbook*, **24,** 1979, pp. 1–9.

GA5 **Gallopin, G. C.,** The abstract concept of environment. *International Journal of General Systems*, **7,** No. 2, pp. 139–149.

GA6 **Gardner, M. R.**, and **W. R. Ashby,** Connectance of large dynamic (cybernetic) systems: Critical values of stability. *Nature*, **228,** 1970, p. 784.

GA7 **Garey, M. R.**, and **D. S. Johnson,** *Computers and Intractability: A Guide to the Theory of NP-Completeness.* W. H. Freeman, San Francisco, 1979.

GA8 **Gause, D.**, and **G. M. Weinberg,** *Are Your Lights On*? Winthrop, Cambridge, Massachusetts, 1982.

GE1 **Gelfand, A. E.**, and **C. C. Walker,** The distribution of cycle lengths in a class of abstract systems. *International Journal of General Systems*, **4,** No. 1, 1977, pp. 39–45.

GE2 **George, L.**, Tests for system complexity. *International Journal of General Systems*, **3,** No. 4, 1977, pp. 253–258.

GE3 **Gerardy, R.**, Probabilistic finite state system identification. *International Journal of General Systems*, 8, No. **4,** 1982, pp. 229–242.

GE4 **Gerardy, R.**, Experiments with some methods for the identification of finite-state systems. *International Journal of General Systems*, **9,** No. 4, 1983, pp. 197–203.

GE5 **Gershuny, J.**, *After Industrial Society*? Humanities, Atlantic Highlands, N.J., 1978.

GI1 **Givone, D. D.**, **M. E. Liebler,** and **R. P. Roesser,** A method of solution of multiple-valued logic expressions. *IEEE Transactions on Computers*, **C-20,** No. 4, 1971, pp. 464–467.

GO1 **Goguen, J. A.**, and **F. J. Varela,** Systems and distinctions: Duality and complementarity. *International Journal of General Systems*, **5,** No. 1, 1979, pp. 31–43.

GO2 **Goldblatt, R.**, *Topoi: The Categorical Analysis of Logic.* North-Holland, Amsterdam, 1979.

GO3 **Goodman, L. A.**, *Analyzing Qualitative/Categorical Data.* Abt, Cambridge, Massachusetts, 1978.

GO4 **Gottinger, H. W.**, Complexity and information technology in dynamic systems. *Kybernetes*, **4,** 1975, pp. 129–141.

GR1 **Greenspan, D.**, *Discrete Models.* Addison-Wesley, Reading, Massachusetts, 1973.

GR2 **Greenspan, D.**, An arithmetic, particle theory of fluid mechanics. *Computer Methods in Applied Mechanics and Engineering*, **3,** 1974, pp. 293–303.

GR3 **Greenspan, D.**, A completely arithmetic formulation of classical and special relativistic physics. *International Journal of General Systems*, **4,** No. 2, 1978, pp. 105–112.

GR4 **Greenspan, D.**, Discrete modeling in microcosm and in the macrocosm. *International Journal of General Systems*, **6,** No. 1, 1980, pp. 25–45.

GR5 **Greenspan, D.**, *Arithmetic Applied Mathematics.* Pergamon Press, Oxford, 1980.

GR6 **Gropius, W.**, *Scope of Total Architecture.* Harper & Row, New York, 1943.

GU1 **Guiasu, S.**, *Information Theory with Applications.* McGraw-Hill, New York, 1977.

GU2 **Guida, G.**, **D. Mandrioli** and **M. Somalvico,** An integrated model of problem solver. *Information Sciences*, **13,** No. 1, 1977, pp. 11–33.

GU3 **Gukhman, A. A.**, *Introduction to the Theory of Similiarity.* Academic Press, New York, 1965.

GU4 **Gusev, L. A.**, and **A. A. Tal,** The possibilities of constructing algorithms for the abstract synthesis of sequential machines using the questionnaire language. *Automation and Remote Control*, **6,** No. 3, 1965, pp. 507–514.

HA1 **Hai, A.**, and **G. J. Klir,** An empirical investigation of reconstructability analysis. *International Journal of Man–Machine Studies*, **22,** 1985.

HA2 **Hajek, P.**, and **T. Havranek,** *Mechanizing Hypothesis Formation.* Springer-Verlag, New York, 1978.

HA3 **Halme, A., R. P. Hamalainen,** and **O. Ristaniemi** (eds.), *Topics in Systems Theory. Acta Polytechnics Scandinavica,* Mathematics and Computer Science Series No. 31, Helsinki, 1979.

HA4 **Hamilton, W. L.,** Reproduction in tessellation structures. *Journal of Computer and System Science,* **10,** No. 2, 1975, pp. 248–225.

HA5 **Hammer, P. C.** (ed.), *Advances in Mathematical Systems Theory.* Pennsylvania State Univ. Press, University Park, Pennsylvania, 1969.

HA6 **Hanken, A. F. G.,** and **H. A. Reuver,** *Social Systems and Learning Systems.* Martinus Nijhoff, Boston, 1981.

HA7 **Happ, H. H.** (ed.), *Gabriel Kron and System Theory.* Union College Press, Schenectady, New York, 1973.

HA8 **Harary, F.,** *Graph Theory.* Addison-Wesley. Reading, Massachusetts, 1969.

HA9 **Hartley, R. V. L.,** Transmirsion of information. *The Bell System Technical Journal,* **7,** 1928, pp. 535–563.

HA10 **Hartmanis, J.,** *Feasible Computations and Provable Complexity Properties.* SIAM, Philadelphia, 1978.

HA11 **Hartnett, W. E.** (ed.), *Systems: Approaches, Theories, Applications.* Reidel, Boston, 1977.

HA12 **Hayes, P.,** Trends in artificial intelligence. *International Journal of Man–Machine Systems,* **10,** No. 3, 1978, pp. 295–299.

HE1 **Heise, D. R.,** *Causal Analysis.* Wiley-Interscience, New York, 1975.

HE2 **Herman, G. T.,** and **G. Rozenberg** (eds.), *Developmental Systems and Languages.* North-Holland, Amsterdam, 1975.

HE3 **Herrlich, H.,** and **G. E. Strecker,** *Category Theory.* Allyn and Bacon, Boston, 1973.

HI1 **Higashi, M.,** A systems modelling methodology; probabilistic and possibilistic approaches. Ph.D. dissertation, School of Advanced Technology, SUNY-Binghamton, 1983.

HI2 **Higashi, M.,** and **G. J. Klir,** Measures of uncertainty and information based on possibility distributions. *International Journal of General Systems,* **9,** No. 1, 1983, pp. 43–58.

HI3 **Higashi, M.,** and **G. J. Klir,** On the notion of distance representing information closeness: Possibility and probability distributions. *International Journal of General Systems,* **9,** No. 2, 1983, pp. 103–115.

HI4 **Higashi, M., G. J. Klir,** and **M. Pittarelli,** Reconstruction families of possibilistic structure systems. *Fuzzy Sets and Systems,* **11,** No. 3, 1983.

HI5 **Himmelblau, D. M.** (ed.), *Decomposition of Large-Scale Problems.* North-Holland, Amsterdam, 1973.

HI6 **Hisdal, E.,** Conditional possibilities, independence and noninteraction. *Fuzzy Sets and Systems,* **1,** No. 4, 1978, pp. 283–297.

HO1 **Holland, J. H.,** *Adaptation in Natural and Artificial Systems.* University of Michigan Press, Ann Arbor, 1975.

HS1 **Hsu, J. C.,** and **A. U. Meyer,** *Modern Control Principles and Applications.* McGraw-Hill, New York, 1968.

IN1 **Ingham, H. S.,** *Discretus Calculus.* Philosophical Library, New York, 1964.

JA1 **Jaynes, E. T.,** Prior probabilities. *IEEE Transactions on Systems Science and Cybernetics,* **SSC-4,** No. 3, 1968, pp. 227–241.

JA2 **Jaynes, E. T.,** Where do we stand on maximum entropy? In: *The Maximum Entropy Formalism*, edited by R. L. Levine and M. Tribus, MIT Press, Cambridge, Massachusetts, 1979, pp. 15–118.

JO1 **Jones, B.,** Determination of reconstruction families. *International Journal of General Systems*, **8,** No. 4, 1982, pp. 225–228.

KA1 **Kapur, J. N.,** On maximum-entropy complexity measures. *International Journal of General Systems*, **9,** No. 2, 1983, pp. 95–102.

KA2 **Karp, R. M.,** Reducibility among combinatorial problems. In: *Complexity of Computer Computations*, edited by R. E. Miller and J. W. Thatcher, Plenum Press, New York, 1972, pp. 85–103.

KA3 **Katz, M. B.,** *Questions of Uniqueness and Resolution in Reconstruction from Projections.* Springer-Verlag, New York, 1978.

KA4 **Kauffman, S. A.,** Metabolic stability and epigenesis in randomly constructed genetic sets. *Journal of Theoretical Biology*, **22,** 1969, pp. 437–467.

KE1 **Kellerman, E.,** A formula for logical cost. *IEEE Transactions on Computers*, **E-17,** No. 9, 1968, pp. 881–884.

KE2 **Kemeny, J. G.,** *A Philosopher Looks at Science.* D. Van Nostrand, Princeton, New Jersey, 1959.

KE3 **Kemeny, J. G.,** *Man and the Computer.* Charles Scribner's Sons, New York, 1972.

KH1 **Khinchin, A. I.,** *Mathematical Foundations of Information Theory.* Dover, New York, 1957.

KL1 **Kleene, S. C.,** General recursive functions of natural numbers. *Mathematische Annalen*, **11,** 1936, pp. 727–742

KL2 **Klir, G. J.,** The general system as a methodological tool. *General Systems Yearbook*, **10,** 1965, pp. 29–42.

KL3 **Klir, G. J.,** *An Approach to General Systems Theory.* Van Nostrand Reinhold, New York, 1969.

KL4 **Klir, G. J.** (ed.), *Trends in General Systems Theory.* Wiley-Interscience, New York, 1972.

KL5 **Klir, G. J.,** *Introduction to the Methodology of Switching Circuits.* Van Nostrand Reinhold, New York, 1972.

KL6 **Klir, G. J.,** A study of organizations of self-organizing systems. *Proceedings of the Sixth International Congress on Cybernetics*, Namur (Belgium), 1972, pp. 165–186.

KL7 **Klir, G. J.,** Identification of generative structures in empirical data. *International Journal of General Systems*, **3,** No. 2, 1976, p. 89–104.

KL8 **Klir, G. J.,** The general systems research movement. In: *Systems Models for Decision Making*, edited by N. Sharif and P. Adulbhan, Asian Institute of Technology, Bangkok,1978, pp. 25–70.

KL9 **Klir, G. J.** (ed.), *Applied General Systems Research.* Plenum Press, New York, 1978.

KL10 **Klir, G. J.,** Architecture of structure systems: A basis for the reconstructability analysis. *Acta Polytechnica Scandinavica*, Mathematics and Computer Science Series, No. 31, Helsinki, 1979, pp. 33–43.

KL11 **Klir, G. J.,** General systems problem solving methodology. In: *Modelling and Simulation Methodology*, edited by B. Zeigler, M. S. Elzas, G. J. Klir, and T. I. Oren, North-Holland, Amsterdam, 1979, pp. 3–28.

KL12 **Klir, G. J.,** Computer-aided systems modelling. In: *Theoretical Systems Ecology*. edited by E. Halfon, Academic Press, New York, 1979, pp. 291–323.

KL13 **Klir, G. J.,** On systems methodology and inductive reasoning: The issue of parts and wholes. *General Systems Yearbook*, **26,** 1981, pp. 29–38.

KL14 **Klir, G. J.,** and **H. J. J. Uyttenhove,** Computerized methodology for structure modelling. *Annals of Systems Research*, **5,** 1976, pp. 29–66.

KL15 **Klir, G. J.,** and **H. J. J. Uyttenhove,** On the problem of computer-aided structure identification: Some experimental observations and resulting guidelines. *International Journal of Man–Machine Studies*, **9,** No. 5, 1977, pp. 593–628.

KL16 **Klir, G. J.,** and **H. J. J. Uyttenhove,** Procedures for generating reconstruction hypotheses in the reconstructability analysis. *International Journal of General Systems*, **5,** No. 4, 1979, pp. 231–246.

KL17 **Klir, G. J.,** and **M. Valach,** *Cybernetic Modelling*. Illiffe, London, 1967.

KO1 **Koestler, A.,** *The Ghost in the Machine*. Macmillan, New York, 1967.

KO2 **Koestler, A.,** and **J. R. Smythies** (eds.), *Beyond Reductionism*. Hutchinson, London, 1969.

KO3 **Kolmogorov, A. N.,** *Foundations of the Theory of Probability*. Chelsea, New York, 1950.

KO4 **Kolmogorov, A.,** Three approaches to quantitative definition of information. *Problems of Information Transmission*, **1,** No. 1, 1965, pp. 1–7.

KR1 **Krantz, D. H., R. D. Luce, P. Suppes,** and **A. Tversky,** *Foundations of measurement, Vol. I: Additive and Polynomial Representations*. Academic Press, New York, 1971.

KR2 **Krippendorff, K.,** An algorithm for identifying structural models of multivariable data. *International Journal of General Systems*, **7,** No. 1, 1981, pp. 63–79.

KR3 **Krohn, D., R. Langer,** and **J. Rhodes,** Algebraic principles for the analysis of biochemical systems. *Journal of Computer and Systems Science*, **1,** No. 2, 1967, pp. 119–136.

KR4 **Krohn, K. B.,** and **J. L. Rhodes,** Algebraic theory of machines. In: *Mathematical Theory of Automata*, edited by J. Fox, Polytechnic Press, Brooklyn, New York, 1963, pp. 341–384.

KR5 **Krohn, K. B.,** and **J. L. Rhodes,** Complexity of finite semigroups. *Annals of Mathematics*, **88,** 1968, pp. 128–160.

KU1 **Kullback, S.,** *Information Theory and Statistics*. John Wiley, New York, 1959. (Reprinted by Dover, New York, 1968.)

LA1 **Lange, O.,** *Wholes and Parts*. Pergamon Press, Oxford, 1965.

LA2 **Langhaar, H. L.,** *Dimensional Analysis and Theory of Models*. John Wiley, New York, 1964.

LA3 **Lasker, G. E.** (ed.) *Applied Systems and Cybernetics*. (six volumes). Pergamon Press, New York, 1981.

LA4 **Laszlo, E.** (ed.), *The Relevance of General Systems Theory*. Braziller, New York, 1972.

LE1 **Lerner, D.** (ed.), *Parts and Wholes*. Free Press, New York, 1963.

LE2 **Lewis, P. M. II,** Approximating probability distributions to reduce storage requirements. *Information and Control*, **2,** No. 3, 1959, pp. 214–225.

LI1 **Lilienfeld, Robert,** *The Rise of Systems Theory: An Ideological Analysis*. Wiley-Interscience, New York, 1978.

LI2 **Lindenmayer, A.,** Mathematical models for cellular interactions in development. *Journal of Theoretical Biology*, **18,** 1968, pp. 280–315.

LI3 **Lindenmayer, A.,** Developmental systems without cellular interactions, their languages and grammars. *Journal of Theoretical Biology*, **30,** 1971, pp. 455–484.

LI4 **Lindenmayer, A.,** Developmental algorithms for multicellular organisms: A survey of *L*-systems. *Journal of Theoretical Biology*, **54,** 1975, pp. 3–22.

LI5 **Lindenmayer, A.,** Developmental algorithms: Lineage versus interactive control mechanisms. In: *Developmental Order: Its Origin and Regulation*, edited by S. Subtelny and P. B. Green, Alan R. Liss, New York 1982, pp. 219–245.

LI6 **Lindenmayer, A.,** and **G. Rozenberg** (eds.), *Automata, Languages, Development*. North-Holland, Amsterdam, 1976.

LO1 **Lofgren, L.,** Complexity of descriptions of systems: a foundational study. *International Journal of General Systems*, **3,** No. 4, 1977, pp. 197–214.

LO2 **Lofgren, L.,** Some foundational views on general systems and the Hempel paradox. *International Journal of General Systems*, **4,** No. 4, 1978, pp. 243–253.

LO3 **Loveland,** D. W., A variant of the Kolmogorov concept of complexity. *Information and Control*, **15,** 1969, pp. 510–526.

LU1 **Lucadou, W., von,** and **K. Kornwachs,** The problem of reductionism from a system theoretical viewpoint. *Zeitschrift für Allgemeine Wissenschaftstheorie*, 1983, pp. 338–349.

MA1 **Maciejowski, J. M.,** *The Modelling of Systems with Small Observation Sets*. Springer-Verlag, New York, 1978.

MA2 **Madden, R. F.,** and **W. R. Ashby,** On the identification of many-dimensional relations. *International Journal of Systems Science*, **3,** No. 4, 1972, pp. 343–356.

MA3 **Makridakis, S.,** and **C. Faucheux,** Stability properties of general systems. *General Systems Yearbook*, **18,** 1973, pp. 3–12.

MA4 **Markov, A. A.,** *The Theory of Algorithms*. National Science Foundation, Washington, D. C., 1961. (Russian original published in 1954.)

MA5 **Martin-Lof, P.,** The definition of random sequences. *Information and Control*, **9,** 1966, pp. 602–619.

MA6 **Mattessich, R.,** *Instrumental Reasoning and Systems Methodology: An Epistemology of the Applied and Social Sciences*. Reidel, Boston, 1978.

ME1 **Mesarovic, M. D.** (ed.), *Views on General Systems Theory*. Wiley, New York, 1964.

ME2 **Mesarovic, M. D., D. Macko,** and **Y. Takahara,** *Theory of Hierarchical Multilevel Systems*. Academic Press, New York, 1970.

ME3 **Mesarovic, M. D.,** and **Y. Takahara,** *General Systems Theory: Mathematical Foundations*. Academic Press, New York, 1975.

MI1 **Mihram, D.,** and **G. A. Mihram,** Human knowledge: The role of models, metaphors, and analogy. *International Journal of General Systems*, **1,** No. 1, 1974, pp. 41–60.

MO1 **Morris, C.,** *Foundations of the Theory of Signs* (two volumes). University of Chicago Press, Chicago, 1938.

MO2 **Morris, C.,** *Signs, Language, and Behavior*. Prentice-Hall, New York, 1946.

MO3 **Morris, C.,** *Signification and Significance*. MIT Press, Cambridge, Massachusetts, 1964.

MO4 **Moshowitz, A.,** Entropy and the complexity of graphs. *Bulletin of Mathematical Biophysics*, **30,** Nos. 1 and 2, 1968, pp. 175–204, 225–240.

NA1 **Naisbitt, J.,** *Megatrends*. Warner Books, New York, 1982.

NA2 **Naylor, A. W.,** On decomposition theory: Generalized dependence. *IEEE Transactions on Systems, Man, and Cybernetics*, **SMC-11,** No. 10, 1981, pp. 699–713.

NE1 **Negoita, C. V.,** *Fuzzy Systems*. Abacus, Tunbridge Wells (U. K.), 1981.

NE2 **Negoita, C. V.,** and **D. A. Ralescu,** *Applications of Fuzzy Sets to Systems Analysis*. Birkhauser, Basel and Stuttgart, 1975.

NE3 **Newell, A.,** and **H. A. Simon,** *Human Problem Solving.* Prentice-Hall, Englewood Cliffs, New Jersey, 1972.

NG1 **Nguyen, H. T.,** On conditional possibility distributions. *Fuzzy Sets and Systems,* **1,** No. 4, 1978, pp. 299–309.

NO1 **Norgerg-Schulz, C.,** *Intentions in Architecture.* MIT Press, Cambridge, Massachusetts, 1965.

OR1 **Oren, T. I., B. P. Zeigler,** and **M. S. Elzas** (eds.), *Simulation and Model-Based Methodologies.* Springer-Verlag, New York, 1983.

OS1 **Ostrand, T. J.,** Pattern reproduction in tessellation automata in arbitrary dimensions. *Journal of Computer and System Science,* **5,** 1971, pp. 623–628.

PA1 **Padulo, L.,** and **M. A. Arbib,** *System Theory.* W. B. Saunders, Philadelphia, 1974.

PA2 **Pask, G.,** *Conversation Theory: Applications in Education and Epistemology.* Elsevier, Amsterdam and New York, 1976.

PA3 **Pattee, H. H.** (ed.), *Hierarchy Theory: The Challenge of Complex Systems.* Braziller, New York, 1973.

PA4 **Patten, B. C.,** Systems approach to the concept of environment. *Ohio Journal of Science,* **78,** No. 4, 1978, pp. 206–222.

PE1 **Pearl, J.,** On the connection between the complexity and credibility of inferred models. *International Journal of General Systems,* **4,** No. 4, 1978, p. 255–264.

PF1 **Pfanzagl, J.,** *Theory of Measurement.* Wiley, New York, 1968.

PH1 **Phillips, D. C.,** *Holistic Thought in Social Science.* Stanford University Press, Stanford, California, 1976.

PI1 **Pippenger, N.,** Complexity theory. *Scientific American,* **238,** No. 6, 1978, pp. 114–124.

PO1 **Poincare, H.,** *Science and Hypothesis.* Dover, New York, 1952.

PO2 **Polya, G.,** Kombinatorische Anzahlbestimmungen für Gruppen, Graphen und Chemische Verbindungen. *Acta Mathematica,* **68,** 1937, pp. 145–254.

PO3 **Polya, L.,** An empirical investigation of the relations between structure and behavior for nonlinear second-order systems. Ph.D. dissertation, School of Advanced Technology, SUNY-Binghamton, 1981.

PO4 **Porter, B.,** Requisite variety in the systems and control sciences. *International Journal of General Systems,* **2,** No. 4, 1976, pp. 225–229.

PO5 **Post, E. L.,** Finite combinatory processes—Formulation. *The Journal of Symbolic Logic,* **1,** 1936, pp. 103–105.

PU1 **Puri, M. L.,** and **D. Ralescu,** A possibility measure is not a fuzzy measure. *Fuzzy Sets and Systems,* **7,** No. **3,** 1982, pp. 311–313.

RA1 **Rapoport, A.,** *et al., Response Models for Detection of Change.* Reidel, Boston, 1979.

RE1 **Renyi, A.,** *Probability Theory.* North-Holland, Amsterdam, 1970 (Appendix: Introduction to information theory, pp. 540–616).

RE2 **Rescher, N.,** *Scientific Explanation.* Free Press, New York, 1970.

RE3 **Rescher, N.,** and **R. Brandom,** *The Logic of Inconsistency.* Blackwell, Oxford, 1980.

RE4 **Rescher, N.,** *The Primacy of Practice.* Blackwell, Oxford, 1973.

RE5 **Rescher, N.,** *The Coherence Theory of Truth.* Oxford Univ. Press, Oxford, 1973.

RE6 **Rescher, N.,** *Plausible Reasoning.* Van Gorcum, Amsterdam, 1976.

RE7 **Rescher, N.,** *Methodological Pragmatism: Systems-Theoretic Approach to the Theory of Knowledge.* New York Univ. Press, New York, 1977.

RE8 **Rescher, N.,** *Scientific Progress.* Blackwell, Oxford, 1978.

RE9 **Rescher, N.,** *Cognitive Systematization: A Systems-Theoretic Approach to a Coherent Theory of Knowledge.* Blackwell, Oxford, 1979.

RE10 **Rescher, N.,** *Induction.* Blackwell, Oxford, 1980.

RE11 **Rescher, N.,** and **R. Manor,** On inference from inconsistent premisses. *Theory and Decision,* **1,** No. 2, 1970, pp. 179–217.

RE12 **Reza, F. M.,** *An Introduction to Information Theory.* McGraw-Hill, New York, 1961.

RO1 **Robertshaw, J. E., S. J. Mecca,** and **M. N. Rerick,** *Problem Solving: A Systems Approach.* McGraw-Hill, New York, 1979.

RO2 **Rogers, H.,** *Theory of Recursive Functions and Effective Computability.* McGraw-Hill, New York, 1967.

RO3 **Roosen-Runge, P. H.,** Toward a theory of parts and wholes: An algebraic approach. *General Systems Yearbook,* **11,** 1966, pp. 13–18.

RO4 **Rosen, R.,** *Dynamical System Theory in Biology,* Wiley-Interscience, New York, 1970.

RO5 **Rosen, R.,** Complexity as a system property. *International Journal of General Systems,* **3,** No. 4, 1977, pp. 227–232.

RO6 **Rosen, R.,** *Fundamentals of Measurement and Representation of Natural Systems.* North-Holland, New York, 1978.

RO7 **Rosen, R.,** Old trends and new trends in general systems research. *International Journal of General Systems,* **5,** No. 3, 1979, pp. 173–184.

RO8 **Rosen, R.,** *Anticipatory Systems.* Pergamon Press, Oxford, 1985.

RO9 **Rozenberg, G.** and A. **Salomaa** (eds.), *L. Systems.* Springer-Verlag, New York, 1974.

RU1 **Rubinstein, M. F.,** *Patterns of Problem Solving.* Prentice-Hall, Englewood Cliffs, New Jersey, 1975.

SA1 **Sage, A. P.,** *Optimum Systems Control.* Prentice-Hall, Englewood Cliffs, New Jersey, 1968.

SA2 **Saucedo, R.,** and **E. E. Schiring,** *Introduction to Continuous and Digital Control Systems.* Macmillan, New York, 1968.

SH1 **Shackle, G. L. S.,** *Decision, Order and Time in Human Affairs.* Cambridge Univ. Press, Cambridge, 1969.

SH2 **Shafer, G.,** *A Mathematical Theory of Evidence.* Princeton Univ. Press, Princeton, New Jersey 1976.

SH3 **Shannon, C. E.,** A mathematical theory of communication. *Bell System Technical Journal,* **27,** July and October 1948, pp. 379–423, 623–656.

SH4 **Shannon, C. E.,** and **W. Weaver,** *The Mathematical Theory of Communication.* University of Illinois Press, Urbana, 1964.

SH5 **Shore, J. E.,** and **R. W. Johnson,** Axiomatic derivation of the principle of maximum entropy and the principle of minimum cross-entropy. *IEEE Transactions on Information Theory,* **IT-26,** No. 1, 1980, pp. 26–37.

SI1 **Simon, H. A.,** The architecture of complexity. *Proceedings of the American Philosophical Society,* **106,** 1962, pp. 467–482. (Reprinted in the *Sciences of the Artificial,* MIT Press, Cambridge, Massachusetts, 1969.)

SI2 **Simon, H. A.,** Complexity and the representation of patterned sequences of symbols. *Psychological Reviews,* **79,** 1972, pp. 369–382.

SI3 **Simon, H. A.,** Does scientific discovery have a logic? *Philosophy of Science,* **40,** Dec. 1973, pp. 441–480.

SI4 **Simon, H. A.,** How complex are complex systems? In: *PSA* 1976, Volume 2, edited by F. Suppe and P. D. Asquith, Philosophy of Science Association, East Lansing, Michigan, 1977, pp. 507–522.

SI5 **Simon, H. A.**, *Models of Discovery*. Reidel, Boston, 1977.

SK1 **Skoglund, V.**, *Similitude: Theory and Applications*. International Textbook Co., Scranton, Pennsylvania, 1967.

SL1 **Sloane, N. J. A.**, *A Handbook of Integer Sequences*. Academic Press, New York, 1973, p. 64.

SL2 **Sloman, A.**, *The Computer Revolution in Philosophy*. The Harvester Press, Hassocks (U.K.), 1978.

SM1 **Smuts, J. C.**, *Holism and Evolution*. Macmillan, London, 1926.

SO1 **Sowa, J. F.**, *Conceptual Structures: Information Processing in Mind and Machine*. Addison-Wesley, Reading, Massachusetts, 1984.

SP1 **Spriet, J. A.**, and **G. C. Vansteenkiste**, *Computer-Aided Modelling and Simulation*. Academic Press, New York, 1982.

ST1 **Streeter, D. N.**, *The Scientific Process and the Computer*. Wiley-Interscience, New York, 1974.

SU1 **Sugeno, M.**, Fuzzy measures and fuzzy integrals: A survey. In: *Fuzzy Automata and Decision Processes*, edited by M. M. Gupta, G. N. Saridis, and B. R. Gaines, North-Holland, Amsterdam and New York, 1977, pp. 89–102.

SU2 **Suppes, P.**, Models of data. In: *Logic, Methodology and Philosophy of Science*, edited by E. Nagel *et al.*, Stanford Univ. Press, Stanford, 1962, pp. 252–261.

SU3 **Suppes, P.**, Some remarks about complexity. In: *PSA* 1976, Volume 2, edited by F. Suppe and P. D. Asquith, Philosophy of Science Assoc., East Lansing, Michigan, 1977, pp. 543–547.

SV1 **Svoboda, A.**, Synthesis of logical systems of given activity. *IEEE Transactions on Electronic Computers*, **EC-12**, 1963, No. 6, pp. 904–910.

SV2 **Svoboda, A.**, Behaviour classification in digital systems. *Information Processing Machines*, **10**, Czechoslovak Academy of Sciences Press, Prague, 1964, pp. 25 42.

SV3 **Svoboda, A.**, A model of the instinct of self-preservation (in French). *Information Processing Machines*, Academia, Prague, 1960, pp. 147–155.

SV4 **Svoboda, A.**, and **D. E. White**, *Advanced Logical Circuit Design Techniques*. Garland STPM Press, New York, 1979.

SZ1 **Szucz, E.**, *Similitude and Modelling*. Elsevier, Amsterdam, 1980.

TA1 **Takahara, Y.**, and **B. Nakao**, A characterization of interactions. *International Journal of General Systems*, **7**, No. 2, 1981, pp. 109–122.

TA2 **Tal, A. A.**, Questionnaire language and the abstract synthesis of minimal sequential machines. *Automation and Remote Control*, **25**, No. 6, 1964, pp. 846–859.

TA3 **Tal, A. A.**, The abstract synthesis of sequential machines from the answers to questions of the first kind in the questionnaire language. *Automation and Remote Control*, **26**, 4, 1965, pp. 675–680.

TE1 **Teller, E.**, *The Pursuit of Simplicity*. Pepperdine University Press, Malibu, California, 1980.

TO1 **Torgenson, W. S.**, *Theory and Methods of Scaling*. Wiley, New York, 1958.

TO2 **Towner, G.**, *The Architecture of Knowledge*. University Press of America, Washington, D. C., 1980.

TR1 **Tranoy, K. E.**, *Wholes and Structures*. Munksgaard, Copenhagen, 1959.

TR2 **Trappl, R.** (ed.), *Cybernetics: Theory and Applications*. Hemisphere, Washington, D. C., 1983.

TS1 **Tsypkin, Y. Z.**, *Adaptation and Learning in Automatic Systems*. Academic Press, New York, 1971.

TU1 **Turing, A. M.,** On computable numbers, with an application to the Entscheidungs problem. *Proceedings of the London Mathematical Society*, Series 2, 1936, **42,** pp. 230–265, **43,** pp. 544–546.

UY1 **Uyttenhove, H. J.,** Computer-aided systems modelling: An assemblage of methodological tools for systems problem solving. Ph.D. dissertation, School of Advanced Technology, SUNY-Binghamton, 1978.

UY2 **Uyttenhove, H. J.,** *SAPS (Systems Approach Problem Solver): An Introduction and Guide.* Computing and Systems Consultants, Binghamton, New York, 1981.

UY3 **Uyttenhove, H. J.,** Systems approach problem solver and the open heart surgery patient. In: *Cybernetics and Systems Research*, edited by R. Trappl, North-Holland, Amsterdam, 1982, pp. 655–661.

VA1 **Valdes-Perez, R. E.,** and **R. C. Conant,** Information loss due to data quantization in reconstructability analysis. *International Journal of General Systems*, 9, No. **4,** 1983.

VA2 **Van Emden, M. H.,** *An Analysis of Complexity.* Mathematical Centre, Amsterdam, 1971.

VA3 **Van Gigch, J. P.,** *Applied General Systems Theory.* Harper & Row, New York, 1974.

VA4 **Varela, F. J.,** *Principles of Biological Autonomy.* North-Holland, New York, 1979.

VA5 **Varela, F. J., H. R. Maturana,** and **R. B. Uribe,** Autopoiesis: The organization of living systems, its characterization and a model. *Biosystems*, **5,** No. 4, 1974, pp. 187–196.

VI1 **Vitruvius Polio, M.,** *Ten Books on Architecture.* Dover, New York.

WA1 **Walker, C. C.,** Behavior of a class of complex systems: The effect of system size on properties of terminal cycles. *Journal of Cybernetics*, **1,** No. 4, 1971, pp. 55–67.

WA2 **Walker, C. C.,** and **W. R. Ashby,** On temporal characteristics of behavior in certain complex systems. *Kybernetik*, **3,** No. 2, 1966, pp. 100–108.

WA3 **Wanser, J. C.,** Systems modelling of an archaelogical site. *General Systems Workshop Project*, Department of Systems Science, SUNY-Binghamton, 1980.

WA4 **Warfield, J. N.,** *Societal Systems.* Wiley-Interscience, New York, 1976.

WA5 **Wartofsky, M. W.,** *Models.* Reidel, Boston, 1979.

WA6 **Watanabe, S.,** *Knowing and Guessing.* Wiley, New York, 1969.

WA7 **Watanabe, S.,** Pattern recognition as a quest for minimum entropy. *Pattern Recognition*, **13,** No. 5, 1981, pp. 381–387.

WE1 **Weaver, W.,** Science and complexity. *American Scientist*, **36,** 1968, pp. 536–544.

WE2 **Wedde, H.,** (ed.) *Adequate Modelling of Systems.* Springer-Verlag, Berlin, 1983.

WE3 **Weinberg, G. M.,** *An Introduction to General Systems Thinking.* Wiley-Interscience, New York, 1975.

WE4 **Weiss, S. M.** and **C. A. Kulikowski,** *A Practical Guide to Designing Expert Systems.* Roman and Allerheld, Totowa, New Jersey, 1984.

WH1 **Whitehead, A. N.,** *Science and the Modern World.* Free Press, New York, 1967.

WH2 **Whitehead, A. N.,** *Modes of Thought.* Free Press, New York, 1968.

WH3 **Whittemore, B. J.,** and **M. C. Yovits,** The quantification and analysis of information used in decision processes. *Information Sciences*, **2,** 1974, pp. 171–184.

WI1 **Wiener, N.,** *Cybernetics.* M. I. T. Press, Cambridge, Massachusetts, 1948.

WI2 **Wilson, K. A.,** MISP: A computer simulation of a model for the instinct of self-preservation. *SAT Document Series* 74/05/07, School of Advanced Technology, SUNY-Binghamton, 1974.

WI3 **Windeknecht, T. G.,** *General Dynamical Processes.* Academic Press, New York, 1971.

WY1 **Wymore, A. W.**, *A Mathematical Theory of Systems Engineering: The Elements*. Wiley, New York, 1969.

WY2 **Wymore, A. W.**, *Systems Engineering Methodology for Interdisciplinary Teams*. Wiley-Interscience, New York, 1976.

YA1 **Yager, R. R.**, A foundation for a theory of possibility. *Journal of Cybernetics*, **10**, Nos. 1–3, 1980, pp. 177–204.

YA2 **Yamada, H.**, Structural and behavioral equivalences of tessellation automata. *Information and Control*, **18**, 1971, pp. 1–31.

YA3 **Yamada, H.**, and **S. Amoroso**, Tessellation automata. *Information and Control*, **14**, No. 3, 1969, pp. 299–317.

YA4 **Yamada, H.**, and **S. Amoroso**, A completeness problem for pattern generation in tessellation automata. *Journal of Computer and Systems Sciences*, **4**, No. 2, 1970, pp. 137–176.

YA5 **Yamada, H.**, and **S. Amoroso**, A completeness problem for pattern generation in tessellation automata. *Information and Control*, **18**, No. 1, 1971, pp. 1–31.

ZA1 **Zadeh, L. A.**, From circuit theory to systems theory. *IRE Proceedings*, **50**, No. 5, 1962, pp. 856–865.

ZA2 **Zadeh, L. A.**, On the definition of adaptivity. *Proceedings of the IEEE*, **51**, 1963, p. 469.

ZA3 **Zadeh, L. A.**, Fuzzy sets. *Information and Control*, **8**, No. 3, 1965, pp. 338–353.

ZA4 **Zadeh, L. A.**, Outline of a new approach to the analysis of complex systems and decision processes. *IEEE Transactions on Systems, Man, and Cybernetics*, **SMC-1**, No. 1. 1973, pp. 28–44.

ZA5 **Zadeh, L. A.**, Fuzzy sets as a basis for a theory of possibility. *Fuzzy Sets and Systems*, **1**, No. 1, 1978, pp. 3–28.

ZA6 **Zadeh, L. A.**, and **C. A. Desoer**, *Linear System Theory*. McGraw-Hill, New York, 1963.

ZA7 **Zadeh, L. A.**, and **E. Polak** (eds.), *System Theory*. McGraw-Hill, New York, 1969.

ZE1 **Zeigler, B. P.**, A conceptual basis for modelling and simulation. *International Journal of General Systems*, **1**, No. 4, 1974, pp. 213–228.

ZE2 **Zeigler, B. P.**, *Theory of Modelling and Simulation*. Wiley-Interscience, New York, 1976.

ZE3 **Zeigler, B. P.**, The hierarchy of system specifications and the problem of structural inference. In: *PSA* 1976, edited by F. Suppe and P. P. Asquith, Philosophy of Science Association, East Lansing, Michigan, 1976, pp. 227–239.

ZE4 **Zeleny, M. (editor)**, *Autopoiesis: A Theory of Living Organization*. Elsevier North-Holland, New York, 1981.

ZE5 **Zeleny, M.**, *Multiple Criteria Decision Making*. McGraw-Hill, New York, 1982.

ZE6 **Zemanek, H.**, Formal definition and generalized architecture. In: *Operations Research* 1972, edited by M. Ross, North-Holland, Amsterdam, 1973, pp. 59–73.

ZE7 **Zemanek, H.**, Abstract architecture. *Paper for the Winter School on Abstract Software Specification*, Danish Univ. of Technology, Copenhagen, 1979.

SUBJECT INDEX

AUTHOR INDEX

About the Author

George J. Klir is a Distinguished Professor of Systems Science and Chairman of the Department of Systems Science, Thomas J. Watson School of Engineering, Applied Science, and Technology, State University of New York at Binghamton. Before joining the State University of New York in 1969, he had been with the Institute for Computer Research and the Charles University in Prague, University of Baghdad, University of California at Los Angeles, and Fairleigh Dickinson University. He has also taught summer courses at several other universities, has worked for IBM and Bell Laboratories, and has been a consultant to various governmental as well as private institutions. During the academic years 1975–76 and 1982–83, he was a Fellow of the Netherlands Institute for Advanced Studies in Wassenaar, and in 1980 he was a Fellow of the Japan Society for the Promotion of Science.

Dr. Klir's main research interests are in the areas of basic and applied systems research, computer architecture and logic design, switching and automata theory, expert systems, information theory, and the philosophy of science. He is the author of more than one hundred articles and has published fourteen books, among them *Cybernetic Modelling, An Approach to General Systems Theory,* and *Methodology of Switching Circuits.* In addition, he holds a number of patents, and has given presentations at numerous conferences in the U.S. and abroad.

Since 1974, Dr. Klir has been Editor of the *International Journal of General Systems.* He has also been actively involved in various professional organizations, particularly the Institute for Electrical and Electronic Engineers (IEEE), the Society for General Systems Research (SGRR), and the International Federation for Systems Research (IFSR). He has served as President of the SGSR and IFSR, and is currently Editor of *IFSR Book Series on Systems Science and Engineering.*

GPSR Compliance

The European Union's (EU) General Product Safety Regulation (GPSR) is a set of rules that requires consumer products to be safe and our obligations to ensure this.

If you have any concerns about our products, you can contact us on ProductSafety@springernature.com

In case Publisher is established outside the EU, the EU authorized representative is:

Springer Nature Customer Service Center GmbH
Europaplatz 3
69115 Heidelberg, Germany

Batch number: 10692442

Printed by Printforce, the Netherlands